Foreword

Good decision-making about how we manage the waste we create is one of the most important contributions humanity can make to reducing its impact on the natural world. The Global Waste Management Outlook (GWMO) is being released at a critical moment, one where the world is considering a new regime to keep global warming to below 2 degrees above pre-industrial temperatures, and, at the same time, discussing what the future development agenda will look like and how it will be funded.

Environmental sustainability is the core issue that will need to be addressed for development to focus on human well-being and yet stay within the limitations of planet's capacity. Environmentally sound waste management is one of the key elements for sustainable development.

Already, science has demonstrated that we are on an unsustainable path, and that urgent and united action is required to correct this. The global population, currently at 7.3 billion, will grow in the coming decades to 9 billion and perhaps 11 billion by the end of the 21st century. Some 80% of this growing population will live in cities, most of which are yet to be built. Of this projected 9 billion people, 3 billion will belong to the middle class, with sufficient disposable income to purchase the consumer goods that others enjoy elsewhere in the world, further draining the planet's already strained natural resources.

Moving to a circular development model – which works to reduce waste before it is produced, but which treats waste as a resource when it is – is essential, and holistic and integrated sustainable waste management will be crucial.

The GWMO is the first comprehensive, impartial and in-depth assessment of global waste management. It reflects the collective body of recent scientific knowledge, drawing on the work of leading experts and the vast body of research undertaken within and beyond the United Nations system. The six chapters inform the reader about trends, provide an analysis on governance and financial mechanisms, and offer policy advice on the way forward. The main document targeting professionals is accompanied by two summary documents, one for decision makers and the other for the public more broadly.

This GWMO offers a profound analysis of the enormous potential better waste management provides to assist in meeting the sustainability challenges ahead.

Achim Steiner
UNEP Executive Director
United Nations Under-Secretary General

David Newman
President
International Solid Waste Association

Acknowledgements

This publication is dedicated to the memory of Matthew Gubb, who passed away during the development of the Global Waste Mangement Outlook (GWMO). As Director of UNEP-IETC, Matthew was heavily involved in the initial planning of the GWMO, but he was sadly unable to participate in its preparation.

Core Team

Editor-in-Chief

David C. Wilson (Independent waste and resource management consultant and Visiting Professor in Waste Management at Imperial College London, UK).

Authors

David C. Wilson, Ljiljana Rodic (Senior Researcher, Wageningen University and Research Centre, Wageningen, The Netherlands), Prasad Modak (Executive President of Environmental Management Centre (EMC) LLP, India), Reka Soos (Director of Resources and Waste Advisory Group and Managing Director of Green Partners, Romania), Ainhoa Carpintero Rogero (Associate Programme Officer, UNEP-IETC), Costas Velis (Lecturer in Resource Efficiency Systems, School of Civil Engineering, University of Leeds, UK), Mona Iyer (Associate Professor at Faculty of Planning, CEPT University, Ahmedabad, India), and Otto Simonett (Director and co-founder of Zoï Environment Network, Switzerland).

Authorship of the six chapters and specific contributions

Chapter 1: Waste Management as a Political Priority
David C. Wilson and Ainhoa Carpintero Rogero

Chapter 2: Background, Definitions, Concepts and Indicators
David C. Wilson and Ainhoa Carpintero Rogero

Chapter 3: Waste Management: Global Status
Prasad Modak, David C. Wilson and Costas Velis

Chapter 4: Waste Governance
Ljiljana Rodic

Chapter 5: Waste Management Financing
Reka Soos, David C. Wilson and Otto Simonett

Chapter 6: Global Waste Management – The Way Forward
David C. Wilson

Case Studies
Mona Iyer and Ainhoa Carpintero Rogero

Academic advisors
Costas Velis, Ljiljana Rodic

Communications advisor
Otto Simonett

Project Coordinator

Ainhoa Carpintero Rogero

Special support from: Kata Tisza (Technical Manager, ISWA) and Rachael Williams Gaul (Technical Manager, ISWA – until February 2015).

Supervision

Surendra Shrestha (Director, UNEP-IETC), Surya Chandak (Senior Programme Officer, UNEP-IETC), David Newman (President, ISWA) and Hermann Koller (Managing Director, ISWA).

Special acknowledgements for either making available databases to, or compiling databases for, the GWMO: AcuComm for their Waste Business Finder database; the Bureau of International Recycling (BIR); Environmental Management Centre LLP towards building the master database for the GWMO; Henry Hickman and Michael Lawless, MEng students at the University of Leeds, UK, whose theses were prepared as inputs to the GWMO.

Steering Committee

Gary Crawford (Vice President – Sustainable Development Technical, Scientific and Sustainable Development Department, Veolia Environmental Services, France), Johannes Frommann (Senior Advisor, Deutsche Gesellschaft fuer Internationale Zusammenarbeit (GIZ), Germany), Christian Holzer (Director General Federal Ministry of Agriculture, Forestry, Environment and Water Management, Austria), Hermann Koller (Managing Director, ISWA, Austria), Raquel Lejtreger (Vice Minister, Subsecretary, Ministry of Housing, Spatial Planning and Environment, Uruguay) Stanley Motsamai Damane (Director, National Environment Secretariat, Ministry of Tourism, Environment and Culture, Lesotho), Nadzri bin Yahaya (Director General, National Solid Waste Management Department, Ministry of Housing and Local Government, Malaysia), Carlos RV Silva Filho (CEO ABRELPE, Brazil), Surendra Shrestha (Director UNEP-IETC, Japan).

Peer reviewers and contributors

P. Agamuthu (Malaya University, Malaysia), Vivek Agrawal (Centre for Development Communications, India), Marcos Alegre (National Cleaner production Centre, Peru), Riad Alkabekli (Society for the Protection of Environment and Sustainable Development, Syria), Roberto Azofeifa (Ministry of Agriculture and Livestock, Costa Rica), Conrad Martin Bader (Hitachi Zosen Inova Ltd., Switzerland), Fritz Balkau (Co-Chair, ISWA Working Group Hazardous Waste, France), Margaret Bates (University of Northampton, UK), Michael Betts (Integrated Skills Ltd., UK), Khanendra D. Bhardwaj (APO), Shriya Bhatia (Environmental Management Centre LLP, India), Valentina Baiamonte (International Green Cross), Peter Borkey (OECD), Emmanuelle Bournay (Zoï Environment Network, Switzerland), Nadezhda Boyarkina (Independent Consultant, UK), Manuel Bräu (MEng student, University of Leeds, UK), Alberto S. Capra (Secretary of Environment and Sustainable Development, Argentina), Paddy Carless (ATKINS, UK), Jérémie Cavé (Independent consultant, France), Dalson Chung (National Environment Agency, Singapore), Kimberly Cochran (U.S. Department of State, U.S.), Ho De Leon (Waste Management Association of Malaysia, Malaysia), Christof Delatter (Interafval, Belgium), Leila Devia (National Cleaner production Centre, Argentina), Sonia Maria Dias (WIEGO, Brazil), Hassan El Bari (University of Ibn Tofail, Morocco), Olabode Esan (UAE Government, UAE), Nicolas Escalante M. (Universidad de los Andes, Colombia), Peter Faircloth (Cranford Economics Ltd., UK), Roland Ferth (Federal Ministry of Agriculture, Forestry, Environment and Water Management), Edmund Fleck (MARTIN GmbH, Germany), Vicenzo Gente (Directorate-General for Research and Innovation, EU), Derek Greedy (CIWM, ISWA), Sue Grimes (Imperial College London, UK), Asma Gul Qazi (PhD Student, Fatima Jinnah Women University, Pakistan), Sanjay K. Gupta (Skat Consulting Ltd., Switzerland), David Haynes (SPREP), Johannes Heitmann (Remondis, Germany), Henry Hickman (MEng student University of Leeds, UK), K. M. Nurul Huda (Bangladesh Municipal Development Fund, Bangladesh), Alodia W. Ishengoma (Independent Consultant, Tanzania), Yuyun Ismawati (International POPs Elimination Network, UK), Ratna Jain (City of Kota, India), Keith James (WRAP, UK), Michel A. Jenny (Caterpillar Industrial & Waste Group, Switzerland), Aida Karazhanova (UN-ESCAP), Sofia Khalid (Fatima Jinnah Women University, Pakistan), Alex Kirby (Zoï Environment Network, Switzerland), Dmitri Kolganov (PWC Moscow, Russia), Maria Libert (Zoï Environment Network, Switzerland), Markus Luecke (SWEEP-Net, GIZ, Tunisia), Lios McGilchrist (Living Earth Foundation, UK), Amar B. Manandhar (Society for Environmental and Economic Development Nepal), Jose A. Martinez Sepulveda (Universidad EAN, Colombia), Antonis Mavropoulos (D-Waste, Greece), Martin Medina (U.S. National Oceanic and Atmospheric Administration, U.S.), Diane Menard (Zoï Environment Network, Switzerland), Feliz Mil-Homens (IWM Business Services Limited, UK), Chris Murphy (Chartered Institution of Wastes Management, UK), Chidi Nzeadibe (University of Nigeria, Nigeria), Godfrey Oluka (Division of Health and Environment Department, Kampala, Uganda), Oladele Osibanjo (Basel Convention Coordinating Centre, Nigeria), Dhundi Raj Pathak (Ministry of Urban Development, Nepal), Wolfgang Pfaff-Simoneit (KfW Entwicklungsbank, Germany), Flaviu Pop (Resources and Waste Advisory Group and Green Partners, Romania), Pierre Portas (Waste Environment Cooperation Centre/WE 2C, France), Joachim Quoden (Extended Producer Responsibility Alliance, Belgium), Mark Reiter (Institute of Scrap Recycling Industries, USA), Natalia Reyna (PhD student, Imperial College London, UK), Esther Richard (Secretariat of the Pacific Regional Environment Programme, SPREP), Diana M. Rodriguez Velosa (Ministry of Environment and Sustainable Development, Government of Colombia), Amiya Kumar Sahu (National Solid Waste Association of India, India), Lorenzo Santucci (UN-ESCAP), Atilio Savino (Solid Waste Association, Argentina), Anuja Sawant (Environmental Management Centre LLP, India), Anne Scheinberg (WASTE, The Netherlands), Klaas Schwartz (UNESCO-IHE Institute for Water Education, Delft, The Netherlands), Julia Stegemann (University College London, UK), Alison Stowell (Lancaster University Management School, UK), Ruth Stringer (Health Care without Harm), Denise Soesilo (Zoï Environment Network, Switzerland), Christian Susan (UNIDO), Vishwa Trivedi (Environmental Management Centre LLP, India), Wolfang Teubner (International Council for Local Environmental Initiatives, Germany), Baskut Tuncak (Center for International Environmental Law, U.S.), Stijn Van Ewijk (University College London, UK), Mariel Vilella (Global Alliance for Incinerator Alternatives, UK), Rob Visser (Former Senior Advisor, UNEP, France), Chettiyappan Visvanathan (Asian Institute of Technology, Thailand), Verele de Vreede (WASTE, The Netherlands), Waste Management Society of Nigeria (Nigeria), Roy Watkinson (RWEC Ltd., UK), Peter Wessman (Directorate -General for Environment, EU), Michiel Westerhoff (Circulus, The Netherlands), Andrew Whiteman (Wasteaware/ RWA, Bulgaria), Noriko Yajima (EcoLomics International, Canada/Switzerland), Shunta Yamaguchi (OECD), Victoria Romero (ATKINS, UK), Anja Zimmermann (Zoï Environment Network, Switzerland).

UNEP peer reviewers and contributors
Meriem Ait Ali Slimane, Cristina Battaglino, Matthew Billot, Garrette Clark, Ludgarde Coppens, Christopher Corbin, Charles Davies, Heidelore Fiedler, Carla Friedrich, Mijke Hertoghs, Arab Hoballah, Shunichi Honda, Matthias Kern, Birguy Lamizana, Mushtaq Memon, Llorenç Milà i Canals, Desiree Montecillo Narvaez, Jordi Pon, Pierre Quiblier, Heidi Savelli, Ibrahim Shafii, Elisa Tonda, Mick Wilson, Kaveh Zahedi.

Thanks to our UNEP colleagues, in particular to: Fuaad Alkizim, Utako Aoike, Junko Fujioka, Azumi Nishikawa, Solange Montillaud-Joyel, Mayumi Morita, Moira O'Brien-Malone, John Peter Oosterhoff, Michiko Ota, Sheeren Zorba.

Thanks to ISWA Members/colleagues, in particular to: Jos Artois (Indaver, Belgium), Josef Barth (Informa, Germany), Gunilla Carlsson (Sysav, Sweden), Adam Read (Ricardo-AEA), Marco Ricci-Jürgensen (CIC Italian Composting and Biogas Consortium, Italy) and the ISWA General Secretariat team.

ISWA International Solid Waste Association

International Solid Waste Association
General Secretariat
Auerspergstrasse 15/41, 1080 Vienna, Austria
Tel: +43 1 2536 001   Fax: +43 1 253 600199
E-mail: iswa@iswa.org   Web: www.iswa.org

Editor: Tara Cannon

Cover illustration: High Moon

Design and layout: UNESCO

Table of Contents

List of Tables

List of Figures

List of Boxes

CHAPTER 1

WASTE MANAGEMENT AS A POLITICAL PRIORITY

Why is waste management of concern? Why would a politician choose to give it priority? Why has UNEP chosen to prepare this Global Waste Management Outlook (GWMO)? This introduction aims to provide a concise overview which answers these and similar questions. Chapter 1 and Chapter 6 in combination provide an extended summary of the GWMO.

The chapter begins by looking at the public health and environmental consequences when wastes are not collected and managed properly. The first priority is to bring wastes under control (1.1). But that on its own is not enough – it is also necessary to move from waste management in a linear economy to resource management within a circular economy (1.2). Tackling waste management helps to address more than half of the high-level sustainable development goals within the Post-2015 Development Agenda and promises significant early reductions in greenhouse gas emissions throughout the economy (1.3). A readers' guide to the GWMO is provided (1.4) and the Chapter is followed by two Topic Sheets, on Waste and Climate and on the World's 50 Biggest Dumpsites.

© Natalia Reyna

1.1 THE CHALLENGE OF SUSTAINABLE WASTE MANAGEMENT

Waste is a global issue. If not properly dealt with, waste poses a threat to public health and the environment. It is a growing issue linked directly to the way society produces and consumes. It concerns everyone.

Waste management is one of the essential utility services underpinning society in the 21st century, particularly in urban areas. Waste management is a basic human need and can also be regarded as a 'basic human right'. Ensuring proper sanitation and solid waste management sits alongside the provision of potable water, shelter, food, energy, transport and communications as essential to society and to the economy as a whole. Despite this, the public and political profile of waste management is often lower than other utility services.

Unfortunately, the consequences of doing little or even nothing to address waste management can be very costly to society and to the economy overall. In the absence of waste regulations and their rigorous implementation and enforcement, a generator of waste will tend to opt for the cheapest available course of action. For example, household solid waste may be dumped in the street, on vacant land, or into drains, streams or other watercourses, or it may be burned to lessen the nuisance of accumulated piles of waste.

- **Public health:** Not having a solid waste collection service has a direct health impact on residents, particularly children. The uncontrolled burning of waste creates particulate and persistent organic pollutant emissions that are highly damaging locally and globally. Accumulated waste and blocked drains encourage vectors to breed, resulting in the spread of cholera, dengue fever and other infectious diseases and are a major contributing factor to flooding. Uncontrolled dumpsites, and in particular the mixing of hazardous and other wastes, can cause disease in neighbouring settlements as well as among waste workers. *Box 1.1* presents selected case studies illustrating both the public health problems of uncollected waste as well as the solutions.[1]

On a larger scale, when significant quantities of municipal or industrial solid waste are dumped or burned in the open, the adverse impacts on air, surface and groundwater, soil and the coastal and marine environment, and thus indirectly on public health, can be severe.

- **Environmental pollution:** Dumpsites on land can pollute both surface and groundwater. These sites are often alongside rivers or the sea, and therefore may directly pollute them as well as the coastal environment. Coastal dumpsite erosion is one source of marine litter. Other potential damage costs include losses resulting from decreases in tourism due to polluted beaches and losses incurred through damage to fisheries. Former dumpsites, particularly those that have received hazardous waste, are a major category of contaminated site. *Box 1.2* presents selected case studies illustrating both the environmental impacts of uncontrolled disposal and also environmentally sound management.[2]

By definition, uncontrolled waste is not 'managed' and thus not measured, making it difficult to estimate either the size of the problem or the scale of the associated costs. However, the evidence suggests that in a middle- or low-income city, the costs to society and the economy are perhaps 5-10 times what sound solid waste management (SWM) would cost per capita. It is dramatically cheaper to manage waste now in an environmentally sound manner than to clean up in future years the 'sins of the past'.

This Global Waste Management Outlook (GWMO) aims to make the case that sound waste management is much more than merely desirable. Sound waste management is absolutely essential.

1 The evidence linking waste and health is reviewed in Table 5.2 in Chapter 5; selected references are included in Annex A under Chapter 5. Further research in this area is recommended (see Table 6.8).

2 Topic Sheet 2 following Chapter 1 summarizes information from a recent compilation of the world's 50 biggest dumpsites.

BOX 1.1 PUBLIC HEALTH ISSUES FROM UNCOLLECTED WASTE

Collection crisis – waste piling up in the streets[3]

Case: Naples, Italy, 1994-2014

The Naples metropolitan area has had long-running problems with municipal solid waste management. The inability to find a satisfactory permanent solution for waste treatment and disposal has led to periodic 'crises' in which the absence of anywhere to take the waste has meant that waste could not be collected. Left to pile up in the streets, the waste became breeding grounds for vector-borne diseases, representing a public health risk for the population. Emergency solutions have sometimes included new uncontrolled dumpsites near the city, sparking further citizen protests. Such crises made national and international headlines in 1994, 1999, 2003, 2008, 2010 and 2011. The situation has caused substantial friction among the different stakeholder groups and has been further complicated by the active involvement of criminal organizations in waste transport. The new municipal government in 2011 appointed a leading anti-waste protester to head the waste management agency, and waste was thereafter transported to other Italian regions and other EU countries while a sustainable local solution was sought.

© lapas77

In 2013 an incinerator in Accerra, near Naples, started operating with a processing capacity of 650,000 t/y. Waste recycling has also increased. This has eased the city's situation considerably. However waste is still shipped to other regions, notably organics, due to the lack of compost plants in the Naples region, and waste is also transported to the Netherlands for incineration. An unresolved issue is what to do with the 8 million tonnes of waste that accummulated in 'temporary' sites in the the area surrounding Naples, as this issue requires both money and political will, both currently lacking.

© Inviaggiocommons

Major floods exacerbated by plastic bags blocking drains

Case: Accra, Ghana

© Modern Ghana

Every year Ghana suffers from seasonal flooding. In Accra, as in many other places, drains blocked by plastics and other wastes are an important factor in this. Due to a lack of organized collection, many people simply dump their solid wastes, some of which gets washed into the drains. Litter on the streets and in the gutters is also widespread, much of it from the indiscriminate disposal of commonly used plastic sachets holding drinking water.

Floods in 2011 incurred loss of life and damaged or destroyed livelihoods and economic value. Fourteen people were killed, 43,000 were affected, and 17,000 lost their homes, with damage to roads, waterways and bridges also reported. In addition 100 incidents of cholera were identified a week after the flooding occurred.[4]

3 D'Alisa, G., D. Burgalassi, H. Healy, M. Walter (2010). Conflict in Campania: Waste emergency or crisis of democracy. Ecological Economics, 70 239-249. Additional information from David Newman, President of the Italian National Waste Association, ATIA-ISWA Italia, until 2014.

4 UNEP/OCHA (2011). Rapid disaster waste management assessment: 26 October Flash Flooding, Central Accra, Ghana. https://ochanet.unocha.org/p/Documents/Ghana_ Flash%20Flooding_DWM%20Assessment.pdf

Epidemic resulting from uncollected waste blocking drains and creating breeding grounds for disease vectors

Case: Plague-like epidemic in Surat, India, 1994

© Sanjay Gupta

In 1994, the city of Surat suffered an outbreak of plague-like disease caused by major flooding as a consequence of uncollected waste blocking the drains. In the preceding years, the city had experienced a growth in population not matched by the infrastructure necessary to provide adequate solid waste management services – services under the responsibility of the municipality. At the time municipal solid waste collection was limited to 40% of the city[5] and the remaining waste was left uncollected and often came to be disposed of in drains and water bodies.

The disease outbreak resulted in 693 cases reported and 56 deaths. Additionally, the country may have suffered as much as USD 2 billion in economic losses,[6] including approximately USD 420 million in lost export earnings (for example, the United Arab Emirates suspended all cargo trans-shipments from India). The disease outbreak occurred just before the Deepavali festival and over 45,000 people cancelled trips to India.[7]

Over the following 18 months the city administration, Surat Municipal Corporation, transformed the city into one of the cleanest cities in the region. The initiatives carried out included monitoring, infrastructure development in slums, engagement of the private sector in waste collection and transportation, capacity building and coordination among municipal employees, awareness raising among the public and the introduction of complaint handling systems.[8] They also created the Littering Detection Squad, an initiative under which people work to maintain the cleanliness of specific areas considered most vulnerable to litter and generate revenue by penalizing residents or shopkeepers who continue littering despite previous warnings.[9] One of the ongoing initiatives, a private sector waste treatment facility, is discussed later in the GWMO *(Box 5.12)*.

© Rahulogy/CC BY-SA 3.0/wikipedia.org

5 According to a 1995 study mentioned in "Surat Solid Waste Management Project under JNNURM" – Jawaharlal Nehru National Urban Renewal Mission – July 2013 http://jnnurm.nic.in/wp-content/uploads/2013/08/City-Report-on-Surat-SWM-Project-under-JnNURM_Final.pdf

6 Brahmbhatt, M. and A. Dutta (2008). 'On SARS-Type Economic Effects during Infectious Disease Outbreaks.' *Policy Research Working Paper 4466*. World Bank, Washington, D.C.

7 Pallipparambil, G. R. (2005) "The Surat Plague and its Aftermath" http://entomology.montana.edu/historybug/YersiniaEssays/Godshen.htm

8 Surat Solid Waste Management Project under JNNURM - Jawaharlal Nehru National Urban Renewal Mission – July 2013

9 Sanjay Gupta, Case Study Report for Surat, India, prepared for GIZ Operator Models project. [Online] via http://www.giz.de/en/mediacenter/publications.html, search the database for 'operator models', then select Annex 5.

BOX 1.2 ENVIRONMENTAL IMPACTS FROM LARGE SCALE UNCONTROLLED DUMPING AND BURNING

Damage to the tourist industry from waste littering and dumping

Case: Barbados

Tourism is one of the largest industries in the world and it accounts for a large proportion of revenue and employment in many nations. For example, Barbados, a Caribbean island nation, each year receives over a million tourists whose economic contribution amounts to almost 50% of GDP and total employment. Many tourists are attracted to this destination by its pristine coastal and marine environment.

Marine and coastal litter[10] is an issue of particular concern to Barbados because of the biological and aesthetic impacts and the costs, both public and private, of coastal and marine clean up, and because of the link between litter and decreased tourism demand. A study[11] carried out in Barbados with the support of the Ministry of Tourism of Barbados and The Caribbean Tourism Organization concluded that beach litter potentially creates economic costs because of its adverse effects on tourists' probability of returning.[12]

Local health crisis resulting from historic uncontrolled disposal of hazardous waste

Case: Love Canal, USA – late 1970s

Love Canal was originally planned as a community settlement powered by a canal connecting the Niagara River with Niagara Falls. Only one mile of the canal had been dug when the project came to a halt. The site was sold to Hooker Chemical Company and the canal became a municipal and chemical dumpsite where some 20,000 tonnes of toxic chemicals[13] were dumped over the years, from the 1920s to the 1950s. In 1953 the canal, having reached full capacity, was covered with earth and sold. Homes and a school were constructed there at the end of the 1950s. In the 60s and 70s problems with odours and residues were reported, as the water table rose and brought contaminated groundwater to the surface. In 1978, a record amount of rainfall resulted in contaminated leachate containing toxic chemicals migrating into the yards and basements of the houses and the school and contaminating the air, as the drums containing the waste had deteriorated over time. More than 100 cases of chemically induced illnesses among children caused a huge public outcry. Investigations found 82 different compounds, of which 11 are suspected carcinogens. Emergency financial aid of USD 10 million was approved to relocate some of the families and in 1980 the US 'Superfund' law[14] was enacted in the wake of discovering toxic waste dumps such as Love Canal. Cleanup activities at such sites have been ongoing. In 1995 according to a consent decree, the chemical company agreed to reimburse the federal government's past costs for response actions, natural resource damages claims and oversight costs, amounting to over USD 129 million.[15] In 2004 the site was deleted from the National Priorities List (NPL).[16]

© BuffaLutheran/CC0 1.0/Wikipedia.org

10 In Barbados 70-80% of the marine litter comes from shoreline/recreation activities and 20% from ocean and waterway activities.
11 Schuhmann, W. P. (2011). Tourist perceptions of beach cleanliness in Barbados: Implications for return visitation http://etudescaribeennes.revues.org/5251
12 Refer to Table 5.2.
13 Including halogenated organics, pesticides, chlorobenzenes and dioxin.
14 The Comprehensive Environmental Response, Compensation, and Liability Act (CERCLA), commonly known as 'Superfund', was signed into law by President Jimmy Carter on December 11, 1980.
15 Love Canal EPA ID:NYD000606947 http://www.epa.gov/region2/superfund/npl/0201290c.pdf
16 The NPL is the list of the US' national priorities among releases of hazardous substances, pollutants or contaminants. The US Environmental Protection Agency uses the NPL as a guide in determining which sites require further investigation and clean-up.

Landslide at a major uncontrolled municipal dumpsite

Case: Payatas. Quezon City, Philippines, 2000[17]

© Merry et al.

In July 2000, after ten consecutive days of heavy rain, a slope of the Payatas municipal dumpsite collapsed on a slum community. The 18 hectare facility received an average of 1,500-1,800 tonnes of waste per day and was home to waste pickers who constructed their houses up to and onto the waste slope. The landslide killed nearly 300 people and left hundreds of families homeless and without a source of livelihood. The slope failure was the result of raised leachate level (due to the heavy rain, absence of soil cover and high infiltration), landfill gas pressures, low waste density (due to limited compaction) and unstable slopes at too acute an angle.[18] Although the dumpsite was closed in the aftermath of the tragedy, the impact on the cleanliness of the city forced local authorities to re-open it in November.

In 2001 the City government started work to transform the dumpsite into a controlled waste disposal facility and an eco-park. Work on the site included slope re-profiling and terracing, stabilization and greening, leachate collection and recirculation, drainage system improvement, fortified roadways and access to the site, gas venting and material recovery.

In 2004 a 110kW Pilot Methane Power plant was set up at the site and in 2007/8 a Biogas Emissions Reduction Project started. The project was the first Clean Development Mechanism (CDM) project in solid waste management in the Philippines and in Southeast Asia and registered under the Kyoto Protocol of the United Nations Framework Convention on Climate Change (UNFCCC). GHG emissions have been reduced by an estimated 116,000 tonnes CO_2 per annum. In addition, employment has been created, local capacity of the technology and know-how have been cultivated and financial resources from the sale of Certified Emission Reductions, or carbon credits, have been obtained. The project is considered a showcase of best practices.

Photo courtesy: Luis Sabater

17 The Payatas Dumpsite: From Tragedy to Triumph – http://www.quezoncity.gov.ph/index.php?option=com_content&id=207%3Athe-payatas-dumpsite-from-tragedy-totriumph&Itemid=235&showall=1

18 Based on an analysis carried out by Kavazanjian, E., S. Merry, and colleagues. See http://faculty.engineering.asu.edu/kavazanjian/wp-content/uploads/2011/07/17_Merry-et-al-ASCE-Payatas.pdf and http://casehistories.geoengineer.org/volume/volume2/issue3/IJGCH_2_3_3.pdf

1.2 MOVING FROM WASTE MANAGEMENT TO RESOURCE MANAGEMENT

Many developed countries have made great strides in addressing waste management, particularly since the environment came onto the international agenda in the 1960s, and there are many good practice examples available for the international community to learn from. However, the initial focus was on waste after it had been discarded, whereas at present attention has moved upstream, addressing the problem at its source through, for example, designing out waste, preventing its generation, reducing both the quantities and the uses of hazardous substances, minimizing and reusing, and, where residuals do occur, keeping them concentrated and separate to preserve their intrinsic value for recycling and recovery and prevent them from contaminating other waste that still has economic value for recovery. The goal is to move the fundamental thinking away from 'waste disposal' to 'waste management' and from 'waste' to 'resources' – hence the updated terminology 'waste and resource management' and 'resource management', as part of the 'circular economy'.[19] In this regard, the GWMO interfaces with the earlier Global Outlook on Sustainable Consumption and Production (SCP) policies.[20]

Low- and middle-income countries still face major challenges in ensuring universal access to waste collection services, eliminating uncontrolled disposal and burning and moving towards environmentally sound management for all waste. Achieving this challenge is made even more difficult by forecasts that major cities in the lowest income countries are likely to double in population over the next 20 or so years, which is also likely to increase the local political priority given to waste issues. Low- and middle-income countries need to devise and implement innovative and effective policies and practices to promote waste prevention and stem the relentless increase in waste per capita as economies develop.

1.3 WASTE MANAGEMENT AS AN ENTRY POINT FOR SUSTAINABLE DEVELOPMENT

Waste management is a cross-cutting issue impacting on many aspects of society and the economy. It has strong linkages to a range of other global challenges such as health, climate change, poverty reduction, food and resource security and sustainable production and consumption. The political case for action is significantly strengthened when waste management is viewed as an entry point to address a range of such sustainable development issues, many of which are difficult to tackle.

Four entry points are introduced below. These correspond to the three 'pillars' or 'domains' of sustainability and to their integration into the Sustainable Development Goals (SDGs) within the Post-2015 Development Agenda.

Environment and climate change ➡ Environmental Domain

Waste is a cross-cutting environmental issue in which a great many different strands interact. The local environmental impacts of waste have already been discussed – the focus here is rather on the global impacts of climate change.[21]

The IPCC estimates that solid waste management accounted for around 3% of global greenhouse gas (GHG) emissions in 2010, with most of that attributable to methane emissions from landfill sites. However, the potential contribution of better waste and resource management to climate change mitigation exceeds that by far. More than 50% of worldwide municipal solid waste in 2010 was generated in OECD countries, and by that time most OECD waste was properly managed, with landfill gas being collected and either flared or utilized in energy recovery. Some countries had already succeeded in diverting significant percentages of waste away from landfill disposal. For example, Germany attributed 24% of its total savings in GHG emissions between 1990 and 2006 to solid waste management, even though by 1990 West German standards of landfill gas control were already high.

19 This is expanded on Section 2.2.2 and in more detail in Box 4.1.

20 UNEP (2012), as listed in Annex A, Chapter 1, first subsection. SCP is also covered in Topic Sheet 3, found after Chapter 2.

21 See Topic Sheet 1 on Waste and Climate, following Chapter 1, for a more detailed discussion of these issues, and the rationale for the various estimates included below.

In addition, upstream actions to address the problem at source, such as designing out waste, preventing its generation and reusing products, have the potential for much greater carbon savings, as these displace greenhouse gas emissions across a wide range of sectors which would otherwise be incurred to provide the displaced products. Similarly, recycling replaces virgin materials at a much lower carbon cost and thus reduces emissions across the economy.

So a focus on waste and resource management has the potential to achieve substantial, short-term mitigation of climate change, amounting to perhaps 15-20%, across a number of sectors.

Good governance ➡ Social Domain

A clean city in which solid waste management service is clearly effective is a healthful and pleasant place for residents, attractive for tourists and a good place to do business and attract inward investment – in other words, a successful city. The GWMO cites example studies in which the loss of tourist income from poor solid waste management was high, and likely more than the cost of implementing a modern SWM system. But the benefits go much further by promoting a sense of community and security, of belonging and of well-being.

For the city to appear clean, the city's SWM system needs to be working well. Achieving that requires attention to a wide range of governance issues. These include user inclusivity (e.g. reaching previously unserved or underserved areas, involving citizens in decision making, changing attitudes and behaviours), provider inclusivity (e.g. mobilizing all service providers, from both the municipal and non-municipal (including the formal private, community and 'informal') sectors, tackling corruption issues), financial sustainability (e.g. knowing relevant costs, providing affordable services, improving revenue streams, accessing finance) and putting in place effective policies and institutions at both the national and local levels (e.g. implementing legislation, assigning clear roles and responsibilities, building professional capacity, instituting management information systems). It has thus been suggested that the cleanliness of a city can be used as proxy indicator for good governance, which is both a major objective of development programmes and extremely difficult to measure.

Enterprise & creating sustainable livelihoods ➡ Economic Domain

Good waste and resource management also brings many positive benefits to society and the economy. Many recent reports have attempted to quantify the business benefits to industry of reduced resource consumption, improved resource efficiency and reduced costs of waste management. Estimates usually exceed USD 1,000 billion globally, with a large proportion of this attributed to waste prevention. There is also much potential for job creation in the green economy, and for contributions to increased resource and energy security. The potential of the waste and resource sector for creating sustainable livelihoods has been demonstrated for both high- and low-income economies.[22]

Sustainable Development Goals ➡ Integration

At the time of writing the GWMO, the action plan for the Post-2015 Development Agenda was in final stages of approval at the UN. Waste management is well embedded within the Sustainable Development Goals (SDGs), being included either explicitly or implicitly in more than half of the 17 goals (*Table 1.1*). So a strong argument can be made for the strategic importance of improving waste management, insofar as actions here will contribute to progress towards a range of SDG targets. Setting and monitoring global targets for waste management will thus contribute significantly to attaining the SDGs.

22 See Table 5.3 on wider benefits to society and the economy from sustainable waste and resource management.

Table 1.1 Proposed Global Waste Management Goals within the Post-2015 Development Agenda

<table>
<tr><th colspan="2">GLOBAL WASTE MANAGEMENT GOALS</th><th colspan="2">RELATED REFERENCES WITHIN THE SUSTAINABLE DEVELOPMENT GOALS</th></tr>
<tr><td colspan="4">Primary goals</td></tr>
<tr><td>W.1</td><td>By 2020, ensure access for all to adequate, safe and affordable solid waste collection services</td><td>11 Make cities inclusive, safe, resilient and sustainable</td><td>11.1 By 2030, ensure access for all to adequate, safe and affordable housing and basic services, and upgrade slums
11.6 By 2030, reduce the adverse environmental impact of cities, including by paying special attention to air quality and municipal and other waste management
11.b By 2020, increase … the number of cities and human settlements adopting and implementing integrated policies and plans towards inclusion, resource efficiency, mitigation and adaptation to climate change [and] resilience to disasters …</td></tr>
<tr><td rowspan="2">W.2</td><td rowspan="2">By 2020, eliminate uncontrolled dumping and open burning</td><td>3 Ensure healthy lives and promote well-being for all at all ages</td><td>3.9 By 2030, substantially reduce the number of deaths and illnesses from hazardous chemicals and air, water, and soil pollution and contamination</td></tr>
<tr><td rowspan="4">12 Ensure sustainable consumption and production patterns</td><td>12.4 By 2020, achieve the environmentally sound management of chemicals and all wastes throughout their life cycle, in accordance with agreed international frameworks, and significantly reduce their release to air, water and soil in order to minimize their adverse impacts on human health and the environment</td></tr>
<tr><td>W.3</td><td>By 2030, ensure the sustainable and environmentally sound management of all wastes, particularly hazardous wastes</td><td>12.5 By 2030, substantially reduce waste generation through prevention, reduction, recycling and reuse</td></tr>
<tr><td>W.4</td><td>By 2030, substantially reduce waste generation through prevention and the 3Rs (reduce, reuse, recycle) and thereby create green jobs</td><td>12.a Support developing countries to strengthen their scientific and technological capacity to move towards more sustainable patterns of consumption and production</td></tr>
<tr><td rowspan="2">W.5</td><td rowspan="2">By 2030, halve global per capita food waste at the retail and consumer levels and reduce food losses in the supply chain</td><td>12.3 Halve per capita global food waste at the retail and consumer levels, and reduce food losses along production and supply chains, including post-harvest losses</td></tr>
<tr><td>2 End hunger, achieve food security and improved nutrition and promote sustainable agriculture</td><td>2.1 By 2030, end hunger and ensure access by all people, in particular the poor and people in vulnerable situations, including infants, to safe, nutritious and sufficient food all year round</td></tr>
</table>

<table>
<tr><th colspan="2">GLOBAL WASTE MANAGEMENT GOALS</th><th colspan="2">RELATED REFERENCES WITHIN THE SUSTAINABLE DEVELOPMENT GOALS</th></tr>
<tr><td colspan="4">Secondary goals</td></tr>
<tr><td rowspan="2">W.6
(also linked to W.4)</td><td rowspan="2">Create green jobs through the circular economy
Build sustainable livelihoods by integrating the informal sector into mainstream waste and resource management in the poorest cities</td><td>1 End poverty in all its forms everywhere</td><td>1.1 By 2030, eradicate extreme poverty for all people everywhere, currently measured as people living on less than USD 1.25 a day
1.2 By 2030, reduce at least by half the proportion of [people] living in poverty ...</td></tr>
<tr><td rowspan="2">8 Promote sustained, inclusive and sustainable economic growth, full and productive employment and decent work for all</td><td rowspan="2">8.2 Achieve higher levels of economic productivity ... including through a focus on ... labour-intensive sectors
8.3 Promote development-oriented policies that support productive activities, decent job creation, entrepreneurship, creativity and innovation, and encourage the formalization and growth of micro-, small- and medium-sized enterprises, including through access to financial services
8.7 Take immediate and effective measures to secure the prohibition and elimination of the worst forms of child labour, eradicate forced labour and, by 2025, end child labour in all its forms ...
8.8 Protect labour rights and promote safe and secure working environments for all workers, including migrant workers, in particular women migrants, and those in precarious employment
8.4 Improve progressively, through 2030, global resource efficiency in consumption and production and endeavour to decouple economic growth from environmental degradation ...</td></tr>
<tr><td rowspan="2">W.7
(also linked to W.4)</td><td rowspan="2">By 2030, reduce industrial waste generation at source through waste prevention, resource efficiency and greater adoption of clean and environmentally sound technologies and industrial processes</td></tr>
<tr><td>9 Build resilient infrastructure, promote inclusive and sustainable industrialisation and foster innovation</td><td>9.4 By 2030, upgrade infrastructure and retrofit industries to make them sustainable, with increased resource-use efficiency and greater adoption of clean and environmentally sound technologies and industrial processes, all countries taking action in accordance with their respective capabilities
(Note also 9.1 and 9.3)</td></tr>
<tr><td colspan="4">Other Linkages</td></tr>
<tr><td rowspan="3">(Linked to W.2)</td><td rowspan="3">(By 2020, eliminate uncontrolled dumping and open burning)</td><td>6 Ensure availability and sustainable management of water and sanitation for all</td><td>6.3 By 2030, improve water quality by reducing pollution [and] eliminating dumping and minimizing release of hazardous chemicals and materials ...</td></tr>
<tr><td>14 Conserve and sustainably use the oceans, seas and marine resources</td><td>14.1 By 2025, prevent and significantly reduce marine pollution of all kinds, in particular from land-based activities, including marine debris and nutrient pollution</td></tr>
<tr><td>15 Protect, restore and promote sustainable use of terrestrial ecosystems, sustainably manage forests, combat desertification, and halt and reverse land degradation and halt biodiversity loss</td><td>15.1 By 2020, ensure the conservation, restoration and sustainable use of terrestrial and inland freshwater ecosystems and their services, in particular forests, wetlands, mountains and drylands, in line with obligations under international agreements</td></tr>
<tr><td rowspan="2">(Linked to W.3)</td><td rowspan="2">(By 2030, ensure environmentally sound management of all wastes)</td><td>13 Take urgent action to combat climate change and its impacts</td><td>(See discussion earlier in this section, and Topic Sheet 1)</td></tr>
<tr><td>7 Ensure access to affordable, reliable, sustainable and modern energy for all</td><td>7.2 By 2030, increase substantially the share of renewable energy in the global energy mix</td></tr>
</table>

NOTE: The text and target dates for each element of the Global Waste Management Goals are already explicit or implied in the cross-referenced targets within the draft SDGs. The exception is goal W.2, highlighted in italics, which is a stepping stone en route to the goal in W.3.

This chapter has made a clear case for taking action on waste management and indeed for making waste management a priority alongside other pressing and perhaps more obvious priorities such as poverty, health and food. The GWMO sets out to look at what is happening worldwide on waste management, at what needs to be done and at the tools to make that happen. The technologies required to 'solve' the waste problem are largely already available and have been much written about. The GWMO has chosen rather to focus primarily on the 'governance' issues which need to be addressed to establish a sustainable solution – including the regulatory and other policy instruments, the partnerships and, crucially, the financing arrangements – and to provide a 'toolkit' to be used in developing a solution appropriate to the local situation.

Guide to the Global Waste Management Outlook

The GWMO is organized into 5 further chapters:

2 **Background, definitions, concepts and indicators**
Providing a full introduction to the GWMO

3 **Waste management: global status**
Covering both waste generation and management practices

4 **Waste governance**
Focusing in depth on regulatory and policy instruments, institutional arrangements and stakeholder inclusivity

5 **Waste management financing**
Addressing the alternative financing models available

6 **Global waste management – the way forward**
Synthesizing the overall findings, drawing conclusions and making recommendations for actions towards sustainable waste and resource management

A number of topics cut across this chapter structure. Their key elements are presented through stand-alone 'topic sheets' and case studies which can be found both within and between chapters.

TOPIC SHEET 1

WASTE AND CLIMATE

Contributing to greenhouse gas mitigation through waste and resource management[1]

Using a lifecycle approach, it has been estimated that a 10 to 15% reduction in global greenhouse gas emissions could be achieved through landfill mitigation and diversion, energy from waste, recycling, and other types of improved solid waste management. Including waste prevention could potentially increase this contribution to 15 to 20%.

The potential contribution of waste management to climate change mitigation

- Waste and resource management offers a number of opportunities for mitigating greenhouse gas (GHG) emissions across a wide range of industrial sectors. However, there are peculiar challenges in estimating and accounting for GHG emissions from waste and resource management.
- The approach adopted by many countries is to report GHG emissions in accordance with the UN Framework Convention on Climate Change (UNFCCC), with emission sources broken down by sector in keeping with recommendations issued by the Intergovernmental Panel on Climate Change (IPCC). To avoid double counting, the economy is divided into sectors, each with its own emission sources. The waste sector includes 'solid waste disposal on land' (landfills), 'wastewater handling' (anaerobic digestion), 'waste incineration' without energy recovery and 'other' (which is effectively limited to composting).
- Other components of waste management, including transport, recycling, agricultural use of compost and waste incineration with energy recovery, are reported under other IPCC sectors.
- Alternative methods often employ a life-cycle approach in order to allow the inclusion of the effects of waste management on other parts of the economy.
- Whatever GHG accounting method is used for waste management, the upstream-operating-downstream (UOD) generic framework[2] allows the system boundary to be specified for a particular measurement. Indirect upstream accounts for emissions associated with energy and material inputs. Direct operating emissions include all unit operations of waste management, from collection and transport through recycling to treatment and disposal. Indirect downstream includes savings and emissions from energy substitution, material substitution and carbon binding (sequestration).
- A significant body of work has applied lifecycle assessment (LCA) to GHG accounting for waste management.[3] LCA methods allow for the inclusion of the benefits of materials recycling, organics recycling and energy recovery, as such processes displace the use of alternative sources of virgin materials or energy, and thus also the GHG emissions associated with their production and distribution (which the IPCC would allocate to other sectors).
- LCA methods can in principle be extended further in order also to include the benefits of waste prevention and reduction, such as the benefits of avoiding producing products that would end up as waste. However, in extending the methods this way, delineating a system boundary that is consistent becomes quite challenging.

1 Topic Sheet prepared by David C. Wilson, with input from Natalia Reyna, Imperial College London and Wolfgang Pfaff-Simoneit, KfW.

2 Gentil, E. et al. (2009). As with all references within this Topic Sheet not specifically cited in footnotes, see Annex A, Chapter 1, Waste and Climate.

3 An early compilation of such papers is contained within Christensen (ed.) (2009).

Figure 1 Simplified schematic of waste management and climate change (excluding prevention)[4]

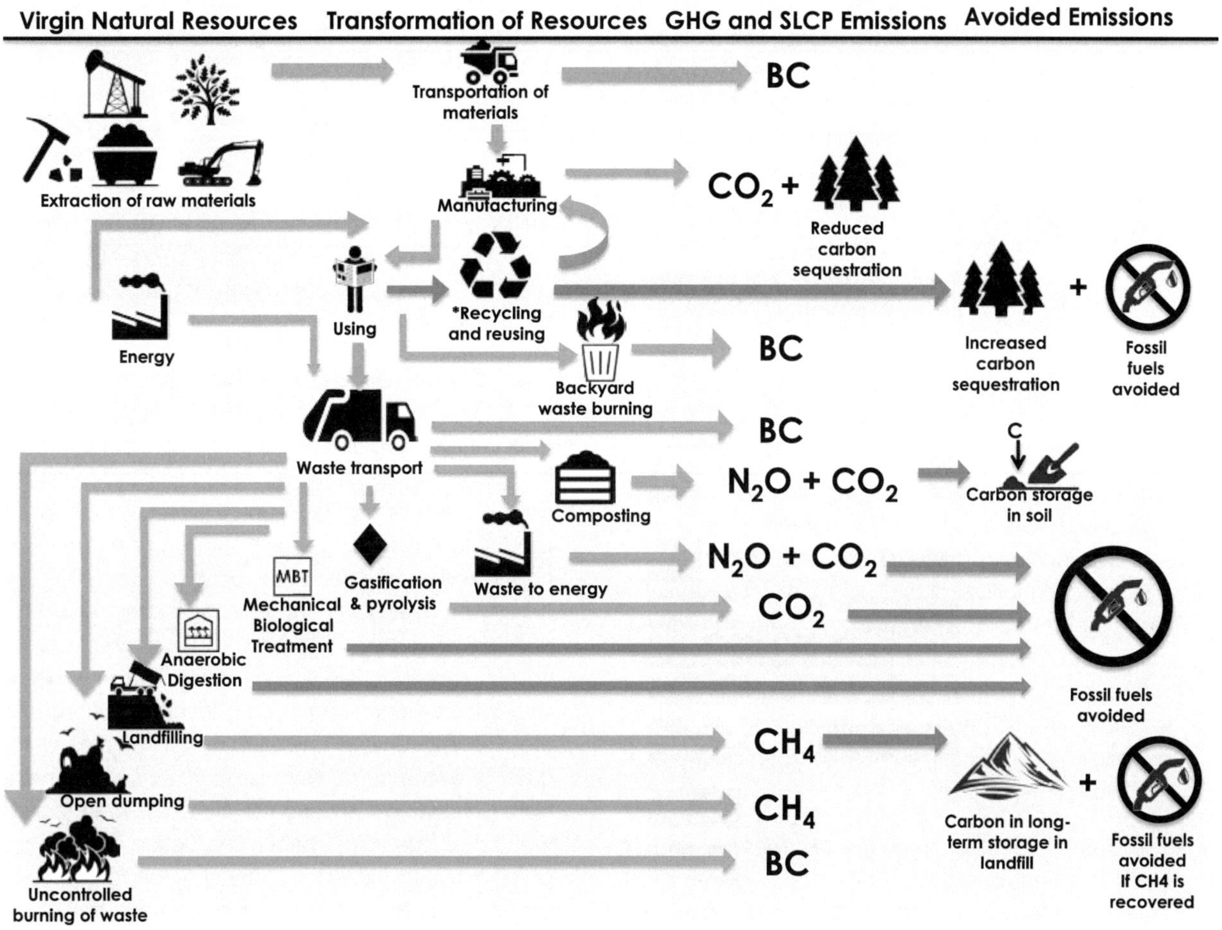

- Each LCA analysis is conducted on the basis of a particular set of detailed assumptions. This naturally restricts the applicability of the conclusions to a very specific local situation. Even small variations in the assumptions, such as in the source of the electricity that is displaced by energy or materials recovery, can change the results significantly. This makes it very difficult to generalize the results or even report 'generic' estimates. So any summaries based on LCA analyses will necessarily quote ranges of figures while emphasizing both the uncertainty of any estimates and the need to conduct a specific LCA for each individual, local situation.

Emissions from solid waste management

- *Figure 1* shows schematically the wide range of waste management and recycling's potential impact upon GHG emissions and mitigation. For clarity, waste prevention is excluded from consideration.

- The IPCC has estimated the direct contribution of the solid waste and wastewater sectors to GHG emissions at 3 to 5%, based on 2010 data. Of this total, 97% is due to methane emissions, split roughly equally between methane from landfills and from wastewater.[5] This result reflects both the limited scope considered (emissions for landfill, incineration without energy recovery and composting only), the high global warming potential (GWP) of methane compared to CO_2, and the lack of data on other emissions shown in *Figure 1*, including black carbon (BC) and nitrous oxide (N_2O).

- Table 1 shows the relative global warming potential (GWP) values of different kinds of emissions. In GHG emission mitigation, waste management is of particular interest, both because its primary emissions have a much higher GWP than CO_2 and because both methane and black carbon are short lived climate pollutants (SLCPs).

- SLCPs are currently receiving much attention. Even if emissions of long-lived GHGs ceased today, atmospheric levels would not decline significantly for a long period, as removal processes require considerable time. In contrast, reductions in SLCPs offer an opportunity to influence the rate of near-

4 Figure prepared by Natalia Reyna, Imperial College London. Inspired by an earlier figure published by USEPA, available at http://www.epa.gov/region9/climatechange/waste.html

5 IPCC (2013).

term global warming. Moreover, polar regions are especially sensitive to the effects of black carbon as the deposition of black carbon onto snow and ice has an additional warming effect.[6] What's more, black carbon is a component of fine particulate matter (PM2.5), and methane contributes to ground-level ozone formation, both of which have adverse impacts on human health and ecosystems. Thus reductions of these SLCPs are expected to bring broad-based benefits.

- Methane emissions from landfill have received the most attention in terms of waste and climate. Black carbon is much less documented and its impacts are less well understood. Estimates of global black carbon emissions suggest that 40% comes from open burning of biomass, including forest and grassland fires and open burning of both urban and agricultural waste, while 60% comes from energy sources, including power plants, industry, transport and residential fuel use. The mass emissions of black carbon from biomass burning are an order of magnitude less than the methane generated from waste, but the much higher GWP of black carbon in terms of CO_2 equivalent makes black carbon potentially a significant contributor, particularly over a 20-year time horizon.
- Therefore, control of methane from landfill and of black carbon from open burning both require urgent attention. Obtaining better data on black carbon emissions from open burning is a key priority.

Greenhouse gas reductions through waste and resource management

- The IPCC estimate of around 3% contribution from SWM to global GHGs in 2010 focused on methane emissions from landfill. Efforts to control these emissions through upgrading open dumps to controlled landfills, through engineering measures to collect and either burn or generate energy from the methane, and through diversion of waste from landfill, have been ongoing in high-income countries since the 1970s, so this figure does not include savings that have already been made.
- The IPCC estimate also does not account for savings achieved through recycling. Using secondary raw materials in industrial production to displace virgin materials significantly reduces GHG emissions, both by reducing direct energy consumption in the production process – e.g. in glass production by 35%, paper and steel over 50%, plastics over 70% and aluminium over 90% – and by the indirect upstream avoidance of mining, processing and transport of primary raw materials. The net savings from recycling vary widely depending on the local conditions and the assumptions in the LCA. For example, in the case of waste paper, very large GHG savings result from expanding the scope of the analysis to include wood displaced from paper production as a source of energy production substituting for fossil fuels.[8] Similarly, textile recycling can demonstrate large GHG savings when factoring in the energy intensive nature of cotton production.

Table 1 Relative global warming potentials and lifetimes of various emissions[7]

CONTAMINANT	LIFETIME	GLOBAL WARMING POTENTIAL (GWP) TIME HORIZON		
		20 years	100 years	500 years
Carbon Dioxide	**Variable (long)**	1	1	1
Methane	12.4 years	86	34	8
Black Carbon (BC)	3-11 days	1,200-3,200	300-1000	100-400
Nitrous Oxide	121 years	270	300	150

6 Bond, T. et al. (2013).
7 IPCC (2013) and other sources

8 A series of papers in Christensen (ed.) (2009) examines GHG emissions and the contribution to climate change mitigation from recycling paper, glass, plastic, metals and wood.

- Similarly, the IPCC estimate does not include the indirect downstream impacts of organic recycling and energy from waste.[9]

- A German study has used LCA to examine the GHG savings already achieved in Germany and other EU countries and their potential for further savings.[10] The results show that by diverting organic waste from landfill to recycling and also to energy, Germany has moved since 1990 from the waste sector being a net polluter emitting annually the equivalent of 38 million tonnes of CO_2 (38 mt CO_2 eq/a), to contributing a net saving of 18 mt CO_2 eq/a. The overall reduction from the waste sector between 1990 and 2006 amounts to 56 mt CO_2 eq/a, or 24% of the total decrease in GHG emissions achieved in Germany over that period. This represents a reduction of around 5% of total GHG 1990 emissions already achieved by 2006. These gains were achieved despite the fact that West Germany had already made significant progress in improving its SWM practices and reducing landfill gas (methane) emissions prior to 1990.

- The same study also modelled the potential for GHG savings from improved waste management practices in three upper middle income countries with high levels of uncontrolled disposal or landfill (Mexico, Tunisia and Turkey). The results show a potential 10 to 15% contribution to reducing total national GHG emissions from improved SWM, including landfill mitigation and diversion, recycling and energy from waste. GIZ and KfW have together developed a SWM-GHG Calculator that allows developing countries to compare the GHG emissions of different SWM options.[11]

- A 2004 study[12] examined the available technological options to stabilize global GHG emissions at the then-current levels, rather than the doubling that would be forecast under business as usual. To help focus policy efforts, the total reduction required by 2050 to achieve such stabilization was divided into seven equal parts, each of 1 billion tonnes of carbon per year, known as 'stabilization wedges'. It has since been argued that reductions achieved through improvements in integrated solid waste management could roughly equal the reduction goal of one of these wedges.[13] This suggests the potential to reduce current total GHG emissions by around 14%, roughly in agreement with the independent German estimate.

- All of these estimates exclude waste prevention. To take just one example, ongoing work is focusing on the environmental footprint of global food waste. Early results suggest that the carbon footprint of food produced but not consumed is 3.3 billion tonnes CO_2 eq/a, representing around 9% of total global GHG emissions.[14] Were such emissions generated by a country, that country would be the world's third-largest GHG emitter, behind only the US and China.

- As noted above, it has been estimated that waste management has the potential to contribute a 10 to 15% reduction in global GHG emissions. Extending that to waste and resource management in a broader sense would significantly increase the potential, to 15 to 20% or perhaps even more, if waste prevention was also included.

9 Christensen (ed.) (2009) addresses GHG emissions and the contribution to climate change mitigation from incineration and co-combustion, composting and compost utilization and anaerobic digestion and digestate use. Landfill and uncontrolled disposal are also addressed.

10 Dehoust, G. et al. (2009).

11 Giegrich, J. and R. Vogt (2009).

12 Pacala, S. and R. Socolow (2004). Stabilization wedges: Solving the climate problem for the next 50 years with current technologies, *Science*, 305, 968-972.

13 Bahor, B. et al., Waste Management & Research (2009), 27: 839-849. In Hansen (ed.) (2009).

14 FAO-UNEP (2013). Food Wastage Footprint. Impacts on Natural Resources, Technical Report.

TOPIC SHEET

2

THE WORLD'S 50 BIGGEST DUMPSITES[1]

The world's 50 biggest active dumpsites have been profiled by the Waste Atlas Partnership,[2] an initiative in which teams of academics and waste professionals on four continents relied on 59,000 crowd-sourced files from 25 countries to compile a list of the '50 biggest sites' where waste is currently being dumped. As shown in Figure 1, most of the sites are in Africa, Asia and Latin America/Caribbean, with two in Europe. The sites are historically and physically different. They also differ in size, receive various amounts of waste and host different numbers of people either working at the dumps or living in the surroundings. However these 50 sites all have at least one thing in common: the serious threat they pose to human health and the environment. They cannot be considered simply as a local problem. They affect the daily lives of 64 million people, a number similar to the population of France. Eliminating all such dumpsites around the world must be a priority for the global community.

Figure 1 World's 50 biggest dumpsites

Source: Developed from the Waste Atlas Partnership (2014). *Waste Atlas: The World's 50 Biggest Dumpsites, 2014 Report.*

1 Waste Atlas Partnership (2014), listed in Annex A, Chapter 3, Municipal solid waste management.

2 Waste Atlas is an online database produced in collaboration between D-Waste, ISWA, WtERT, SWEEP-Net, SWAPI and the University of Leeds. The aim is to collate and standardize the wealth of municipal solid waste data available on the Internet.

Jam Chakro in Pakistan is one of the largest dumpsites in the world, extending over 202 hectares. The dump 'serves' one of the biggest informal sector communities with more than 5,000 members and affects the life and health of an additional 5 million people living within 10 km of the site.[3] Among the most impactful e-waste dumpsites is the Agbogbloshie dump in Accra, Ghana which receives around 192,000 tonnes of e-waste annually and pollutes soil, air and water and causes serious health impairments in the 10,000 scavengers gaining their livelihood from sorting and recycling. The Bantar Gebang dump in Bekasi, Indonesia takes in 230,000 tonnes of municipal solid waste every year, with the total amount of disposed waste estimated to be 28 to 40 million tonnes.

The Vinca dumpsite in Belgrade, Serbia also qualified for the top 50 list. It receives 700,000 tonnes of waste annually, including medical waste and e-waste. The site is located only 2 km from the Danube river and the nearest settlement.

Drawing conclusions from the study of these dumspsites, the Waste Atlas establishes the (potential) correlation between the size of the sites and the number of people living within a 10 km radius (*Figure 2*), shows the types of waste being received at the sites (Figure 3) and provides the characteristics of a 'typical' site, based on the median values of the 50 studied sites (*Figure 4*).

Figure 2 Population living within a 10 km radius of the 50 dumpsites

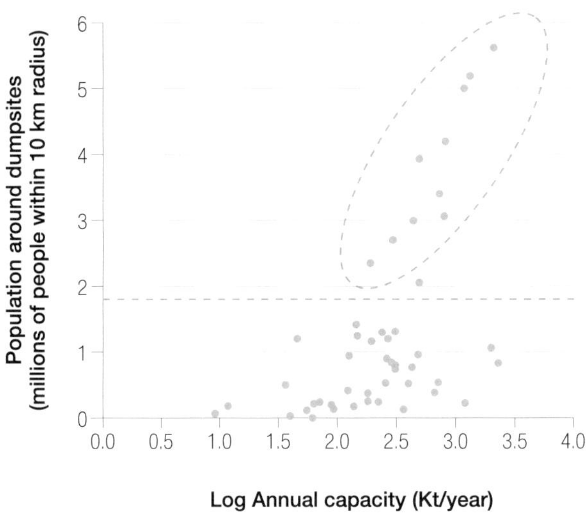

For populations of more than 2 million people living within 10 km around dumpsites, the logarithm of the annual capacity of the site (amount of waste by weight it accepts per year) has a slight linear relationship with the number of people. This indicates that bigger dumpsites can, in general, be associated with greater numbers of people living around them.

Source: Waste Atlas Partnership (2014). *Waste Atlas: The World's 50 Biggest Dumpsites, 2014 Report.*

Figure 3 Number of the 50 dumpsites accepting different combinations of waste types

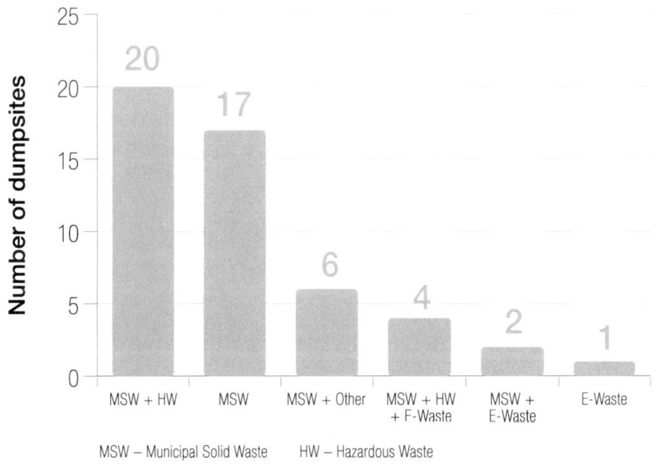

This figure classifies each of the 50 sites according to the main types of waste it receives: municipal solid waste (MSW), hazardous waste (HW), e-waste and other waste. All sites but one receive municipal solid waste either with or without hazardous waste; e-waste was identified as an important input at seven of the sites.

Source: Waste Atlas Partnership (2014). *Waste Atlas: The World's 50 Biggest Dumpsites, 2014 Report.*

3 The Jam Chakro site was built as a sanitary landfill, but within five years of opening it reverted to being a dumpsite. Rouse, J.R. (2006). Seeking common ground for people: Livelihoods, governance and waste. Habitat International 30(4): 741-753.

Figure 4 The typical 'monstrous' dumpsite

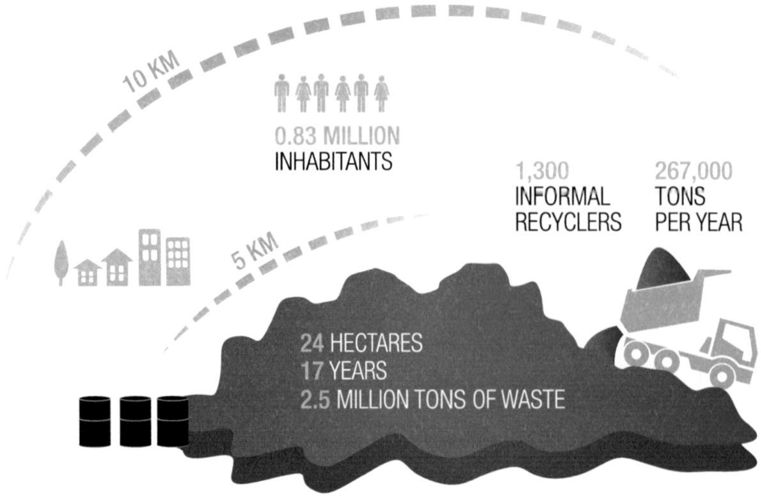

The figures show the median values for each variable across the 50 dumpsites. This means that, for example, the nearest settlement is less than 0.5 km away for half of the sites, and farther than 0.5 km away for the other half.

Source: Amended from Waste Atlas Partnership (2014). *Waste Atlas: The World's 50 Biggest Dumpsites, 2014 Report.*

BACKGROUND, DEFINITIONS, CONCEPTS AND INDICATORS

CHAPTER 2

Following the introduction to the importance of waste management in Chapter 1, this Chapter provides a full introduction to the GWMO. It begins by setting out the mandate for the GWMO together with its aims, objectives, audience, outcome, development process and plans for regional Outlooks (2.1). It then explains what the GWMO means by 'waste' and the concept of waste as a resource to be managed, followed by an overview of the scope and coverage of the GWMO (2.2). A historical perspective is provided through consideration of the various drivers for waste and resource management (2.3). The principal analytical tools used in the GWMO, including integrated sustainable waste management, lifecycle analysis and the waste management hierarchy, are briefly introduced (2.4), as are waste related data and indicators (2.5).

© Sanjay K. Gupta

2.1 THE GLOBAL WASTE MANAGEMENT OUTLOOK (GWMO)

In 2013, the **UNEP Governing Council**,[1] in **decision GC 27/12** on Chemicals and Waste Management, requested UNEP *"to develop a global outlook of challenges, trends and policies in relation to waste prevention, minimization and management ... to provide guidance for national policy planning."* This is to be done *"taking into account the materials life-cycle, subject to the availability of extra-budgetary resources and in consultation with Governments and stakeholders, building on available data, best practices and success stories, taking into account the Global Chemicals Outlook and any other relevant initiatives."*

Responding to this mandate, UNEP's Executive Director appointed the International Environmental Technology Centre (IETC), in collaboration with the International Solid Waste Association (ISWA), to develop this Global Waste Management Outlook (GWMO).

2.1.1 Overall aims

The GWMO has been developed to provide an overview of waste management around the world – how it has developed over time, what the current status is and where the current thinking on "state of the art" indicates that the future lies over the medium term. To reflect UNEP's mandate, the GWMO presents waste management as essential for ensuring public health and environmental protection once waste has been generated, and also, in the wider context of 'waste and resource management', highlights the need to consider the entire life-cycle of materials and products in order to prevent and minimize waste.

The Outlook aims to provide the rationale and the tools for taking a **holistic approach towards waste management** and for **recognizing waste and resource management[2] as a significant contributor to sustainable development and climate change mitigation.**

2.1.2 Specific objectives

Within these overall aims, a number of specific objectives have been defined:

- **Position waste management as an area requiring urgent action,** and call for policy and decision makers to take such action.
- **Expand the concept of 'waste management' to become 'waste and resource management'**, including waste prevention and minimization and also aspects of resource efficiency and sustainable consumption and production (SCP).
- **Demonstrate the relation of waste and resource management to other global challenges** such as sustainable development, water and energy balance and security, sound chemicals management, climate change, food security, resource scarcity and security and poverty alleviation; establish the links to wider health and environmental policy challenges.
- **Identify policies and governance strategies for sound waste management**, considering the varying levels of economic and human development between countries, their needs and the practices in use; provide a critical overview of what instruments have been deployed towards which goals and have worked under which circumstances.
- **Examine the available approaches to waste management financing and set out a framework for establishing a sustainable financing model in a particular local situation**; consider the direct costs and revenues, the costs of inaction and the indirect benefits of environmentally sound waste management; examine how to raise sufficient revenue to cover the net costs of service provision, and examine investment financing.

1 At the first universal session of the UNEP Governing Council held in February 2013, Member States recommended to the UN General Assembly that the Governing Council, the 58-member governing body of UNEP in place since 1972, be renamed the United Nations Environment Assembly of the United Nations Environment Programme with universal membership.

2 'Resource management' is used in this document to refer to the use of materials and energy that can be recovered from waste and to the prevention of the use of resources. 'Waste management' implies a focus on what to do with wastes once they have been generated, including reuse, recycling and recovery. 'Waste and resource management' broadens the focus to include also 'upstream' measures to prevent, reduce and minimize wastes prior to their generation. (See also Section 2.2.2, including *Figure 2.1*.)

- **Assemble[3] sets of standardized performance indicators** on waste management that allow benchmarking exercises and facilitate better analysis of the state of waste management around the world and provide a standardized means for monitoring progress.

Widespread experience around the world has shown that solving society's waste management challenges requires both the technical aspects (the physical infrastructure) and the governance and financial aspects to be in place.

While the GWMO has a very broad scope, it also aims to be a relatively concise, descriptive and readable document. The GWMO complements more technical publications on waste management by focusing in particular on governance and finance. However, a brief summary of the technical approaches and their status is also included.

2.1.3 Audience and outcome

The GWMO aims to be **relevant to all countries**, regardless of their current state of development in terms of waste and resource management.[4] That said, it places **particular focus on lower-income countries** which are often struggling to bring their waste under control.

The GWMO is targeted at a relatively high but non-technical level for a wide range of professionals and decision makers both in central and local governments and other groups of stakeholders, including NGOs, civil society, business, the manufacturing sector, the waste industry, financial institutions and academia.

To expand its reach and cater to the wide-ranging needs of such a broad audience, the GWMO has been developed as a set of three products comprising:

- **This publication, targeting mainly a professional audience**, to increase the understanding of possible paths to integrated sustainable waste and resource management, taking due cognizance of the particular situation under consideration. Chapters 1 and 6 are aimed also at high-level decision makers, as a call for action and to set out a potential way forward.
- **A summary for high-level decision makers** responsible for developing policy instruments or allocating funding and thus facilitating effective waste management. Such decision makers might include policy makers at the national and municipal levels and within international organizations, financial institutions and other funding agencies as well as business leaders.
- **A short non-technical summary aimed at the general public** to establish and disseminate the linkages between local waste management actions and their impacts at the global level, as well as their contributions to global and national sustainable development goals, and to raise awareness of the role of the civil society in waste management.

2.1.4 GWMO development process

The GWMO has been developed through a multi-stakeholder process.

A highly experienced **editorial team** was established to work on the main content and additional **contributions** have been integrated from stakeholders around the world.

The GWMO has been subject to **consultations** at different stages of its development:

- An initial consultation meeting was held at the beginning of the development process to ensure that the Outlook would be designed to meet the expectations of the different stakeholder groups. This meeting brought together experts from national and local governments, intergovernmental organizations, the private sector, academia and public interest groups. Their contributions assisted in the refinement of the scope and initial outline of the proposed Outlook and in the design of the development process, including provisions for stakeholder consultations.[5]

3 The GWMO aims to gather existing indicators, rather than develop them.
4 See further discussion in Section 2.3.
5 Additional information on the initial consultation meeting can be found at http://www.unep.org/ietc/InformationResources/Events/GlobalWasteManagementOutlookGWMO/tabid/106373/Default.aspx

- Peer reviews of the annotated outline and draft of the GWMO in the form of broad c consultations aiming to reach practitioners from around the world across various fields. The expanded group of stakeholders actively provided data, additions to the bibliography, case studies and substantial comments.

Additionally a **Steering Committee** was established to act as the quality assurance body of the GWMO.

2.1.5 Future plans

The GWMO is the first of a series of Outlooks. This publication providing the general overview will be complemented by **Regional Waste Management Outlooks** which are currently planned for Asia, Central Asia, Africa, Latin America and the Caribbean (LAC), Mountain Regions and Small Island Developing States (SIDS). The regional Outlooks will allow deeper analyses of the state of waste management specific to each region, reflecting its concrete situation and proposing recommendations better tailored to each region.

Different member countries have already requested the development of the Waste Management Outlook for their specific region. For example, in the case of LAC, during the XIX Meeting of the Forum of Ministers of Environment of Latin America and the Caribbean (Los Cabos, Mexico, 11-14 March 2014), LAC countries requested UNEP to "*develop a regional outlook on the main challenges, trends and policies related to integrated waste avoidance, minimization and management, so that it can be used as guidance for the design and implementation of national policies, plans, programmes and projects.*" Additionally, the Interstate Commission on Sustainable Development (ICSD) in Central Asia requested IETC's assistance to put together an Outlook for their region, and the Asian Development Bank (ADB) and the Association of Southeast Asian Nations (ASEAN) have asked for the development of the Regional Outlook for Asia.

2.2 DEFINING THE SCOPE AND COVERAGE OF THE GWMO

Building on a discussion of what the GWMO means by 'waste', this section sets out the 'system boundary' for the GWMO, with a summary provided in Table 2.1 at the end. This is explained within the context of the life-cycle of materials and products (*Figure 2.1*).

2.2.1 What does the GWMO mean by waste?

'Waste' is a very broad concept. The GWMO focuses on one group of the many usages of the word 'waste': unwanted or discarded materials *'rejected as useless, unneeded or excess to requirements'.*[6] Waste can be viewed as the combination of four wrongs - a wrong substance, in a wrong quality, in a wrong place at a wrong time.[7]

Even this usage of 'waste' is still very broad, as it includes such unwanted outputs of human activity as gases, liquids and solids as well as discharges to the three environmental receiving media of air, water and land. The UN Statistics Division uses the term 'residuals' rather than 'waste' in this broad context, where they comprise one of six components of a comprehensive set of environmental statistics.[8] 'Residuals' are then subdivided into three parts: emissions to air, generation of wastewater and *generation of wastes*.

It is in this rather narrower context that internationally agreed definitions of 'waste' exist, for instance within the Basel Convention on the Control of Transboundary Movements of Hazardous Wastes and their Disposal, which has 183 Parties[9] and is thus nearly universal. The Basel Convention defines 'wastes' as '*substances or objects which are disposed of or are intended to be disposed of or are required to be disposed of by the provisions of national law*'. This includes substances or objects which are subject to disposal operations which either lead to or do not lead to the possibility of resource recovery, recycling, reclamation, direct re-use or alternative uses.[10] A colloquial summary of this definition of 'wastes' might simply be 'stuff people throw away'.

6 Taken from the authoritative Chambers 21st Century Dictionary. http://www.chambers.co.uk/search.php
7 Huda (2008), listed in Annex A, Chapter 1, Waste management.
8 UNSD (2013), listed in Annex A, Chapter 2, Data and indicators. The six components are: 1. Environmental conditions and quality; 2. Environmental resources and their use; 3 Residuals; 4. Extreme events and disasters; 5. Human settlements and environmental health; and 6. Environmental protection, management and engagement.
9 Current as of August 2015. The updated status of ratifications is available at http://www.basel.int/Countries/StatusofRatifications/PartiesSignatories/tabid/4499/Default.aspx. Further information on the Basel Convention and other Multilateral Environmental Agreements is given in Box 4.13.
10 For the list of such operations, reference is made to Annex IV of the Basel Convention, available from http://www.basel.int/Portals/4/Basel%20Convention/docs/text/BaselConventionText-e.pdf

One way of visualizing the difference between 'wastes', as defined by the Basel Convention, and emissions to the atmosphere or to receiving water bodies, is that wastes are collected, stored and managed. The impacts are thus delayed and perhaps moved elsewhere. In contrast, emissions may first be treated and are then discharged locally into the atmosphere or receiving water bodies, where they are generally dispersed relatively quickly - the impacts are thus more immediate. As a result, legislation and technologies for environmental control of 'residuals' have traditionally been subdivided largely by the receiving media - air pollution control, water pollution control and waste management (often termed solid waste management, and broadly synonymous with what might be termed 'land pollution control').

So the GWMO is broadly following the accepted international concept and definition of 'wastes'. However, the GWMO is also taking a holistic approach: environmental management works best if policy and regulatory control, and the agencies in charge of each, are integrated across pollution control (to air, water and land) and waste management. While these are often separate control regimes, the interfaces between them need to be strong. Control of air pollution to meet emissions standards concentrates the contaminants as air pollution control (APC) residues; the treatment of both human sewage and other wastewaters to meet discharge standards concentrates the contaminants in sludges; both APC residues and wastewater treatment sludges then become 'solid wastes' requiring some form of handling and/or disposal, often to land. In this sense, the huge progress around the world over the last 50 years in controlling air and water pollution has further increased the challenges of (solid) waste management, which was in many ways a less immediate and thus a less high-profile issue. But unless (solid) waste management is tackled alongside air and water pollution control, the pollution is merely shifted from one receiving medium to another. Former uncontrolled dumpsites, particularly those that have received hazardous waste, are a major category of contaminated site.

The function of the kidneys might be a useful analogy for waste management - each removes contaminants from the materials flowing through the system and concentrates them into 'waste' which then needs to be collected and managed. The existence of environmentally sound waste management is thus a prerequisite for the effective control of hazardous chemicals.

2.2.2 Waste as a resource

The mandate for the GWMO is: '*To develop a global outlook of challenges, trends and policies in relation to waste prevention, minimization and management ...*'. So the GWMO is about waste *and* resource management, looking upstream at waste prevention and all of the components of a circular economy as well as downstream at the management of wastes after they have been discarded (*Figure 2.1*). Using resources more efficiently and generating less waste reduces business costs while providing other upstream benefits such as lower vulnerability to unreliable supplies, decreased dependency on global markets for critical materials and reduced depletion of natural capital stocks (or natural resources expenditure). It also reduces risks to public health and the environment.

Looking upstream in *Figure 2.1*, the scope of the GWMO firmly includes waste prevention, which interacts strongly with sustainable consumption and production (SCP),[11] which looks at the implications of, among other aspects, the design, manufacturing and packaging of products in the market, as well as consumption patterns and lifestyles. Waste prevention also has strong relevance in other complimentary policy areas such as the 'green economy', 'innovation for sustainability' and indeed the 'circular economy'.

11 UNEP (2012d). Listed in Annex A, Chapter 1, Precursors of the GWMO. Sustainable consumption and production is also covered in Topic sheet 3, found after Chapter 2.

Figure 2.1 Moving from waste management to resource management within a circular economy

Where we are coming from: The linear economy and waste management

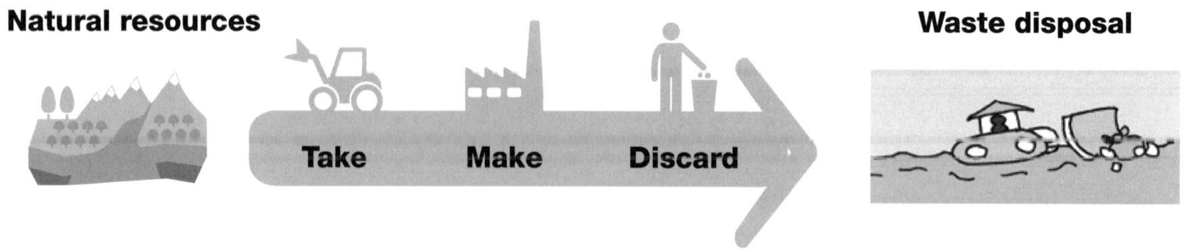

Where we need to get to: Resource management within a circular economy

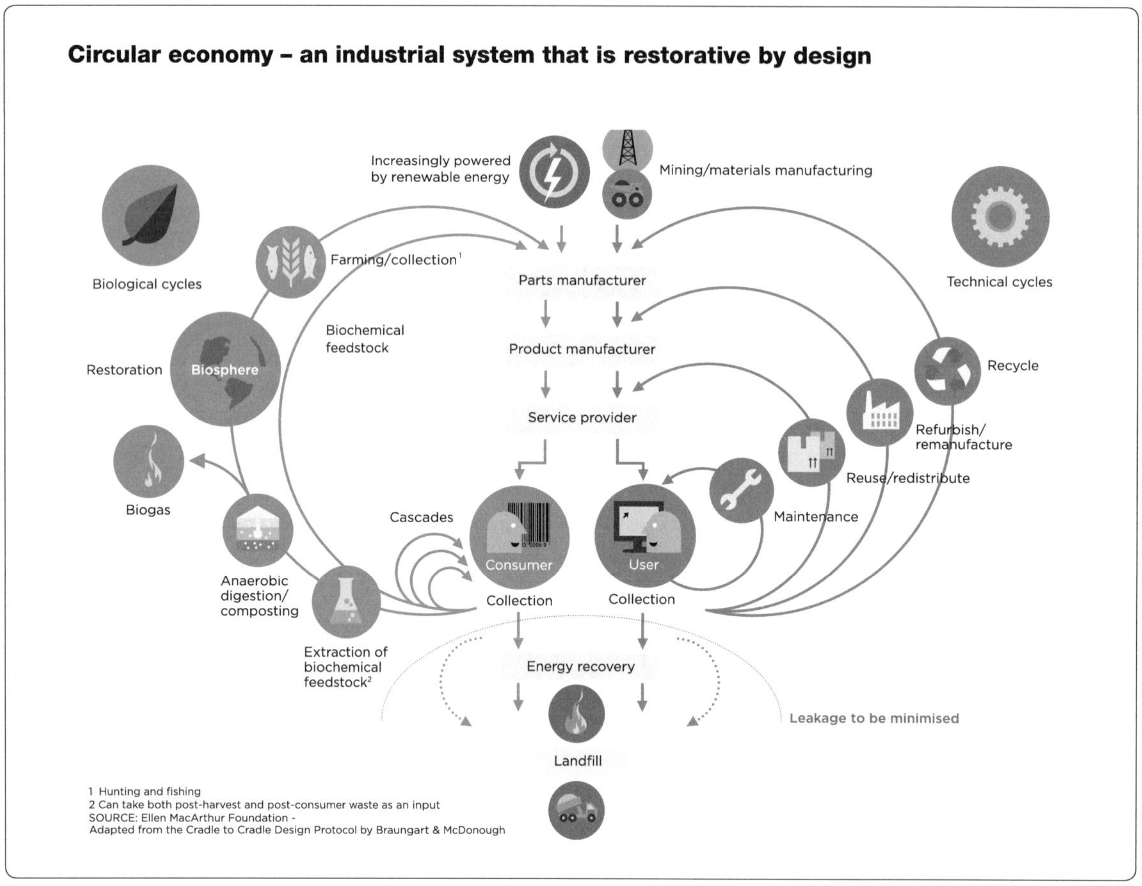

Source: Ellen MacArthur Foundation (2013) (http://www.ellenmacarthurfoundation.org/circular-economy/circular-economy/interactive-system-diagram).

Note: Both parts of this Figure are a simplification. Waste is generated at all steps in the life-cycle of materials and products, in procuring the raw materials through agriculture or mining, manufacturing the materials and the products, and distribution and retail, not just when the consumer discards a product at end-of-life. In the circular economy, these wastes from each step can also enter numerous feedback loops, or be utilized for energy recovery, or go to landfill. Each step in the life-cycle also generates other residuals, including emissions to air and water. Air pollution control and wastewater treatment concentrate these contaminants into additional waste. The energy recovery step could also be represented with a number of additional feedback loops, as both recovered energy and potentially recycled metals and ash are fed back into the system.[12] To show all of this detail would be to make for a rather complex graphic.

The 'feedback loops' depicted in *Figure 2.1* indicate that recycling and recovery are also firmly within the scope of the GWMO. So the GWMO considers not only what is commonly called the 'waste management sector', but also the recycling or secondary raw materials industry (the industrial value chain) and also the agricultural value chain. It is clearly important for national and international legislation to differentiate between wastes for disposal and materials for recycling and recovery, and indeed between 'wastes' and 'non-wastes'; for the GWMO, all of these distinctions are covered in the discussions.

12 A recent variant elaborating on the energy recovery step has been developed. See ISWA (2015a) listed in Annex A, Chapter 4, Sustainability.

2.2.3 Coverage of the GWMO

A life-cycle approach

The GWMO needs to look at the waste arising throughout the entire life-cycle of materials and products (*Figure 2.1*). These include mining and quarrying (extraction); agriculture and forestry; industry (materials, parts and product manufacturing); construction and demolition; commerce and institutions (distribution and services); and consumption (households).

Focusing on 'higher-risk' wastes

However, 'waste management' as it currently exists has taken a pragmatic approach, having in fact tackled some of the 'higher risk' waste first.[13] The criteria for selecting the waste to be treated include proximity to people and the characteristics of the waste. It was public health concerns that resulted in municipal solid waste (MSW) in urban areas being controlled during the 19th century. Then, when environmental controls were first introduced in the 1970s, these included not only MSW but also commercial and industrial (C&I) and construction and demolition (C&D) wastes. Specific additional controls were focused on hazardous waste, due to its higher risks.

Responsibility for the environmentally sound management of waste belongs with the waste generator. The necessity to maintain public health in cities led to city authorities being given legal responsibility for the safe collection and disposal of MSW. As a consequence of this, the city auhority becomes in effect a 'proxy-generator'. While the definition of MSW varies widely between countries, household wastes are always included, as are some C&I and C&D wastes from smaller businesses and institutions. In principle, C&I and C&D wastes from larger waste generators remain the direct responsibility of the waste generator, rather than passing on to the city authorities. However, these distinctions are often 'fuzzy' in developing country cities - by default, the city often manages all the waste generated in the municipal area, including C&I and C&D. The integrated nature of the management of these 'urban' wastes is further highlighted by the fact that for the physical management of their waste, both cities and business waste generators rely on 'the waste industry' - which depending on the country may be mainly private sector or part public and part private; moreover, it may be mainly 'formal sector' or part formal and part informal.

So while the GWMO may have its main focus on MSW and on hazardous waste, it is necessary also to include both general C&I and C&D waste firmly within the scope. Similarly, both the public and private, and formal and informal sectors, are within its scope.

A number of more specific waste types are also highlighted at specific points in the GWMO, particularly using Topic Sheets, Case Studies and Boxes. Examples include food waste and e-waste, which may arise at many different stages of the life-cycle, and so need more focused attention, and also disaster waste and marine litter.

Agricultural and forestry, and mining and quarrying (extraction) residues and wastes

All this leaves two very large sources of waste. Traditionally most agricultural and forestry residues and wastes (crop residues, animal faeces and urine, wood residues and waste) have been managed by the farmers and forestry managers as part of the agricultural value chain and the nutrient cycle. With the intensification of agriculture, the question of whether or not to include such materials in national waste control regimes has become a higher profile issue as increasing quantities of crop residues are finding an alternative use as biomass for energy generation. A substantial portion of mining and quarrying residues and waste are surplus rock or over-burden, which are high volume, low risk materials that are of necessity managed by the mining companies as near as possible to the point of generation. Again here, such materials have often been outside of national waste control regimes. A major exception is mine tailings - which are arguably not an 'extraction' waste but rather an industrial waste from the processing of ore in order to concentrate the metals. **In order to keep the GWMO manageable, we have followed most national waste legislation and left the bulk of agriculture and forestry and mining and quarrying residues and wastes aside for this first edition. The Outlook does draw attention to mine tailings** and recommends a follow-up study specifically on that topic. It also recommends that the subsequent Regional Outlooks revisit their scope to determine the appropriate priority for such wastes in each region.

13 The drivers and history of waste management are expanded on in the next section.

2.2.4 Geograhical scope

The GWMO focuses primarily on policy planning at the national level, but in order to do so, it addresses waste management at local (city), regional and national levels. Most waste is generated in or near cities, and public health and environmental risks are greater when in proximity to people, so for MSW at least, it makes sense to focus first on cities. Many high-income countries have already gradually extended MSW management to rural areas, but for most developing countries this remains a long-term aspiration rather than a short-term priority.

Internationally, the GWMO aims to be relevant to all countries, whatever the income-level, in all the major world regions and in the context of different climates and local geographies. Clearly, the level of detail cannot be great – hence the need for follow-up Regional Outlooks and for specific Topic Sheets, Case Studies and Boxes to draw attention to examples of more specific issues (e.g. small island developing states [SIDS]). A number of global issues are also highlighted, incuding global markets for secondary raw materials, global trafficking of waste and marine litter.

A summary of the scope of the GWMO is provided in *Table 2.1*.

Table 2.1 The scope of the GWMO – setting the 'system boundary'

NO.	CATEGORY	MAIN FOCUS WITHIN THE GWMO	ALSO CONSIDERED	OUTSIDE THE SCOPE
1.	**Receiving environmental media**	A holistic approach, but with a focus on 'solid waste' to land	Residues from air and water pollution control Gaseous and liquid hazardous wastes	Gaseous emissions to air and wastewater discharges
2.	**Waste as a resource**	Scope includes waste prevention, reduction, reuse, recycling and recovery – including both industrial and agricultural value chains	Related aspects of sustainable consumption and production (SCP) Both 'waste' and 'non-waste'	N/A
3.	**'Source' of waste**	Some focus on municipal solid waste (MSW), including waste from households and smaller business and institutions	Mining and quarrying (M&Q) wastes and agricultural and forestry (A&F) wastes	N/A
4.	**Responsibility for waste**	Commercial and industrial ('C&I') wastes, and construction and demolition (C&D) wastes, from larger waste generators		
5.	**Properties of waste**	Non-hazardous waste Hazardous wastes (including hazardous healthcare and household hazardous wastes)	N/A	N/A
6.	**Specific types of waste**	Includes plastics, food waste, e-waste, packaging, emerging waste streams, post-disaster wastes, marine litter	N/A	Radioactive (nuclear) waste
7.	**Public and private sectors**	Waste managed by both public and private sector operators. Private sector includes: waste generators, producers and distributors, waste industry, industrial value chain recyclers and agricultural value chain	N/A	N/A
8.	**Formal and informal sectors**	Both formal and informal sectors, including both waste management and recycling	N/A	N/A
9.	**Geographical scope**	Urban waste Considers local, regional and national levels, with primary focus on national policy Regional differences around the world Small island developing states (SIDS) Global markets for materials for recycling	Waste generated in rural areas	N/A

2.3 DRIVERS FOR WASTE AND RESOURCE MANAGEMENT

Particular drivers have contributed to the evolution of solid waste management systems over the last thousand years. This section introduces the three broad groups of drivers attributed to what are commonly referred to as 'developed' countries; together with a broad brush summary of the origins of modern waste and resource management systems in developed countries.[14] This is followed by an examination of the current drivers operating in emerging and developing countries.

2.3.1 Historical drivers in developed countries

- **Resource value of the waste.** The earliest driver was a financial one, with real economic demand providing an incentive for the repair and reuse of products and recycling of materials. Food and garden waste entered the agricultural supply chain, either as animal feed or being composted and used to improve the soil. The Industrial Revolution concentrated people in the cites, increasing waste quantities while interfering with natural cycles back to the soil. Large numbers of the urban poor made their livelihoods as informal 'street buyers' of source-separated recyclable and reusable materials, or as 'waste pickers' retrieving any saleable items from mixed waste. A combination of 'free' municipal waste collection, rising living standards and widespread availability of increasingly complex goods from cheap mass production led to a steady decline in MSW recyling rates, although the intrinsic resource value continued to be a main driver for the recycling of 'home scrap', clean materials produced within industrial processes, so that a significant secondary materials industry (the industrial value chain) has operated continuously since the 19th century.

  Both rising disposal costs since the 1980s and increased public opposition to new sites have led developed countries to 'rediscover' recycling for MSW, driven not so much by the inherent commodity value of the materials, but rather by recycling markets offering a competitive 'sink' (see *Figure 2.2*). These two resource drivers have recently been joined by a third – growing concerns about the depletion of virgin raw materials and resource scarcity, leading to a focus on waste prevention and resource efficiency. This is often described as a (potential) shift from 'waste management' to 'resource management', with complementary policy initiatives such as sustainable consumption and production (SCP), the 'circular economy' and the 'green economy'.

- **Public health.** Municipal solid waste management (MSWM) as we know it today has its origins in the cholera epidemics which struck the newly industrializing countries of Europe and North America from the 1830s.[15] A causal link was made to decaying organic wastes, and the public health revolution focused primarily on improving sanitary conditions and collecting municipal solid wastes.

- **Environmental protection.** Until the emergence of the environmental movement in the 1960s and 1970s, the norm was for uncontrolled disposal or burning of wastes, sometimes on a large scale at a centralized site. The local environmental impacts through air pollution, surface and groundwater pollution and land and marine contamination were often severe, leading to numerous scandals caused in particular by abandoned dumpsites which had received hazardous waste. However, even on a small scale, through littering, the amenity loss can be significant (particularly in touristic areas). Comprehensive environmental legislation was introduced, with waste controls being implemented in stages. The initial focus was on phasing out uncontrolled disposal, both on land and by burning; environmental standards were then gradually tightened to control local pollution from landfill leachate, landfill gas, air emissions from incinerators and the environmental impacts of other treatment and disposal facilities.

  A more recent driver is impacts upon the *global environment*, notably anthropogenic climate change. As noted in Topic Sheet 1, the potential contribution of waste and resource management to climate mitigation is well in excess of the direct contribution from reducing methane generation from the anaerobic decomposition of organic waste in landfills and eliminating the generation of black carbon from uncontrolled waste burning. For this reason, climate change is an additional driver for eliminating open

14 This discussion is very much a summary and by its nature an over-simplification. A more detailed discussion of a wider range of drivers is provided by Wilson (2007), and later expanded on for developing countries by Marshall & Farahbakhsh (2013), while the historical overview is expanded on by Scheinberg, Wilson, Rodic (2010), pp 19ff, listed in Annex A, Chapter 1, Waste management. A short bibliography on waste history and drivers is provided in Annex A, Chapter 2, Waste history and drivers.

15 It can be argued that MSWM was recognized as an essential utility service in a number of earlier human civilizations, of which perhaps the earliest was in the city of Mahenjo-Daro in the Indus Valley around 2000 BC. See Worrell and Vesilund (2012), listed in Annex A, Chapter 1, Waste management.

dumping and burning, for diverting waste from landfill and for both recycling and waste prevention, all of which significantly reduce greenhouse gas emissions across a number of economic sectors.

In summary, and as shown in *Figure 2.2*, the primary drivers for SWM in most OECD and high-income countries have thus changed over time, from resource value up to the mid-19th century, to public health up to about 1970, then to (local) environmental protection, and more recently to a more complex mix including global environmental issues and the extension of 'waste management' to 'waste and resource management'.

Figure 2.2 Schematic timeline showing evolution of the primary drivers for municipal solid waste management in developed countries

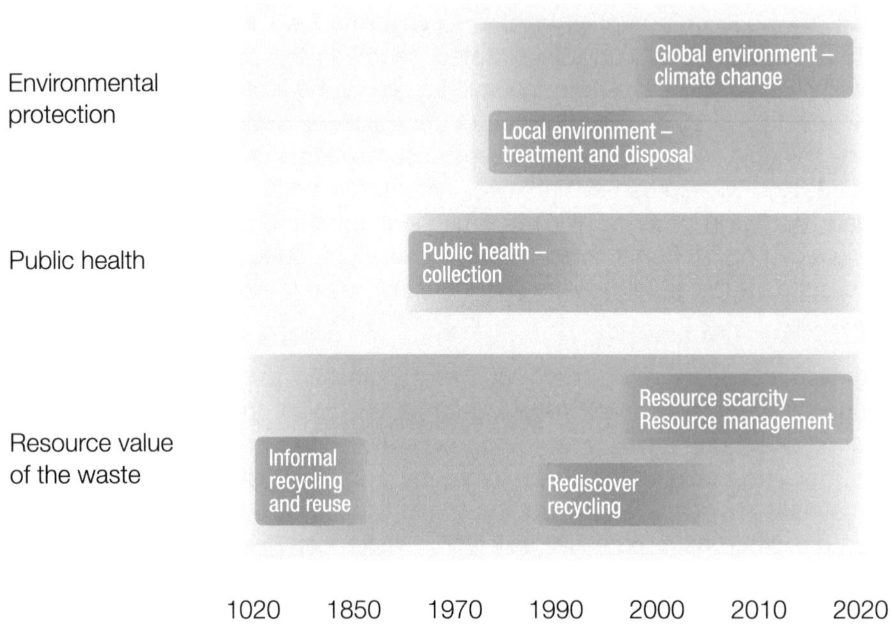

Note: The vertical axis arranges individual drivers into one of the three groups introduced in the text. Individual boxes show how particular drivers have come into play at different periods in the original OECD countries – the ends of the boxes are intended to be 'fuzzy', as the timings are not exact, and many of the drivers have continued to be important well beyond the time periods shown

Source: David C. Wilson (n.p. 2015)

2.3.2 Current drivers around the world

The current situation in the highest income countries represents one end of a spectrum. Near the opposite end of the spectrum would be some of the least developed countries, for whom the early drivers shown in *Figure 2.2* are still very relevant. In the absence of control systems, uncontrolled dumping and open burning of wastes remain common practices, having both direct and indirect health as well as environmental impacts. Direct health impacts affect for example children coming into contact with uncollected wastes or breathing the fumes if those wastes are burned. Indirect health impacts occur for example via blocked drains and watercourses, which provide breeding grounds for vectors causing infectious diseases such a cholera, dengue fever and plague, and also cause flooding. The rise of new infectious and auto-immune diseases like AIDS, avian flu and Ebola adds new urgency to the public health driver if the waste generated is not treated adequately.

So, in some of the least developed countries, public health is still a major driver, as not all MSW is collected, resulting in serious impacts, both directly on human health and indirectly through flooding. The local environmental driver is also still strong, with uncontrolled disposal widespread and hazardous wastes often dumped alongside MSW. In terms of the resource driver, the informal recycling sector is thriving in many cities as a means of livelihood for the urban poor and often achieves moderately high recycling rates, albeit at a social and environmental cost.[16]

Each country will be in a different position between these two ends of the spectrum of realities around the world, with most of the drivers shown in *Figure 2.2* coming into play to differing extents. As rural populations continue to migrate to the cities and living standards improve, urban waste quantities in developing countries are growing rapidly. In response to this, governments are seeking to learn from experiences elsewhere and to 'leapfrog' the slow, step-by-step progress that most developed countries made over the last century and a half. Many developing and emerging countries have already made substantial progress in developing their solid waste management systems. In many cases, legislation and even a national strategy may be in place, but implementation and enforcement are often lacking.

The GWMO aims to be relevant to all countries wherever they fall on this spectrum of current drivers and regardless of their state of development in terms of waste and resource management. It aims to provide assistance in identifying and implementing the appropriate policies and actions for the next steps in developing their own specific waste and resource management systems.

It is however essential to highlight from the beginning that every situation is different. Thus there are no inherently 'right' or 'wrong' solutions, nor is it possible to provide a simple 'user manual' that will solve every problem. Rather, the GWMO intends to illustrate what can be achieved, and provide some interesting examples and case studies from which each country can draw lessons relevant to their own situation. **Regardless of their starting point, all countries can continue to improve.**

2.4 AN ANALYTICAL FRAMEWORK FOR THE GWMO

A number of assessment methods have been developed for or adapted to waste management systems. The principal analytical tools used in the GWMO are briefly introduced here.

2.4.1 Integrated sustainable waste management (ISWM)

Developing a waste management system is complex. Experience suggests that, for a system to be sustainable in the long term, consideration needs to be given to:

- All the **physical elements** (infrastructure) of the system, from waste generation through storage, collection, transport, transfer, recycling, recovery, treatment and disposal.
- All the **stakeholders** (actors) involved, including municipalities; regional and national governments; waste generators/service users (including industry, business, institutions and households); producers (those who put products on the market which become waste at the end of their life, including manufacturers, brand owners, importers and others in the supply chain); service providers (whether public or private sector, formal or informal, large or small); civil society and non-governmental organizations (NGOs) (which play a variety of roles, including facilitating the participation of other parties); international agencies; etc.
- All the **strategic aspects**, including the political, health, institutional, social, economic, financial, environmental and technical facets.

The term *integrated waste management* has been widely used with a variety of meanings,[17] but often refers only to integration across the physical elements. The concept of *integrated sustainable waste management*

16 For data on recycling rates, see *Figure 3.13* in Section 3.5. For further discussion of informal sector recycling, including common occupational health risks, see Topic Sheet 14 on the Informal Waste Sector, following Section 4.7.

17 Integrated Solid Waste Management was mentioned in the UNEP's Governing Council Decision GC 24/5 (2007) and again in GC 25/8 (2009). Past usages of the terms 'integrated waste management' and 'integrated solid waste management' are classified in Table 1 of Wilson et al. (2013). See http://www.icevirtuallibrary.com/content/article/10.1680/warm.12.00005

(ISWM),[18] which explicitly brings together all three dimensions, is gradually becoming the norm in discussion of solid waste management in developing countries. In the GWMO, the primary analytical framework used is a simplified form of ISWM, first developed for UN Habitat's *Solid Waste Management in the World's Cities (2010)*. This is shown schematically in *Figure 2.3* as two overlapping 'triangles'.

The first triangle in *Figure 2.3* comprises the three primary physical components (elements), each linked to one of the key drivers identified in *Figure 2.2*. These provide the necessary infrastructure for solid waste management:

1. *Waste collection*: driven primarily by *public health*;
2. *Waste treatment and disposal*: driven primarily by *environmental protection*; and
3. *The 3Rs – reduce, reuse, recycle*: driven by the resource value of the waste and more recently by 'closing the loop' in order to return both materials and nutrients to beneficial use.

The second triangle focuses on the 'softer' aspects of ISWM – the governance strategies:

4. *Inclusivity of stakeholders:* focusing in particular on service users and service providers;
5. *Financial sustainability*: requiring the system to be cost-effective, affordable and well financed; and
6. *Sound institutions and proactive policies*: including both the national policy framework and local institutions.

An integrated and sustainable waste managment system must address all technical (infrastructure) and governance aspects to allow a well-functioning system that works sustainably over the long term.

As previous publications have tended to have a more technical focus, **the GWMO has chosen to focus primarily on issues of governance and finance**.

Figure 2.3 The integrated sustainable waste management (ISWM) framework used in the GWMO

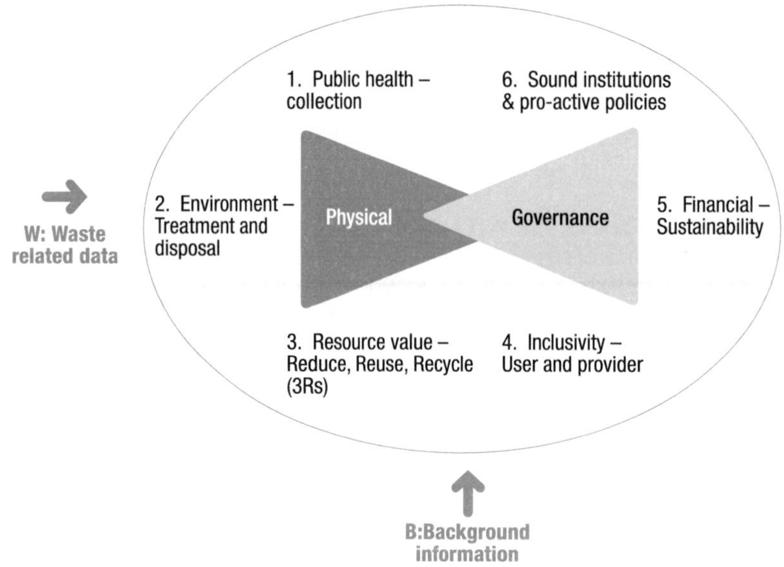

Notes: This is a simplified version of the original ISWM concept and was first devised for UN-Habitat.[19] The numbers and letters in this version of the Figure[20] cross-refer to the 'Wasteaware' benchmark indicator set, which is built around this framework and introduced in Section 2.5.3 below.

2.4.2 Life-cycle analysis (LCA) and other assessment tools

Traditional 'end-of-pipe' waste management focuses on just one segment of the life-cycle of materials and products, namely, after the point of discard. In contrast, the GWMO takes into account the wider issues of waste and resource management across the product life-cycle. Life-Cycle Thinking (LCT) is a well-established concept which aims to provide a holistic view of all the environmental, social and economic impacts that could

18 ISWM was first developed in the 1990s by the Dutch NGO WASTE and the Collaborative Working Group on Solid Waste Management in Low- and Middle-Income Countries (known as CWG). See Annex A, Chapter 2, Integrated sustainable waste management.

19 See Scheinberg, Wilson, Rodic (2010) in Annex A, Chapter 1, Waste management.

20 Wilson et al. (2013).

possibly occur during a product's lifetime.[21] Life-Cycle Assessment or Analysis (LCA) is a set of tools to quantify these impacts through the entire lifecycle. The traditional use of LCA was to compare products, with a relatively limited focus on end-of-life impacts. Recently, there has been a push to extend the scope systematically to explore resource and waste management.[22]

LCA is just one of an ever-growing set of assessment methods[23] to support decisions regarding waste and resource management. Both LCT and ISWM could be seen as specific applications of a systems approach. Another tool used in the GWMO is Materials Flow Analysis (MFA), which can be used, for example, to produce national-level accounts of material and resource flows through the economy, and to prepare a mass balance of a city's waste and resource management system.

The results of an LCA can be used to arrange the available management options for a specific type of waste into a priority order. A simple and often used rule of thumb is the **waste management hierarchy**, which provides a *generalized* priority order for waste management options and technical approaches. Waste policy initiatives are often framed in terms of 'moving waste management up the hierarchy', which as a general principle has widespread acceptance. The hierarchy can be stated in many different ways, and the specifics of the generalized priority order are themselves hotly debated.[24] The version appearing in *Figure 2.4* was agreed by the parties to the Basel Convention,[25] who further stated that the hierarchy *'encourage[s] treatment options that deliver the best overall environmental outcome, taking into account life-cycle thinking'*.

A detailed comparison of options in any specific context requires a detailed LCA, because the option considered 'better' can vary depending on the precise questions asked and the particular local circumstances at play. It follows that different versions of the hierarchy could be constructed for use in different contexts. For example, in developing countries a major priority is often to phase out uncontrolled disposal (open dumping or burning) in order to get onto the hierarchy in the first place. *Figure 2.4* recognizes this by showing 'uncontrolled disposal' below the hierarchy, and an intermediate step of 'controlled landfill' below the bottom rung of the hierarchy itself, which in the EU would be taken to equate to 'state-of-the-art' landfill meeting high technical standards.

Also, the hierarchy leaves out some of the early steps in the waste management chain, including waste storage and collection. Extending collection to the entire population for the protection of public health continues to be a major priority in many developing countries.

So, the GWMO does make use of the waste management hierarchy, but at the same time, it recognizes its limitations.

Figure 2.4 Waste management hierarchy

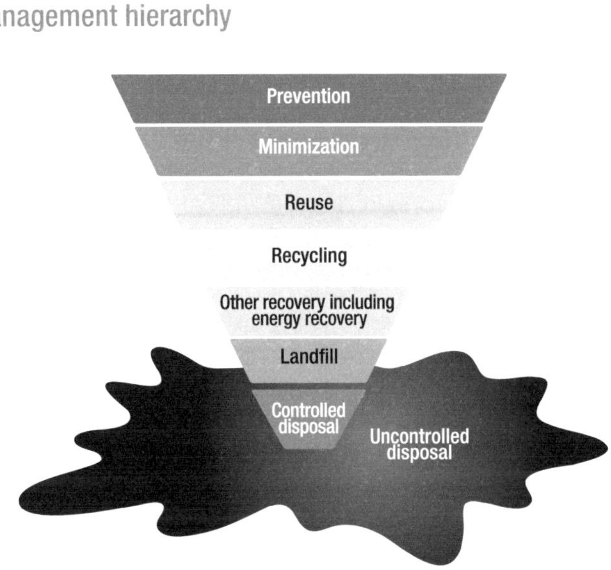

Note: This version of the hierarchy prepared for the GWMO. Please see the caveats in footnote 23 below. The sequence of steps is based on that agreed by the parties to the Basel Convention (see footnote 24)

21 http://www.lifecycleinitiative.org/starting-life-cycle-thinking/what-is-life-cycle-thinking/
22 See Annex A, Chapter 2, LCA and other assessment tools for waste and resource management.
23 See Annex A as above, which includes recent reviews by Allesch & Brunner (2014) and Zurbrügg et al. (2014).
24 The applicability of the waste hierarchy has been questioned in a world where resource recovery involves global value chains, as it refers only to environmental aspects and not to public or occupational health, financial or other considerations such as materials criticality. Some current research efforts aim to supplement or replace it with more sophisticated tools.
25 Basel Convention, COP10 2011 - Decision BC-10/2: Strategic framework for the implementation of the Basel Convention for 2012–2021. Annex Item II, Guiding Principles, point a.

2.5.1 Introduction

'Data are the lifeblood of decision-making and the raw material for accountability. Without high-quality data providing the right information on the right things at the right time, designing, monitoring and evaluating effective policies becomes almost impossible.'[26]

It is useful to distinguish clearly between some related terms:[27]

- Environmental **data** are large amounts of unprocessed observations and measurements about the environment and related processes. These can then be processed and structured to produce information in the form of **statistics**, using an agreed set of statistical methods, standards and procedures.
- Environmental **indicators** are used to synthesize and present complex statistics (which may be either quantitative or qualitative) in a form that is designed to summarize, simplify and communicate information and to turn that into knowledge. A special case is where a set of indicators is combined into an environmental **index**, in which each individual indicator is weighted so that they can be combined into a single number.
- In summary, statistics aim to convert raw **data** into useful **information**; indicators then help to transform that **information** into **knowledge**, which can then be used to make **wise decisions.**

One of the key objectives of the GWMO includes: *'Assemble a set of standardized performance indicators on waste management that allow benchmarking exercises and facilitate better analysis of the state of waste management around the world and provide a standardized means to monitor progress'.* The collation of comprehensive, credible, and updated data on waste and resources is essential in achieving this objective.

Performance indicators provide a good basis for assessing the existing situation, carrying out a comparison and tracking changes or progress made over time. For indicators to be useful as a tool for decision makers and politicians, they need to simplify the potential mass of data by being selective, by focusing on the important elements rather than trying to cover all aspects. By doing so, the information the indicators present will be relatively easy to use and understand.

2.5.2 Quality and availability of waste-related data

Unfortunately, compiling high quality data on waste and waste treatment has long been a challenge. The available estimates are diverse, not verified or reliable, and often rather dated. Thus transforming waste data into reliable waste statistics has proven difficult. Some of the major areas of concern are:

1. **Lack of standard definitions and classifications.** The definitions of waste streams such as MSW, C&I waste, C&D waste, agricultural and forestry wastes and mining and quarrying wastes vary widely among countries, including within the European Union (EU).[28] Arguably, it is unrealistic to suggest that a single global definition of waste could or indeed should be applied in all countries. At the same time, it is realistic to ask that each country's definitions be clear and applied consistently.
2. **Absence of measurement and of standard methodologies for measurement.** The weighing of wastes is both relatively recent[29] and still not universally practised. Many cities in the developing world still rely on estimates of MSW based on the volume of the vehicles used for collection and disposal. Here, as with many other aspects of waste assessment, there is a lack of standard methodologies. For example, it is often unclear at what point estimates or measurements have been made and whether the data refer to MSW or to all waste in the city, or whether it is waste as generated, or as collected, or as delivered to a disposal site (has some separation of materials for recycling already taken place before the measurement was made?). Measurement and assessment also tend to be limited to the official or formal waste management system. Activities outside of that system, including uncontrolled (and often illegal) dumping or burning, as well as recycling by the informal sector, are neither measured nor reported. Waste composition data are even more uncertain, even in high-income countries, as measurement tends to be occasional and not carried out on a comprehensive or consistent basis.

26 This is the opening summary statement of the UN Secretary General's Independent Expert Advisory Group on a Data Revolution for Sustainable Development report published in November 2014. See IEAG (2014), listed in Annex A, Chapter 2, Waste data and indicators.
27 UNSD (2013).
28 See Section 2.2.3.
29 For example, universal weighing of all collected municipal solid waste at the transfer or disposal facility was achieved in the UK in 1993.

3. **Lack of standard reporting systems.** Statutory reporting systems for waste management in a standard format are still the exception. While tracking systems for the transboundary movement of hazardous wastes are mandatory under the Basel Convention, data collection and reporting of total quantities generated are only advisory and thus patchy. National data collection systems do exist for MSW; however for other waste streams including C&I and C&D wastes, the reporting systems are not uniform even among high-income countries.[30] So although data are reported to and collated by both the EU (Eurostat) and the OECD, there are both gaps in the data and questions over their inter-comparability. Double counting is one issue, as often when waste is processed, the output from the treatment facility is counted again as a 'new' waste. As a result, not only tallying the total quantities but also tracking a particular item of waste from its origin to its final destination is difficult. For developing countries, the availability of any data beyond MSW, never mind reliable data, is rather uncommon.

Given these many interrelated challenges, there is an urgent need for clear and consistent methodologies for waste assessment. Any approach needs to be underpinned by carrying out as much fieldwork 'on the ground' as possible. It is also essential to gather information from as wide a range of 'actors' in the city's waste and resource management system as possible. Two additional pieces of advice would be to go as far as is practicable to establish at least a rough mass balance, including estimates of unmeasured 'losses' from the system, and to document carefully all of the assumptions and estimates made and the 'rules of thumb' used, so that the assessment process is transparent and able to be audited. Some guidance is already available on the more detailed 'how' of data collection in order to establish the baseline situation in a developing country city.[31]

2.5.3 Waste management indicators

Converting environmental data into statistics and then into indicators, and potentially using those as an input to indices, requires a level of confidence in the underlying data that statisticians do not have in the case of waste management data.[32] This is illustrated starkly by the Environmental Performance Index (EPI), which publishes rankings of the comparative environmental performance of countries every two years. The EPI has been developed over 15 years and six iterations and its component indicators measure performance across 20 environmental issues. A further nine issues are explicitly listed as important, but are currently excluded from the EPI[33] as 'global data are still lacking'. These include municipal solid waste management, recycling rates and toxic chemical exposures. Hazardous waste and any other waste indicators are also excluded.

The UN has had guidance on environmental statistics in place since 1984, and this guidance was updated in 2013.[34] The basic set related to the generation and management of waste comprises the total waste collected, the amount treated by different methods, the number of treatment and disposal facilities and their capacity for MSW, hazardous waste and other/industrial waste; the amount of recycled waste; and imports and exports of both waste and hazardous waste.

Once waste statistics have been standardized, a selection of these statistics can be used as indicators. For example, the definition and use of quantitative 'key performance indicators' has been relatively common within individual, mainly high-income countries.[35] The typical set of benchmark indicators in the 1990s was: waste generated per capita; proportion of waste being managed by different methods; and proportion of households with regular collection service. Similar indicators are still used today as part of composite sustainable development indicators in cities.[36] A common addition is the per cent of recyclables actually recycled. It is difficult to achieve consistent comparisons here, so a three-year European collaboration has developed a standard methodology to improve benchmarking.[37]

Recently, much activity has focused on developing indicators for benchmarking the performance of waste management systems in individual cities or countries in greater detail. An overview appears in *Table 2.2*. The majority of this work has been academic research, often as PhD projects. The scope is often quite specific,

30 An interesting exception, where data for the broad range of wastes from industry, including from mining and mineral extraction, is much more widely available than data for MSW, is the Russian Federation and some of its neighbouring countries - comprehensive reporting from (formerly state-run) enterprises has been in place since Soviet times, and detailed annual reports are published. See http://www.mnr.gov.ru/regulatory/list.php?part=1101 (in Russian; waste data ('ОТХОДЫ') are on pages 64-67 of the 2013 report).

31 For example, see Scheinberg, Wilson and Rodic (2010; UN-Habitat publication) listed in Annex A, Chapter 1, Waste management; also Wilson et al. (2001) (World Bank publication), IJgosse et al. (2003) and UNEP (2009c), listed in Annex A, Chapter 1, Strategic planning.

32 Ronconi (2001), listed in Annex A, Chapter 2, Data and indicators.

33 http://epi.yale.edu/our-methods

34 UNSD (2013), listed under Annex A, Chapter 2, Data and indicators.

35 For example, KPIs were defined and used in England up to 2010. Ireland published an annual waste benchmarking report from 2006-2010 (Forfás, 2010). A review of four international and eight national indicator sets is provided by Munizaga & Garcia (2012), including those used in Chile and Colombia.

36 Global City Indicators Facility (2014). See http://www.cityindicators.org/

37 Regions4Recycling (2014). See http://www.regions4recycling.eu/R4R_toolkit/R4R_methodology

focusing on a particular aspect of the waste management system, such as selective collection or comparing technologies. Some proposals are untested, while others have been applied to one (local) case study.

Table 2.2 Overview of recent initiatives to develop waste management performance indicators

The focus here is on for benchmarking the performance of waste management systems in individual cities or countries. Indicators for comparing the performance of particular technologies have not been included. The sequence is alphabetical, in order of the first authors. Both the selection and each summary are based on a review of the published literature. In the rows marked with an asterisk (*), specific references have been chosen to represent multiple similar but unrelated initiatives.[38]

No.	PURPOSE/DESCRIPTION	REFERENCE	ORIGIN	APPLICABILITY	EXTENT OF USE
1.	Various methodological proposals for a SWM indicator set*	Armijo et al. (2011)	University research	This one - Brazil *Others - Europe	None One case study each
2.	Evaluation of programs for selective collection of MSW including social inclusion*	Bringhenti et al. (2011)	University research	Brazil	Indicators validated but not tested
3.	Performance Assessment System for urban water, sanitation and solid waste	CEPT University (2010)	Action research funded by Gates Foundation	India	400 urban local bodies in two Indian states
4.	Comparison and ranking of SWM programs in the U.S.	Greene & Tonje (2014)	University research	US	10 municipalities in New York State
5.	Monitoring progress of 3R efforts towards a green economy. Discussion paper and factsheets on performance indicators in the 3Rs & resource efficiency	Hotta et al. (2014)	Asia Resource Circulation Policy Research Group (international collaboration)	Low-, middle- and high-income countries	None (Discussion paper and factsheets)
6.	Assessment of the performance of SWM collection*	Huang et al. (2011)	University research	People's Republic of China (PRC)	307 local governments
7.	Evaluation of MSWM services using a Balanced Score Card*	Mendes et al. (2012)	University research	Europe	1 municipality in Portugal
8.	Assessment of the performance of recycling, treatment and disposal as a component of sustainable SWM systems*	Menikpura et al. (2013)	University research	Thailand	1 municipality in Thailand
9.	Service level benchmarks for water supply, sanitation and solid waste management in Urban Local Bodies	Ministry of Urban Development of India (2010)	Government of India	India	Widespread but results not yet published
10.	Garbometer: Evaluation on MSWM systems	Munizaga & Garcia (2013)	University research	Spain	1 city in Spain
11.	Sustainability indicators for SWM	Polaz & Teixeira (2009)	University research	Brazil	1 city in Brazil
12.	Benchmarking performance of a national hazardous waste management system	Romualdo (2014)	University research	Low-, middle- and high-income countries	Proposed indicator set tested in 7 countries (4 in Europe plus 1 each in Africa, Asia and Latin America)
13.	'Wasteaware' ISWM benchmark indicators to compare performance of SWM in cities	Wilson et al. (2015) (Also Scheinberg, Wilson, Rodic (2010); Wilson, Rodic et al. (2012); Soos et al. (2013a)	International community of practice. Parts of the 6-year programme funded by UN-Habitat and GIZ	Low-, middle- and high-income countries	Tested in 39 cities in 6 continents; being used to monitor progress in 19 cities in Egypt. An adaptation has been used to benchmark performance across 9 countries (SweepNet, 2014).
14.	Measuring progress in national waste prevention programs	Wilts (2012)	University research	Europe	1 case, Germany
15.	Development of a zero waste index for measuring performance of SWM	Zaman & Lehmann (2013)*	University research, Zero Waste South Australia	High-income cities	3 cities (Adelaide, San Francisco, Stockholm)

38 A more extensive listing of references is provided in the paper by Wilson et al. (2015).

Some of the initiatives in *Table 2.2* stand out in terms of both their origins and the extent of their use. Both the 'Wasteaware' (row 13) and the two Indian (rows 3 and 9) initiatives have had institutional input to develop comprehensive indicators to benchmark performance of cities' SWM systems. The work of the Asia Resource Circulation Policy Group (row 5) is also notable for its broad stakeholder input, although their proposals remain at the discussion stage. The 'Wasteaware' and Indian indicators are expanded upon here as examples. The results from their application are used later in the GWMO alongside other available data.

The 'Wasteaware' ISWM benchmark indicators were developed over six years, building on work for UN-Habitat and GIZ. The framework provides a self-assessment and diagnostic tool for a city to benchmark the performance of its MSWM system and raise the 'waste awareness' of decision makers. The indicators have been designed such that they are applicable to cities in high-, middle- and low-income countries. They have so far been applied to more than 50 cities in all six inhabited continents. An earlier version was adapted and applied to benchmarking the state of MSWM systems across nine countries (see row 13). A recent research project extended the methodology and built on other published work to propose an indicator set to benchmark the performance of national hazardous waste management systems (row 12).

The 'Wasteaware' indicator set combines relatively well-established quantitative indicators for waste generation, composition and the three main physical components or infrastructure, with a corresponding, qualitative, composite indicator for the 'quality' of service provision for each physical component as well as five qualitative, composite indicators assessing performance for the three main governance aspects (see *Figure 2.3*). *Table 2.3* provides an example to illustrate this, for the city of Maputo in Mozambique.

In India, two related indicator sets have been designed to provide service level benchmarks for Urban Local Bodies (cities and municipalities), to be used for monitoring performance of water supply, sanitation, solid waste services and storm water drainage. The first of these is a National Service Level Benchmark (SLB) for four key sectors, which has been rolled out across India by the Ministry of Urban Development (MoUD) as part of its programme to facilitate critical reforms in the urban sector (row 9).[39] The SLB has now been included as an eligibility criterion for performance grants from the Central Finance Commission, with the result that over 1400 ULBs notified the central government of their service levels for the year 2010-11. These were published by MoUD as an SLB Data Book.[40]

A parallel initiative is the Performance Assessment System (PAS) (row 3),[41] which aims to develop an assessment system at the local and state levels and link the planning and fund allocation process to performance. It includes indicators related to the access and coverage, equity, service levels and quality, efficiency and financial sustainability of service provision. There is five years of annual data now available for 400 ULBs in two states. At the state level, the information is used to monitor ULB performance and assess investment requirements and also used for benchmarking. At the ULB level, it assists in planning, target setting and tariff determination. *Table 2.4* illustrates the application of the PAS to the city of Navi Mumbai.

39 To encourage and facilitate adoption of the SLB framework, the MoUD launched an SLB Pilot Initiative in February 2009. The Initiative involved provision of technical support for implementation of the framework in 28 pilot cities across 14 states and one union territory. The Pilot Initiative was undertaken under a partnership arrangement with the involvement of various development agencies, viz. Water Sanitation Program - South Asia, JICA, GTZ, CEPT (supported by the Bill & Melinda Gates Foundation) and PROOF.

40 http://moud.gov.in/servicelevel

41 PAS is an action research programme initiated by the CEPT University, Ahmedabad, with funding from the Bill and Melinda Gates Foundation since 2009. The seven-year programme works with all levels of government: national, state and local. The performance indicators developed under PAS are aligned with the MoUD framework. PAS was tested in over 400 ULBs in the two states of Gujarat and Maharashtra over five years. The PAS system is now mainstreamed in state governments through state-level SLB cells. Other states in India have also begun to implement this system.

Table 2.3 The WasteAware ISWM benchmark indicators - worked example for Maputo, Mozambique[42]

No.	CATEGORY	BENCHMARK INDICATOR (DATA POINT)	RESULTS	'TRAFFIC LIGHTS'
Background information		**Data point**		
B1	Country income category	Gross National Income (GNI) per capita	$470	
		World Bank income category	Low Income	
B2	Population of city	Total population of the city	1,131,149	
B3	Waste generation	Total municipal solid waste (MSW) generation (tonnes/year)	508,000	
Key Waste-related data		**Data point**		
W1	Waste per capita	MSW per capita kg per year	316	
		MSW per capita kg per day	0.9	
W2	Waste composition:	Summary composition of MSW for four key fractions – all as % by weight of total waste generated		
W2.1	Organic	Organics (food and green wastes)	65%	
W2.2	Paper	Paper	8.5%	
W2.3	Plastics	Plastics	8%	
W2.4	Metals	Metals	2.5%	
Physical components		**Benchmark indicator**		
1	Public health – Waste collection	1.1 Waste collection coverage (% households who have access to a reliable waste collection service)	82% (Medium)	
		1.2 Waste Captured by the System (% of MSW generated that is handled completely by the waste management and recycling system)	75% (Medium)	
1C		Quality of waste collection service	Medium/High	
2	Environmental control – waste treatment and disposal (T&D)	Controlled treatment & disposal (% of MSW for T&D which goes to at least a 'controlled' site)	0% (Low)	
2E		Quality of environmental protection in waste treatment and disposal	Low/Medium	
3	Resource Value – 3Rs: Reduce, Reuse, Recycle	Recycling rate (% of total MSW generated that is recycled as materials or as organic products)	<5% (Low)	
3R		Quality of 3Rs – Reduce, reuse, recycle – provision	Low/Medium	
Governance factors		**Benchmark indicator**		
4U	Inclusivity	User inclusivity	Medium	
4P		Provider inclusivity	Medium	
5F	Financial sustainability	Financial sustainability	Medium/High	
6N	Sound institutions, proactive policies	Adequacy of national solid waste management framework	Low/Medium	
6L		Local institutional coherence	Medium	

Key for abbreviations

B	Background information	4U	User inclusivity
W	Waste information	4P	Provider inclusivity
1, 1C	Public health	5F	Financial sustainability
2, 2E	Environmental control	6N	National framework
3, 3R	Resource value – 3Rs	6L	Local institutions

Key for colour coding

Low
Low/Medium
Medium
Medium/High
High

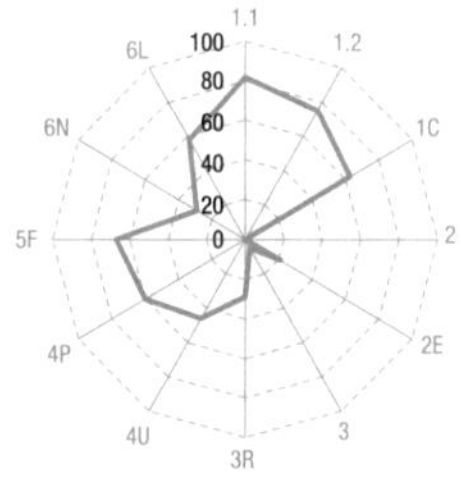

For details of how the colour coding has been assigned to the different indicators, see the original source paper

42 Wilson et al. (2015).

Table 2.4 Performance assessment system (PAS) for urban water and sanitation in India: SWM indicators for the city of Navi Mumbai, Maharashtra 2013-2014[43]

GENERAL INFORMATION

Parameters	2014
Area (Km²)	110
Population (1000s)	1244
Households (1000s)	293
Slum population (1000s)	280
Slum households (1000s)	47
Establishments (1000s)	7.3
Financial Information	
Total revenue income (Million Rs) 1 USD ≈ 60 Rs	11896
Total revenue expenditure (Million Rs)	14612
Total Capital receipts (Million Rs)	250
Total Capital expenditure (Million Rs)	7642

Solid Waste Management

Parameters	2014
Waste generated (metric tonnes per day (TPD*)	778.5
Waste collected/ transported to disposal site (TPD)	755.0
Waste at all type of processing facilities (TPD)	583.3
Waste disposed at compliant landfill sites (TPD)	150.0
Waste disposed at open dump sites (TPD)	ND
Door to door collection – Households and establishments (1000s)	294.8
Door to door collection – Slum households (1000s)	27.3
Financial Information	
Revenue expenditure on SWM* (Million Rs)	890
Revenue receipts from SWM (Million Rs)	582
Capital expenditure on SWM (Million Rs)	47

Service Level Benchmarking (SLB) Indicator Values

Coverage and Equity

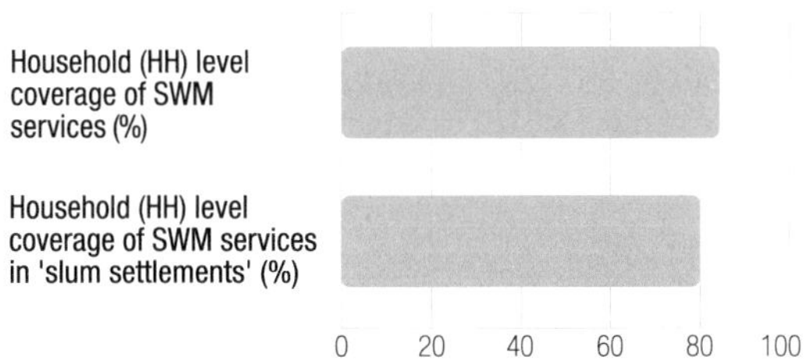

Service Levels and Quality

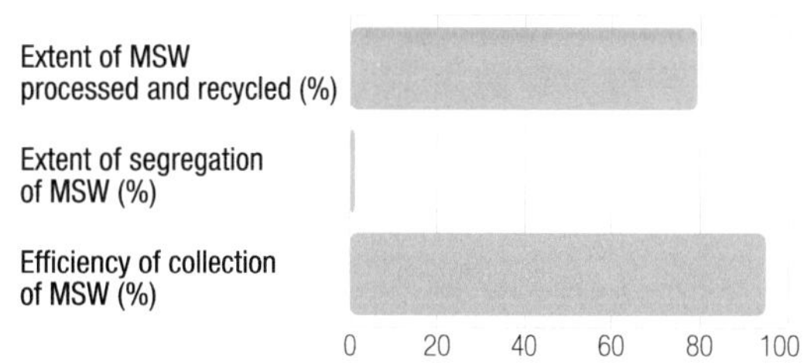

Efficiency in Service Operations

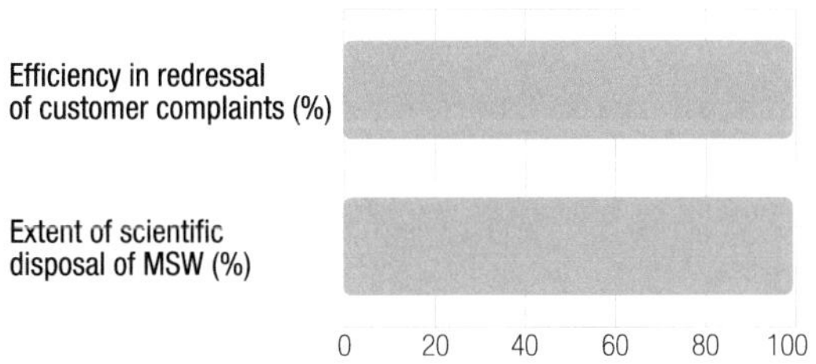

Financial Sustainability

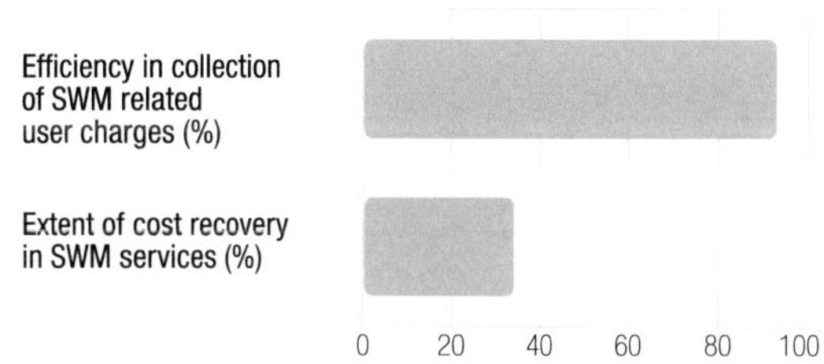

* SWM – solid waste management TPD – tonnes per day MSW – municipal solid waste

43 CEPT University – PAS (2010). See www.pas.org.in

2.5.4 Resource management indicators

Some of the indicator work listed in *Table 2.2* goes beyond waste management, moving more towards resource management. The Asia Resource Circulation Policy Research Group's discussion paper focused in particular on performance indicators in the 3Rs (reduce, reuse, recycle) and resource efficiency (row 5). Rows 14 and 15 present work on measuring progress in national waste prevention programs, and on a 'zero waste index' designed for use in high-income cities striving for zero waste.

Looking more generally, there has been extensive recent work to extend well-established national and international statistics on the financial flows associated with manufacturing and trade (i.e. the economy) to resource and waste flows, for example using Material Flow Accounting (MFA).[44] MFA is now regularly practised by some member states of the EU and detailed national calculations have been available for several years for countries such as Austria, Denmark, Finland, Germany and the UK. In comparison with efforts in the EU, efforts to establish comprehensive MFA for the United States are more recent and to date less institutionalized. In the Asia-Pacific region, Australia, PRC, Japan and the Republic of Korea have been pioneers in developing an MFA system, and considerable work in the region has been done by UNEP.[45] *Box 2.1* and *Figure 2.5* illustrate the material flow indicators in Japan. It can be observed that the final waste disposal rates have been reduced drastically in Japan since 2000, with a steady increase in the indicators for 'resource productivity' and 'cyclical use rate'.

Presently, the focus of MFA is still on individual substances (e.g. cadmium flows), specific materials, or bulk material flows (e.g. steel and steel scrap flows within an economy). The global recycling trade of plastic wastes has been documented using MFA in various studies.[46] Using the concepts of MFA may also help in throwing some light on the present situation of transboundary movements of wastes. It is likely to take many years however for full national and international accounts that show the mass flows of both virgin and secondary raw materials.

Two further applications of MFA should also be mentioned:

- The concept of *material input per unit of service* (MIPS) was developed by the Wuppertal Institute[47] to measure material resource productivity. This is particularly interesting for SWM because MIPS includes two types of material flows, the latter of which is quite unique in the literature:
  - Direct materials inputs (the natural resource commodities that enter the economy for further processing to become a material for products); and
  - "Hidden" material flows, or "*ecological rucksacks*", which are materials removed from the environment along with the desired material, and the material moved or disturbed in resource extraction or in building and maintaining infrastructure. These would include huge amounts of mine tailings and the like.[48]
- The Urban Metabolism (UM) approach measures the materials flowing into a system, the stocks and flows within it, and the resulting outputs from the system to other systems in the form of pollution, waste or exports.[49] This approach has been applied to assess and describe urban material flows and associated impacts, using different indicators and tools.[50]

44 National economic accounts were first proposed in the 1920s and became institutionalized around 1950. National resource accounts were first proposed in the 1990s, and progress so far has been similarly slow.
45 UNEP and CSIRO (2011), UNEP and CSIRO (2013), West et al. (2013)
46 Velis (2014) and Minter (2013), listed in Annex A, Chapter 3, Global secondary materials, discuss the issues related to transboundary movements of plastic wastes, market shifts, complexities and impacts attached.
47 Schmidt-Bleek (1994), Bringezu (1997), Hinterberger et al. (1997)
48 Cleveland & Ruth (1998)
49 Baccini & Brunner (2012); Pincetl et al. (2012).
50 Loppolo et al. (2014).

BOX 2.1 MATERIAL FLOW INDICATORS IN JAPAN

Japan's 2003 strategic plan for establishing a Sound Material-Cycle Society introduced material flow indicators as a tool for monitoring Japan's material flows on a regular basis. The material flows monitored refer to:

- **Inputs** – domestic and imported natural resources such as non-metallic mineral resources (e.g. rock, earth and sand), fossil resources, metallic resources (e.g. molybdenum, zinc, copper, rhodium, platinum, nickel, palladium, iron, aluminium, gold and other metals) and biomass resources, plus imported products;
- **Input of water** – included in waste and the like (sludge, animal manure, human waste, waste acid, waste alkali) and sediment and the like associated with economic activities (sludge from mining, building and water works and tailing from mining);
- **Circulative resources** – paper, scrap iron and steel, slag, non-ferrous scrap metal, scrap plastic, vegetable oil cake and grain residues; and
- **Outputs** – MSW and industrial waste for final disposal, and sediments.

Targets were set for 2010 (short-term), 2015 (medium-term) and 2020 (long-term), using a 2000 baseline.

MATERIAL FLOW INDICATORS AND TARGETS USED IN THE FUNDAMENTAL PLAN FOR ESTABLISHING A SOUND MATERIAL-CYCLE SOCIETY IN JAPAN[52]

Measuring	Indicator	Calculated as	Baseline value in 2000[53]	Target 2010	Target 2015	Target 2020
Inputs	Resource productivity (10,000 yen/ton)	GDP / Natural resources input	25	37[54] (originally 39) (50% improvement from 2000)	42 (+70 improvement from 2000)	46 Approximately 80% improvement from 2000)
Circulation	Cyclical use rate (%)	Amount of cyclical use (i.e. reuse and recycling) / (Amount of cyclical use + natural resources input)	10	14 (40% improvement from 2000)	14-15 40%-50% improvement from 2000	17 (70% improvement from 2000)
Outputs	Final disposal amount (million tons)	Amount of waste disposed	56	28 (50% reduction from 2000)	23 (60% reduction from 2000)	17 (70% reduction from 2000)

Figure 2.5 Material flow indicators in Japan[54]

In 2010, the resource productivity amounted to 374,000 yen/tonne, the cyclical use rate improved to 15.3%, and the final disposal amount declined to 19 million tons. If compared with the target set for that year it can be observed that the cyclical use rate and final disposal amount have not only achieved those targets but also surpassed the targets set for 2015.

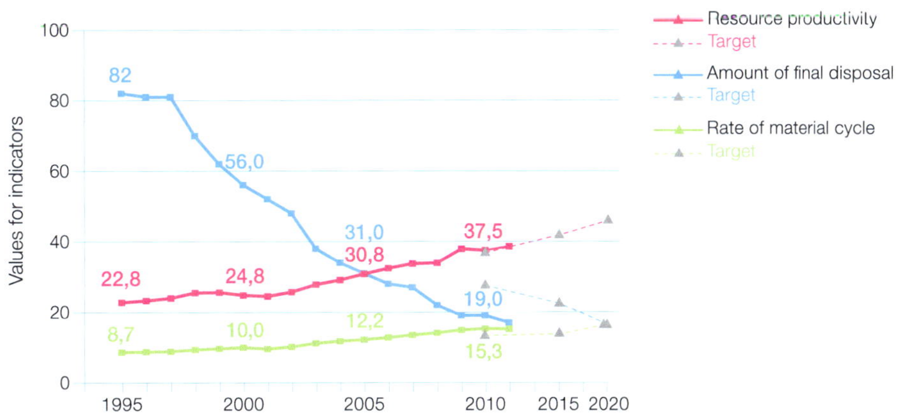

51 Fundamental Plan for Establishing a Sound Material-Cycle Society – Third Progress Report of the Plan (2013) http://www.env.go.jp/en/recycle/smcs/3rd-f_plan.pdf

52 In Japan, the government's fiscal year runs from 1 April to 31 March.

53 Changed from the first Fundamental Plan due to the revision of the method for calculating GDP. See Fundamental Plan for Establishing a Sound Material-Cycle Society - First Progress Report of the Plan (2005) at https://www.env.go.jp/en/recycle/smcs/f_plan.pdf

54 See http://www.env.go.jp/doc/toukei/data/2014_4.02.xls (in Japanese).

TOPIC SHEET 3

SUSTAINABLE CONSUMPTION AND PRODUCTION[1]

While symptomatic of and supporting economic growth, consumption has been directly linked with a range of global environmental problems – in highly industrialised countries consumption has reached rates that clearly cannot be sustained within the limits of the Earth ecological systems. Populous BRIC nations (Brazil, Russia, India and China) with economies in transitions are heading in the same direction – everyone aspires to good life and prosperity. At the same time, a large portion of human population is still in dire need of economic and human development. The UN Sustainable Development Goal No. 12 focuses on ensuring sustainable consumption and production patterns.[2]

As the UN Agenda 21 states, there is a need for new concepts of wealth and prosperity. At this crucial juncture our civilization is redefining what constitutes a good quality of life and informing our understanding of 'who gets how much', in order to enable sustainable development where it is most needed, and secure living within ecological limits. To that end, policies are being introduced to address sustainable production as well as consumption around the world. Initially, the focus was on technological innovations to achieve sustainable production by industries; however, historical evidence suggests that, notwithstanding its merits, this is not enough if we are to achieve sustainable lifestyles. Therefore, consumption has been added as one of the key concepts for sustainable development, to complement the efforts in production, as stated in the 10-year Framework of Programmes on Sustainable Consumption and Production adopted at the Rio+20 Conference.[3] In essence, sustainable consumption calls for citizens and governments to consider the environmental consequences of their consumption patterns. Accordingly, business, people and government constitute the three groups on which the change towards sustainable consumption and production depends.

1 Topic Sheet prepared by Ljiljana Rodic, with inputs from David Wilson.

2 http://sustainabledevelopment.un.org/sdgsproposal

3 See, for example, http://sustainabledevelopment.un.org/index.php?menu=204

After the 1992 UN Conference in Rio, the Network of National Cleaner Production Centres (NCPCs) was established as collaboration between UNIDO and UNEP.[4] UNEP's 2012 *Global Outlook on Sustainable Consumption and Production Policies*[5] provides a wealth of interesting cases of concrete policy measures and initiatives that aim at mandating, promoting and supporting sustainable consumption and production (SCP) in countries around the world. In addition, the UNEP's Global SCP Clearinghouse was launched in 2013.[6]

Sustainable Production

Sustainable production addresses a broad range of production features, including (a) function as the starting point for design, rather than a material product, (b) design of products for durability, repair and disassembly, (c) properties of materials used for the product and in the production processes, (d) water and energy consumption, (e) responsible sourcing, including environmental impacts such as those on climate change, on land and biodiversity, (f) social impact on workers and local communities. A useful systematic overview of issues that play a role in sustainable production is provided in the UNEP's 2009 document 'Design for Sustainability'.[7]

Sustainable production can be directly linked with waste management; at the production stage, major gains can be achieved in both quantitative and qualitative waste

4 http://www.unido.org/ncpc.html

5 Reference to this and other UNEP publications on this topic are listed in Annex A, Chapter 2, section SCP.

6 http://www.scpclearinghouse.org/. Also, there is a European Round Table on Food Sustainable Consumption and Production, http://www.food-scp.eu.

7 Reference to this and other UNEP publications on this topic are listed in Annex A.

Illusion of dematerialization: 'out-sourcing' of waste generation and management

A shift to the 'information age' is often seen as a trend of dematerialization of society, as technical innovations result in ever smaller electronic products such as computers and telephones, accompanied by incorporation of electronic components into electrical appliances, vehicles and industrial machinery. Researchers have been warning however against the prevailing 'gut' feeling that this gain in resource efficiency (amount of resources used *per unit product*) inexorably leads to decrease in the resource use and reduced environmental impact. The evidence suggests the opposite often to be true: technological efficiency improvements push consumption higher, resulting in higher use of resources *in total*. For example, technological gains in energy efficiency have been counteracted by increased energy consumption. Similarly, even though our electronic devices are getting smaller and smaller, in order to satisfy the market demand, the use of scarce materials such as rare metals for their production is sky-rocketing. This phenomenon is known as the rebound effect or Jevons paradox.[7]

The 'service-based economies' that have characterized development in industrialized countries are also seen as significant contributors to dematerialization. In reality, some of the service sectors, such as recreation and leisure, are responsible for large amounts of emissions. In addition, products for consumption in these 'modern' economies are largely manufactured elsewhere, with all the ensuing resource use, generation of huge amounts of waste from raw material extraction (mining) and from manufacture, and emissions to air and water. So it is not only production that is outsourced, but also the accompanying waste generation and pollution and the need for its adequate management and control. Environmentally sound waste management and pollution control in developing countries is thus an issue of both international social justice and corporate social responsibility (CSR).

prevention through design and selection of materials and production processes which eliminate or reduce use of hazardous substances. In addition, choice of materials and component assembly methods largely determine the recycling potential at the end of the use period. Addressing production is also important because most waste does not come from households but from other sources, such as mining, agriculture, construction, manufacturing industry and others.[9]

Sustainable Consumption

Consumers can be citizens, businesses or governments. Various education campaigns are launched and diverse channels and forms are used to provide individual consumers with information that will guide their future purchasing decisions towards (more) sustainable choices and thereby support them in developing a sense of environmental citizenship and empower them to take a key role as social agents in sustainable development.

However, the approaches to promoting sustainable consumption will probably be ineffective if they fail to address people's more pressing, overarching social concerns about equity, power, privilege and justice which they see in their everyday life both in their communities and in the society as a whole. Also, telling people to consume less when so many people can hardly afford to consume enough, telling people what to do and how to live when governing institutions do little to address their own consumption issues, will not be received well.

Moreover, consumption is not just a way to satisfy human material needs for shelter, food and clothing. For better or for worse, the act of consumption serves a multitude of other purposes; it is a way of expressing one's identity, demonstrating status and social aspirations, as well as creating lifestyle and servicing personal relationships. Therefore, in order to be effective, proposed sustainable alternatives ought to be adequate replacement not only in terms of product's functionality but also in terms of these symbolic values and meanings.

Finally, how is it possible for an individual to live sustainably when society is not geared towards sustainable living? Products are designed with planned obsolescence as a requirement and often cannot be repaired. Furthermore, it is difficult to make sustainable, 'green' individual choices within social structures and culture that promote materialistic consumerism, supported by the messages from the media and the government about the need to increase consumption in support of the economy. Simplistic approaches that ignore the currently dominant societal context of consumption will not be successful. More profound changes, nothing short of a paradigm shift, are needed.

Businesses, as one of the key actors in the system with direct influence on consumption patterns, have a major role to play in facilitating and promoting sustainable consumption and production. This can be achieved, among others, through their own purchasing choices from the suppliers, the extent to which they engage with them and other actors in their supply chain, as well as corporate culture and workers' habits on the premises.[10]

A new discourse is being created through diverse forms of interaction between the public, civil society organizations and the governments around the world, including a constructive dialogue about joint efforts towards

8 It is particularly important to examine this effect in relation to decoupling of economic growth and resources use, as discussed in 2011 UNEP document on decoupling, cited in Annex A, Chapter 2, section SCP.

9 See data presented in Section 2.3.

10 More detailed information can be found in the 2010 UNEP/Wuppertal Institute document cited in Annex A, Chapter 2, section SCP.

sustainable communities, within planetary ecological boundaries. Inclusion of sustainability considerations in formal education, starting from elementary school onwards, is generally seen as an indispensible basis for such efforts. In addition, building strong and authentic leadership, in combination with governmental institutions taking responsibility for their own day-to-day consumption practices, in other words, 'walking the talk' and abiding by the principles of fairness and equity, will have positive effects on public perception and behaviour.

Government as a 'green' customer

Around the world, governments procure trillions of dollars' worth of works, goods and services every year in a wide range of sectors, with construction, health, transport and education being the predominant ones. While

in many cases the focus is still on tendering rules and opting for the lowest price, the purchasing power of government and the wider public sector is increasingly being used to support businesses that offer more sustainable products and services. Furthermore, being at least as effective as some of the (frequently used) research and development subsidies in fostering innovation, public procurement provides a major opportunity to further the policy goals in the areas of sustainability and innovation, often in combination with support to SMEs as an engine of growth.

In the EU, public procurement has recently received renewed attention and is now being used as a potentially powerful policy instrument to promote – both trigger and respond to – innovation.[10]

In order to prepare so-called 'green purchasing affirmative procurement programs', US Federal agencies use the EPA's guidelines on Environmentally Preferable Purchasing (EPP).[11]

Source: http://www.epa.gov/epawaste/conserve/tools/cpg/

11 The 2011 proposal for a EU Directive on public procurement is listed in Annex A, Chapter 2, section SCP.
12 The U.S. EPA guidelines are listed in Annex A, Chapter 2, section SCP.

TOPIC SHEET

4

WASTE PREVENTION[1]

According to an old saying stating that 'an ounce of prevention is worth a pound of cure', waste prevention is the most desirable option in the waste management hierarchy and, as such, an important goal and guiding principle of future waste strategies.[2] Moreover, waste prevention has a prominent place within broader sustainable consumption and production (SCP) considerations. The awareness is increasing that our civilization is reaching the limits of the finite environment that supports life on Earth – as ecologist Bill McKibben[3] puts it, 'we're running out of planet.' In terminology of systems thinking, we are running out of sources and sinks. Despite local efforts to restore the balance, we are using natural resources at unprecedented rates resulting in depletion.[4] We are also reaching the assimilative capacity limits of planetary sinks – places where emissions, discharges and waste can be safely deposited, absorbed and processed. Waste prevention can deliver on both – save resources and decrease the need for sinks.

While waste prevention can be understood in different ways, fundamentally, two aspects of prevention can be distinguished: quantitative and qualitative. Quantitative prevention comprises measures that reduce the quantities of substances, materials and products that become waste. This can be so-called strict avoidance – avoidance of unnecessary consumption, acceptance of fruit and vegetables with less-than-perfect shape or deviating from a size standard, product design and production processes that use lesser amounts of materials, as well as design for durability and dissasembly. Or, once a product is purchased, it can be diverted from becoming waste through re-use by the owner or someone else, possibly after repair and refurbishment to extend the product's life span. Qualitative prevention seeks to reduce or eliminate use of specific hazardous substances in materials and products that become waste, thereby directly improving environmental performance of the products or production processes concerned.

1 Topic Sheet prepared by Ljiljana Rodic, with inputs from David Wilson.

2 For example, the Cartagena Declaration adopted by the tenth meeting of the Conference of the Parties to the Basel Convention, which was held in Cartagena, Colombia, in October 2011, calls on the active promotion and implementation of more efficient strategies to achieve prevention and minimization of the generation of hazardous waste and other wastes

3 McKibben B (2007) 'Deep Economy. Economics as if the World Mattered.' Oneworld Publications.

4 Not only non-renewable resources are being depleted, we are also using renewable resources (such as fish stocks in the oceans or top layers of agricultural soil) at rates beyond their natural regenerative capacities, resulting in depletion of these stocks.

Waste prevention is also a climate issue. Much of the world's greenhouse gas emissions are associated with the production and distribution of the food we eat and the products we consume or use – waste prevention has the potential for substantial carbon savings across many sectors of the economy, including agriculture, forestry, mining, construction, manufacture, transport and logistics and others.[5]

Accordingly, waste can be prevented at almost any point along the life cycle of a product, from extraction of a raw material to its processing into a functional material to manufacturing of a product, further into packaging, distribution and retail, to the use stage and the end-of-use stage (*Figure 1* below). Accordingly, waste prevention policies and specific instruments are devised to address individual stages in the life cycle, and targeted at the respective key stakeholders involved.

5 For more details see Topic Sheet 1 on waste and climate.

Figure 1 Waste prevention at different stages in product life cycle

Waste Prevention

	Sustainable production		Sustainable consumption	Waste management	
Life cycle stage	Production	Transport/ Purchase	User/ Consumption	End-of-use	Collection
Key stakeholder	Manufacturer/ Importer	Distributer/ Retailer	Consumer	Consumer/ Community	Local authorities/ Waste companies
Qualitative prevention	Elimination or reduction of hazardous substances content	Product-service systems; Eco-labels	'Green' purchasing choices by citizens, businesses and governments	Segregation at source	Education & information programmes; second hand shops
Quantitative prevention	Design for repair, reuse, disassembly & recycling; Reduced packaging; Remanufacture	Product-service systems; Reduced packaging, refilling	Purchasing choices; no food waste; repair & reuse	Reuse; donation; exchange (Freecycle, Swapit); buying second hand, also online	Reverse logistics for remanufacture; reuse awareness campaigns; second hand shops and events; swap events

Note: Inspired by Cox et al (2010)

For example, in addition to reuse, the concept of cascading use offers possibilities to extend the use of materials and prevent waste. Raw materials from renewable resources such as wood lend themselves well for this: wood is first used as timber for high-level applications like furniture, then as cardboard, and only then it is incinerated for energy recovery.[6]

In discussions on waste management priorities, worrying about waste prevention may be dismissed as a 'luxury' of the rich countries that have addressed their waste-related public health and environmental protection concerns. However, taking waste prevention seriously is at least as important – if not more – in the countries with developing economies, due to sheer amounts of waste to be handled in their cities on daily basis. As many a waste practitioner from an African capital will confirm, waste amounts generated by their growing urban populations – accompanied by the changes in waste composition towards more plastics and e-waste – are putting additional pressure on already strained municipal services. And with the current growth rates, waste quantities in many fast growing cities in the 'South' are likely to double over the next decade or two.[7] Under such circumstances, focusing only on the technical, end-of-pipe solutions of extending collection and building environmentally sound treatment and disposal capacities will forever be 'running to stand still'. Probably SIDS know this better than other nations – faced with constant imports of goods for their populations and stream of tourists (on which they depend for income), and the limited space for waste disposal, they face unique waste management challenges.[8] Waste prevention is therefore of paramount importance to curtail ever increasing waste quantities.

Prevention may however ma appear difficult to envisage in low- and middle-income countries because, for the first time, their populations have the opportunity to increase personal consumption: promoting prevention could be interpreted as restricting access to markets and products, and therefore needs to be explained

6 The idea is not new – respectful use of natural materials, including wood, is part of traditional practices around the world. Equally, the latest modern-day strategies focus on innovation to support and advance sustainable use of biomass for industrial purposes, as is the case in the EU. For more information see 'Innovating for Sustainable Growth: A Bioeconomy for Europe', at http://ec.europa.eu/research/bioeconomy/pdf/201202_innovating_sustainable_growth.pdf

7 Waste generation trends are discussed in section 3.3.3.

8 Specific SIDS issues in waste management are discussed in Topic Sheet 5.

carefully and placed in the local context. Moreover, these countries often know more about waste prevention than high-income countries, and have more and better waste prevention skills and systems (e.g. for repair and reuse) in place. Indiscriminate transfer of technologies and 'best practices' from North to South eliminates and delegitimizes these well-anchored waste prevention practices. As emphasized in section 4.2.2 on Strategic planning, it is better to embrace such existing strengths and build upon them.

Deconstruction (instead of demolition) of buildings in the United States

Estimated more than 250 million tons[9] of waste materials are generated annually from the construction, renovation, and demolition of buildings in the United States;[10] more than half of this amount ends up in landfills. Across the country, deconstruction of buildings – a systematic disassembly by hand to salvage the materials – is emerging as an economically viable alternative to demolition. Deconstruction saves natural resources, reduces waste disposal, and creates job opportunities.

Deconstruction is the core programme of the Finger Lakes ReUse Center in Ithaca, NY. The ReUse Center was established as part of the Tompkins County's 20-Year Solid Waste Management Plan of 1995; its mission is about 'Enhancing Community, Economy, and Environment through ReUse'.[11]

9 US short tons are used here following the data source used (1 short ton = 0.907 metric tonne).

10 The data on C&D waste generation are not systematically measured throughout the U.S.A. and reported to the U.S. EPA. A 2003 estimate of building related C&D waste was 170 million tons/year. (http://www.epa.gov/epawaste/conserve/imr/cdm/pubs/cd-meas.pdf, p. 17) The 2009 World Waste Survey reported a 2006 estimate of 243-288 million tons/year (Chalmin and Gaillochet, 2009; as listed in Annex A, under Chapter 3, Collated data sources).

11 http://www.fingerlakesreuse.org

SOLID WASTE MANAGEMENT IN SMALL ISLANDS DEVELOPING STATES (SIDS)[1]

Good waste management is a vital component of sustainable development in SIDS. Poorly managed waste has negative impacts on public health, fragile terrestrial, coastal and marine ecosystems, and on important economic sectors such as tourism and fisheries. Waste has been recognised as one of the areas for priority attention for SIDS where the lack of regulations and poor enforcement, under-developed infrastructure, limited recycling opportunities due to economies of scale; general poor public attitudes and practices, and barriers to moving waste from one country to another due to different legal regimes and definitions of waste, all contribute to inadequate waste management practices.

SIDS, as with most countries around the world, are experiencing an increase in waste generation due not only to common factors such as increasing population, urbanization, and change of consumption patterns, but most acutely due to the large quantities of imported material and packaging, and the excess amount of waste produced by tourism[2], including cruise-ship generated wastes. Additionally the complexity and hazard of particular waste streams such as e-waste, pesticides, asbestos, used oil, items containing heavy metals and also biomedical wastes is also adding pressure to local waste management systems, since facilities for their treatment and disposal are often not in place.

In many SIDS the most prevalent method of disposal continues to be open and uncontrolled dumping, which leads to human health problems, as well as risks to the marine ecosystems, especially to mangroves, sea grasses and coral reefs; and other sensitive land areas and water courses. While in other SIDS progress with regards to sanitary landfilling has been made, it is recognized that the selection of appropriate locations for waste disposal sites poses challenges in the land use planning in the nation. The land area to be used as landfill sites is constrained by the lack of appropriate flat land close to the major urban centres and custom ownership of land; or in some cases by the small land area of the island itself (e.g. some atoll nations).

Another barrier, which is characteristic of SIDS in general, is the constraints to greater recycling. Segregation of waste streams in SIDS is still uncommon and recycling is generally not well developed in most of the islands. The constraints are mostly related to the small size and population of SIDS and to their relative geographic isolation: specifically, the resultant low quantities of recyclable waste mean that economies of scale cannot be achieved; their small size restricts local recyclables

Marine litter in the Caribbean

1 Topic Sheet prepared by Ainhoa Carpintero, with inputs from Prasad Modak and Chris Corbin.

2 Tourism generates substantial amounts of solid waste in some SIDS with tourists generating twice as much as solid waste per capita as local residents in the Caribbean. Cruise ship passengers are estimated to produce as much as four times the amount of garbage per day compared to local residents.

markets; other recycling markets require expensive transportation.[3] The limited human capacity and lack of incentives to encourage recycling, including the absence of legal and regulatory provisions for recycling, economic instruments for the citizens and businesses or voluntary agreements with the private sector, are additional constraints to recycling. Despite this, a 2013 survey showed that one in five Pacific island countries achieve recycling rates of more than 50%, and a further two in five of more than 15%.[4]

Therefore, to respond effectively to the SIDS challenges, there is an urgent need to promote an integrated waste management approach based on where each SIDS stands. In many low income SIDS, as with other low income countries, the first priority is to tackle public health and environmental problems by extending waste collection to all the citizens, and by phasing out uncontrolled disposal and burning. In these efforts attention should be given to careful land use planning and to the development of cost effective systems for waste collection and disposal. There is an extensive evidence base to show that the costs of inaction in SIDS, through impacts on tourism, fisheries and health and environmental impacts, greatly exceed the direct financial costs of environmentally sound waste management for both municipal solid wastes and hazardous wastes.[5]

In parallel, emphasis should be placed on waste reduction, reuse, recycling, recovery and return approaches in accordance with national capacities and priorities[6] with particular reference to the use of waste as a resource, the use of appropriate technologies relevant for SIDS that will allow the further use of local materials (e.g. local resources used for landfill cover) and the required capacity building and awareness raising to the public.

Kolonnawa compost Plant, Sri Lanka

Keep Samoa clean

Mechanisms to facilitate the development of partnerships with clear win-win opportunities could also be identified, including regional strategies and approaches for advocacy in specific areas e.g. extended producer responsibility, deposit refunds, shipping sector to offset the high cost of transport of recyclables, incentives at regional level required to encourage greater private sector involvement in recycling or regional infrastructure for the treatment and disposal of hazardous waste. For example in the Pacific the sustainable management of waste is guided by a Regional Solid Waste Management Strategy (2005-2015) implemented by the Secretariat of the Pacific Regional Environment Programme (SPREP). This strategy is currently being revised to include management of all waste and pollution in the region. A critical component of the strategy is the Japanese Government sponsored J-PRISM Programme which provides hands on technical support to improve solid waste management in the region. The objective of the J-PRISM Partnership is to strengthen and develop national waste management systems and human capacity in Pacific SIDS as well as to promote the regional sharing of knowledge, understanding and expertise.

3 UNEP (2014) GEO SIDS- Small Island Development States Outlook listed in Annex A under Chapter 2, SIDS.
4 JICA (2013), Data Collection Survey on Reverse Logistics in the Pacific Islands, Japan International Cooperation Agency.
5 See Section 5.2, Table 5.2.
6 SIDS Accelerated Modalities of Action (S.A.M.O.A.) Pathway http://www.sids2014.org/index.php?menu=1537

Source segregation, recycling and composting in Santa Cruz Island[7]

The island of Santa Cruz, the second largest islands of the Galapagos Archipelago, Ecuador, with a land area of 986 km² and a population of 12,000, has been implementing separation at source and differentiated waste collection since 2006.

Citizens segregate their waste into recyclables, non-recyclables and organics, and place it in colour-coded containers (green for organic waste, blue for recyclable materials, black for other waste (residuals)). Hospitals and clinics use red containers for toxic and hazardous waste.

The municipality has established the frequency and the schedule of the collection by city sector and by type of waste; therefore the citizens place out the containers on the public roads at the specified times. The vehicles used for the collection are equipped with a system of digital radios and GPS that allows tracking the location of the vehicles and the monitoring and supervision of the differentiated collection.

The organic waste and the recyclables are sent to the Fabricio Valverde Recycling Centre. The organic waste is composted, and the product is sold and used by the municipality as fertiliser in municipal green areas. The recyclables are manually sorted, compacted and/or shredded to be sent to companies on the mainland. The residual waste is sent to the sanitary landfill.

The system has been supported by continuous education campaigns undertaken over consecutive years and focused on different target groups and key audiences, including the commercial and tourist sectors and through visits to households, training sessions for students, organizing visits to the Recycling Centre, distribution of informative materials, advertisements on the radio and the television and contests involving the local community.

In 2012, approximately 50% of the overall waste generated on Santa Cruz Island was recycled.

7 Castillo M. and Hardter U., (2014) Integrated Solid Waste Management in Island Regions. Ed. Irma Larrea Oña. WWF and Toyota, Galapagos-Ecuador.

Moving away from dumpsites to sanitary landfill and then towards recycling in Mauritius[7]

In Mauritius, until the late 1980s, waste was disposed of in open dumpsites. The dumpsites were often on fire and no records were kept of the incoming wastes. In the early 1990s the Government of Mauritius started efforts to improve the situation by adopting landfill as the disposal method for the medium-term and improving the collection services. Operation of the sanitary landfill started in 1997. Five transfer stations were set up across the island to provide for cost effective transportation of waste to the landfill. A gradual increase in collection was established until the total coverage of the island was achieved; and public awareness activities were undertaken on solid waste management.

Over the years since then, the infrastructure and services have been strengthened, in parallel with developing the legal framework and building the institutional capacity, in order to provide sanitary conditions. Private initiatives have been undertaken for the removal of recyclables such as paper, plastic, batteries and waste motor oil at the transfer stations. In 2011 a private operator invested in a composting plant, where 35,000 tonnes are processed out of the 450,000 tonnes of waste generated annually (2014).

Driven in particular by the limited land resources to further develop the landfill site, the Government is deploying much efforts to establish an integrated waste management system which will uphold the 3Rs principle. As such, infrastructure and measures for diversion of both, organic and dry recyclables from the landfill are to be implemented from 2015.[8]

A transfer station in Mauritius

Landfill in Mauritius

8 Text prepared by D. Dookee, Principal Project Officer, Solid Waste Management Division, Ministry of Environment, Sustainable Development, Disaster & Beach Management.

Partnerships and financing in Kiribati

In the Republic of Kiribati, the largest SIDS in terms of ocean territory, a beverage container deposit system has been operational since 2004. Under this system PET bottles and aluminium cans have an AUD 0.05 (5 cent) levy paid on them at importation and collected by the Ministry of Finance and Economic Development (MFED). This cost is passed on through the commercial system to the consumer – who upon returning the empty beverage container at a collection point, receives AUD 0.04 (4 cents) back. One cent as 'handling fee' is for the recycling operator to make the operation viable[9]. In this way, these easily recyclable elements are removed from the waste stream by the use of a simple economic tool and it is reducing the burden to the landfill. The deposit system was set up as part of the project *Kaoki Maange!* – which means *'Return the Rubbish!'*[10] – an innovative scheme to recycle different recoverable materials run by the Foundation for the Peoples of the South Pacific Kiribati (FSPK). The container deposit system was supported by the Special Funds (Waste Material Recovery) Act 2004 passed by the Kiribati government[11].

A Green Bag scheme for source segregation and collection also started in early 2000. Under this scheme the people were encouraged to separate their waste and use the Green Bags for non-compostable and non-recyclable waste – waste to be sent to the Nanikai landfill. In two years the results showed a decrease of about 60% quantity of household waste deposited in the landfill.[12] The Green Bag scheme was discontinued and in 2012 was re-introduced as a user-pays system where the people purchase the Green bags (20¢ per bag) and get a "free" weekly collection (the cost of the collection is built into the price of the bag).[13] The collection is undertaken by private contractors. This low-cost scheme encourages the reduction of waste, and seems fair since those who generate more waste are to pay more.

Truck collecting green bags for disposal

9 The private recycling operator claims the entire deposit amount of 5 cents per can from the Ministry of Finance and Economic Development (MFED) on a monthly basis. Generally the aluminium cans are exported to Australia every month and PET bottles are exported to Hong Kong.

10 Government of Kiribati and UNICEF (2005) Kiribati Islands: A situation analysis of children, women & youth. http://www.unicef.org/pacificislands/Kiribati_Sitan.pdf

11 ADB (2014) Solid Waste Management in the Pacific - Kiribati Country Snapshot http://www.adb.org/sites/default/files/publication/42671/solid-waste-management-kiribati.pdf

12 Leney A (2006) The impact of the Greenbag on waste generation in South Tarawa, Kiribati. IWP-Pacific Technical Report (International Waters Project) no. 22 http://www.sprep.org/att/publication/000518_IWP_PTR22.pdf

13 ADB (2014) Solid Waste Management in the Pacific – Financial Arrangements http://www.adb.org/sites/default/files/publication/42656/solid-waste-management-financial-arrangements.pdf and http://www.environment.gov.ki/?page_id=37

CHAPTER 3

WASTE MANAGEMENT: GLOBAL STATUS

Chapter 3 provides a context or setting to the GWMO by presenting data as an evidence base regarding waste generation and its management across the world.

The Chapter starts by looking at the relative quantities of different types of wastes from various sources (Section 3.2). The focus then turns to municipal solid waste (MSW) where data on quantities and composition as well as past trends and future projections are presented (Section 3.3). Section 3.4 overviews the status of MSW management across income groups and regions. It focuses first on the protection of public health by ensuring that all wastes are collected, and then on environmental protection by phasing out uncontrolled disposal and open burning of waste.

The later sections focus on resource recovery (Section 3.5), looking at collection for recycling, the importance of source segregation, and available technologies for resource recovery. This is followed by an examination of the global industry in secondary materials (Section 3.6).

The chapter is followed by a series of Topic Sheets focusing on waste streams of particular interest, including construction and demolition waste, hazardous waste, e-waste, plastic waste and marine litter, disaster waste and food waste.

© Ainhoa Carpintero

3.1 SUMMARY OF THE CHAPTER – KEY MESSAGES ON THE GLOBAL STATUS OF WASTE MANAGEMENT[1]

- A best 'order of magnitude' estimate of the total global arisings of municipal solid waste (MSW) is around 2 billion tonnes per annum. A broad grouping of 'urban' wastes, including MSW, commercial and industrial (C&I) waste, and construction and demolition waste (C&D), is estimated at around 7 to 10 billion tonnes per annum.
- Although generation rates vary widely within and between countries, MSW generation per capita is strongly correlated with national income. In high-income countries, MSW generation rates are now beginning to stabilize, or even show a slight decrease, which may indicate the beginning of waste growth 'decoupling' from economic growth. However as economies continue to grow rapidly in low- and middle-income countries, one can expect per capita waste generation to increase steadily.
- Waste generation is growing rapidly in all but the high-income regions of the world, as populations rise, migration to cities continues, and economies develop. In 2010, the traditional high-income countries accounted for around half of all waste generation. That is forecast to change quickly, with Asia overtaking these countries in terms of overall MSW generation by around 2030 and Africa potentially overtaking both later in the century.
- Organic fractions comprise a greater percentage of the MSW arisings in low-income countries (where organic waste is typically 50 to 70% of all MSW) than in high-income countries (where organics account for typically 20 to 40%). The percentage of paper appears to be proportional to income levels (23% of MSW in high-income, 19% to 11% in middle-income and 7% in low-income countries). Plastic levels generally appear high across the board (8% to 12%), not showing as much dependence on income level as other waste types. 'Dry recyclable' materials (metals, glass and textiles) range from 12% of MSW in high-income to 12% and 9% in middle-income and then 6% in low-income countries. Household hazardous waste (HHW) is estimated to make up less than 1% of all MSW across all income ranges, but its presence makes certain management options much more difficult.
- Extending MSW collection to 100% of the urban population is a public health priority. Evidence suggests that significant progress has been made in many middle-income countries over the past few years, particularly those with gross national income (GNI) per capita above USD 2500 per year. At the same time, median collection coverage is still around 50% in low-income countries and figures are much lower in some countries. It also drops sharply in the more rural areas of many countries. It is estimated that at least 2 billion people worldwide still lack access to solid waste collection.
- Eliminating uncontrolled disposal is a priority for protecting the environment. Evidence suggests considerable progress has been made. However, the 100% and 95% controlled disposal rates in high- and upper-middle income countries respectively are in stark contrast with rates that are often well below 50% in low-income countries, and 0% controlled disposal is still relatively common in rural areas in many countries. In lower-income countries, waste disposal is often in the form of uncontrolled dumpsites with open burning. It is estimated that at least 3 billion people worldwide still lack access to controlled waste disposal facilities.
- Recycling may provide a source of income, help conserve scarce resources and reduce the quantities of waste requiring disposal. However the success of recycling depends critically on materials being kept separate and clean and being found in sufficiently high concentrations. Recycling rates in high-income countries have progressively increased over the last 30 years, driven largely by legislative and economic instruments. In lower-income countries, the informal sector is often achieving recycling rates of 20 to 30% for MSW.
- The secondary materials industry operates globally, with active international 'commodity' markets for ferrous and non-ferrous metals, paper, plastics and textiles. Most secondary materials come from industry and most are utilized inside national boundaries, but a sharp increase in the availability of materials from MSW recycling since the 1990s, together with the relocation of much of the world's manufacturing industry to Asia in general and to the People's Republic of China (PRC) in particular, has led to an increasingly transboundary and even global market. The PRC accounts for 60% by weight of global imports of

1 Please refer to Annex B for details of the data sources used to compile the evidence presented in this chapter.

aluminium scrap, 70% of recovered paper and 56% of waste plastics. Other Asian countries are also major importers, with Turkey the recipient of 30% of the total world trade in steel scrap.[2]

- Resource recovery from waste includes processes both for the recovery and recycling of the organic fraction and for energy recovery. Global activity in new waste processing facilities is high. Over the past two years, waste processing investment projects worth more than 300 billion USD have been active, of which 85 billion USD was directed to MSW processing (although not all of these projects will be built). Most of this investment activity is in the high-income countries, including energy from waste projects utilizing biomass and so on.
- Some waste streams require particular focus. Topic sheets are provided in the GWMO for large-volume C&D waste; for high-risk hazardous waste and e-waste; for plastic waste and for its associated problem of marine litter, which is receiving global attention; for waste from disasters; and for food waste, the scale of which is huge when considered alongside food scarcity and global starvation.
- Mining and quarrying and agriculture and forestry residues and wastes are generally managed close to source, with most agriculture and forestry wastes either being returned to the soil as soil improvers and nutrients or used as biomass fuel. These large volume streams are thus generally outside of national waste control regimes and data are not reported. A very rough, 'order of magnitude' estimate is that each of these major sectors generates 10 to 20 billion tonnes per annum of residue and waste. Mine tailings merit further attention as waste due to their potential for health and environmental impacts.
- The definitions of waste categories vary widely; waste quantities are often not measured; national reporting systems are often weak. As a result, it is not surprising that international data on MSW generation, composition and management lacks adequate breadth and depth and is weak and unreliable. The international data situation is still worse for other waste types. Use is made here of recent work that developed indicators to benchmark the performance of a city's MSW management system on a consistent basis. It is recommended that waste and resource management data are actively included within wider international action as part of the data revolution to improve data for sustainable development, that a globally recognized and internationally agreed methodology be developed for collecting and reporting waste data at the local (municipal) and national levels, and that the available performance indicators be subjected to widespread testing, with the results used to inform further work to develop standardized indicators.

Textile waste

Compost

2 Excluding intra-EU trade.

3.2 OVERVIEW OF GLOBAL WASTE GENERATION

Providing a global overview of total waste generation would appear to be a fundamental element of the GWMO, but in reality it is almost impossible to do so with sufficient accuracy.[3] Where data exist, they generally refer to MSW,[4] hence that is the focus for most of this chapter. A wider compilation of data on waste from different points in the material and product life cycle exists mainly in higher-income countries, especially in OECD countries. Therefore these data have been used as a 'proxy', to show the relative quantities of waste from different sources.

Figure 3.1 suggests that the three major waste streams of construction and demolition (C&D), commercial and industrial (C&I – appearing as two segments in *Figure 3.1*) and municipal solid waste (MSW) predominate. In the higher-income OECD countries from which these data are taken, MSW is generally managed by municipalities and C&I and C&D waste by the waste generators themselves through the waste industry (through business to business [B2B] arrangements).[5] However, even in these cases there is overlap between the definitions and considerable variation between countries. The distinctions between these three major waste types are even more 'fuzzy' in developing country cities.[6]

Figure 3.1 Relative quantities of waste from different sources in the material and product life cycle

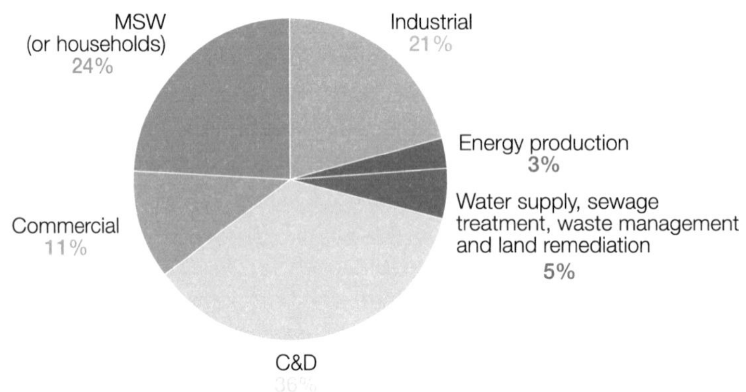

Notes: Data is for the OECD countries as a proxy, due to limitations on availability of data from the rest of the world. All data exclude agricultural and forestry and mining and quarrying wastes. Where there are significant gaps in the OECD database for a particular waste arising in a specific country, other sources have been used (using the EMC Master database [2014, n.p.] compiled for the GWMO), or an estimate has been made. Estimate of waste from a broad range of municipal, commercial and industrial sources (total waste quantity generated in the OECD countries, including construction and demolition (C&D) but excluding agricultural and forestry and mining and quarrying): **3.8 billion tonnes per annum**.

Figure 3.1 distinguishes two further types of waste, which are reported separately in the OECD database. Wastes arising from water supply, sewage treatment, waste management and land remediation represent around 5% of the total, while waste from power generation represents around 3%. These sources are interesting, as they represent the best measure available of those residues which have been removed from emissions to air and water, and concentrated as 'solid waste'.[7]

In principle, it is possible to attempt to extrapolate from the OECD data in *Figure 3.1* to estimate total worldwide waste arisings. Such extrapolation is facilitated by the availability of waste data for some non-OECD countries, in particular Russia and the PRC.[8] Extrapolating from the EMC database prepared for the GWMO to estimate 2010 worldwide MSW arisings results in an estimate of around 2 billion tonnes per annum, which is roughly twice the MSW figure for the OECD.

For the other waste streams, extrapolation is even more challenging. Based on the available information, the best 'order of magnitude' estimate of total arisings worldwide for the broad grouping of 'urban' wastes (municipal, commercial and industrial wastes, including C&D waste) comparable to the data indicated for the OECD in *Figure 3.1* is in the range of 7 to 10 billion tonnes per annum. However, more reliable, measured data are urgently needed: a major recommendation from the GWMO is to ensure that waste and resource

3 See Section 2.5.2 on the quality and availability of waste-related data.
4 See Section 3.3 for an overview of municipal standard waste generation, its composition and its properties.
5 See Chapter 5, in particular Sections 5.3 and 5.5.
6 See Section 2.2.3.
7 See Section 2.2.1.
8 See Annex B.

management data are actively included within wider international action as part of the data revolution to improve data for sustainable development.

A decision was taken early in the development of the GWMO to focus on the 'higher risk' grouping of wastes included in *Figure 3.1.* The two major sectors 'Agricultural and Forestry' and 'Mining and Quarrying' had been set aside, as these sectors' residues and wastes are generally managed close to source, with most agricultural and forestry residues either being returned to the soil as nutrients or used as biomass fuel; are often outside of national waste control regimes; and data for them are generally not reported.[9] The quantities are potentially very large, as these wastes include crop residues, animal manure and wood residues from agriculture and forestry as well as rock, over-burden and processing residues from mining and quarrying. Based on data from the few countries which collect and report them, and on estimates based on production data and assumptions concerning residues per unit of production, it is possible to make rough, 'order of magnitude' estimations of total worldwide arisings of residues and wastes, which are in the range of 10 to 20 billion tonnes per annum for each of the two sectors. The main component of interest in the GWMO, due to its potential impacts on public health and the environment, is mine tailings, on which a specific follow-up study is recommended.

3.3 OVERVIEW OF MSW GENERATION

3.3.1 MSW generation

MSW generation rates vary widely within and between countries. The generation rates depend on income levels, socio-cultural patterns and climatic factors. *Figure 3.2* shows the relationship between waste per capita[10] and income levels per capita for 82 countries. Despite the 'scatterplot', there is a strong positive correlation, with the median generation rates in high-income countries being about six-fold greater than in low-income countries. There is also considerable variation within countries. For example, Brazil's national database shows state waste generation per capita in 2012 ranging from a low of 310 kg per capita per annum to a high of 590 kg per capita per annum.[11]

Figure 3.2 Waste generation versus income level by country

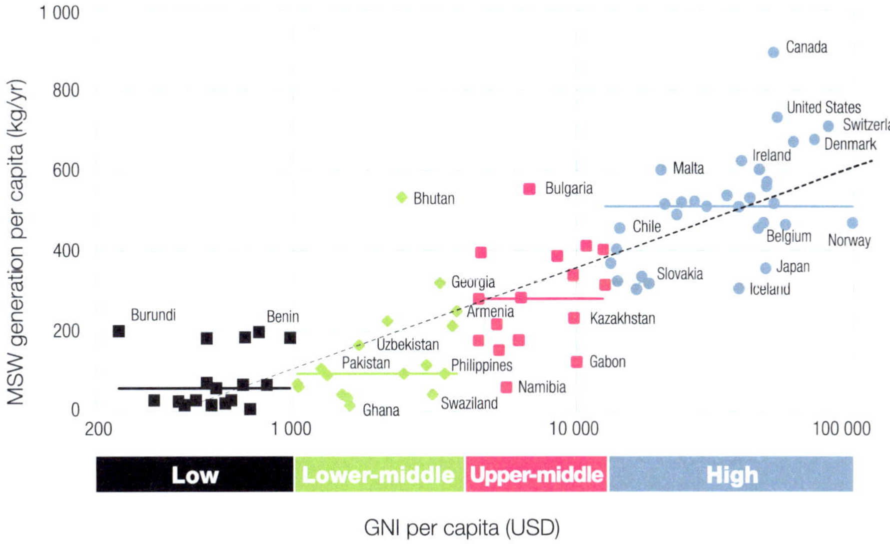

Notes: Based on data from 82 countries using the latest available data within the period 2005-2010. For 12 countries, the latest available data was older than 2005.

Regression: $y = 109.67\ln(x) - 651.45$, $R^2 = 0.72$

Data sources: EMC's Master Country Database (n.p., 2014) using primarily data from the EU, OECD and World Bank; Lawless (2014), Waste Atlas: Recycling and resource recovery around the world (Unpublished master's thesis). University of Leeds, Leeds, UK. Both were prepared for the GWMO (see Annex B, under Waste databases).

9 See Section 2.2.3.

10 It is important to note that data reported for many countries is likely to be MSW collected rather than generated. This not only affects interpretation of the waste generation data but also the data on waste composition.

11 Annual reports on Brazilian waste statistics (in Portuguese). See www.abrelpe.org.br

3.3.2 MSW Composition and Properties

In spite of the high variability and low reliability of source data, a comparison of average compositions relative to the countries' income level shows some interesting patterns (see *Figure 3.3).*

- One major difference is in organic fractions, which are significantly higher in middle- and low-income countries (averaging 46 to 53%) than in high-income countries (averaging 34%). Yet in fact these averages might be understating the differences. One comparative study contrasts an average of 67% across many middle- and low-income cities with 28% for cities in Europe, North America and Australia.[12] Also, the nature of the organic waste differs. In middle- and low-income countries, most organic waste is 'unavoidable', as it is the organics left over after the preparation of fresh food – organic matter that could not have been eaten. In contrast, in high-income countries there is a great deal of avoidable food waste – that is, food that could have been eaten.[13]
- The percentage of paper waste appears to be proportional to income levels, rising steadily from 6% in low-income countries, through 11% to 19% in middle-income and 24% in high-income countries. These figures are in line with data on the annual per capita consumption of paper worldwide, which ranges from 240 kg in North America, through 140 kg in Europe, to 40 kg in Asia and 4 kg in Africa. There has long been speculation that per capita consumption of printing and writing paper and newsprint in high-income countries has been falling due to electronic readers. The world average per capita consumption had shrunk by 4% in 2012 compared to the peak recorded in 2007.[14]
- While plastic levels appear generally high, they perhaps do not show as much dependence on income level as might be expected, with the averages for all income categories having a fairly narrow range of 7 to 12%. However, these averages do hide considerable variation between countries, with much higher values being reported in certain countries. For example, a regional comparative report indicated high levels for both Jordan (about 16%) and Mauritania (about 20%).[15]
- Levels of other 'dry recyclable' materials, which include metals, glass, and textiles, are all relatively low. Taken in aggregate, there is a small but steady increase in this type of waste as incomes rise, from 6% in low-income countries, through 9% and 12% in middle-income to 12% in high-income nations.
- MSW now increasingly contains relatively small amounts of hazardous substances. Often known as household hazardous waste (HHW), typical sources may include mineral oils such as motor oil; asbestos products such as roofing and heating blankets; batteries; waste electrical and electronic equipment (WEEE or e-waste); paints and varnishes; wood preservatives; cleaning agents such as disinfectants; solvents such as nail varnish; pesticides such as rat poison; cosmetics such as hair dyes; and photo lab chemicals such as developer. Statistics are unavailable on the percentage of household hazardous waste in MSW on a global basis. Estimates suggest a percentage of household hazardous waste in MSW of less than 1%, but up to 5% if e-waste is included.[16]

Waste composition affects the physical characteristics of the waste, including density, moisture content and calorific value, which in turn affect waste management and the choice of technology for collection, treatment and the 3Rs. For example, the ash content of MSW in high-income countries has decreased over the last 50 years, while the content of paper, plastics and other packaging materials has increased, significantly reducing the bulk density and increasing the calorific value. Reduced density has increased the need for compaction during collection to achieve higher and more economic vehicle payloads, while increased packaging content and rising calorific values make both recycling and energy from waste (EfW) more attractive. Conversely, the higher levels of organic waste in lower-income countries means that the waste is wetter, denser and has a lower calorific value, so there is less need for compaction during collection and the MSW may not burn without auxiliary or support fuel.

Some plastic wastes, in particular PVC, can result in air emissions of toxins such as dioxins and furans if unmanaged wastes are subjected to open burning, or if the thermal treatment and pollution control at EfW

12 Wilson et al. (2012). Comparative Analysis of SWM in 20 cities. 13 of the 15 'Southern' middle- and low-income countries are within the range 48–81% (average 67%); while the five cities in Europe, North America and Australia (i.e. the four high-income cities plus Varna in Bulgaria) report 24–34% (average 28%). Source listed in Annex A, under Chapter 1, Waste management.

13 See Topic Sheet 11 on Food Waste and Case Study 3 on reducing food waste, both found after Chapter 3.

14 Bureau of International Recycling (2014). Recovered paper market in 2012 (2014 report), listed in Annex A, Chapter 3, Global secondary materials industry.

15 Sweepnet (2014a), listed in Annex A, Chapter 2, Waste data and indicators.

16 Slack et al. (2007), listed in Annex A, Chapter 3, Municipal solid waste management.

facilities are inadequate. In light of this, it is important to establish a reliable database on waste composition and characteristics and monitor the trends.

Figure 3. 3 Variation in MSW composition grouped by country income levels

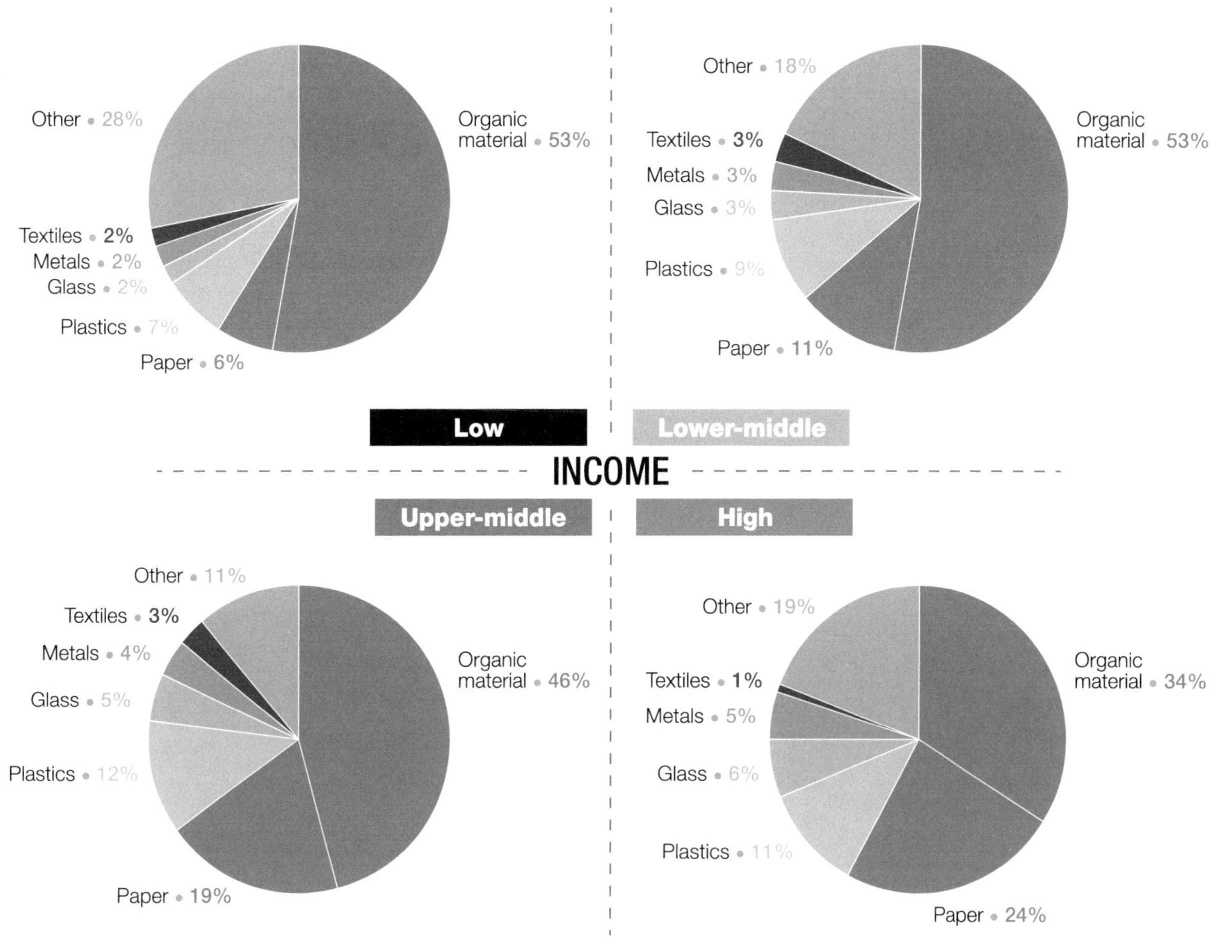

Notes: Based on data from 97 countries (22 in Africa; 14 Asia-Pacific; 35 Europe; 19 Latin America/Caribbean; 2 North America; 5 West Asia). Dates of the data vary between 1990-2009. "Other" means other inorganic waste.

Source: EMC's Master Country Database (n.p., 2014) using primarily data from the UN and World Bank and Hoornweg & Bhada-Tata (2012)[17]

3.3.3 Trends in MSW generation

Waste generation per capita has risen markedly over the last 50 years and shows a strong correlation with income level. *Figure 3.4* shows data for the last 20 years in some high-income countries. This figure also suggests that MSW generation rates are beginning to stabilize in high-income countries, or even show a slight decrease. This is often cited as evidence for the beginning of waste growth 'decoupling' from economic growth, as the trend became apparent before the 2008-09 financial crisis. However, the previous rising trend may resume if economic growth returns to previous levels. Also, a contributing factor may be the shifting of manufacturing industries to emerging economies. This shift would not be such a major factor in MSW generation, but would be expected to have a larger impact on industrial waste quantities.

17 Listed in Annex A, Chapter 3, Collated data sources.

Figure 3. 4 Trends in MSW generation since 1995 in selected high-income countries

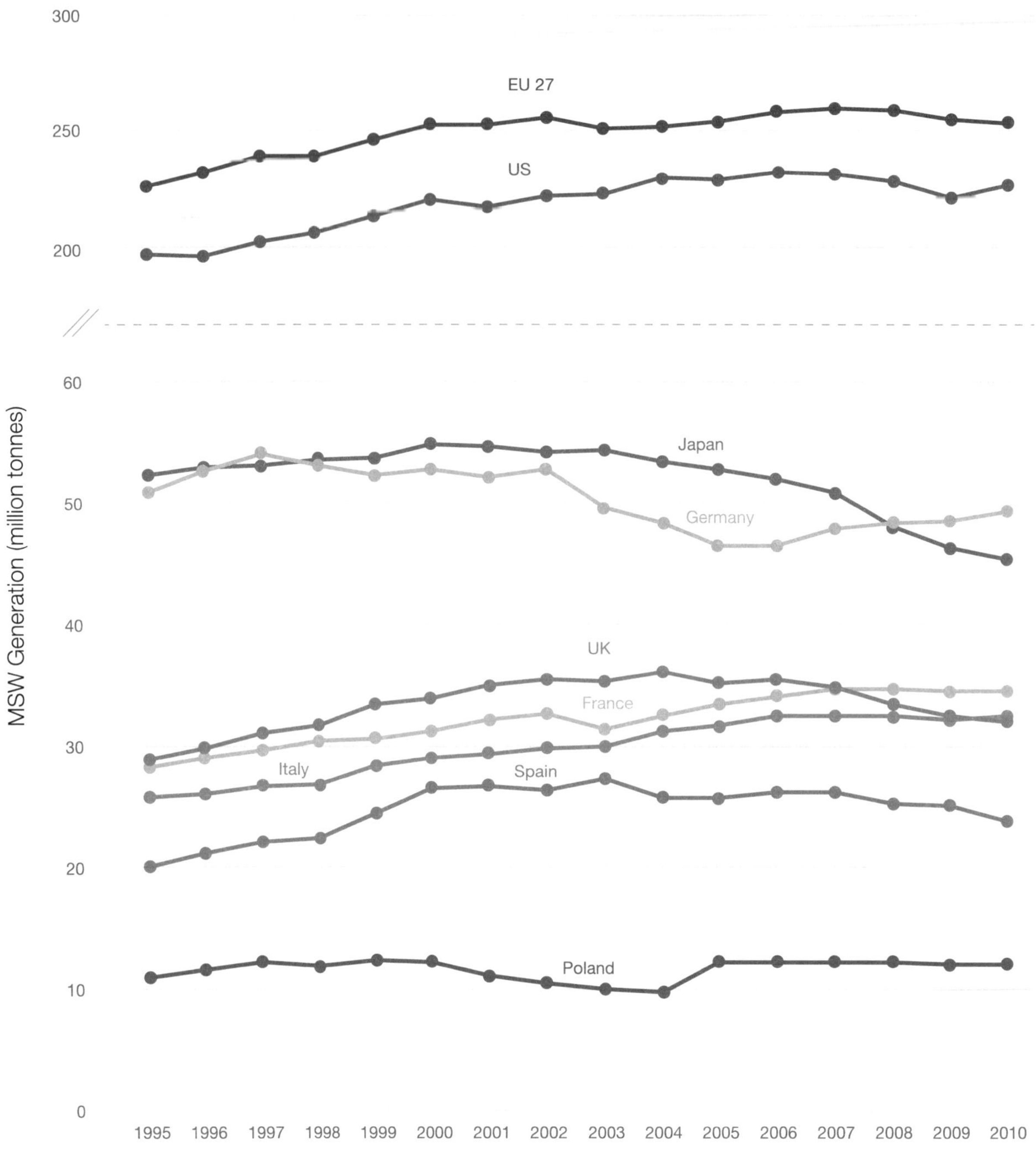

Data source: EMC's Master Country Database (n.p., 2014) using data from Eurostat and OECD

The best available data on current total world generation of MSW come from a combination of 'real' national statistics, where waste arisings have been systematically measured, recorded and reported; and calculated figures where population data have been combined with estimates for MSW generation per capita. Forward projections of both population and waste per capita data are needed to project these figures into the future and forecast future changes in MSW arisings.

Forecasting population has been a major focus for the world's statisticians. The UN's World Population Prospects[18] publishes a range of scenarios for future population growth through to 2100, although the scenarios begin to show an increasingly broad range in their forecasts beyond the next 30 years. *Figure 3.5* shows estimated and projected world population by region from 1950 to 2100 for the 'medium variant'. The general

18 UNDESA. World Population Prospects. http://esa.un.org/wpp/

trend is an initial rise, followed by a levelling out and then either a stabilization or a fall. Under this scenario, Asia is forecast to reach its peak population around 2050 while Africa continues to grow through to 2100.

Figure 3. 5 Estimated and projected world population by region

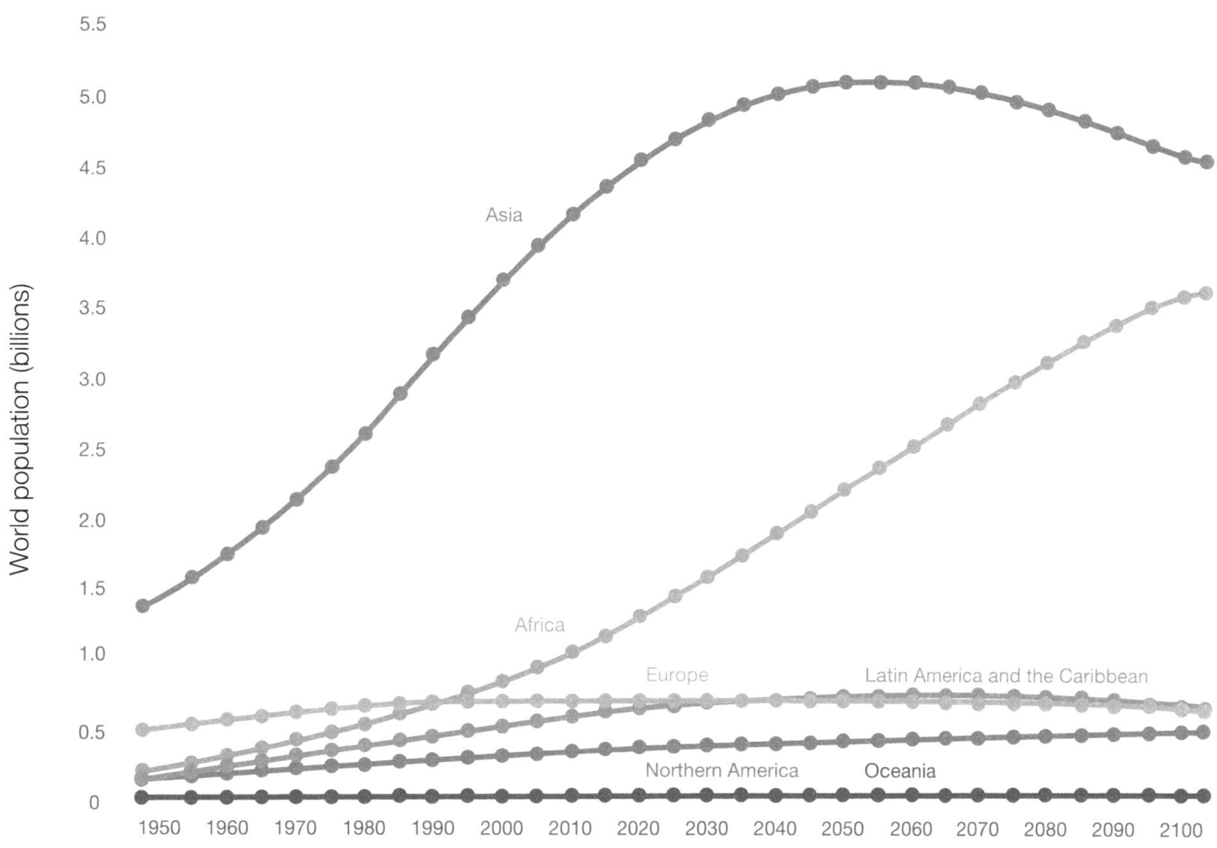

Notes: Estimates: 1950-2010; Medium variant: 2015-2100

Source: UNDESA, Population Division (2013)[19]

Since waste generation is significantly greater in urban than in rural areas, forecasting the split between urban and rural populations is also important. *Figure 3.6* presents UN data showing the percentage of people living in urban areas by country, and also the location of cities in three size ranges above 1 million people, for four 'snapshots' in time. The shift from rural to urban areas since 1970 has been marked, and the projection for 2030 reinforces the trend. The only three megacities with a population over 10 million in 1970 were in Japan and the US; by 2014, there were 28 megacities, of which 20 were in the global 'South'; by 2030, it is forecast that there will be 12 more megacities, all in the 'South'. Urban populations are already at or approaching 80% in much of the Americas, Europe, Japan and Australia; the trend of migration to the cities still has a long way to run in Asia and particularly in sub-Saharan Africa – which coincides with the regions where total population is also forecast to continue growing most strongly.

Using these data to forecast waste arisings in individual cities in the fastest growing regions provides quite startling results. To take one example, Kinshasa in the Democratic Republic of the Congo had a population of less than 4 million in 1990, had risen to 11 million by 2014 and is forecast to reach 20 million by 2030. Allowing for increases in waste per capita with development, the total MSW generation in the city now is more than three times that in 1990, and will likely have doubled again by 2030. The challenge of providing basic MSW management services to such rapidly growing cities which are already under-served is enormous.

Unlike world population and urbanization trends, there are no authoritative UN forecasts of future waste generation per capita, and filling that gap is one of the GWMO's recommendations for future work. As shown in *Figure 3.2*, there is a clear link between waste per capita and income level; so unless specific waste prevention

19 UNDESA, Population Division (2013). World Population Prospects: The 2012 Revision. New York.

measures are taken, one can assume that per capita waste generation levels in the current low- and middle-income countries will increase as their economies continue to develop and gross national income (GNI) levels rise.

Box 3.1 shows the results of a recent research project which has attempted to project MSW generation forward to 2100. It is worth repeating here the caveat that any projection beyond 2050 becomes extremely speculative. It should be interpreted as a scenario of what might happen under a particular set of assumptions, rather than a forecast of what is likely to happen.

Figure 3.6 Percentage of urban population and locations of large cities, 1970 – 2030

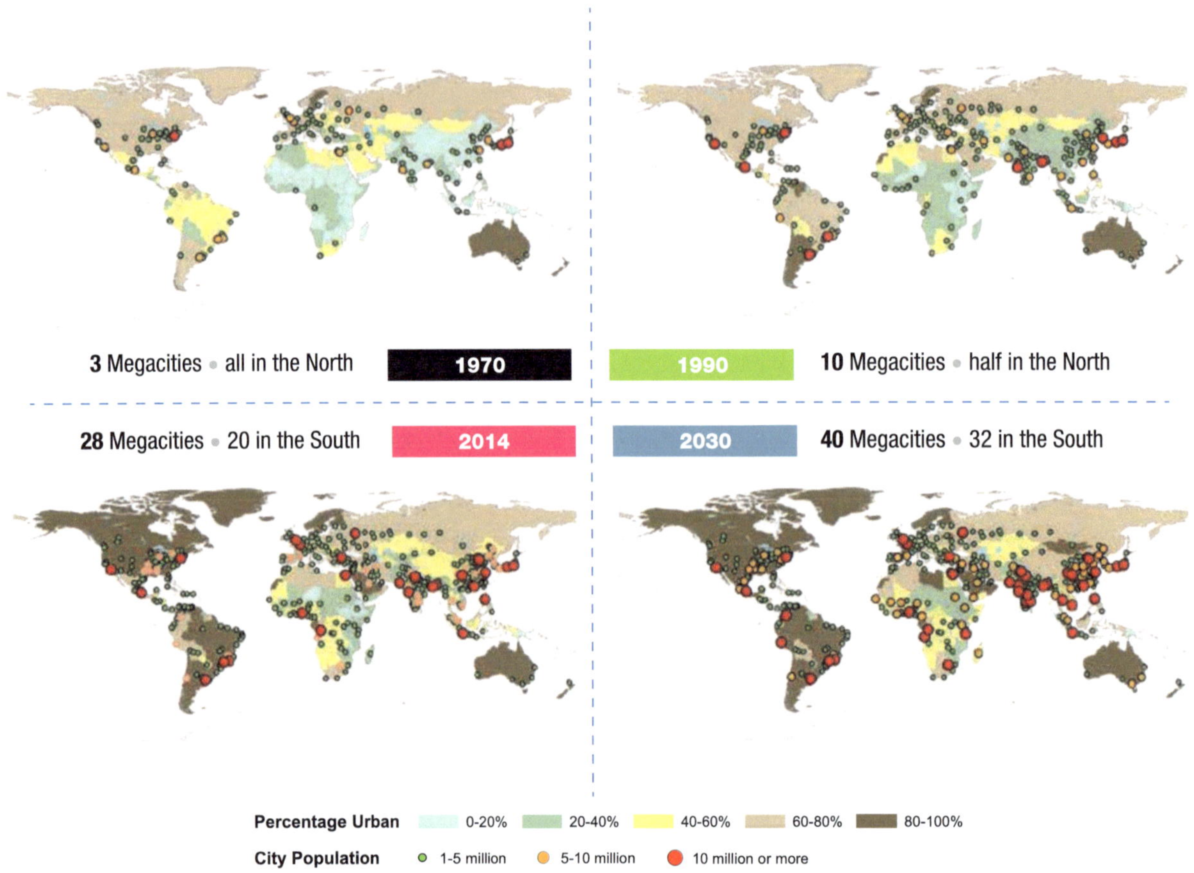

UN disclaimer: Designations employed and the presentation of material on this map do not imply the expression of any opinion whatsoever on the part of the Secretariat of the United Nations concerning the legal status of any country territory or area, or of its authorities, or concerning the delimitation of its frontiers or boundaries.

Source: UNDESA, Population Division (2014). World Urbanization Prospects, the 2014 Revision. New York. http://esa.un.org/unpd/wup/

BOX 3.1 PROJECTION TO 2100 OF MSW GENERATION, BY WORLD REGION[20]

Figure 3.7 Total MSW generation by region

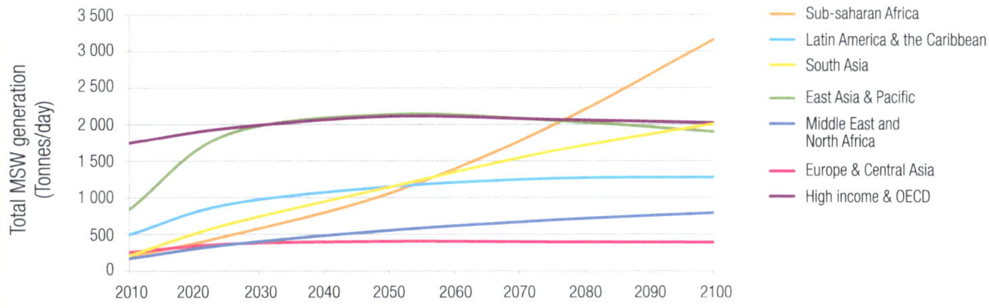

Figure 3.8 MSW generation per capita by region

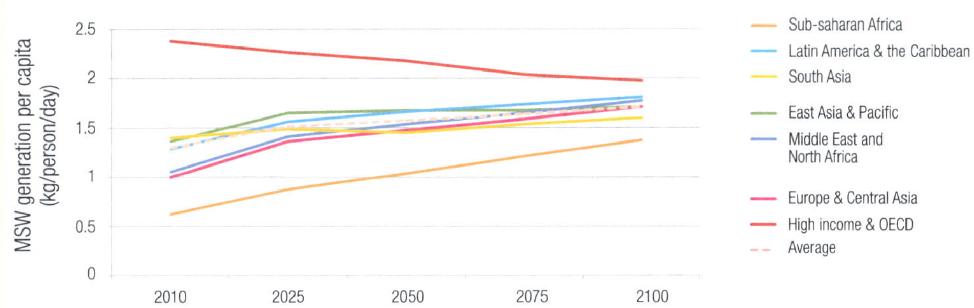

Notes: In the research study from which the above graphs were taken, the authors used the five scenarios developed for the Intergovernmental Panel on Climate Change (IPCC) that relate climate change and socioeconomic factors such as population expansion, urbanization, economic and technological development.[21] The five scenarios on the shared socio-economic pathways (SSP) are: "SSP1 – Low challenges; SSP2 – Intermediate challenges, business as usual; SSP3 – High challenges; SSP4 – Adaptation challenges dominate; SSP5 – Mitigation challenges dominate." The scenario used for the projections shown in Figures 3.7 and 3.8 is SSP2, defined as "middle of the road, or business as usual," in which the current trends continue and the world makes some progress towards sustainability. In that scenario, by the year 2100, population is around 9.5 billion and slightly declining. Waste per capita is linked to GNI per capita by a series of linear relationships, the gradient of which is assumed to decline over time. Five lines are used, from 2010, 2025, 2050 and 2075.

Source: Hoornweg et al. (2015). Peak Waste: When Is It Likely to Occur? Journal of Industrial Ecology, 19 (1), 117-128. http://onlinelibrary.wiley.com/enhanced/doi/10.1111/jiec.12165/ Listed in Annex A, Chapter 3, MSW management.

While projections of outcomes so far into the future are speculative, particularly beyond 2050, they do provide some interesting insights. *Figure 3.7* shows that in the "high income and OECD" group of nations, waste generation first rises only slowly, then stabilizes and declines. As a percentage of the world total, it is declining rapidly, initially as the contribution of the two Asia regions increases rapidly, before they too stabilize. As would be expected from the previous discussion on population and urbanization, the contribution of Africa, and particularly sub-Saharan Africa, starts as relatively small, and begins to rise very quickly after 2050. What is both surprising and speculative is the forecast that Africa may become the dominant region in terms of total waste generation. *Figure 3.8* provides the corresponding data for waste per capita. It should be noted that this is part input data on how waste per capita is assumed to change as GNI per capita levels rise in individual countries, and part output, back-calculated from the results in *Figure 3.7* and population projections.

20 This box summarizes and provides commentary on a research project led by Daniel Hoornweg of the University of Ontario Institute of Technology. See Hoornweg et al. (2013, 2015), listed in Annex A, Chapter 3, Municipal solid waste management.

21 Moss, R.H., J.A. Edmonds, K.A. Hibbard, M.R. Manning, S.K. Rose, D.P. van Vuuren, T. R. Carter, et al. (2010). The next generation of scenarios for climate change research and assessment. *Nature* 463 (7282): 747–756.

3.4 CURRENT STATUS OF MSW MANAGEMENT: PROTECTION OF PUBLIC HEALTH AND THE ENVIRONMENT

MSW management is an essential utility service. The first steps in ensuring sound MSW management are providing a reliable collection service to all citizens and eliminating uncontrolled dumping and open burning. The world's progress towards this target is the focus of this section.

3.4.1 Collection coverage

A resident handing over waste, India

Providing a regular and reliable waste collection service to 100% of the urban population has been a public health objective since at least the mid-19th century. Data compiled for the GWMO from 125 countries gives the average collection coverage in low-income countries as 36% (the World Bank provides an average of 43%), lower-middle income countries 64% (World Bank 68%) and upper-middle income countries 82% (World Bank 85%), with higher income countries showing collection coverage approaching 100%. On a regional basis, collection coverage has the following ranges: Africa (25% to 70%); Asia (50% to 90%); Latin America and Caribbean (80% to 100%), Europe (80% to 100%) and North America (100%). Although these estimates are quoted as country-wide data, some incorporate the entirety of the population, both urban and rural, while others focus on urban areas. Many countries show great variation in degree of coverage among local areas or regions. For example, the national database for Brazil gives a national average of 90.2% in 2011, but the State averages range from 60% to 99.2%.[22] Similarly, in India, ministry data for 105 major cities shows collection coverage ranging from 40 to 100%.[23] Because rural areas typically have lower rates of collection coverage than urban areas, national averages on collection coverage are likely to be lower than the averages of urban areas alone.

In order to assess the status of collection coverage just at the city level, *Figure 3.9* shows data on 39 cities for which Wasteaware ISWM indicators are available.[24] *Figure 3.9* appears to fall into two parts: at lower income levels, collection coverage appears to increase with increasing income, while above a certain threshold, collection reaches 'saturation' as levels approach 100%. If one apparent outlier (Canete, Peru) is set aside, then the threshold income level for this transition appears to lie at a GNI per capita in the range 2,000 to 3,000 USD per year. It needs to be borne in mind that data for entire cities may conceal a gap between the 'haves' and 'have-nots', in which often, the central business district and affluent neighbourhoods have near 100% coverage, while low-income and unlawful settlements often have none.

Supporting evidence comes from the 2014 comparative report for member countries of the SWEEP-Net consortium in North Africa and the Near East.[25] The consortium reports collection coverage for nine countries at an average of 63%, with an average across urban areas of 75% (range 30 to 100%), and for rural areas of 40% (four at 0%, and the others at 35%, 70%, 70%, 90% and 100%).

The World Bank assessment of collection coverage quoted on their website, that "30 to 60% of all the urban solid waste in developing countries is uncollected and less than 50% of the population is served,"[26] appears to be more of a reasonable historical baseline applying up to 2000 than a current estimate. If that is indeed the case, then the data presented here, both from the GWMO and from the World Bank's own project data,[27] suggests that there has been significant progress since that time, particularly in cities in those countries with an income above about 2,500 USD per capita per year (which represents approximately the mid-point in the

22 Annual reports on Brazilian waste statistics (in Portuguese). www.abrelpe.org.br
23 India, Ministry of Urban Development (2012), listed in Annex A, Chapter 3, Collated data sources.
24 Wilson et al. (2015), listed in Annex A, Chapter 2, Waste data and indicators. See also Section 2.5.3 for more information on indicators.
25 SWEEP-Net (2014), listed in Annex A, Chapter 2, Waste data and Indicators.
26 World Bank (n.d.). Urban Solid Waste Management. http://go.worldbank.org/A5TFX56L50
27 Hoornweg & Bhada-Tata, *What a Waste* (2012)

range of lower-middle income countries). However, it is clear that many low-income cities still have collection coverage in the range of 30 to 60%, and that the figures may be much lower in some countries, and also in the more rural areas of many countries. If the figures here for collection coverage are combined with the 2014 data for world population by country income groups,[28] then **it can be estimated that at least 2 billion people worldwide still lack access to solid waste collection**.

Figure 3.9 Collection coverage for selected cities by income level

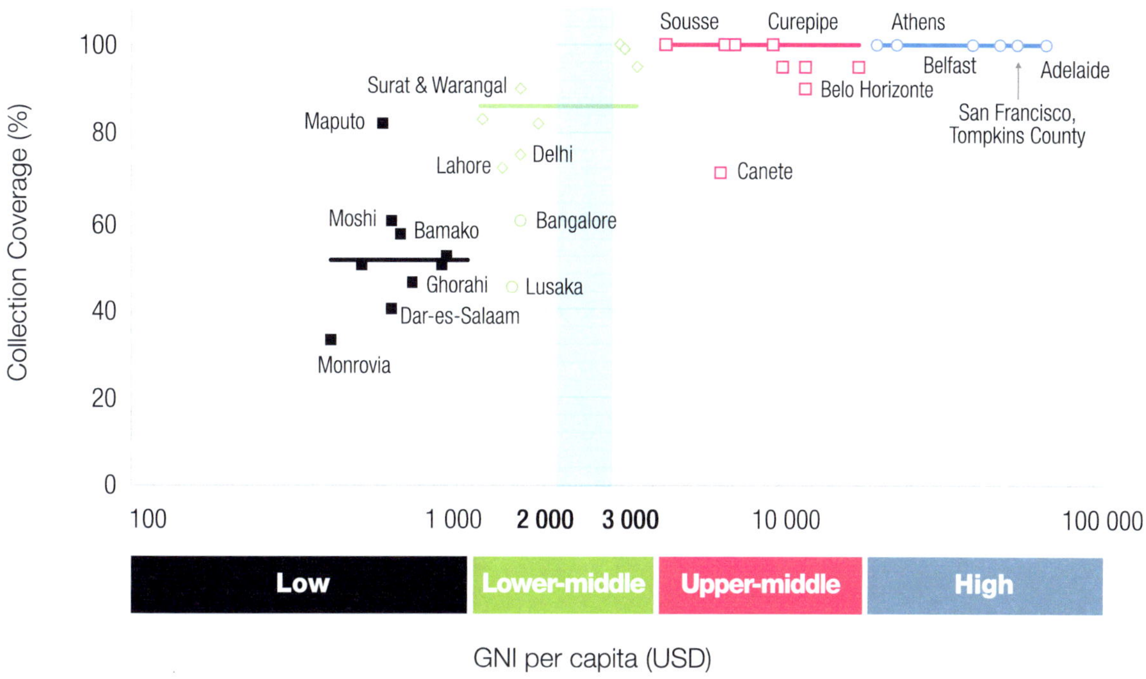

Notes: Figure shows collection coverage versus gross national income (GNI) per capita on a logarithmic scale. The cities are those for which Wasteaware indicators[29] were available in May 2014. The blue vertical bar represents an apparent 'collection coverage threshold' (GNI of 2000-3000 USD per capita), as it was first discussed in Wilson et al. (2012). Below that threshold, collection coverage increases with income level; above the threshold, collection coverage reaches 'saturation' as levels approach 100%. The four coloured horizontal lines show the median collection coverage for each income group.

Source of data: Wasteaware – University of Leeds.

Waste collection services come in a wide variety of shapes and forms (*Box 3.2*). Services may be delivered by the formal sector, through either public- or private-sector operators, or by the community or 'informal' sector, through for example community based organizations (CBOs), non-governmental organizations (NGOs) or micro- and small enterprises (MSEs).[30] Services may be on a relatively small scale, providing primary collection to local neighbourhoods, or on a larger scale, providing either secondary collection or an integrated collection service across the city. Pickup is carried out by a range of vehicle types, such as bicycles, tricycles, tractor and trailer, tipper trucks or purpose-build compaction vehicles, and sometimes by pushcarts or animal powered carts.[31] To optimize collection systems, the use of GPS and GIS, or even route optimization software, may be relevant for large municipalities or substantial collection coverage areas.

28 World Urbanization Prospects, 2014 Edition. http://esa.un.org/unpd/wup/
29 See Section 2.5.3 as well as Wilson et al. (2015).
30 See Section 5.6 on a financing model for delivering services.
31 Coffey & Coad (2010), listed in Annex A, Chapter 3, Municipal solid waste management

BOX 3.2 EXAMPLES OF DIVERSITY IN MSW COLLECTION PRACTICES

Scale of operation

SMALLER-SCALE PRIMARY COLLECTION	LARGER-SCALE SECONDARY COLLECTION

© KKPKP

Cycle cart, India

© Photo courtesy: Odeniyi Ra

Close truck collection, Nigeria

© Ainhoa Carpintero

Push cart, Vietnam

© GIE Salambougou by Erica Trauba

Open collection, Mali

© Ainhoa Carpintero

Small truck collection, China

© Petri Rogero

Truck collection, Spain

© KKPKP

Primary collectors delivering waste to a secondary refuse collection vehicle, India

3.4.2 Controlled disposal

Uncontrolled disposal (through open dumping and open burning) was the norm everywhere until the 1960s,[32] and according to the World Bank is still the norm in most developing countries.[33] This practice gives rise to substantial public health and environmental risks. These risks are significantly increased in cases in which hazardous waste is delivered to a dumpsite alongside MSW.[34]

The high-income countries have learned that 'cleaning up the sins of the past' can be significantly more expensive than disposing of waste in an environmentally sound manner (ESM).[35] Legislation phasing out uncontrolled disposal was first introduced in high-income countries in the 1970s, and the standards required for ESM facilities have since been gradually raised.[36]

Figure 3.10 shows progress around the world in achieving the first step of eliminating open dumps and achieving controlled disposal, as measured by the Wasteaware controlled disposal indicator.[37] This novel indicator is the percentage by weight of the residual waste remaining after collection for recycling that is received at a controlled treatment or disposal facility. 'Controlled' disposal involves adequate treatment of waste and operation of secured facilities so as to meet defined compliance requirements. However, a controlled facility does not necessarily have to meet the latest EU or US standards. It can also for example be an 'intermediate' engineered landfill or an upgraded dumpsite.[38]

Figure 3.10 Controlled disposal for selected cities by income level

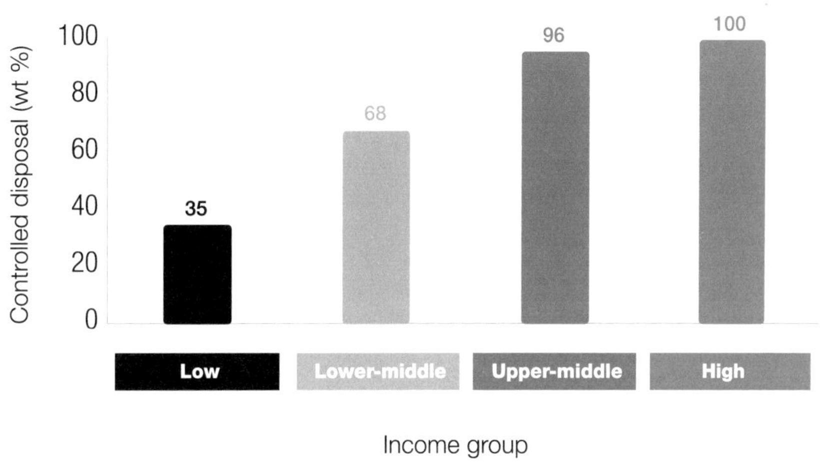

Notes: Graph shows average percentage of controlled disposal for each of the four standard World Bank income level categories. 'Controlled disposal' is the primary quantitative environmental indicator defined in the Wasteaware ISWM indicator set. Data are for the 39 cities for which the indicators were available in May 2014.

Source of data: Wasteaware – University of Leeds

Phasing out uncontrolled disposal practices is one of the first objectives in improving MSW management in developing countries. Besides the 100% controlled disposal generally achieved in high-income countries, the rates in upper-middle income cities (with an average of 95%) and in lower-middle-income cities (with an average of 70%) are still substantially better than the historical '0% norm'. Even the average 35% in the lower-income cities is better than the historical norm. Evidence to support this apparent recent progress is given by other sources. The Brazilian national database divides disposal into three categories: sanitary landfill, which makes up 57% of the nation's disposal on average; other landfill, at 24%; and uncontrolled dumping, at 18%. This means that in Brazil, despite relatively high controlled disposal rates, the waste from around 35 million people is dumped in an uncontrolled manner, amounting to some 15 million tonnes annually. It also suggests a comparable controlled disposal indicator likely around 80% on average. The averages for controlled disposal at

32 See Section 2.3 on drivers for waste and resource management.
33 World Bank (n.d.). Urban Solid Waste Management. http://go.worldbank.org/A5TFX56L50
34 See Box 1.2 in Chapter 1. Also, Topic Sheet 2, found after Chapter 1, provides information on 'the 50 biggest dumpsites in the world'.
35 The costs of inaction are documented in Section 5.2.3.
36 See Sections 2.3 on waste history and Section 4.3.4 on environmental legislation.
37 See Section 2.5.3 on waste management indicators.
38 Rushbrook & Pugh (1999) and Hoornweg & Bhada-Tata (2012), listed in Annex A, Chapter 3, Collated data sources.

an individual state level range generally from 40% to 90%, with one outlier below 20%. The 2014 comparative report for the SWEEP-Net consortium in North Africa and the Near East gives controlled disposal rates in the range of 10 to 70% across nine countries, with an average of 24%. As of 2010 there were 7,518 waste disposal sites officially reported in Russia, of which around 23% were MSW landfills, 7% industrial waste disposal sites, and 70% unauthorized dumps.[39] In 2011, the PRC achieved a national average controlled disposal rate of around 90%.[40]

A case study of the successful elimination of open dumping in a small town in Colombia is shown in *Box 3.3*.

In summary, the status as assessed in 2015 using the latest available data appears to be significantly better than mere dumping as the norm across developing countries. The Wasteaware data suggest that significant progress is being made by some cities in middle-income countries, with controlled disposal rates often in the range of 70 to 95%, although there is a lot of variation both within and between countries. Such achievements are impressive and compare well with the early take-up of controlled disposal in Europe in the 1970s and 1980s. The situation is much worse in low-income countries, where controlled disposal rates are often well below 50% overall and 0% in rural areas. If the figures here for controlled disposal are combined with the 2014 data for world population by country income groupings,[41] then **it can be estimated that at least 3 billion people worldwide still lack access to controlled waste disposal facilities**.

BOX 3.3 VERSALLES, COLOMBIA: AN EXAMPLE OF INTEGRATED MUNICIPAL SOLID WASTE MANAGEMENT[42]

In Versalles, a small town in Colombia, open dumping was a common sight until 1997. Through technical support from Suna Hisca, a non-profit Colombian organization, and financial support from Corporación Autonoma Regional del Valle (CVC), an integrated municipal solid waste management plan was devised. The implementation of this plan enabled Versalles to stop the contamination of its water resources and avoid potential health impacts from this practice.

The objectives of the Plan of Integrated Management of Solid Waste (PIMSW) were: (a) to achieve adequate collection, transport and disposal of municipal solid waste; (b) to engage the active participation of the stakeholders (users, the utility, the municipal administration, recyclers); (c) to get the community to practice source separation into three fractions: organics (food waste), recyclables (plastic, cardboard, metal, etc.) and sanitary waste (items contaminated with blood, urine or excreta such as sanitary towels, wound dressings, nappies, or incontinence pads); (d) to build an Integrated Solid Waste Plant to process the solid waste; (e) to create a public utility; (f) to generate employment; and (g) to improve municipal environmental sanitation. A public utility called Cooperativa Campo Verde was responsible for implementing the plan and is responsible for the collection and transportation of the waste, as well as for the operation of the plant.

As a result of the plan's successful implementation, the rate of separation at source in 2015 was above 80%, with recoverable materials marketed and organic matter transformed into compost for sale. Of the 42 tonnes of waste generated by the community per month, 27 tonnes of organic matter and 7 tonnes of recycled materials are recovered and transformed. Overall, the town has reduced by 83% the amount of waste it would have otherwise sent to landfill. *Figures 1 and 2* show the town's new weighbridge and a new vehicle for the separate collection of solid waste.

39 IFC (2014), listed in Annex A, Chapter 3 Collated data resources
40 China Statistical Yearbook 2014, listed in Annex B.
41 World Urbanization Prospects, 2014 Edition. http://esa.un.org/unpd/wup/
42 Information and text provided by Leonardo E. Navarro J. of Suna Hisca, consultant to the Housing Ministry of Colombia

3.3 RESOURCE RECOVERY

Sitting alongside the public health driver for waste collection and the environmental driver to phase out uncontrolled disposal is the resource value driver for the '4Rs' – reduce, reuse,[43] recycle and recover. The focus here is on recycling and recovery, of MSW in particular but not exclusively.

3.5.1 Collection for recycling

Most 'recycling rates' for MSW refer to the waste collected for recycling.[44] Corrections are sometimes made for subsequent 'rejects' – materials not passed on up the materials value chain for eventual recycling – but it is difficult to audit how far corrections have been done, especially in globalized value chains of secondary materials, such as in the case of waste plastics.[45] The data presented here include the collection of materials for both 'dry recycling' (e.g. paper, plastics, metals, glass, textiles) and organic recycling. The downstream processing of the collected waste materials for recycling is dealt with in subsequent sections.

Local innovation to implement the container deposit system.
The basket for 500 cans avoids manual counting at the time of deposit and refund, Kiribati

Official data for MSW recycling often come from municipal governments, which in many developing countries focus on managing the MSW they collect (or which is collected on their behalf by the 'formal sector', leaving collection of materials for recycling often to the 'informal sector'). Official data, either at the city level or compiled from city data by national governments, are thus likely to be under-reporting recycling rates. This was indeed one of the motivations behind the methodology developed to collect the data for the city-level Wasteaware ISWM indicators, that the system being studied should be the complete waste and recycling system for the city.[46] Recycling rates were calculated with the assistance of a material flow analysis (MFA) developed for each city. Waste flows were estimated and cross-checked against each other using the MFA.[47]

Figure 3.11 shows the Wasteaware recycling rates from a sample of 39 cities across various income groups. This figure shows no clear relationship between recycling rates and income levels. While recycling rates are indeed highest in the high-income countries, some low- and lower-middle income countries do collect quite reasonable percentages of their total MSW for recycling (20 to 40%). Interestingly, there is some evidence that recycling rates are lower in some of the more developed, upper-middle income countries, perhaps reflecting the history in the developed world where formalization of solid waste management as a municipal service displaced pre-existing informal recycling systems as standards of living rose, prior to the more recent 'rediscovery' of recycling and a resurgence in recycling rates in the high-income countries.[48] More research would be required to confirm this hypothesis.

43 Both 'reduce' and 'reuse' are addressed in the Topic Sheet 4 on waste prevention, which follows Chapter 2.
44 Velis & Brunner (2013), listed in Annex A, Chapter 3, Recycling.
45 Velis (2014), listed in Annex A, Chapter 3, Global secondary materials industry.
46 See Section 2.5.3 and Table 2.3, both on indicators.
47 See Section 2.4.2 on life cycle analysis.
48 See Section 2.3.1 on historical drivers in developed countries.

Figure 3.11 Average recycling rates for 39 cities by income level

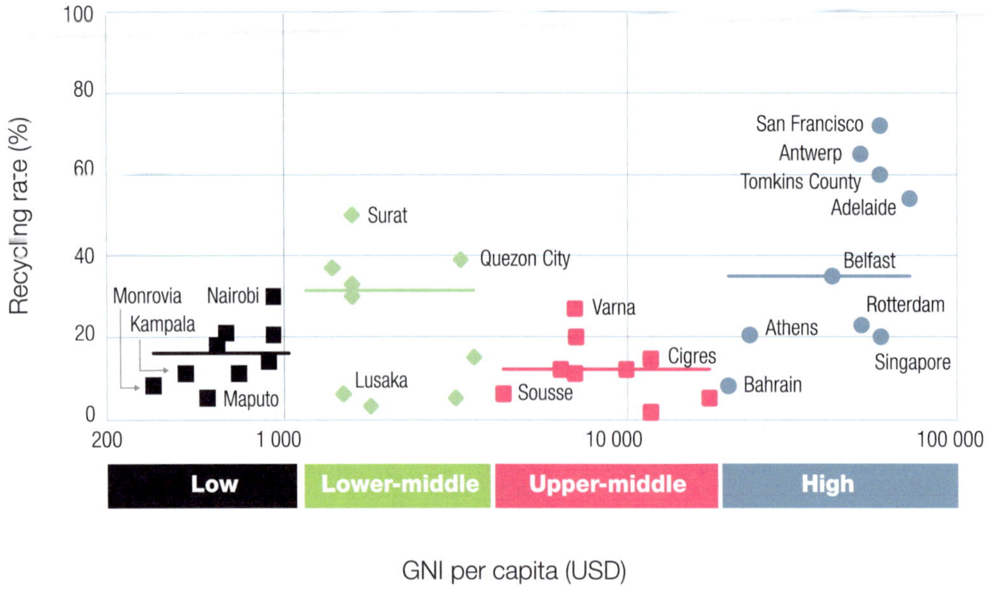

Notes: Verified and consistent data from application of the Wasteaware indicators[49] to benchmark recycling performance. In calculating recycling rates, consideration has been given to both the formal municipal SWM system and any informal/community recycling being carried out in parallel (whereas 'official' statistics often understate recycling by ignoring the latter). The cities are those for which Wasteaware indicators were available in May 2014. Data covers both dry recyclables and organics. Where possible, correction has been made for materials collected for recycling, but ultimately disposed of after initial processing (reject fraction).

Source of data: Wasteaware – University of Leeds.

Information on recycling rates in the EU countries is now collected regularly and systematically but inconsistencies in the definition still exist (e.g. regarding counting collection for recycling vs. counting outputs of MRFs and composting plants, regarding whether or not to count metals and aggregates obtained as the output of EfW combustion plants) and the level of data reliability still differs among the EU countries. *Figure 3.12* provides statistics on recycling rates in the EU countries. It may be observed that the recycling rates in the EU have increased substantially between 2001 and 2010, as the lower-performing countries have worked towards meeting the EU-wide targets.[50]

Figure 3.12 Municipal solid waste recycling in the European Union

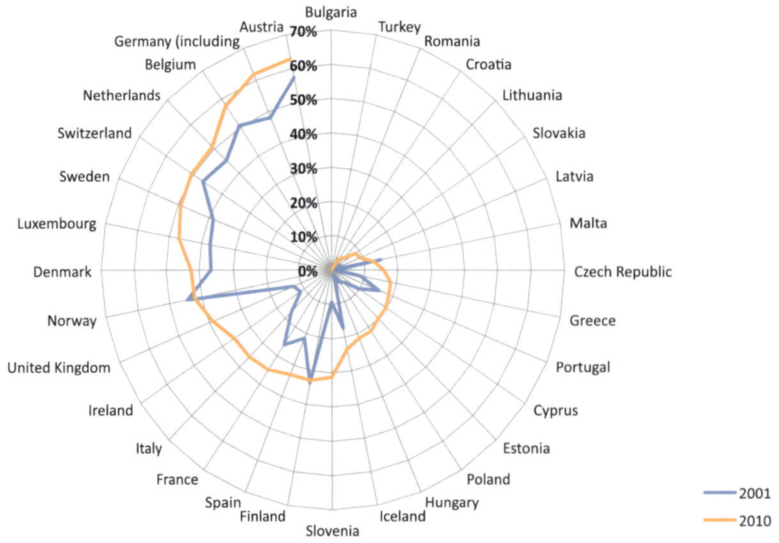

Source: http://www.eea.europa.eu/data-and-maps/figures/municipal-waste-recycling-rates-in

49 See Section 2.5.3 as well as Wilson et al. (2015)
50 Legislation and policies to promote recycling are discussed in Sections 4.3.5 and 4.4.

3.5.2 The importance of segregation

Recycling depends critically on two aspects of 'segregation'. The first is the degree of mixing of different elements or materials within a product, or the concentration at which the element is present, which can be addressed through design for recyclability. The second is to keep different 'wastes' separate at the point of generation, to ensure that they remain clean and uncontaminated by other waste streams. This can be addressed through segregation at source. These two aspects are elaborated here in turn.

Design for recyclability

Figure 3.13 shows a plot of recyclability versus the degree of material mixing for a wide range of consumer products. This clearly shows that products with lower degrees of material mixing are easier and more economical to recycle than others, with the degree of mixing at which recycling is feasible increasing as the value of the recycled materials rises. Products with a lower degree of mixing and higher values of the component materials are economic to recycle, while those with higher degrees of mixing and lower values are not.

One way to 'manipulate' this relationship is to address recyclability explicitly in the design process. For example, automobile manufacturers have recently focused on designing their products to facilitate both future dismantling (design for dismantling – DfD) and recycling (design for recycling – DfR).

Figure 3.13 Single product recycled material values

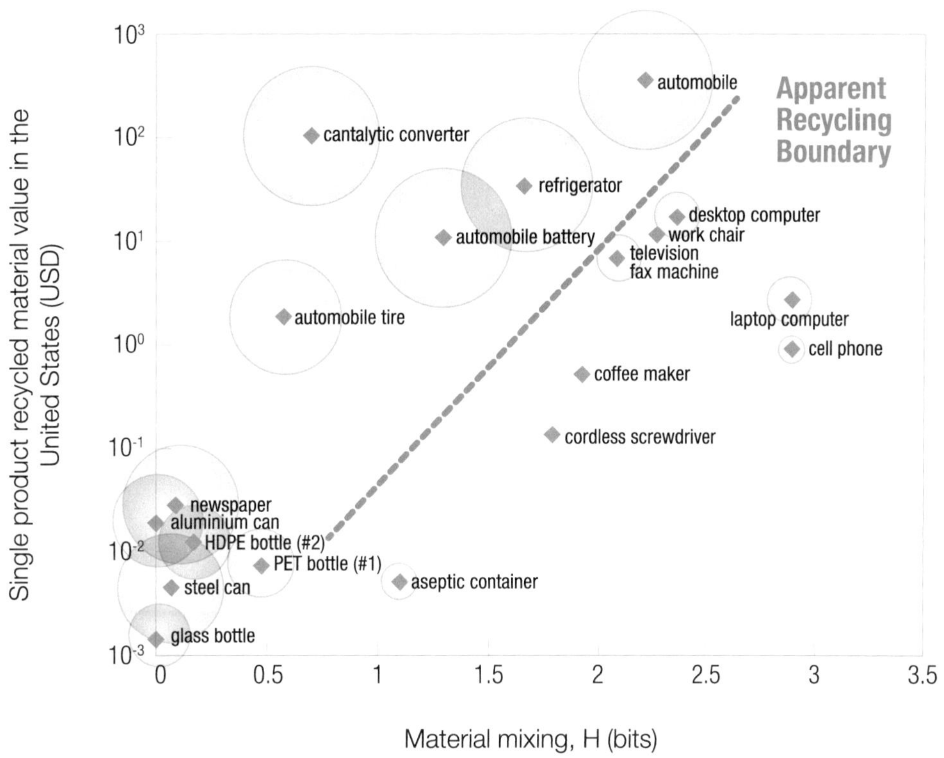

Note: Material mixing (H) and recycling rates (for 20 products in the USA. Recycling rates are indicated by the size of the spheres: for example, automobile and catalytic convertor recycling rates are 95%, newspapers 70%, PET bottles 23% and televisions 11%. The 'apparent recycling boundary' separates the graph into two regions: where recycling tends to take place and where it tends not to take place. Material mixing incorporates the number of components as well as the concentrations of the components in a product, expressed as 'H', which is the average number of binary separation steps needed to obtain any material from the mixture (i.e. the product). For a simple product consisting of a single material only (e.g. a glass bottle), H is zero. H is therefore proportional to the complexity of the product. For instance, automobiles contain many materials and are thus more complex, so they therefore have a very high value of H. However, this graph does not indicate the actual recyclability via liberation of items by manual, mechanical or thermal processing means. Initial complexity is one thing, but the ability to liberate material is another, more critical feature.

Source: Dahmus & Gutowski (2007). What gets recycled: an information theory based model for product recycling. *Environmental Science & Technology* 41: 7543–7550.

Recycling rates depend both on the degree of mixing and on the concentration of the target material or element. In a study of 60 metals (*Figure 3.14*), only one third have recycling rates greater than 50%. These include aluminium, titanium, chromium, manganese, iron, cobalt, nickel, copper, zinc, rhodium, palladium, silver, platinum and gold, all of which are either used in high concentrations and/or have a high value. Although more than half the metals have very low recycling rates of less than 1%, many of them are regarded as 'critical materials', including indium and gallium, or are rare earth metals including lanthanum, cerium, praseodymium, neodymium, gadolinium and dysprosium. These metals are all used in a wide range of electronic products including screens, chips and speakers and microphones and also in the magnets that are critical in many renewable energy technologies. The problem with recycling these critical metals is that the concentrations are often very low while the degree of mixing with other elements is very high. A major challenge moving forward is to ensure that design for dismantling and design for recyclability is prioritized in these rapidly growing industrial sectors.

Figure 3.14 End of life recycling rates for 60 metals

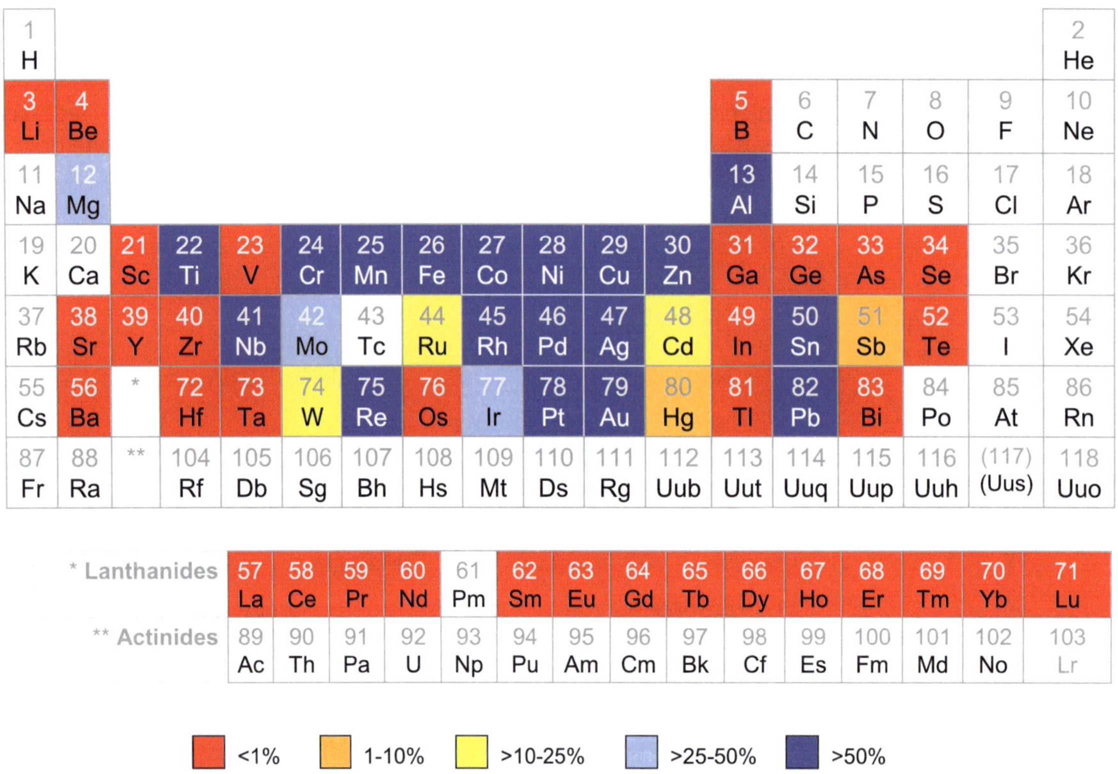

Notes: The figure uses the periodic table to show the global average end-of-life (post-consumer) functional recycling for sixty metals. Functional recycling is recycling in which the physical and chemical properties that made the material desirable in the first place are retained for subsequent use. Unfilled boxes indicate that no data or estimates are available, or that the element was not addressed as part of the study. These evaluations do not consider metal emissions from coal from power plants.

Source: UNEP (2011b). *Recycling Rates of Metals: A Status Report.* http://www.unep.org/resourcepanel/Portals/24102/PDFs/Metals_Recycling_Rates_110412-1.pdf

The presence of hazardous components is particularly important: for recycling to be economically feasible, recycling streams should ideally be contaminant free. Household hazardous waste (e.g. spent batteries), if not segregated, can contaminate the organic fractions and result in compost that is contaminated by toxic heavy metals.[51] Another example is that of waste paper containing polychlorinated biphenyls (PCBs), a persistent organic pollutant (POP) that is released when some older carbonless copy papers are recycled. A 2014 study on paper and board collected from Danish household waste suggested presence of measurable quantities of PCBs that could potentially have health and environmental consequences.[52]

51 Velis & Brunner (2013)

52 Pivnemko, K., E. Eriksson and T.F. Astrup (2014). Polychlorinated biphenyls (PCBs) in waste paper from Danish household waste. 5th International Conference on Engineering for Waste and Biomass Valorisation, Rio de Janeiro, 25-28 August 2014. Despite being banned since 1993, measurable levels of PCBs were found in wastepaper samples.

Segregation at source

Segregation of MSW at source is critical to ensure that the waste is separated into organic and dry recyclable fractions. Segregation of MSW at source, by separating organic and dry recyclable fractions, is critical to avoid cross-contamination and to maintain the quality of the materials, which will lead to more effective recycling and divert waste from landfill. Further, segregated waste reduces health and safety related risks to waste pickers and to the ecosystems around the waste treatment and disposal sites.

Despite the advantages of segregation, source separation prior to recycling is of relatively recent origin in formal MSW management systems. Referring back to *Figure 3.13*, the high recycling rates in high-income countries are now almost all based on segregation at source, resulting in relatively clean fractions being collected for recycling. Some examples of the use of such systems in middle-income countries are shown in *Box 3.4.*

BOX 3.4 EXAMPLES OF SEGREGATION INFRASTRUCTURE AS PART OF FORMAL SWM SYSTEMS

Separate collection of food waste, Yangshuo, PRC

Separate collection of plastics in Siddhipur, Nepal

Recycling containers in Phitsanulok, Thailand

Recycling containers in Benalmadena, Spain

Waste Segregation in Oslob, Cebu, Philippines

The informal recycling that is often dominant in many developing countries is generally from mixed MSW, although there can be a significant contribution from 'itinerant waste buyers', who collect and pay for source separated materials accumulated by householders or domestic servants. Community initiatives may also collect source-separated materials to raise funds for local charities.[53] Increasing segregation at source is a critical component of any programme to include the informal sector into mainstream waste management and would both improve their working conditions and improve their livelihoods by improving the quality of the recycled materials.[54]

Persuading citizens to segregate their MSW at source and to present it separately for collection requires a focus on changing behaviours.[55] Building design is also important, in order to ensure adequate space to store several separated fractions pending collection.

3.5.3 Technologies for resource recovery

A number of technologies are used for the processing and recovery of resources from waste. A summary for material separation technologies, including MRF, sorting centres and MBT is provided in *Box 3.5*, for organic recovery, including composting, anaerobic digestion and animal feed in Box 3.6 and in *Box 3.7* for energy recovery technologies. A complementary broad comparative analysis of some of the technologies that are commonly used, focusing on MSW, is presented in *Table 3.1*. This comparison is only indicative, addressing various elements such as applicability, advantages, relative costs and key factors for success. A listing of key resource materials to obtain more details of each technology is provided in Annex A, Further Resources.[56]

The selection of technologies appropriate to a particular local situation is as much of a governance issue as a technical matter: a key starting point is the waste composition and resultant waste properties, which need to be considered alongside local governance issues and the goals of the waste strategy. Selection of technologies is thus discussed later in the GWMO both under strategic planning (*Section 4.2.2*) and how to select an appropriate set of policy instruments that will be most effective in a particular situation (Section 4.9.2), where *Box 4.37* explicitly considers the selection of appropriate technologies for a developing country.

Emptying underground waste containers, Netherlands

53 An example is the city of Salem in Tamil Nadu, India. One kg of plastic waste is exchanged for a pencil, and 10 kg for a notebook. The community in turn sells plastic at Rs 2.50/kg to the market. Jars are kept outside city temples where worshipers are encouraged to bring used glass bottles. Glass is then sold at Rs 0.5/kg and the money collected is used to whitewash the temples.

54 See Section 4.7 on Including stakeholders and Topic Sheet 14 on the informal waste sector.

55 See Section 4.5 on Economic instruments Figure 4.4 on the '4Es' framework for designing behaviour change initiatives.

56 See Annex A, Chapter 3, Technologies for resource recovery.

Figure 3.15 uses a triangle chart to show the percentages of disposal sent to landfill, recycling and composting, and combustion with energy recovery from waste (EfW) in different countries. Countries can be grouped into two main clusters: in one, rates of disposal to landfill range from 50 to 100%, with (collection for) recycling rates of 0 to 40%; in the other, rates of disposal to landfill are less than 10%, with both recycling and EfW in the range of 30 to 70%.

Figure 3.15 Proportions of recycling and composting, energy from waste (EfW) and disposal to landfill in European and non-European countries

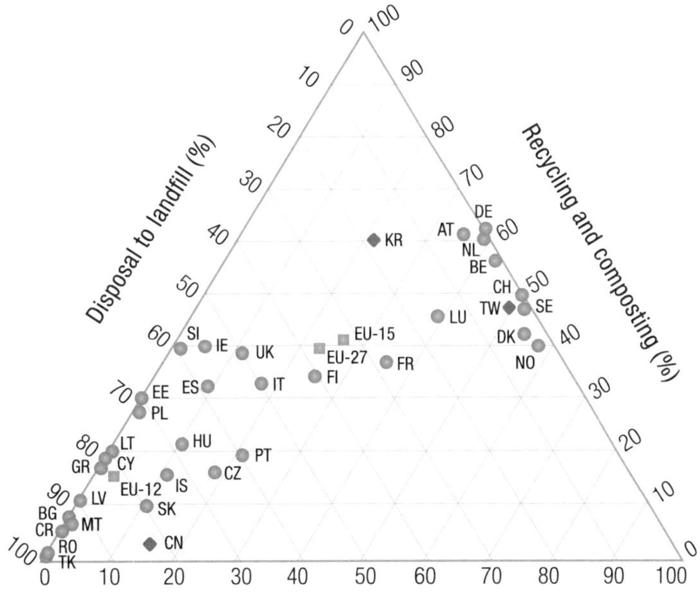

Notes: Proportions of collection for recycling and composting, combustion with energy recovery (EfW) and disposal to landfill in European countries (circles) and non-European countries (rhombuses), along with European averages (cubes), in 2011.

Source: Bartl (2014). Moving from recycling to waste prevention: A review of barriers and enables. *Waste Management & Research.* 32(9) Supplement 3–18

Separation of wastes in an MRF/MBT plant in Madrid, Spain

BOX 3.5 ALTERNATIVE TECHNOLOGIES FOR RESOURCE RECOVERY FROM WASTE

(1) MATERIALS RECOVERY AND SORTING FACILITIES, INCLUDING MBT

Material recovery facilities (MRFs)

- *'Clean' MRFs.* 'Clean' material recovery facilities further separate clean, source-segregated dry materials for recycling and/or produce a prepared fuel. They may use either hand or automated sorting systems, or some combination of the two. They are used extensively in developed countries alongside source separation of mixed recyclables.
- *'Dirty' MRFs.* 'Dirty' material recovery facilities accept mixed waste (MSW or from other sources), from which dry recyclable materials are separated out from the organic fraction. These can be similar to the mechanical part of an MBT plant. Cross contamination results in lower quality outputs. These are more common than clean MRFs in developing countries.
- *Specific purpose MRFs.* Specialized material recovery facilities focus on specific waste streams, such as e-waste, C&D waste, or plastic waste.

Waste sorting centres

- *Waste sorting centres.* 'Waste sorting centre' is the term used mainly in developing countries to cover a range of options. For example, the city of Pune city in India has set up a number of mainly manual waste sorting centres with the informal sector to integrate them into the mainstream waste management system. Centres which involve the informal sector but use a mix of manual and mechanical sorting are common in Brazil and some other countries. The United Nations Economic and Social Commission for Asia and the Pacific (UNESCAP) has been promoting decentralized and Integrated Resource Recovery Centres (IRRC) in seven secondary cities across five countries in the Asia-Pacific region as a demonstration project.[57] In practice, waste sorting centres may overlap with MRFs, and in the case of IRRCs, with MBTs (explained below).

Mechanical biological treatment facilities (MBTs)

- Mechanical biological treatment facilities use a group of technologies and accept either MSW, or residual MSW after source separation of recyclables. MBTs are extensively used, particularly in Europe.
- MBTs use a range of combined mechanical and biological processes to treat and further separate the waste into recyclable, organic-rich and fuel-rich fractions. Each facility is designed with a particular purpose, using a specific input waste stream to prepare outputs to meet certain specifications. All can separate dry recyclates and/or refuse-derived fuel (RDF).
- Aerobic biological unit processes may be used to 'stabilize' the organic fraction to reduce its biodegradability, and therefore its capacity to generate methane, as a pre-treatment prior to landfill ('biostabilization' being the simplest option). Similar, but more complicated, is the production of compost-like output (CLO) for low-value on-land applications.
- A high-value configuration is to use biodrying (aerobic decomposition with high aeration) along with extensive mechanical processing to prepare a quality-controlled manufactured fuel (solid recovered fuel, or 'SRF'). This is particularly useful for treating high moisture organic wastes. SRF can be stored and transported for use by industry in thermal processing energy generation facilities (see 'Co-combustion in an industrial facility' in *Box 3.7* below).
- Anaerobic biological unit processes (or anaerobic digestion, AD) can produce biogas from the mechanically separated organic fraction of MSW.
- As the input to MBT plants is mixed waste, most of the solid outputs – including low-quality RDF and CLO – are all still regulated in most high-income countries as waste, so the products can only be used if the receiving facility or site obtains a waste management licence. Some higher quality SRF fuels, as well as dry materials separated for recycling, may be able to meet an 'end-of-waste' protocol so that the material can be traded as a product, and the using facility will then not require a waste management licence.[58]

---

57 See Box 4.5 in Section 4.2.2. Also see Storey et al. (2013), listed in Annex A, Chapter 3, Technologies for resource recovery.

58 See Sections 4.3.2 and 4.3.5. Italy was the first country to adopt an end-of-waste protocol for SRF.

BOX 3.6 ALTERNATIVE TECHNOLOGIES FOR RESOURCE RECOVERY FROM WASTE

(2) ORGANICS RECYCLING/RECOVERY

Composting

- Compost is the output of a biological process that converts biodegradable waste to a humus-like material. The principal use is to improve soil quality, as compost improves its biological and physical properties, for example enhancing water retention and resistance to erosion, which is particularly valuable in arid climates. It also has some value as fertilizer.
- Composting is applicable to a wide range of organic wastes. Residence times are typically longer for lignin-rich, 'hard', woody wastes.
- Contamination of compost due to household hazardous waste is an issue. In developed countries, regulations allow the use of waste-derived composts for food production, only if clean source-separated feedstock is used. Such materials may be able to meet an end-of-waste protocol.[59] This is the reason for the use of the term 'compost-like output' (CLO) when mixed waste is used as the feedstock (such as from an MBT plant). This will remain a 'waste' and be restricted to non-food applications at sites which obtain a license as waste facilities.
- Composting requires good process control, to ensure sufficient temperature and retention time to eliminate pathogens and to destroy weed seeds. Open heaps or windrows is the simplest and cheapest method. In-vessel composting uses a variety of proprietary technologies, which claim faster processing times and must be used (including under EU regulations) if the feedstock contains animal by-products. However, open-air or covered windrows are often used for the maturation of the output from in-vessel units.
- Typically 50 to 70% by weight of the MSW generated in developing countries is organic materials suitable for composting. Composting can be facilitated through segregation at source. Decentralized composting systems have been found to work well in many cities in low- and middle-income countries.[60] Home composting is also widely practised all around the world. Vermicomposting, which uses worms, is a popular option,[61] particularly in India.

Anaerobic digestion

- Anaerobic digestion (AD, also known as biomethanization) is considered a reliable source of energy in the form of biogas. AD works best for wet wastes, so is most widely used for sewage sludge and for livestock wastes. In 2013, the majority of the 13,800 AD plants in Europe and the 2,200 AD plants in the US treated those two types of waste.
- In developing countries, AD is widely applied at both the small and the community scale, for domestic or community use of the bio-gas. For example, in 2013 there were more than 40 million AD plants in the PRC, nearly 5 million in India and 300,000 in Nepal. [62]
- Application of AD to MSW is challenging. The high solid content, large particle size and inhomogeneous nature of the waste makes process control difficult.[63] It is particularly difficult to apply AD to lignin-rich, woody wastes. The digestate remaining can in principle be used as a soil conditioner. This usually requires a relatively long maturation (composting) stage prior to application to land.
- Contamination can both disrupt the AD process as well as make the digestate unsuitable for use as a soil conditioner. For MSW, the use of clean source-segregated feedstock is preferable, and essential if the (composted) digestate is to meet end-of-waste criteria and be used for food production. For an organic fraction separated mechanically from either mixed MSW or from residual MSW, such as from an MBT plant (see *Box 3.5*), contamination is a major issue. The digestate can subsequently go through a composting (maturation) step and be used as a compost-like output (CLO) for low-value on-land applications or be dried for use as a low calorific value RDF. RDF, CLO and digested sewage sludge will typically remain as 'wastes'. This means that all handlers and users need to obtain waste management licences.

Animal feeding[64]

- For clean, source-segregated food waste, direct reuse as animal feed is an important option.
- Japan makes extensive use of this option, using central processing plants to sterilize the waste to destroy any pathogens which may carry animal diseases.[65]

59 See Topic Sheet 12 for examples of compost quality protocols in selected EU countries.
60 Eawag/Sandec and Waste Concern (2006), listed in Annex A, Chapter 3, Technologies for resource recovery.
61 See http://vermicomposting.com/ and http://www.calrecycle.ca.gov/Organics/Worms/
62 REN21 (2014)
63 In Section 4.2.2 on waste planning, Box 4.3 provides an example of a failed investment in an AD plant, attributed to a failure to adapt the design to local conditions.
64 See Topic Sheet 11 on Food Waste, found after Chapter 3.
65 See Section 4.3.5, Box 4.8 for more information on food waste recycling in Japan through processing the waste into animal feed.

BOX 3.7 ALTERNATIVE TECHNOLOGIES FOR RESOURCE RECOVERY FROM WASTE

(3) FUEL AND ENERGY RECOVERY FROM WASTE STREAMS

Combustion with energy recovery as electricity and/or heat

- Combustion with energy recovery has been widely used for MSW for many years. This process has been used to destroy the hazardous components of many 'difficult' wastes (such as POPs) and produce renewable and carbon-neutral energy from the biogenic part of the waste (roughly 65% of MSW in high-income countries).
- This option requires a high level of process control and of gas cleaning. Modern plants can achieve very high environmental protection standards. If not performed correctly, there is the potential for generating air emissions of particulates, acid gases, metals and incomplete combustion products such as dioxins. The process must be controlled through multi-stage gas cleaning to meet high standards.
- This technology can achieve high levels of energy efficiency. The EU threshold for energy efficiency, using a policy formula considering conversion of waste to both electricity and heat, is 65%.
- Worldwide, an estimated 765 'energy from waste' (EfW) plants exist for MSW, with an annual capacity of 83 million tonnes. These include 455 plants in the EU, 86 in the US (2011-12) and 150 in the People's Republic of China (PRC; 2014).[66]
- A list of questions regarding suitability in a particular situation, particularly in a developing country where the waste may be of low calorific value and financial sustainability may be an issue, is provided in Box 4.37.

Co-combustion in an industrial facility

- A prepared fuel (e.g. a solid recovered fuel or 'SRF') can be used in a range of industrial facilities, including cement kilns, industrial boilers and power plants. Attention needs to be paid to emission controls in the user facility. Cement kilns are attractive here, as they operate at high temperatures and already have air pollution control systems in place. However, these may be typically of lower standards than purpose-built EfW combustion plants.
- Co-combustion in an industrial facility is widely used for prepared fuels from MBT plants, particularly quality controlled SRF, but also RDF.
- Many cement kilns have been adapted to accept a high calorific value fuel blended from liquid hazardous wastes. This practice is already widespread in developed countries and is becoming more so in developing countries like Brazil, Ecuador, Malaysia, Pakistan, the PRC, Sri Lanka, Tanzania and Vietnam. Most cement production worldwide is controlled by a small number of multinational companies who are well placed to transfer the technology.[67]
- Prepared fuel products need to be of a high and consistent quality if they are to meet end-of-waste criteria. Otherwise, handlers and users will need to obtain waste management licenses.

Gasification

- Gasification was developed for the more efficient recovery of energy from solid fuels such as coal, and to generate a synthetic gas for combustion or as a chemical feedstock. The lack of oxygen reduces the generation of products of incomplete combustion such as dioxins. Gasification is adapted to a range of biomass fuels and to wood wastes.
- Some technologies utilize an RDF or SRF product from MBT pre-processing of MSW.
- Commercial scale gasification of MSW and industrial wastes has been carried out since the 1990s in Japan and the Republic of Korea. A variety of proprietary technologies have been demonstrated at full scale in North America and Europe since the 1970s, but these have all faced both high costs and operational challenges.

Pyrolysis

- Thermal degradation in the complete absence of oxygen can produce a liquid fuel (but also gaseous flow and solid residues). This option is most suitable for feedstock with a high calorific value and low moisture content such as wood waste and plastic waste.
- Some technologies utilize an RDF or SRF product from MBT pre-processing of MSW, but this application is not yet in widespread use.

Landfill gas utilization

- Methane is produced in landfill sites through the decomposition of organic wastes under anaerobic conditions. Uncontrolled release of methane from landfills is a potential major contributor to greenhouse gas emissions. The migration and accumulation of methane may also pose an explosion and fire risk to the surrounding community.
- Landfill gas collection is thus a routine part of a controlled landfill operation. The gas may be utilized, either in a gas engine to generate electricity and/or heat, or it may be cleaned and the pressure increased for injection into a natural gas grid or for direct utilization as a transport fuel.
- Energy recovery from landfill gas has come to be widely implemented in developing countries through climate funding under the Clean Development Mechanism (CDM) of the Kyoto Protocol. This provided an important funding mechanism for many cities, as the payments made for carbon credits from a previous year paid for the current operating costs of the landfill site[68].

66 Planning Commission of India (2014)
67 Huang et al. (2012)
68 See Section 5.7.6 on other revenue sources as well as the Philippines case in Section 1.1, Box 1.1. Also see Terraza & Willumsen (2009) and USEPA/ISWA (2012).

Table 3. 1 Technology sheet: Comparing technologies for resource recovery from MSW

TECHNOLOGY	COMPOSTING	ANAEROBIC DIGESTION (AD)
What is it?	Aerobic decomposition of organic wastes.	Biodegradation of (readily degradable) organic wastes in the absence of oxygen, with anaerobic microorganisms. 'Wet' or 'dry' variations.
Selling points?	Addresses organic fraction, which is a large percentage of MSW. Produces compost with value as soil improver and fertilizer. Completes biological material cycle.	Able to handle wet waste. Produces biogas for direct use after upgrading, or for conversion to electricity/heat.
Input wastes	Separated organic fraction of MSW, or food waste, e.g. from restaurants and canteens.	
	Other solid organic waste. Can treat material high in lignin (woody).	Animal/human excreta. Liquids and sludges. Less suitable for high in lignin (woody) material.
Main outputs and their markets	Compost. Serves as soil conditioner, mitigates erosion and is used in land reclamation and as a final cover for landfills.	Biogas. Digestate can be composted for use as soil conditioner. Digestate can be dewatered and used as low calorific value RDF.
	Use as a soil conditioner depends on control of inputs and the process, and regulatory permits. In food production, MSW-derived outputs can often only be used when the inputs were source-separated organic fractions.	
Volume reduction (%)[1]	50-70%	45-50%
Sophistication of pollution control required	Low-medium	Low-medium
Cost per tonne (USD)[2]	25-70	65-120
Conditions for success	Temperature sensitive. Long residence time. Regular aeration required. Odour control.	Good process control – microbial processes can easily be disrupted. Works best with clean, homogeneous and consistent inputs – so MSW a difficult feed
	Clean input material; market for compost/digestate; contamination sensitive	
Appropriate scale of plants	Household (home composting) and community (backyard, vermicomposting). Centralized level, large-scale (windrow, aerated static pile, in-vessel).	Decentralized small scale digesters, including on-farm Larger scale for the organic fraction of MSW.
Extent of use	Widespread in high-income countries. Asia has a long tradition of making and using compost.	Widespread mainly for non-MSW. Increased interest in high income, and for small scale low-tech in low-income, countries.
Applicability in developing countries[3]	High potential, particularly in developing countries with a high organic fraction in MSW. Not yet widespread due to operating costs and need for source separation.	Small-scale anaerobic digesters are used to meet the heating and cooking needs of individual rural communities.

ENERGY FROM WASTE (EFW)			MECHANICAL BIOLOGICAL TREATMENT (MBT)
COMBUSTION WITH HEAT AND ENERGY RECOVERY	**GASIFICATION**	**PYROLYSIS**	
Direct combustion of waste in the presence of excess air (oxygen) to recover the energy content of the waste as heat energy, which can be used directly for heating or as a means of generating power (e.g. via steam turbine generators), or both (combined heat and power [CHP])	Partial oxidation of the wastes in the presence of less air (or other oxidant) than required for complete combustion.	Thermal degradation in the complete absence of air or other oxidizing agent	Combination of mechanical processing with biological reactors in the same plant. Generic term for many different technologies. Bioreactors can be biodrying or composting or AD.
Produces electricity and/or heat, e.g. for district heating systems. Completely sterilizes, destroys organic com pounds including hazardous wastes. Main output is a sterile ash.	Theoretical capability to use syngas in much more efficient gas engines in comparison to boiler and steam turbine. Potentially lower emissions of pollutants.	Wastes can be readily converted into liquid fuel products.	Advanced management of material flows, versatility and modularity. Actual benefits depend on the type of MBT and the main outputs of the plant.
Mixed MSW or prepared fuel (RDF). Versatile with feedstocks, if they are combustible.	Prepared waste. More suitable for treating the RDF or SRF produced by MBT rather than MSW. Also applicable to a range of other relatively homogeneous organic waste, such as wood waste, agricultural residues, sewage sludge, and plastic waste.		Mixed MSW or after source separation of dry recyclables ('residual MSW').
Heat only, electricity only, or both (CHP). Energy efficiency ranges from up to 30% (electricity only) to up to 95% (CHP). Secondary products: Fe and non-Fe metals and aggregate recycling. Potentially also precious metals.	Synthetic gas (syngas). Further combustion or conversion to chemical feedstock.	Liquid fuel. Further combustion or conversion to chemical feedstock.	Depending on plant type: SRF, RDF, compost-like output (CLO), biogas, reduced biodegradability output for landfill ('stabilized biowaste') SRF can be used in cement kilns, industrial boilers and power plants. Also dry recyclables.
75-90%	90%	50-90%	Variable – depends on plant configuration
High	Medium	Medium	Low-medium (depending on legislative requirements)
95-190	95-190	95-190	20-70
Good process control. Market needed for steam/hot water. Cold climate with heat demand (hot climate with cooling needs is possible but less prevalent). Waste to be within the combustible area of the Tanner diagram.	Pretreatment of waste required for removal of non-combustible materials and feedstock homogenization. Less versatile that combustion EfW. Market needed for synthetic gas.	Market needed for liquid fuels.	Market needed for outputs. Plant design to match process objectives. Not all plant configurations have a sufficient track record.
Centralized large scale is the more common and preferred option. Economies of scale allow for higher standards of emission control and higher energy efficiency.	Small, medium, and large scale configurations are available.	Small, medium, and large scale configurations are available.	Small, medium, and large scale configurations are available. Typically modular, more flexible than thermal processing.
Widely applied, with an established track record in Europe, Japan, the PRC and the US. Increased interest in rapidly developing economies.	Japan and the Republic of Korea have had commercial facilities for gasification of MSW for 20 years. Interest in Europe for small/ medium scale.	Not widely established for MSW.	Very widespread in Europe. Strong interest around the world.
MSW often too wet to burn without auxiliary fuel. Recovering the costs of an MSW EfW plant in low- to medium-income countries is difficult.	Potential for wood gasification technology. India has one of the world's largest programmes for small gasifiers.	Low – not established yet, even in developed countries.	Configurations are available at different levels of cost and sophistication suitable for developing countries.

1 Volume reduction will vary widely with the specific technology used. The rough estimates here are compiled from a variety of literature sources.

2 Estimated total cost per tonne in USD (net of operation and investment costs, less revenues from resource recovery), depending on income of the country. Assumes centralized facilities on a moderately large scale. See Section 5.2.2, Table 5.1 for more on comparative cost data. Source: Pfaff-Simoneit (2013), listed in Annex A,
Chapter 5, General reading on financing and economics.

3 See Sections 4.2.2 and 4.9.2 on strategic planning and the selection of policy instruments. Also see Box 4.37 on selecting appropriate technologies.

3.5.4 Investment worldwide in waste processing technologies

This section presents data on the development of new waste treatment and recovery facilities around the world. It should be noted that these data include all active facility development projects over the 2-year period 2013 to 2014, including projects at all stages of development, from feasibility and planning through construction. The total project value of 309 billion USD overestimates the degree of investment, as not all of the projects will actually be built. However, the data are very useful in providing a sense of the size of the current facility development projects (average value of 113 million USD per project), and the relative levels of activity, by waste type, by technology type and by geographic region. A summary of some of the basic data is provided in the figures that follow.

Analysing facility development projects by waste type, MSW accounts for 28% of all the projects by value (85 million USD). Wood and plant biomass (largely agricultural and forestry wastes) and organic waste together account for another 35% of the total (see *Figure 3.16*). Looking at MSW in particular, *Figure 3.17* shows a breakdown by technology. The largest contributors are various waste-to-energy technologies, particularly combustion with energy recovery (EfW). In terms of distribution by geographic area, the UK and the US show major investments, accounting for 24% and 11% of global MSW investment activity by value respectively, while the most active developing countries are the PRC (10%) and India (5%).[69]

Figure 3.16 Percentage of total facility development projects values by feedstock type

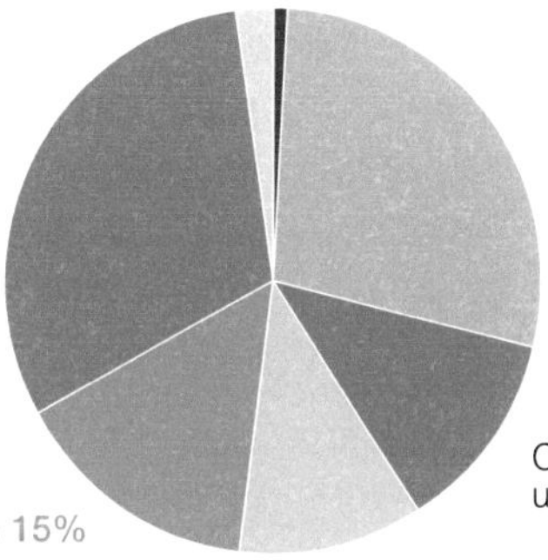

Figure 3.17 Percentage of total project values for MSW by facility type

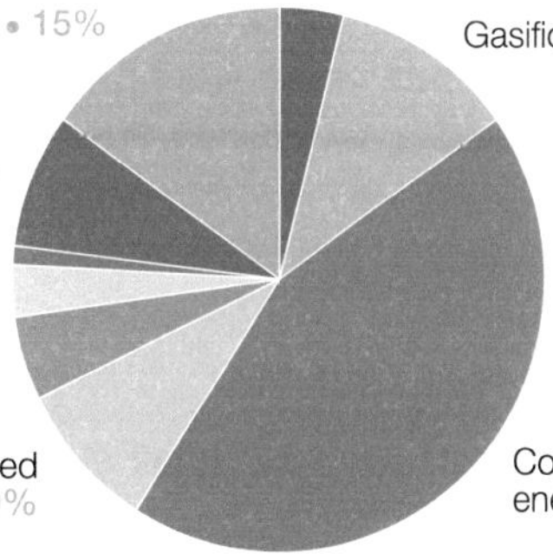

Notes to Figures 3.16 and 3.17: Data covers 2723 facility development projects active between January 2013 and December 2014, across 93 countries in all. The total value across all projects was 309 billion USD, of which 85 billion USD was for MSW projects. The data cover all waste types, including agricultural and forestry (wood) waste and cover all technology types, but may favour larger, higher technology projects while underestimating others, such as activity in landfill investment. Average project value was 113 million USD. Projects at all stages of development, through feasibility, planning and construction, quality as 'active'. Therefore the figure for total active project value is higher than the ultimate amount invested.

Source: Extracted from AcuComm's Waste Business Finder database. http://acucomm.net/

69 For a discussion of the selection of appropriate technologies for developing countries in particular, see Sections 4.2.2 and 4.9.2 (*Box 4.37*).

The secondary materials industry has been important since the industrial revolution. Early in the 20th century, this industry relied mainly on relatively clean industrial waste, but the quantity of material separated from municipal solid waste has increased since the 1980s. This section focuses in particular on the transnational trade of this global industry.[70]

3.6.1 The globalization of secondary materials markets

Separation and collection for recycling only makes economic sense if the material is actually recycled, which depends on there being a market for the material. The waste industry depends closely on the secondary materials industry to provide that market. Some markets are relatively local, for example for compost as a soil conditioner or for aggregates from C&D waste. Others may be national or regional, such as for glass, processed fuels made from MSW (SRF may be more suitable for longer distance transport than RDF[71]) or wood waste. The focus in this section is primarily on those secondary materials which are globally traded commodities, including ferrous and non-ferrous metals, paper and board ('recovered paper' or 'recovered cellulose fibre'), plastics and textiles. The use of recycled materials competes with and displaces the use of primary materials and helps reduce the extraction of virgin material resources and reduce greenhouse gas emissions.[72]

In 2010, 700 to 800 million tonnes of "waste" were recycled as "secondary commodities",[73] derived from MSW as well as other waste streams. In terms of both tonnage and value, recycling markets are dominated by ferrous scrap (steel). In tonnage terms this is followed by paper and board, whereas in terms of value non-ferrous metals rank second, with aluminium and copper dominating this market. Based on the estimates made in Section 3.2, it appears that the main traded secondary materials represent around 10 to 15% of overall world waste generation, excluding construction and demolition, agricultural and forestry and mining and quarrying wastes.[74]

However, as reported below, only a relatively small proportion of the total 700 to 800 million tonnes (likely less than 25%) is traded across national boundaries. Asia makes up the most dynamic and arguably the most important global recycling market. Labour-intensive manufacturing industries and raw material extraction have been increasingly outsourced from developed to developing countries over a number of decades. The import of materials for recycling from high-income countries therefore represents an essential resource for fast-growing Asian economies, such as India, Indonesia, the PRC, Thailand and Turkey. Developing regions of Asia have relatively low labour and operating costs for industry, have different manufacturing quality standards and sometimes not so stringent environmental regulations compared to developed countries such as EU, the U.S., Canada and Japan, so the trend is not without its downside. The PRC is slowly beginning to raise its environmental standards, for both industry in general and the waste processing and recycling sectors in particular. As the costs of meeting environmental compliance are high in developed countries, there are potential large profits to be made from following what has been described as the 'least environmental pathway',[75] by exporting wastes to developing countries with lower levels of control and enforcement. This can be done either illegally for simple dumping, or possibly legally for recycling, but without proper controls on public and occupational health and environmental pollution. The latter applies to some extent to the global market for recycled commodities such as plastics and paper, and arguably even more so for some hazardous wastes, including e-wastes and end-of-life ships.

Perhaps the most notable characteristic of secondary material markets is their price volatility. Secondary materials have traditionally been used to 'top up' a relatively stable supply of primary materials (made from virgin raw materials) in response to short-term variations in market demand, so their prices have tended to be even more volatile than those of the related primary commodities. Both primary and secondary material prices appear to be increasing in volatility: data are provided in *Figure 5.1* in Chapter 5. For example, major price

70 See Sections 2.3 on drivers for waste and resource management and 5.3.3 on the resource recovery business.

71 An example of a regional market for a minimally-processed RDF is the rise in exports from the UK to the Netherlands, Germany and Sweden after 2010 - the trade rose from 0 to more than 2 million tonnes per annum, driven both by a rise in landfill tax in the UK and by surplus capacity in European EfW plants, which needed to meet long term contracts for heat and power. See also Case Study 6 on energy from waste. http://www.ciwm.co.uk/CIWM/MediaCentre/Current_pressreleases/Press_Releases_2013/ciwm_news_310713.aspx

72 See Topic Sheet 1 on waste and climate, following Chapter 1.

73 Bureau of International Recycling (2010, 2011). World Markets for Recovered and Recycled Commodities. See Annex A, Chapter 3, Global secondary materials industry.

74 10 to 15% is an underestimate of total recycling, as it is selective in the listing of recycled materials (for example, it excludes organics recycling), but more comprehensive in its inclusion of waste types.

75 Crang et al. (2013) and Velis (2015). See also Sections 4.3.7 and 5.3.4 for a discussion of waste trafficking and waste crime.

crashes occurred during the world recession of 2009 and due to the slump in oil prices late in 2014 and early in 2015. Such market instability is a major threat to the sustainability of recycling programmes around the world.

Widespread trade in materials recovered from waste depends on the materials being classified as no longer being waste, so that they can be traded freely as a product without the need for handlers and users to obtain waste management licences.[76] The EU is in the process of developing 'end-of-waste' criteria or protocols, with criteria already defined for iron, steel and aluminium scrap, and under preparation (2015) for copper scrap, recovered paper, glass cullet and biodegradable waste/compost.[77]

3.6.2 Ferrous metals

World production of iron and steel is rising steadily, increasing by 40% from 2005 to reach over 1.67 billion tonnes in 2014. According to the industry, every tonne of ferrous metal scrap that goes back into production reduces the use of iron ore by 1,400 kg, of coal by 740 kg, and of limestone by 120 kg.[78] *Figure 3.18* shows a steady increase in scrap use from 2001-2014, although this has not kept pace with steel production, as the ratio of steel scrap to crude steel has steadily decreased over the period. By 2011-2014, total steel scrap use was approaching 600 million tpa, representing rather less than 40% of total steel production. Scrap can be grouped into the three sources of (i) post-consumer (old) scrap; (ii) new scrap (e.g. production off-cuts) purchased by steel mills from industrial users; and (iii) own arisings, directly recycled within the steel mills. The quantities of both new and own scrap are relatively stable, with the quantities of old scrap varying over the 14 years between 180 and 260 million tonnes per annum (*Figure 3.18*). Scrap is the main raw material for electric arc furnaces, while it can only be used as a small percentage of the total feedstock for traditional blast furnaces using coke and iron ore.

Figure 3.18 The use of steel scrap for steelmaking (global totals)

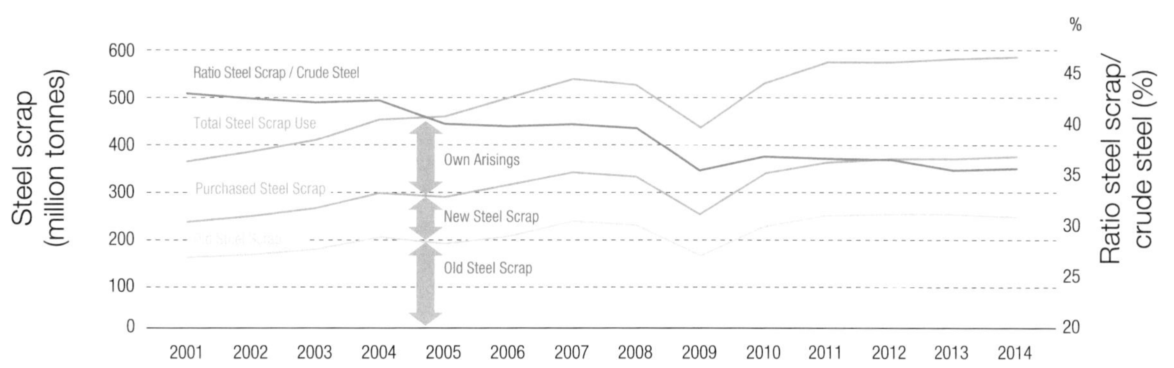

Notes: Figure shows time series data for steel scrap. New scrap is scrap from steel processing and old scrap is from products after their use. Own arisings are rejects from melting, casting and rolling.

Source: Bureau of International Recycling (2015). World Steel Recycling in Figures 2010 – 2014. http://www.bir.org/publications/brochures/

External trade across national boundaries accounts for less than 20% of total steel scrap usage (*Figure 3.19*), but increased in tonnage terms from 73 million tonnes in 2001 to around 100 million tonnes from 2004-2014. Excluding trade within the EU, the major importer in the period from 2010 to 2014 was Turkey (about 30% of the total trade outside of the EU); Turkey is unusual in using electric arc furnaces for around 70% of its steel production, and relying on scrap for around 90% of its raw materials. Other significant importers include India (10%), the People's Republic of China (7%), the Republic of Korea (14%) and Republic of China (7%). Imports to these Asian countries account for two thirds of external trade outside of the EU. Other significant importers

76 See Sections 4.3.2 on legal classifications and 4.3.5 on legislation for resource recovery.

77 http://ec.europa.eu/environment/waste/framework/end_of_waste.htm. 'End-of-waste' criteria are the conditions that need to be met for materials to no longer be classified as 'waste' but rather as a 'product' or a 'secondary raw material.' Box 4.9 in Chapter 4 shows examples of existing protocols for compost in several EU countries, and Topic Sheet 12, found after Section 4.3.5, shows how end-of-waste can be implemented in practice for compost.

78 World Steel Association (2012). Figures based on blast furnace production.

are the U.S., EU, Indonesia, Malaysia, Canada and Thailand. The dominant exporters are the US (about 28%), the EU (25%) and Japan (12%). Russia's share of the export market fell from around 20% in 2005 to 4% in 2010 but had risen again to 9% in 2014. Other significant exporters were Canada, Australia and South Africa.

Figure 3.19 Volume of external global steel scrap trade

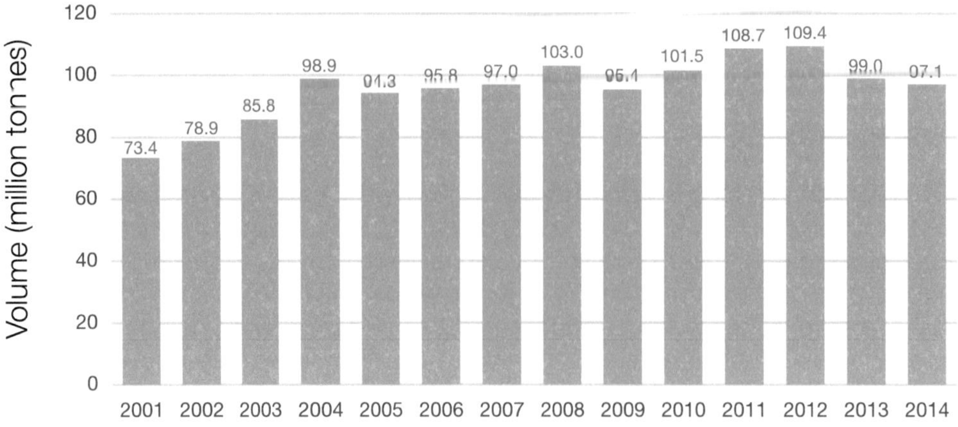

Notes: Figure shows temporal data for the external trade in steel scrap.

Source: Bureau of International Recycling (2014, 2015). World Steel Recycling in Figures 2009 – 2013 and 2010 – 2014. Available from http://www.bir.org/publications/brochures/

3.6.3 Non-ferrous metals

Global production of the main non-ferrous metals is rising fast, as is demand for scrap for recycling (*Table 3.2*). The prices of these commonly-used metals are relatively high, so scrap is in heavy demand. However, as shown earlier in *Figure 3.14*, it is only for 20 out of 60 metals that more than half is recycled when products reach their end of life.

Aluminium is both the most heavily used and the fastest growing of the non-ferrous metals, so that is taken as an example here. The story is dominated by the PRC. In 2000, the PRC produced 2.9 million tonnes of aluminium (12% of the global total); by 2011, this had risen to 19.4 million tonnes (43% of a global total which had grown by 80%). As shown in *Figure 3.20*, of the 7.4 million tonne per annum increase in demand for aluminium scrap over the same period, the PRC accounts for 6.3 million tonnes (85%); consumption also increased in the rest of Asia and in Europe, but decreased in the US. Despite the increase in scrap consumption, supply from scrap has failed to keep pace with the increase in aluminium production: the ratio of scrap used in aluminium production has fallen slightly over the period from 31 to 29%.

As with steel, only a proportion of aluminium scrap is traded externally across national boundaries. *Figure 3.21* shows that quantities have increased fourfold from 2000, reaching around 4 million tonnes (around 20% of the total) in 2011. The PRC accounts for 60% of imports and the rest of Asia 30%, while most exports originate in North America and Europe.

Table 3.2 Global demand for the largest volume non-ferrous metals and global scrap consumption

COMMODITY	GLOBAL DEMAND FOR METAL*			GLOBAL SCRAP CONSUMPTION		
	2000 (Million tonnes)	2011 (Million tonnes)	Percentage growth 2000-2011*	2000 (Million tonnes)	2011 (Million tonnes)	Percentage growth 2000-2011
Aluminium	25	45	82%	11	18	68%
Copper	15	19	30%	7.0	10	45%
Lead	9	12	30%	3.7	5.8	57%
Zinc	7	10	40%	0.8	1.1	34%
Nickel	1	1.1	10%	0.6	0.9	42%
Steel	1144 (2005 data)	1607	(40%)	401	573	43%

Notes: Global demand for primary metal has been rising quickly, as has global scrap consumption. The last row for steel is shown for comparison. The non-ferrous metal tonnages are 35 to 1,000 times lower.

Source: Bureau of International Recycling (2011). Global non-ferrous scrap flows 2000-2011. Available from http://www.bir.org/publications/brochures

Figure 3.20 Global aluminium scrap demand from 2000-2011 by region

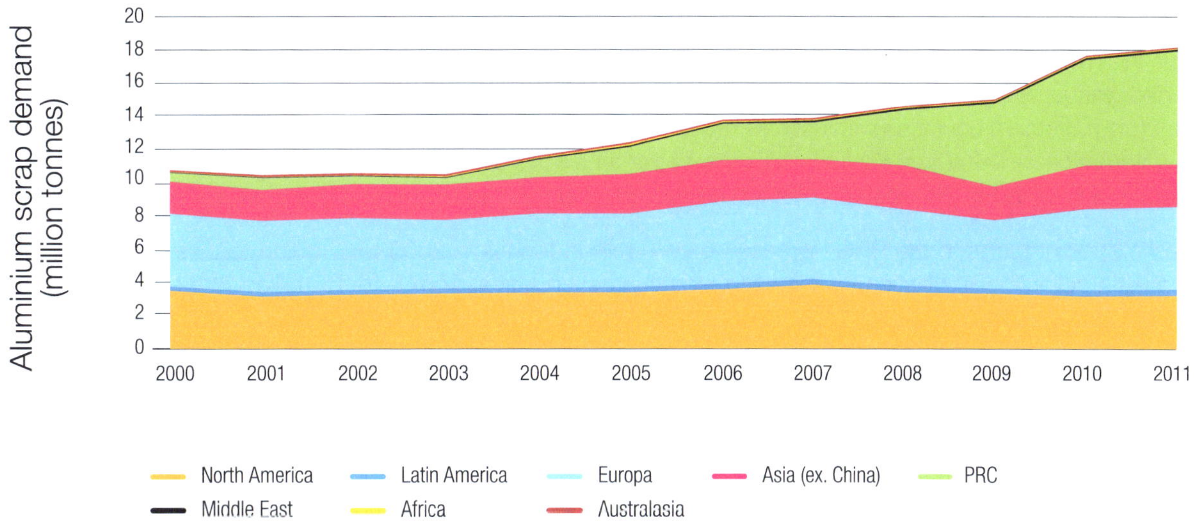

Notes: 85% of the increase is accounted for by the PRC, with the rest of Asia and Europe also showing increases.

Source: Bureau of International Recycling (2014, 2015). World Steel Recycling in Figures 2009 – 2013 and 2010 – 2014. Available from http://www.bir.org/publications/brochures/

Figure 3.21 Aluminium scrap Imports from 2000-2011

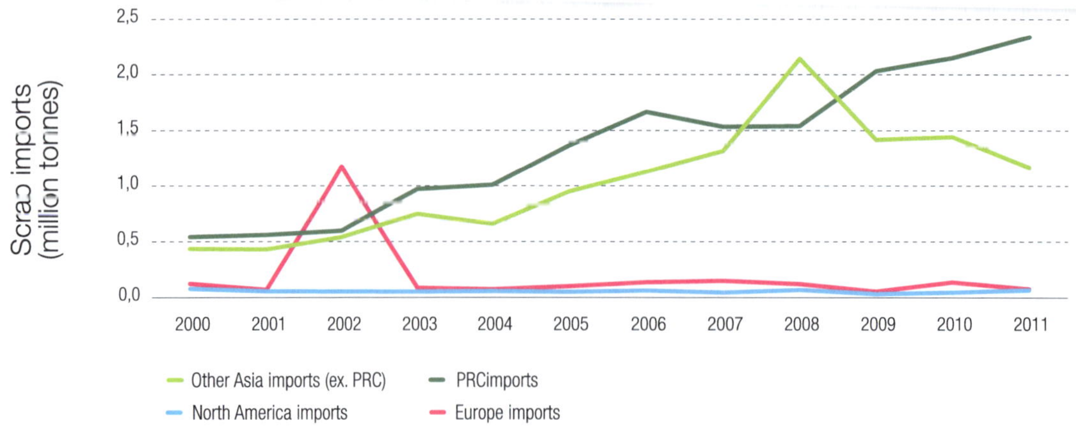

Notes: The total traded quantity has increased approximately fourfold, from around 1 million to around 4 million tonnes per annum. The largest increase is in the PRC, followed by the rest of Asia.

Source: Bureau of International Recycling (2014, 2015). World Steel Recycling in Figures 2009 – 2013 and 2010 – 2014. Available from http://www.bir.org/publications/brochures/

3.6.4 Plastics

International trade in used plastics is prospering. With global production of plastics skyrocketing, from 1.5 million tonnes in 1950 to 204 million tonnes in 2002 and 299 million tonnes in 2013,[79] and a continuing shift of production from the West to Asia (more than 40% by weight of world production in 2013), the annual volume of transnationally traded waste plastics at 15 million tonnes represents just 5% by weight of new plastics production. Plastic scrap flows from Western countries with established recycling collection systems mainly to the PRC, which dominates the international market (see *Figure 3.22*), receiving around 56% wt. of global imports. Europe (EU-27) collectively exports almost half of the plastics collected for recycling, at least 87% of which goes to the PRC.[80]

Plastic scrap imports to the PRC increased from 6 million tonnes in 2006 to 8 million tonnes in 2011, but the domestic collection of plastics for recycling increased even faster, from 7 to 15 million tonnes over the same period. This is expected to rise further as domestic recycling rates increase. However, it is speculated that the poor quality of domestic post-consumer recyclates necessitates quality imports for capital-intensive better quality manufacturing, while the inferior imports and domestic recycled plastics end up at either low-tech, unregulated facilities and maybe also EfW plants. While the PRC government is actively working to increase the quality of imported plastics and reduce the numbers of unregulated facilities (as witnessed by the 2013-14 Green Fence Operation, as an example), the environmental benefits from plastic exports to the PRC are questionable given the dominance of uncontrolled reprocessing/manufacturing with very low environmental standards.[81]

A recent report[82] asked the question: Is dependence on a single importing country a risk to the exporting countries which need to meet high, statutory recycling targets for plastics? The conclusion drawn was 'Yes', from two perspectives. First, the PRC may in the medium- or long-term become self-sufficient in high-quality secondary plastics from domestic sources and may not import. Second, the aim of achieving high recycling rates in exporting countries such as the EU was to achieve sustainable resource recovery, meeting high standards of environmental protection and achieving clean material cycles and resource utilization; this is questionable when almost half of the plastics collected in the EU for recycling are exported to countries with lower environmental standards. The long-term solution likely requires a balance between developing domestic capacity within the EU for recycling and relying on international markets: for example, quality, segregated polymers, such as clean PET from bottles, are increasingly sought-after commodities on the global market, with manufacturers in the US, Europe and the PRC competing for a limited supply. At the same time, continuing efforts are required to ensure a 'level playing field' in terms of environmental standards.

79 http://www.plasticseurope.org/Document/plastics-the-facts-20142015.aspx?FolID=2
80 All of the data in this section is taken from Velis (2014a), listed in Annex A, Chapter 3, Global secondary material markets. Much of it is based on research undertaken by Sihui Zhou at Imperial College London in 2012.
81 Minter (2013).
82 Mavropoulos et al. (2014), listed in Annex A, Chapter 1, Waste management. Based on the detailed work of Velis (2014a).

Figure 3.22 Global flow of plastics to the People's Republic of China in 2010

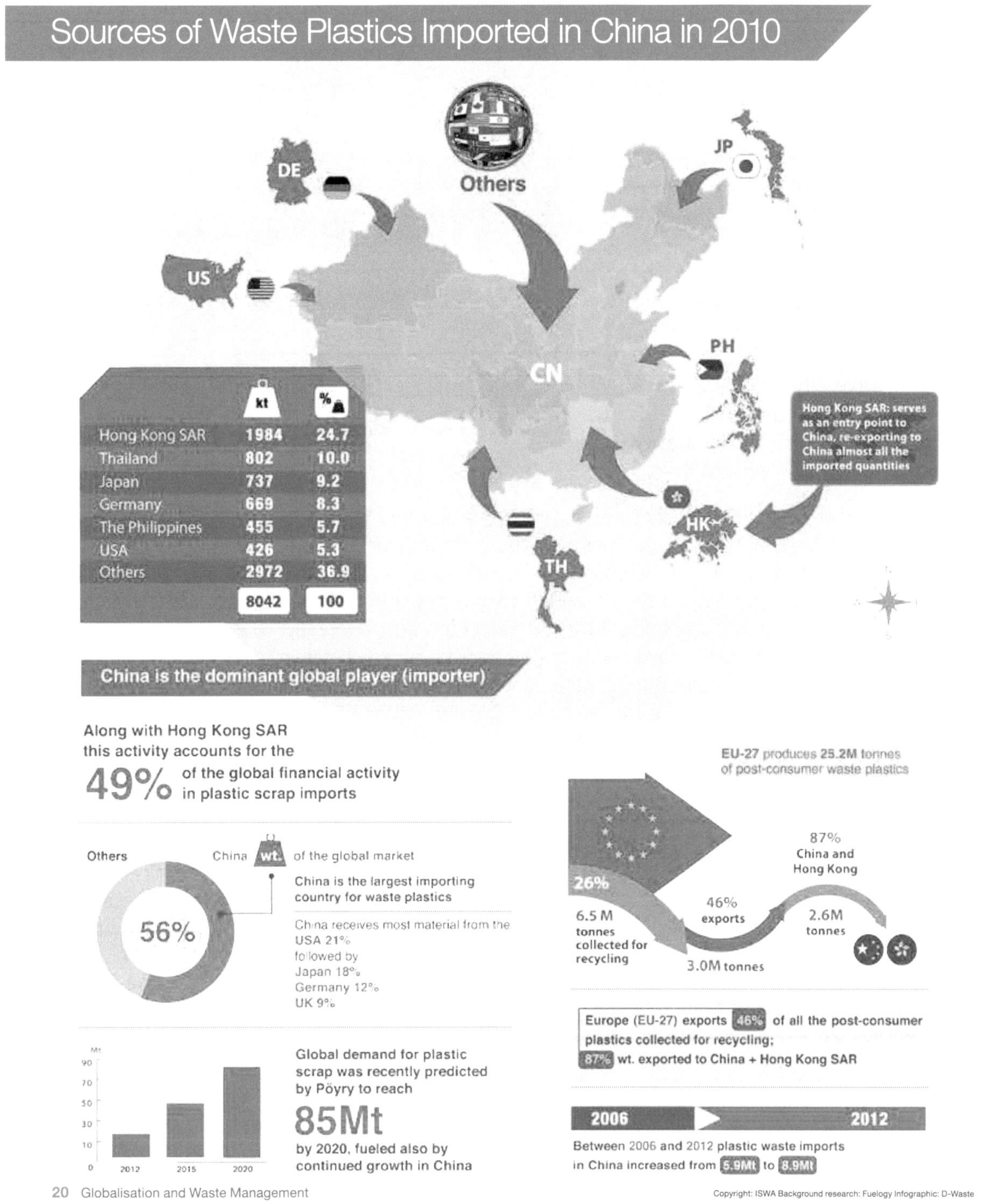

Source: ISWA, reproduced directly from Velis (2014). Global recycling markets – plastic waste: A story for one player – China. Infographic prepared by D-Waste on behalf of International Solid Waste Association – Globalisation and Waste Management Task Force. ISWA, Vienna. http://www.iswa.org/fileadmin/galleries/Task_Forces/TFGWM_Report_GRM_Plastic_China_LR.pdf

3.6.5 Paper

Total world production of paper and paperboard in 2012 was around 400 million tonnes, of which 45% was in Asia, 26% in Europe and 21% in North America.[83]

Recycled paper and paperboard (known in the industry as 'recovered paper' or 'recovered cellulose fibre' (RCF) has always been a major raw material used in the paper industry. In 1990, recovered paper accounted for 40% of the total pulp used in the European paper industry, and by 2013 this had risen to 53%. At the same time total production in Europe had risen by around 50%.[84] This increase in 'recycled content' was driven mainly by the 'rediscovery' of municipal solid waste recycling and thus an increase in recovered paper supply, but the increase in MSW recycling rates from around 8% in 1990 to approaching 50% in 2012 meant that supply was outstripping regional demand.[85]

© David C. Wilson

Table 3.3 shows world collection and consumption of recovered paper by region, distinguishing those individual countries handling more than 1 million tonnes per annum. Total quantities are around 230 million tonnes. Of this, around 80% is consumed within the country where the paper is collected for recycling. All of the named countries in *Table 3.3* consume more than 1 million tonnes per annum of recovered paper in their national paper industry. Transboundary trade totals 40 to 50 million tonnes per annum. This trade is now dominated by the PRC, which despite collecting almost as much paper for recycling within the country (45 million tonnes per annum) as does the U.S., still imported a net quantity of 30 million tonnes in 2012. Other Asian countries, notably India and Indonesia, accounted for a further 8 million tonnes of net imports. Other net importing countries included Mexico, Austria, Germany, Sweden and Spain. The US accounts for 20 million tonnes of net exports, followed by Japan and the UK at 4 to 5 million tonnes each. The paper market is also influenced by regional variations in the availability and price of wood pulp, which is e.g. relatively lower in Latin America.

© Ainhoa Carpintero

Table 3.3 also shows that, although global trade in recovered paper has a long history, the quantities shipped around the world have increased rapidly over the last two decades, in response to both increased collections of MSW for recycling in Europe, North America and Japan, and to the rise of the PRC as the dominant world paper producer. Between 1997 and 2012, net exports from the US and Europe have risen around fourfold, as have net imports to the 'rest of Asia'; Japan has moved from being a small net importer to the second largest net exporter; and the PRC has increased its imports by a factor of 20.

Quality is likely to be key to the future of the global market for recovered paper. Since 2000, large paper companies in the PRC have either acquired or established paper merchants operating in high income countries, in order to increase their control over their recovered paper supply chain. The European paper industry has been making the case that the EU's end-of-waste criteria[86] for waste paper should be set 'high', to ensure that recovered paper delivered to paper mills is pre-sorted and of high quality.[87]

83 Most of the data in this section comes from the BIR 2014 report, Recovered Paper Market in 2012.
84 CEPI Key Statistics 2013 http://www.cepi.org/system/files/public/documents/publications/statistics/2014/Final%20Key%20statistics%202013.pdf
85 See Sections 2.3 and 5.5.1
86 'End-of-waste' criteria are the conditions that need to be met for materials to no longer be classified as 'waste' but rather as a 'product' or a 'secondary raw material.'
87 CEPI press release, 20 September 2013. http://www.cepi.org/topic/recycling/pressrelease/endofwaste

Table 3.3 Leading countries collecting & consuming recovered paper and regional totals (2012)

Unit: Million tonnes

Region		Country	Collections of recovered paper and board	Consumption of recovered paper	Net flows: positive = imports negative = exports	Regional total net flows	
						2012	1997
North America		United States	46.3	26.3	-20.0	-22	-6
		Canada	4.4	2.6	-1.8		
		Regional subtotal	*50.6*	*29.9*	-21.8		
Latin America		Brazil	4.5	4.5	0.0	**1**	
		Mexico	**3.9**	**4.8**	**0.8**		
		Regional subtotal	*12.2*	*13.1*	*0.9*		
Europe		**Germany**	**15.3**	**16.2**	**0.9**	-7	-1.6
		United Kingdom	8.2	3.8	-4.4		
		France	7.3	5.0	-2.3		
		Italy	6.2	4.7	-1.6		
		Spain	**4.6**	**5.1**	**0.5**		
		Netherlands	2.6	2.1	-0.4		
		Belgium	1.9	1.2	-0.7		
		Poland*	1.6	1.3	-0.3		
		Austria	**1.5**	**2.4**	**1.0**		
		Sweden*	**1.3**	**1.9**	**0.6**		
		Switzerland	1.2	1.0	-0.2		
		Russia	2.6	2.2	-0.4		
		Regional subtotal	*62.0*	*54.8*	*-7.2*		
Asia	**Japan**	Japan	21.7	16.8	-4.9	-5	**0.05**
	PRC	**PRC**	**44.7**	**75.0**	**30.3**	**30**	**1.6**
	Rest of Asia	**Republic of Korea**	**8.8**	**9.6**	**0.8**	**8**	**2**
		Indonesia	**3.6**	**5.9**	**2.3**		
		India	**3.4**	**5.7**	**2.3**		
		Republic of China	**3.1**	**3.8**	**0.8**		
		Thailand	**2.7**	**3.6**	**1.0**		
		Malaysia*	**1.2**	**1.6**	**0.4**		
		Turkey*	**1.0**	**1.1**	**0.1**		
		Regional subtotal	*99.4*	*130.9*	*31.5*		
Australasia		Australia*	3.4	2.0	-1.4	-1	
		Regional subtotal	*3.5*	*1.9*	*-1.6*		
Africa		South Africa*	1.0	1.0	0.0	0	
		Regional subtotal	*2.8*	*2.6*	*0.2*		
Totals		**Named countries**	**207.7**	**211.0**	**3.3**		
		World totals	**230.5**	**233.2**	**2.8**		

Notes: Unless otherwise noted, data for collection and consumption is for 2012, and is taken from the Bureau of International Recycling (BIR, 2014): Recovered Paper Market in 2012.

- 'Collections' shows national totals of recovered paper and board collected by the secondary paper industry.
- 'Consumption' shows national consumption of recovered paper by the paper industry (domestic deliveries plus imports)
- 'Net flows' shows national consumption less national collections: a positive figure denotes a net importing country (highlighted in bold); a negative figure denotes a net exporter. These figures do not total exactly zero, as some stocks are carried forward between years. Note that some countries may be both a significant importer and a net exporter. Examples include the Netherlands and Belgium, where the ports of Rotterdam and Antwerp handle exports on behalf of a number of countries.
- The BIR report provides data for most of the leading collecting and consuming countries in 2012. A few other countries collecting or consuming more than 1 million tonnes per annum, for which 2012 data are missing, have been added, using 2009 data from the FAO Recovered Paper Survey. These have been indicated with an asterisk.
- The regional sub-totals, taken from the BIR report, show data for all countries in the region, not just the named countries.
- The last two columns summarize net flows by region, using the 2012 data from the table. The 1997 data provided for comparison is taken from Kojima and Michida (2011). This is available only for selected regions, for which the definitions differ from those used for the 2012 data: the 'Europe' data is for the EU-15; the 'Rest of Asia' data is for the ASEAN-6 (Association of Southeast Asian Nations) countries of Brunei Darussalam, Indonesia, Malaysia, the Philippines, Singapore, and Thailand.

3.6.6 Textiles

Used textiles have become a globally traded commodity. Focusing on the second hand clothing economy in particular, this has doubled from 1.26 billion USD in 2001 to 2.5 billion USD in 2009. Textile recyclers sort clothing into reusable garments or recycling grades, the latter including industrial cleaning cloths and reclaimed fibres. The sector has globalized as a result of the growth of supply from the global North, the relocation of sorting operations to Eastern Europe and the global South, and the development of differentiated markets for reuse.[88]

Five high-income countries (Canada, Germany, Republic of Korea, UK and U.S.) account for more than half of all exports of second-hand clothing, most of it originating as donations to charity when it reaches the end of its perceived useful first life. Charities typically select only a small percentage for domestic reuse (estimated at 20% in the UK), often for sale in their own shops. The larger part is sold on to a complex network of global traders, being sorted many times into increasingly differentiated components. Major sorting centres are located in Poland, India and Ghana. Many of the higher quality garments are sold on in Eastern Europe. Lower quality wearable items from Europe and North America tend to go to Africa, while those from Asian countries tend to go to Asian markets (matching the clothing to the users body shape). Fifteen countries account for half of all imports: Angola, Benin, Cambodia, Cameroon, Canada, Germany, Ghana, India, Kenya, Malaysia, Pakistan, Poland, Russia, Tunisia and Ukraine Many of these countries are major re-exporters of sorted fractions.

3.7 OTHER WASTE STREAMS AND EMERGING ISSUES

Some waste materials are of particular interest due to their characteristics and generation patterns, or due to challenges in their management. Some of these are explored in a series of Topic Sheets through the GWMO, a number of which have been placed after Chapter 3. These cover for example construction and demolition waste; hazardous waste; e-waste; plastic waste and marine litter; disaster waste; and food waste.

It is safe to predict that a number of 'new' waste streams of concern will come onto the agenda over the next decade. One such emerging issue already on the horizon is nano-waste. The chemical-physical properties of nano-materials may pose risks to human health and the environment that are not yet entirely known or understood. The Federal Office for the Environment (FOEN) in Switzerland highlighted the risks in its Policy Paper on the Safe and Environmentally Sound Disposal of Nano-waste.[89] NANoREG is a research project funded by the EU that aims to develop a common European approach to the regulatory testing of nano-waste and other manufactured nano-materials.[90]

88 Chang et al. (2013), Norris (2013)

89 Swiss Federal Office for the Environment. Nanowaste. http://www.bafu.admin.ch/abfall/01472/12850/index.html?lang=en

90 See EU Framework 7 Programme, NANoREG. http://nanoreg.eu/

CONSTRUCTION AND DEMOLITION (C&D) WASTE

TOPIC SHEET 6

C&D waste generation

Construction and demolition (C&D) waste is generated during the construction, renovation or demolition of buildings, roads, bridges, flyovers, subways, and so on. These activities typically generate large quantities of waste, although oftentimes data on C&D waste are not collected routinely or consistently, so most published figures are estimates which need to be interpreted with caution. Such estimates include 821[1] million tonnes of C&D waste generated across the EU in 2012,[2] 77 million tonnes in Japan, 33 million tonnes in China and 17 million tonnes in India (all in 2010),[3] and almost 7 million tonnes in each of the fast developing cities of Dubai (2011)[4] and Abu Dhabi (2013)[5]. C&D waste often represents the largest proportion of total waste generated: for example, C&D waste accounts for 34% of the urban waste generated within OECD countries, as shown in *Figure 3.1.* The volume of C&D waste is also sharply increasing, reflecting the pace of infrastructure development across the world.

C&D waste composition

C&D waste not only contains a high proportion of inert materials (e.g. concrete, masonry, asphalt), but also wood, metal, glass, gypsum and plastics as well as hazardous substances such as treated wood, lead paint and asbestos from demolished old buildings. Due to the variety of materials it is important that the C&D waste be segregated at source, with each stream managed as required.

Baseline C&D waste management practices

When waste management controls were first introduced in the 1970s, the illegal dumping of C&D waste was widespread. In many developed countries this continues to be an on-going problem, and in some developing countries such as India and Jordan, illegal dumping of C&D waste is approaching 'epidemic' proportions.[6,7] This is a major problem as such illegal dumps act as 'magnets,' in that their existence increases the likelihood of other waste such as municipal solid waste or even hazardous waste being dumped, which may pose a much more serious pollution risk. For this reason, the proper management of C&D waste is of critical importance within sound waste management overall.

As the bulk of C&D waste is inert, in the 1970s and 1980s this waste was typically landfilled. Over time, C&D

1 821 million tonnes refers to the latest available Eurostat figure, replacing an earlier one of 531 for the EU-27 in 2009 quoted in the BIO Intelligence Services prepared for the European Commission (DG ENV) (cf. Annex A), which states, "Data on C&D waste treatment suffer from the same gaps and inconsistencies as generation data ... C&D waste quantities are therefore likely to range between a total of 310 and 700 million tonnes per year in the EU-27." See http://ec.europa.eu/environment/waste/pdf/2011_CDW_Report.pdf

2 Eurostat (2015). Waste statistics. See http://ec.europa.eu/eurostat/statistics-explained/index.php/Waste_statistics

3 Guah (2013), listed in Annex A, Chapter 3, Other waste streams, under Topic Sheets, Construction and demolition waste

4 Environmental Center for Arab Towns (2013). Waste Management in Dubai. http://en.envirocitiesmag.com/articles/pdf/waste_management_eng_art1.pdf

5 Environment Agency of Abu Dhabi (2013). Towards Integrated Waste Management in Abu Dhabi: Annual Policy Brief. See http://www.ead.ae/wp-content/uploads/2014/03/Waste-PB-Eng.pdf

6 Centre for Science and Environment (2014)

7 SWEEP-Net (2010). Country report on the solid waste management in Jordan. See http://www.sweep-net.org/ckfinder/userfiles/files/country-profiles/Country-reportJordan-En-mai2011.pdf

waste came to be diverted from landfilling to recycling, while more recently the focus has been on reuse and prevention.

C&D waste recycling

For C&D waste, in principle, high recycling rates should be relatively simple to achieve through segregating and crushing inert materials not contaminated with hazardous waste. The materials can be used as aggregate or used in the manufacture of, for example, concrete products or paving blocks. To compete in this marketplace, items produced through the crushing and sieving of C&D wastes need to meet the specifications and criteria set for 'end of waste'.[8]

Other components of C&D waste can be recycled where markets exist for them. When it is clean and of good quality, untreated timber can be recycled into particleboard or fibreboard, recycled to produce wood chips or alternatively used for energy recovery. Metals such as aluminium and steel offer opportunities for recycling and plastics can be reprocessed into many different materials.

The recycling rates of C&D waste range enormously among countries. The 2011 Bios report provides a 'best estimation' of the 2008-09 EU average as in the range of 30 to 60%, with EU countries reporting recycling and recovery rates as high as over 90% and as low as 10%.[9] In the US in 2009 approximately 40% of the C&D waste generated was reused, recycled or sent to EfW plants.[10]

Targets have been set across a number of countries and cities. For example, the current EU target is for a minimum of 70% of non-hazardous C&D waste to be reused, recycled or to undergo other material recovery (including backfilling operations in which the waste substitutes for other materials) by 2020,[11] while the Construction Resource Initiatives Council introduced a series of targets, shown in the figure below, as a "global call to action" for the entire building industry under the international "Mission 2030" initiative.[12,13]

Nevertheless, the primary driver within the construction industry for the increase in reuse and recycling is cost control, since these practices will generally reduce project costs through reduced disposal costs, decreased purchasing costs for new materials, and greater revenue earned from the sale of materials.[14]

Interest in C&D waste recycling is increasing even beyond developed countries. In the Gulf region, where major infrastructure projects are on the rise but baseline recycling levels are low – just 4% in Qatar in 2012[15] – several C&D waste recycling projects have been developed using the public-private partnership (PPP) model. These projects include plants established in Amman (Jordan), Kuwait, and Dubai and Sharjah (United Arab Emirates) which have been developed in the last decade and give revenue to their municipalities while reducing the extraction of natural resources, increasing life of landfills and generating employment.

In countries such as China and India where urban infrastructure development and re-development are expanding rapidly, C&D waste recycling has become a business opportunity for the private sector, but with C&D recycling rates estimated at 5% in China (2013)[16] and 50% in India (2014)[17], there is still some way to go to fulfil that potential.

8 See Sections 4.3.2 and 4.3.5.
9 European Commission (2015). Waste: Construction and Demolition Waste. http://ec.europa.eu/environment/waste/construction_demolition.htm
10 USEPA Office of Solid Waste and Emergency Response (2009). Deconstruction. http://www.epa.gov/oswer/iwg/pilots/docs/ipco_deconstruct.pdf

11 European Commission (2015). Waste: Construction and Demolition Waste. http://ec.europa.eu/environment/waste/construction_demolition.htm
12 Gratton (2013)
13 http://www.cricouncil.com/
14 USEPA (2000)
15 Ayoub, N., F. Musharavati, and H. Gabbar (2014). A Future Prospect for Domestic Waste Management in Qatar. Presented at International Conference on Earth, Environment and Life Science (EELS-2014). 23-24 December 2014, Dubai (UAE).
16 http://www.theclimategroup.org/what-we-do/news-and-blogs/construction-waste-recycling-in-china-the-climate-group-releases-new-report/
17 Ghosh, S., S. Ghosh and A. Aich (2011). Rebuilding C&D Waste Recycling Efforts in India. See http://www.waste-management-world.com/articles/print/volume-12/issue-5/features/rebuilding-c-d-waste-recycling-efforts-in-india.html

India's first C&D recycling plant in New Delhi[18]

The Municipal Corporation of Delhi, working in cooperation with the private sector, established a C&D recycling plant with the aims of diverting waste from landfill and developing the market for C&D waste. The plant, a public-private partnership in operation since the end of 2009, had an original design capacity of 500 tonnes per day that was expanded in 2014 to 2000 tonnes per day. Incoming material is inspected and weighed. Plastics, metals, wood and certain other materials are separated out by both manual and mechanical means. The remaining waste is again separated, this time sorted into whole bricks for internal use and sale, large pieces of concrete and mixed C&D waste that is managed using dry processing to crush and grade the concrete and C&D waste and also wet processing for mineral processing and washing. The plant recovers products such as sand, stone and ready-mix concrete and uses these to manufacture other value-added products such as paving blocks and tiles, kerbstones and bricks. By early 2015, the plant had sold well over a million tonnes of recycled products.

Prevention and reuse of C&D waste

The recycling of C&D waste is not the ultimate objective. Thus, high recycling targets can be counter-productive. By the 2000s, in high-income countries, one common source of waste was the over-ordering of bricks and other materials for building sites, in order to avoid any delays in supply. Recycling targets provided a perverse incentive to crush and recycle any leftover raw materials, rather than to return and reuse them at another building site. This is an example of a simple waste prevention measure which pays for itself very quickly.[19] Another example slightly more complex is the controlled dismantling rather than the wholesale dismantling of buildings, so that more components can be salvaged for reuse. An example of deconstruction can be found in Topic Sheet 4 on waste prevention.

18 Text prepared from information provided directly by IEISL in June 2015, and from published material. See http://www.waste-management-world.com/articles/print/volume-15/issue-1/features/going-live-india-s-first-c-d-recycling-plant.html and http://www.cseindia.org/userfiles/NB%20Mazumdar.pdf

19 Lundesjo, G. (2011) Using Construction Consolidation Centres to reduce construction waste and carbon emissions. Guidance: Construction Logistics. WRAP. http://www.wrap.org.uk/sites/files/wrap/CCC%20combined.pdf

TOPIC SHEET

7

HAZARDOUS WASTE

Hazardous waste poses a greater risk to human health and the environment than non-hazardous waste. Therefore, greater attention needs to be given to hazardous waste and its proper management.

A historical perspective

Before the categorization of waste into 'hazardous' and 'non-hazardous' was introduced as a part of legislative controls, all waste tended to be managed in the same largely uncontrolled way. Hazardous waste was often dumped together with non-hazardous waste in the oceans or on land, leading to serious environmental pollution that often caused public outcry. Such indiscriminate and improper disposal, which has subsequently required large expenditures for clean-up,[1] and the recognition of the potential risks posed by this waste, has led to the introduction of legislative controls[2] and the establishment of environmentally sound management (ESM) systems.

However, a side-effect of introducing such controls in developed countries in the 1970s was that criminal actors looked to avoid the costs of environmentally sound management by exporting hazardous waste to countries where controls were lacking or weak. In order to prevent such illegal activities, the international community negotiated and adopted the Basel Convention on the Control of Transboundary Movements of Hazardous Wastes and Their Disposal in 1989 along with other regional MEAs.[3] In the early years of developing hazardous waste management systems, the focus was on environmentally sound management. This included both treatment to destroy the hazardous components, for example through the high temperature combustion of organic pollutants including polychlorinated biphenyls (PCBs) from electrical transformers and the chemical treatment of cyanide waste. Some resource recovery was practised, including for example the regeneration of spent solvents for recycling and the recovery of metals such as lead and zinc. In more recent years, attention in the industrialized countries has moved further 'up' the hierarchy, focusing in particular on hazardous waste prevention,[4] both through reducing hazardousness (e.g. the substitution of less toxic materials for mercury and cadmium in batteries and other products) and reducing amounts through modifying manufacturing processes to follow "green chemistry," increasing conversion efficiencies and establishing new business models such as chemical leasing.[5]

Data on hazardous waste generation

Data on the generation and management of hazardous waste are lacking or remain weak in many countries. Furthermore comparisons become difficult when the types of hazardous waste covered, the definitions and the methods used are different among countries. As a result, global data on hazardous waste generation, as shown in *Figure 1*, is not exhaustive despite progress made by many countries in this area and efforts by the Secretariat of the Basel Convention to compile data.[6]

© Shutterstock/Brandon Bourdages

1 See Box 1.1 in Section 1.1.
2 See Section 4.3.
3 See Box 4.13.

4 See Topic Sheet 4 on Waste Prevention, found after Chapter 2.
5 See Box 5.4 in Section 5.5.3.
6 The Secretariat of the Basel Convention (SBC) is requested to maintain data on the generation and transboundary movement of hazardous waste by Parties to the Convention. This data is however not complete, as not all Parties to the Convention submit their national reports on time and the submitted information is frequently incomplete.

Figure 1 Global hazardous waste generation in 2009

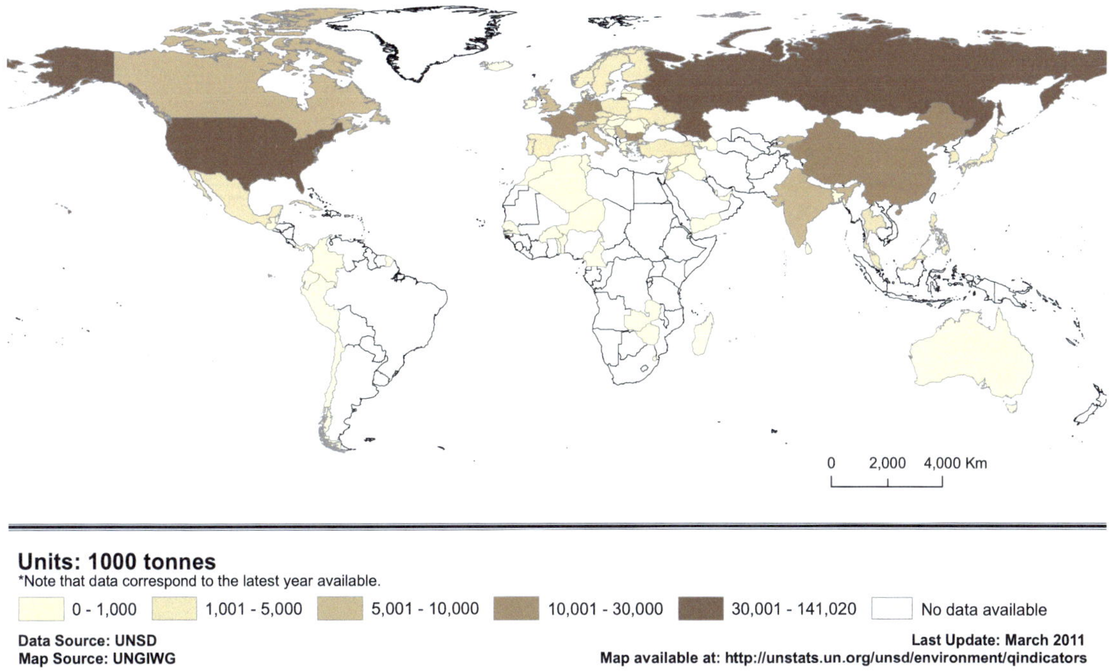

Source: UNSD (2011). Environmental indicators, waste: Hazardous Waste Generation (http://unstats.un.org/unsd/environment/hazardous.htm).

Hazardous waste arises from multiple sources (ranging from large- and small-scale industries to hospitals, offices and households); *Figure 2* shows the distribution of total hazardous waste generation by sector in the U.S. in 2011 (total amount of 34 million [short] tons), as defined in and reported under the U.S. Resource Conservation and Recovery Act.

Figure 2 Sectors in the U.S. generating the largest quantities of hazardous waste

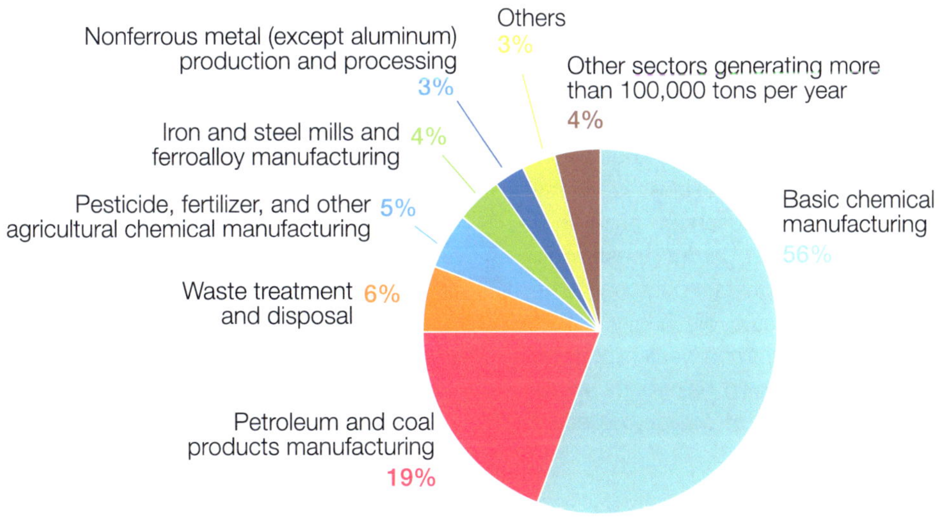

Source: USEPA. National Analysis – The National Biennial RCRA Hazardous Waste Report (Based on 2011 Data) (http://www.epa.gov/epawaste/inforesources/data/br11/national11.pdf).

As shown in *Figure 2*, large volumes of hazardous waste come from resource extraction and processing, heavy industry and product manufacturing. Over recent decades, many of these industries have been moving to emerging economies, where such industrialization and economic growth means that the generation of hazardous waste is increasing. In the BRICS countries (Brazil, Russia, India, China, South Africa) and other rapidly industrializing emerging economies, hazardous waste management systems have been developed and are gradually being implemented, including provision of facilities, as the quantity of industrial hazardous waste in some places already provides the necessary economy of scale. Notwithstanding the progress made, there is still much to be done to ensure that fully compliant ESM facilities are available in all regions within these countries.

Despite such shifts in industry, and despite recessions in much of the developed world, it appears that hazardous waste generation is still increasing in developed countries. For example in the EU, there was an overall 3.3% increase in hazardous waste generated between 2010 (97.5 million tonnes) and 2012 (100.7 million tonnes).[7] That said, changes in definitions may also be a factor here, as the hazard categories are changing.[8]

All countries generate hazardous waste

All countries generate hazardous waste, be it from small-scale industrial processes such as leather tanning, electroplating of metals or photofinishing; from mining and quarrying or dredging activities; from healthcare facilities; or from offices and households. The management of such waste represents a challenge to governments as often this waste is co-disposed with municipal waste or non-hazardous industrial waste. In addition waste not officially categorized as hazardous waste often contains low concentrations of hazardous substances, due in part to the increase in the number of chemicals produced and their use in a broader range of products.

Many countries are starting to pay particular attention to the generation of small quantities of hazardous waste from households, offices and small businesses. These are often termed generically 'household hazardous waste'. Examples include paint, garden pesticides, pharmaceuticals, certain detergents, personal care products, fluorescent tubes, waste oil, heavy metal-containing batteries, print cartridges and waste electronic and electrical equipment. Household hazardous waste should in principle be segregated at source, collected separately and managed within the hazardous waste system. However many developed countries are still in the process of implementing such controls, and the lack of ESM facilities for hazardous waste is a major barrier in low- and middle-income countries.

Another example is healthcare or medical waste, of which around 80% is generally similar to household waste and 20% is hazardous waste such as sharps, materials contaminated with bodily fluids, protective clothing, body parts, chemicals and pharmaceuticals, medical devices and radioactive materials. As per WHO estimates, the average hazardous healthcare waste generated per hospital bed is around 0.2 kg in most low-income countries and 0.5 kg in high-income countries.[9] However, the failure to segregate hazardous from non-hazardous waste in low-income countries increases the quantity of healthcare waste that requires management as infectious or otherwise hazardous waste.

In most developing countries, healthcare waste management is a major issue. There is a lack of awareness of the risks to the patients, healthcare workers and the general public; segregation is poor; the infrastructure for safe collection, handling and treatment or disposal is lacking; legislation often does not exist; and where legislation is in place, enforcement is often poor. Awareness raising, legislation, enforcement and the establishment of common healthcare waste treatment facilities on a city-wide or regional basis have been the strategies to address this challenge. Guidelines and a code of practice have been published by the World Health Organization.[10]

Medical waste at a landfill, Philippines

7 Eurostat (2014). Energy, transport and environment indicators. http://ec.europa.eu/eurostat/documents/3930297/6613266/KS-DK-14-001-EN-N.pdf/4ec0677e-8fec-4dac-a058-5f2ebd0085e4 and Eurostat (2014). Waste Statistics. http://ec.europa.eu/eurostat/statistics-explained/index.php/Waste_statistics

8 As an example, the UK has introduced new guidelines for the classifications of hazardous wastes, to implement new EU requirements: http://resource.co/article/new-waste-assessment-guidance-released-10130

9 http://www.who.int/mediacentre/factsheets/fs253/en/

10 WHO (2014), listed in Annex A, Chapter 3, Other waste streams, under Topic Sheets, Hazardous waste.

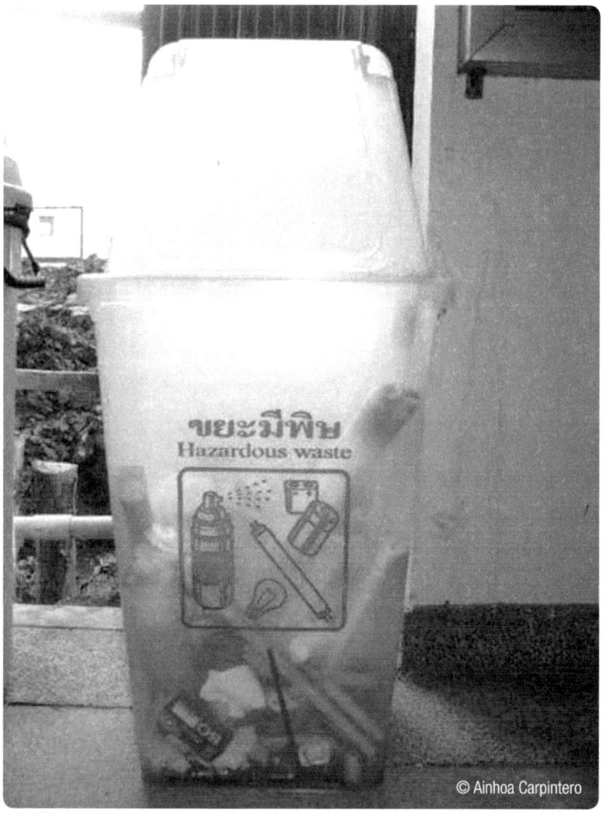

Container for hazardous waste, Thailand

Support to developing countries in managing their hazardous waste

All hazardous waste should be separated at source and managed in an environmentally sound manner. Most developing countries still lack proper treatment and disposal facilities, while at the same time they face the challenge of increasing amounts of hazardous waste, either due to increased imports of products which become hazardous waste at the end of their useful life (e.g. TVs, electronic equipment) or through industrialization (as both production and the associated wastes and pollution are 'out-sourced' by the developed countries). As a result, it could be argued that both multinational companies and developed countries have some responsibility to help developing countries to manage their hazardous waste. Support is needed to raise awareness in government, industry and the general public; to develop appropriate legislation and the institutional structures for implementation and enforcement;[11] to build institutional and human capacity;[12] and to establish the required infrastructure and facilities for environmentally sound management.

A first priority is to stop the uncontrolled mixing of hazardous and non-hazardous waste and to stop the disposal of such waste in an unsafe and uncontrolled manner. This includes the upgrading of existing unsafe recycling practices. For example, the recycling of lead-acid batteries in many developing countries involves draining the sulphuric acid into the gutter and then burning off the rubber casing to obtain molten lead. Similarly, the recycling of e-waste[13] often involves, among other practices, the burning of PVC cladding to recover copper cables. Such practices result in environmental pollution while at the same time poisoning the recycling operators and their neighbours.

11 Several guidelines are published by the Secretariat of the Basel Convention, including model legislation and codes of practice for the management of particular types of waste. For links to these, see Annex A, under Chapter 3, Other Waste Streams, under Topic Sheets, Hazardous waste.

12 A Training Resource Pack (TRP) for hazardous waste management in emerging economies was first published by ISWA, UNEP and the Basel Convention in 2002. The TRP manual and its associated support material is currently being updated. The original 2002 version, and the forthcoming update, entitled "TRP+", can be found at http://www.trp-training.info/

13 See Topic Sheet 8 on E-waste, found after this Topic Sheet.

TOPIC SHEET 8

E-WASTE

Waste electrical and electronic equipment (WEEE), popularly known as 'e-waste', comes from a broad range of electronic products such as computers, televisions, or video games, as well as all kinds of electrical equipment, often divided into large equipment (washing machines, air-conditioners, freezers etc.) and small equipment (hairdryers, electric toothbrushes, vacuum cleaners etc.).

This is the fastest-growing waste stream all around the world due to increased consumer demand, perceived obsolesce, and rapid changes in technology and inventions of new electronic devices. The situation is compounded by the short lifespans of certain products and products not being designed with recycling in mind.

Figure 1 Amounts of e-waste generated in 2014, by type

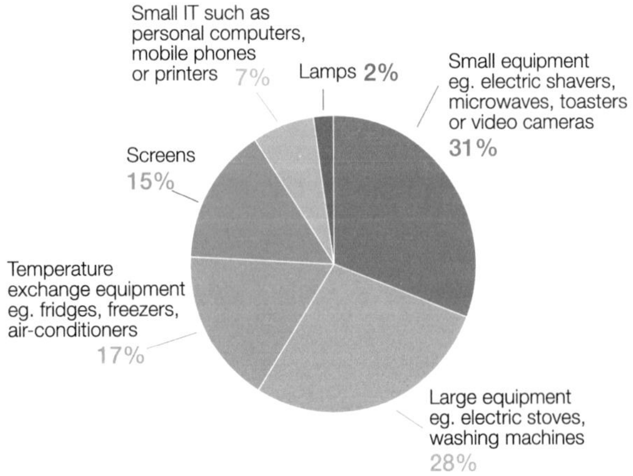

Source: Baldé, C. P., F. Wang, R. Kuehr, J. Huisman (2015). The global e-waste monitor – 2014. United Nations University, IAS – SCYCLE, Bonn, Germany.[1]

1 Available at http://i.unu.edu/media/unu.edu/news/52624/UNU-1stGlobal-E-Waste-Monitor-2014-small.pdf

A 2015 report by the United Nations University (UNU) estimated that 41.8 million tonnes (Mt) of e-waste was generated in 2014, almost 25% more than the 2010 figure of 33.8 Mt.[2] The amounts of e-waste generated by type are shown in *Figure 1.*

Most of this waste was generated in Asia (16 Mt), followed by Europe (11.6 MT), North America (7.9 Mt), Latin America and Caribbean (3.8 Mt), Africa (1.9 Mt) and Oceania (0.6 Mt). However, in e-waste generation per capita, Europe has the highest figure (15.6 kg/person) and Africa the lowest (1.7 kg/person).

Estimated annual generation in the coming years are as high as 50 Mt in 2018.[3]

Composition of e-waste

The composition of WEEE/e-waste is very diverse, as electrical and electronic equipment encompasses such a broad range of categories. In general it is characterized by containing a combination of metals, plastics, chemicals, glass and other substances.[4] Among the substances of particular interest are a very wide range of metals including rare earth metals like lanthanum, cerium, praseodymium, neodymium, gadolinium and dysprosium; precious metals such as gold, silver and palladium; or other metals such as copper, aluminium or iron, which have a high intrinsic value. However, even though 'critical materials' are scarce, they have recycling rates lower than 1%, which represents a threat to resource security over the long term.[5] Others substances which are hazardous or potentially hazardous, such as lead-containing glass, plastics treated with brominated flame retardants or ozone-depleting substances such as CFCs, are of concern due to the potential health and environmental risks they might pose.

2 Baldé, C.P., F. Wang et al. (2015), listed in Annex A, Chapter 3, Other waste streams, under Topic Sheets, E-waste.

3 Baldé, C.P., F. Wang et al. (2015), listed in Annex A, Chapter 3, Other waste streams, under Topic Sheets, E-waste.

4 See Figure 3.13.

5 See Section 3.4.2 and Figure 3.14

© WorldLoop / Hendrickx

E-waste legislation and management

The collection and management of e-waste heavily depends on the legislation in each country. As e-waste could include end-of-life products, it is an obvious candidate for some form of extended producer responsibility (EPR) or product stewardship, with such schemes being either voluntary or mandatory.[6,7] Although various forms of take-back systems are being implemented in both developed and developing countries, the amount of e-waste treated under such systems is reported to be lower than 50% of the total amount generated (40% in Europe, 24 to 30% in China and Japan, 12% in the U.S. and 1% in Australia).[8]

The e-waste not collected under these take-back systems might end up discarded into the general waste stream, or it might be collected by individual dealers or companies who trade the e-waste either for reuse or for recycling its metal, plastic and other substances.

Developing nations often lack the legislation on this type of waste, the infrastructure needed for handling it properly or treatment standards. Informal collection followed by unsafe recovery and recycling methods are commonplace, involving manual disassembly of the equipment, unsafe treatment techniques such as acid leaching to recover precious metals or burning activities to extract valuable metals, for example burning PVC-coated cables to recover copper, and the uncontrolled dumping of the residual waste.[9] These practices expose workers to dangerous working conditions that pose great risks to their health, for example through the inhalation of black soot, carcinogenic polyaromatic hydrocarbons (PAHs) and dioxins during burning activities, and also cause serious environmental pollution. To avoid such practices, initiatives have been launched by different stakeholder groups (including the producers) in many developing countries, aiming at environmentally sound e-waste management and sustainable recycling businesses providing safe and secure livelihoods.[10]

Export/transboundary movement of e-waste

Given the rapid turnover of mobile phones and computer equipment in developed countries, and the high demand for access to such equipment at a more affordable cost in middle- and low-income countries, there is a legitimate international trade in reusable and repairable equipment, a portion of which is facilitated by development charities. The question is, until what point are these materials 'products' which can be traded freely, and from what point do they become 'wastes', subject to control under the Basel Convention.

It is legal under the Basel Convention to export (hazardous) e-waste for recycling and disposal on the

6 An example of the approach being taken to EPR for e-waste in China is given in Topic Sheet 13, found after Section 4.5. (Topic Sheet 13 also addresses the dilemma of how to manage 'informal' sector recycling.)

7 See Section 4.5.2.

8 Baldé, C.P., F. Wang et al. (2015), listed in Annex A, Chapter 3, Other waste streams, under Topic Sheets, E-waste.

9 ISWA (2011). ISWA Position Paper on Waste Trafficking. http://www.iswa.org/index.php?eID=tx_bee4mememberships_download&fileUid=118

10 See Case Study 1 on e-waste, found after this Topic Sheet.

basis of prior informed consent, provided that the waste will be managed in an environmentally sound manner. However, a total ban has been proposed on the export of hazardous waste to developing countries[11] because of concern that many developing countries 'lack the financial, technical, legal, and institutional capacity for monitoring transboundary movements, managing [hazardous] wastes in an environmentally sound way and preventing illegal imports.'[12] In fact the EU and many other developed countries already ban exports of hazardous waste to developing countries.

All this notwithstanding, it is clear that some portion of the export trade in e-waste in particular has in the past 'crossed the line', and could be classified as 'waste trafficking' or 'waste crime', rather than legitimate trade. For example, the Agbogbloshie dump in Accra, Ghana, mentioned in Topic Sheet 2 as one of the 50 biggest dumpsites in the world, and perhaps the world's largest e-waste dump, receives around 192,000 tonnes of e-waste annually, much of it imported, and pollutes soil, air and water and causes serious health impairments in the lives of 10,000 informal sector workers gaining their livelihood from sorting and recycling.

The *modus operandi* for international waste trafficking includes, among other means, deliberate misclassification of waste types, false declarations of waste products or unchecked items as re-usable products and other kinds of fraudulent shipment documents.[13] Controlling such trade entirely at the point of export is impracticable, and as discussed in Section 4.3.7, it is also impossible for the regulator to inspect each container.

Another approach is to work within the product supply chains to prevent inappropriate and illegal trade at source, and to work with the recyclers in developing countries to develop environmentally sound management and sustainable business practices. Several multi-stakeholder partnership initiatives have been facilitated by the Basel Convention or the UN, two of which are highlighted in the box below.[14] The Case Study on e-waste management in Kenya which follows this Topic Sheet showcases two initiatives to turn unsafe recycling practices into environmentally sound management businesses.

11 An amendment banning all exports of hazardous wastes from Basel Convention "Annex VII" countries (primarily OECD and EU member countries) to non-Annex VII countries, including exports intended for reuse, recycling or recovery operations, was adopted in 1995. This Amendment will enter into force upon ratification by three-fourths of the Parties that were Parties to the Convention when the Amendment was adopted in 1995 (i.e. 66 of the 87 Parties to the Convention in 1995).

12 Basel Convention (2002). "Minimizing hazardous wastes: A simplified guide to the Basel Convention," p.16. www.basel.int/pub/simp-guide.pdf

13 See Section 5.3.4.

14 Another collaborative global initiative, led by UNU-IAS since 2007, is the 'Solving the E-waste Problem' (StEP) Initiative, which leads global thinking, knowledge, awareness and innovation in the management and development of environmentally, economically and ethically-sound e-waste resource recovery, re-use and prevention.

E-waste management and support for decision making through partnerships under the Basel Convention (MPPI) and PACE[14]

The Basel Convention has been addressing e-waste Issues since 2002, when the Convention's Mobile Phone Partnership Initiative (MPPI) was adopted.

The overall objective of the MPPI was to promote the objectives of the Basel Convention in the areas of environmentally sound management of end-of-life mobile phones, prevention of illegal traffic to developing countries and building capacity around the globe to better manage e-waste. Under the MPPI, five technical guidelines were developed, including awareness raising and design considerations; collection of used and end-of-life mobile phones; transboundary movement of collected mobile phones; refurbishment of used mobile phones; and material recovery/recycling of end-of-life mobile phones. The guidelines were tested in a facility-type environment and then revised. The final MPPI guidance document[15] was adopted in its entirety in 2011.

In 2006, the Nairobi Declaration on the Environmentally Sound Management of Electrical and Electronic Waste was adopted under the Convention, calling for more structured and enhanced efforts towards achieving global solutions to manage e-waste problems and encouraging Parties to develop further partnerships targeting e-waste.

More recently, the Partnership for Action on Computing Equipment (PACE) was developed as a multi-stakeholder public-private partnership that provides a forum for representatives of manufacturers, recyclers, international organizations, associations, academia, environmental groups and governments to tackle environmentally sound refurbishment, repair, material recovery, recycling and disposal of used and end-of-life computing equipment.

PACE has produced various kinds of materials, including guidelines on environmentally sound testing, refurbishment and recycling of used computing equipment; guidelines on environmentally sound material recovery and recycling of end-of-life computing equipment; and a report on strategies, actions and incentives to promote environmentally sound management of used and end-of-life computing equipment.[16]

PACE
Partnership for Action on Computing Equipment

15 More information can be found at http://www.basel.int/Implementation/TechnicalAssistance/Partnerships.

16 Basel Convention (2012).

17 More information can be found at http://www.basel.int/Implementation/TechnicalAssistance/Partnerships/PACE/Overview/tabid/3243/Default.aspx

CASE STUDY 1

NETWORK FOR E-WASTE RECYCLING IN KENYA AND OTHER AFRICAN COUNTRIES[1]

Background and context

Almost 2 million tons of e-waste exist in Africa,[2] arising from domestic use as well as imports, oftentimes illegal. Most countries in this region lack the expertise, infrastructure and regulations to manage e-waste properly. A push towards greater use of information and communications technology (ICT) picked up momentum and visibility in Kenya in the early 2000s, resulting in a surge in ICT-related e-waste later that decade. In Kenya, an average of 3,000 tons of computers, monitors, printers, mobile phones, batteries and other kinds of e-waste is generated annually.

In response to inadequate e-waste management, initiatives by different stakeholder groups have arisen across Africa to collect, treat and properly dispose of e-waste. This case study highlights two operations in Kenya: the Waste Electrical and Electronic Equipment Centre (WEEE Centre), led by a not-for-profit organization, and the East African Compliant Recycling Company (EACR),[3] led by the private sector. These co-exist in the same region, and both involve multiple stakeholders with different approaches and self-sustaining business models. Their operations helped shape the first legislation on e-waste guidelines managed by Kenya's National Environmental Management Agency (NEMA).

About the initiatives

Waste Electrical and Electronic Equipment Centre (WEEE Centre)

The WEEE Centre is an e-waste recycling organization based in Nairobi, Kenya that is owned and operated by local entrepreneurs with sustained support from various local and international partners. It provides e-waste collection, dismantling and automated processing services in Nairobi and in other major cities in Kenya. To ensure treatment of all e-waste fractions, it participates in the "Best of 2 Worlds" (Bo2W) model,[4] shipping its hazardous and non-valuables[5] through a transparent flow for end-processing by international recyclers and smelters.

This Centre was established to meet the need for safely disposing of end-of-life ICT equipment. The initiative was led by Computers for Schools Kenya (CFSK), a non-governmental organization that since 2002 has facilitated the development of ICT infrastructure and capacity of young Kenyans through the distribution of high-end reconditioned computers. CSFK, together with multiple partners including Computer Aid International, Safaricom Foundation, Close the Gap and WorldLoop,[6] launched the centre as a pilot in 2010. Each partner contributed, and many still do, in different areas. For example WorldLoop provided seed capital and developed business plan support and continues to provide technical expertise and operational support by managing the international recycling and disposal of complex and hazardous fractions as well as offering incentives for the collection of non-valuable fractions. Safaricom helped with community sensitization and runs collection drives. Similarly, other partners provide support as and when needed. Since 2011 the WEEE Centre operates as a formal recycling facility and it is now registered as a separate entity from CFSK.

1 Prepared by Mona Iyer with inputs from WorldLoop and Prof. Margaret Bates, University of Northampton.

2 Baldé, C.P., F. Wang et al. (2015), listed in Annex A, Chapter 3, Other waste streams, under Topic Sheets, E-waste.

3 Originally called the East African Computer Recycling Company.

4 Bo2W is a concept developed by the multi-stakeholder Solving the E-waste Problem (StEP) Initiative hosted by the UN's research arm, the United Nations University (UNU), which provides a network and pragmatic solution for e-waste treatment in emerging economies. It seeks technical and logistic integration of 'best' pre-processing in developing countries to manually dismantle e-waste and 'best' end-processing to treat hazardous and complex fractions in international state-of-the-art end-processing facilities.

5 This includes CRTs, lead-containing glass, printed circuit boards, transformers, mixed scrap and device-specific components containing toxic substances.

6 Besides these, it has both local and international partners. See http://www.weeecentre.com/index.php/about-us/faqs

The WEEE Centre primarily sources e-waste through collection campaigns aimed at individual households and businesses and engages the informal sector as e-waste collectors. In addition, awareness raising initiatives are held at various levels throughout the community, in schools and community centres, through dedicated events and conferences and through meetings with various stakeholders, including the local and national governments. The Centre has two parallel systems, one for repair and refurbishment and the other for dismantling and recycling. The equipment requiring only minor repair or refurbishment are repaired and then donated to resource-deprived community information access centres or learning institutions. End-of-life equipment is recycled at the Centre. Locally recyclable materials, such as aluminium, copper and other metals and hard and soft plastics, are sent to local smelters for future use in locally manufactured products. The fractions that cannot be locally handled are sent to international partner recyclers in Europe who have the capability and capacity to handle the difficult elements and are in compliance with international and domestic regulations and laws governing e-waste disposal.

Safe dismantling at the WEEE Centre

The WEEE Centre currently operates with five full time staff and 13 day labourers, whose work hours vary with the collection volume. All workers receive technical training on best practices and sustainable solutions in the collection, dismantling and recycling of electronic and electrical waste in order to protect the workers and the environment. Additional practitioner and business management training necessary to meet the Centre's marketing, HR and accounting needs, including cash flow management, is also made available throughout the year.

The funding that WorldLoop provides is linked to the Centre's collection targets, since collection is the largest challenge and one of the primary drivers for a financially sustainable e-waste business. An average of 12 to 15 tonnes of e-waste is collected per month. The WEEE Centre has already reached 73% self-sustainability after its first two years of operations based on the resale value of devices tested and eligible for reuse, locally sold raw materials (metals, plastics) and internationally recycled fractions (printed circuit boards, SIM cards and hard drives). WorldLoop's financial involvement decreased incrementally as self-generating revenues increased and has been reoriented to the collection and treatment of hazardous and non-valuable waste until a local financing mechanism and legislation are implemented. As is case with all WorldLoop projects, the financing requirements to cover the collection and treatment costs of non-valuable fractions are currently being addressed through the e-resource certificate programme.[7]

East African Compliant Recycling Company (EACR)

In October 2011, Hewlett-Packard (HP) opened the EACR in Mombasa, Kenya in partnership with Camara Education, an 'Education through ICT' NGO working with disadvantaged communities in Africa.[8] This facility accepted end-of-life IT assets from public and private sector customers, consumers and the informal sector and operated to international health, safety and environmental standards. Professional recycler Reclaimed Appliances (UK) Ltd, took over EACR in 2012 and expanded the scope to 'everything with a plug or battery' and to all types of plastics. In 2013, HP formed a Public Private Partnership with German Bank DEG and Reclaimed Appliances to secure funding for relocation of the EACR to a larger facility in Nairobi.[9] Following this, WorldLoop and EACR entered into a partnership specifically to support the collection and treatment of CRT monitors. EACR restructured in 2015, with ownership passing to local stakeholders.

In addition, members of the E-Waste Solutions Alliance for Africa[10] have assisted in sourcing funding for EACR. This Alliance helps to raise awareness of the need for standards and also developing capacity in government and other key stakeholders for environmentally safe and economically sustainable e-waste management.

The EACR facility sources its e-waste from eight registered collection centres located all over the Kenyan territory with a goal to expand to neighbouring countries where import into Kenya can be approved by the government. Many of the collection centres are sponsored by Dell or HP, covering upfront costs such as personal protective equipment (PPE), weighing equipment and containers. Each collection centre aims to become self-sustaining. A centre typically operates as a microbusiness, contracted to the treatment facility, through a franchise agreement with EACR. Before a contract is signed and

7 http://worldloop.org/get-involved/e-resource-certificates/
8 http://www.eastafricancompliantrecycling.net/about-us
9 E-Waste Solutions Alliance for Africa (2013).
10 E-Waste Solutions Alliance for Africa includes participants like Dell, HP, Nokia, Phillips and Reclaimed Appliances (UK) Ltd. who have developed a set of principles which are critical in the success of managing e-waste in many developing countries.

able to operate a collection point, operators, many of whom were previously in the informal e-waste recycling sector,[11] have to undertake training and pass a test on health and safety, environmental protection, mercury spills and other areas. The standards and methods they now use are clearly stated and enforced by the treatment facility to ensure adherence. No dismantling or recycling takes place at collection points. This is all carried out at the central facility, thereby enabling the provision of the necessary health, safety and pollution control systems. The collection points also work with a network of registered informal sector workers who are trained and provided with the necessary PPE. Environmental and personal protection is ensured throughout the collection process as the contracts, and therefore the payments, require the adherence to standards. There have been no specific awareness raising activities to source e-waste but media interest and support from the ministry and NEMA helped raise awareness of the facility. The franchise model with collectors encourages collectors to undertake local activities to ensure that they get the most material in their area.

The recycling facility receives about 10-15 tonnes of waste per month. EACR collects the containers from collection centres when the operators indicate that they are full. All the equipment is disassembled and the material are separated. Materials are then traded globally. In the long term there are plans to take this to the next stage with the manufacture of other goods and support for local enterprises on the basis of those materials. Non valuables are treated in an environmentally sound manner, including safe export to other facilities.

Safe dismantling at the EACR facility

Lessons and the way forward

Both the WEEE Centre and EACR demonstrate that a sustainable solution to e-waste can operate in developing countries. They also ensure that all fractions, both valuable and non-valuable, hazardous and non-hazardous, are collected with proper enforcement of and incentives for adhering to environmental, health and safety standards. Major lessons learned from the experiences of these two organizations include:

- Local awareness of the hazards in e-waste and the steps required for proper e-waste management is extremely low.
- Projects require not only technical training but also managerial training. Gaps in knowledge include business planning, marketing & communications, corporate accounting and facility design.
- Collection can and should be undertaken through the local informal sector, after proper training and 'formalization'. The collectors do not undertake any dismantling, as dismantling needs to be done in a central facility under close supervision to ensure that both health and safety and environmental standards are met.
- It has been recognized that various fractions (leaded glass from CRT monitors, most batteries, flat panel displays and capacitors) are not able to produce revenue for projects and represent a cost for proper treatment. In order to discourage cherry picking, financing needs to be in place to cover the collection and treatment costs of these fractions.
- These facilities can serve a regional need, either by importing e-waste from neighbouring countries (when regulations are in place), or by replicating there.
- Government engagement and support is a key enabler.

The success of these two facilities validates the claim that a gradually decreasing funding approach can lead to a successful business growth plan. The results of both models demonstrate that a sustainable business model involving local entrepreneurs is possible even when national regulations are not yet in force. However, the regulations are particularly important to ensure that all fractions are collected, including the problematic and hazardous ones.

Sustained interactions with NEMA and the presence of multi-stakeholder participants in these two major e-waste recycling facilities has resulted in the formulation of the draft Environmental Management and Co-ordination (E-Waste Management) Regulations, 2013[12] which will further facilitate the environmentally sound management of e-waste by incorporating extended producer responsibility. On implementation of the legislation, it is expected that municipal support will be made available for collecting and treating non-valuables and for incentivizing the adoption of environmentally sustainable treatment practices.

11 Often carrying out poorly conducted, and potentially dangerous, e-waste recycling and recovery.

12 This regulation has been drafted and gazetted but not implemented.

TOPIC SHEET 9

MARINE LITTER[1]

Marine litter (marine debris) has rapidly emerged as one of the most difficult to resolve global waste and resource management challenges of our times.

According to UNEP, marine litter can be described as *'[...] any persistent, manufactured or processed solid material discarded, disposed of or abandoned in the marine and coastal environment. Marine litter consists of items that have been made or used by people and deliberately discarded into the sea or rivers or on beaches; brought indirectly to the sea with rivers, sewage, storm water or winds; or accidentally lost, including material lost at sea in bad weather.'*[2] The challenge has local, regional, national and genuinely global dimensions. There is a worldwide growing awareness, with thousands of organizations focusing on studying and solving aspects of this major challenge. The problem is almost ubiquitous around the globe to a variable extent, but effective solutions, at a different scale level, are hampered partly by lack of sufficient understanding of the actual science regarding magnitude, formation mechanisms and impacts.

Specifically, the overall magnitude of the marine litter entering the sea annually is unknown. A recent guess stated that around 10% of the current yearly production of plastics could be entering the sea, an amount which was elsewhere quantified to around 18 million tonnes.[3] In another recent study using proxy information to estimate the quantity of waste plastics generated on land and entering the ocean, this was estimated from 4.8 to 12.7 million tonnes per year.[4] That said, there is no conclusive evidence that these numbers are reliable. This order of magnitude has been criticized as hard to believe.[5]

Of vital importance are the various sources and routes via which the used plastics reach the coast and sea. It is often quoted that 80% of marine litter by weight comes from land-based sources (with the remaining 20% by weight from sea-based activities), but this is based on a misinterpretation of a hypothesis first put forward in a

Table 1 Key sources of marine litter: from land and sea

SEA-BASED SOURCES	LAND-BASED SOURCES
• Fishing vessels (e.g. nylon nets)	• Coastal touristic activities (e.g. littering at beaches)
• Merchant ships (e.g. accidents with primary/secondary plastics containers)	• Transport of waste by rivers flows (sources: sewerage, run-off, smaller inland waterways)
• Ferry boats	• Sewage and storm water direct discharge into the sea
• Cruise ships	• Dumpsites (and engineered landfills) sited on the coastline
• Military naval vessels	• Mismanaged residues from industrial facilities
• Pleasure boats	• Wind-blown litter
• Fish farming facilities	
• Offshore oil or gas platforms	

1 Topic Sheet prepared by Costas Velis, based largely, but not exclusively, on text which first appeared as an Editorial in the journal *Waste Management & Research*. Reproduced here with permission. See Velis (2014), listed in Annex A, Chapter 3, Plastic waste and marine litter.

2 UNEP (n.d.) Environment for Development: About Marine Litter. http://www.unep.org/regionaLseas/marinelitter/about/default.asp

3 Velis (2014)

4 Jambeck et al. (2015)

5 Velis (2014)

GESAMP[6] document of 1990 regarding pollution – not solid litter – and has been reproduced since extensively, becoming erroneously a widely accepted assumption.

Used plastics are not the only material entering the ocean and featuring in marine litter, but there is consensus that it is the most challenging material because it degrades slowly, is lightweight and can be transported over long distances, is abundant and has had documented impact on numerous marine organisms and habitats. Global plastics production has grown consistently since 1950, reaching 299 million tonnes in 2013.[7] The People's Republic of China is the leading producer of plastics, manufacturing almost 25% of global plastics by weight. From the data in *Figure 3.3* in Chapter 3, plastics account on average for 7 to 12% of MSW by weight, depending on the income level of the country. While additional research in quantification of the financial losses incurred is necessary, a study indicates that the natural capital cost of the impact of plastics on marine ecosystems is at least 13 billion USD per year.[8]

For example, patches of marine litter exist floating in the oceans, accumulating in the five subtropical ocean gyres as shown in *Figure 1*. The Great Pacific Garbage Patch is one such collection of marine litter in the North Pacific Ocean. Marine litter is not limited to just this area, but is found in accumulation zones in other regions of the North Pacific.

A considerable amount of scientific effort over the last decade focused on attempting to document the multiple aspects of potential harm caused by used plastics to marine fauna and its habitat. Key issues relate to (i) entanglement: for example, sea turtles entangled in abandoned nylon fishing nets. One hundred and fifteen species (44 sea bird species, 9 cetacean species, 11 pinniped species, 31 invertebrate species/taxa, 6 sea turtle species) were reported entangled in marine debris in the US and a total of 200 species worldwide;[10] and (ii) ingestion: for instance, seabird fulmars ingesting plastics, confusing it with food; and even zooplankton, bivalves, etc. may 'feed' on it. Ingestion of plastics could result in a series of problems, from physical harm to bioaccumulation of organic chemical compounds used as additives to the plastic polymers that are known to function as endocrine disruptors such as phthalates, bisphenol A (BPA) and polybrominated diphenyl ethers (PBDEs). Evidence has also accumulated on the absorption into the plastics within the sea of the hydrophobic chemical polluting compounds present in the water, including polychlorinated biphenyls (PCBs) and other persistent organics pollutants (POPs). Ingestion of

Figure 1 Concentrations of plastic debris in surface waters of the global ocean[9]

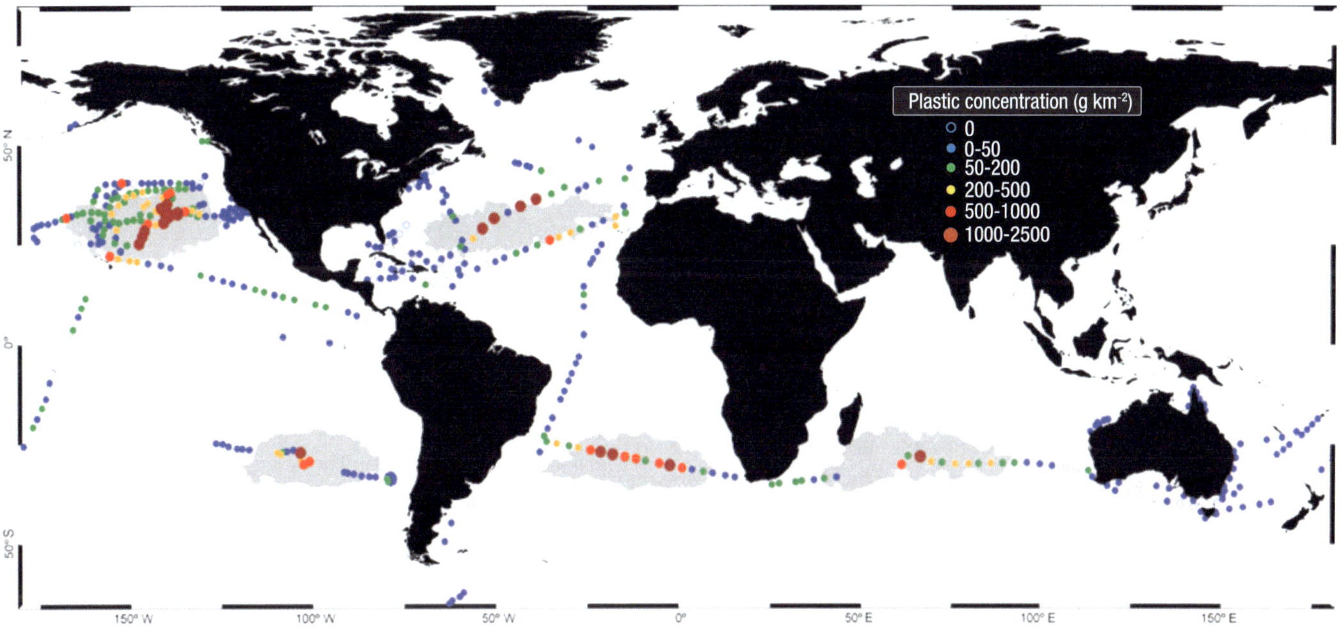

Source: Cózar et al. (2015) Plastic debris in the open ocean. PNAS Vol. 111(28), 10239-10244 http://www.pnas.org/content/111/28/10239.full.pdf

6 The Joint Group of Experts on the Scientific Aspects of Marine Environmental Protection, an advisory body to the UN system.

7 PlasticsEurope (n.d.) Plastics: The Facts 2014/2015 – An analysis of European plastics production, demand and waste data. See http://issuu.com/plasticseuropeebook/docs/final_plastics_the_facts_2014_19122

8 UNEP (2014c).

9 An updated figure showing the plastics accumulation in the Mediterranean Sea can be found at http://journals.plos.org/plosone/article?id=10.1371/journal.pone.0121762 Cozar et al. (2015) Plastic Accumulation in the Mediterranean Sea. PLoS ONE 10(4): e0121762. doi:10.1371/ journal.pone.0121762

10 NOAA (2014).

plastics with accumulated absorbed POPs may result in increased bioaccumulation in fish, in comparison to pure plastics, recent studies suggest. However, not all aspects are sufficiently understood. For instance, a recent study on ingestion of microplastics (<1mm) on small invertebrates (Marine Isopod Idotea emarginata) has shown that "*microplastics, as administered in the experiments, do not clog the digestive organs of isopods and do not have adverse effects on their life history parameters.*"[11]

The sequence of absorption, ingestion and potential bioaccumulation in marine living organisms may, through the food chain, pose risks for human health as well, but this scenario requires further exploration. The list of potential damage of marine plastic litter to the natural and human-made systems also includes the migration of invasive species travelling attached to plastics; harm to reef corals and benthic communities at the seabed; formation of nano-/micro-colloids. Some of these hazards result in direct financial costs, from costly clean-up efforts for beaches of high aesthetical recreational value, to revenue losses from impeded tourism.

© Race for Water 2015 – Peter Charaf

There is considerable data on the size of plastics items found in marine environments, and micro-plastics prevail. However, there are no globally consistent quantification methodologies, and often data collected refer to numbers of items, with mass data needed for quantification of mass flows missing. Increased consideration is given to microplastics, defined as plastics under a certain particle size. These may stem from fragmentation of bigger plastic items (secondary microplastics) or may originate from additives in widely used cosmetics, such as facial exfoliators. Plastics are mainly fossil-derived, but there is increased production of bio-based polymers (bio-polymers). Whereas most of the fossil and bio-polymers are non biodegradable, both fossil and bio-polymers can be manufactured to biodegrade, but their degradability depends on the exact environmental conditions to which they are exposed.[12] Great uncertainty relates to the decomposition and fragmentation of plastics. Sinking mechanisms and rates are also underexplored. Due to microorganisms and algae accumulating on the plastics debris over time ('biofouling'),[13] plastics may sink instead of float.

Hence, on a generic level, considerable evidence has been accumulated on the potential harm caused by plastics present in marine litter. There is consensus that this evidence suffices to lead to immediate action. Given the complex nature of the challenge, any solutions have to equally derive from a systems and multi-stakeholder approach, involving: (i) multiple materials (types of polymers and additives), products (from fishing nets, to plastic carrier bags, to sanitary products such as nappies, to hard thermosets in durable goods such as toys), and points of initial entry and transport into the marine environment (sea, rivers, sewerage systems, wind-blown, seashores – global flows in huge oceanic gyros); (ii) the various scales of impacts, from aesthetical and amenity loss to wildlife species health damage and entire ecosystem disruption; and (iii) the multiple human activities involved in causing the problem, from the innovation of chemical compounds, to product design, to modes of retail, consumption and lifestyles, such as littering of fast food packaging, to unsound solid waste disposal methods, such as in uncontrolled dumpsites. Marine litter is one of many reasons why the elimination of open dumping in non-landlocked low- and middle-income countries needs to be a political priority.[14]

Despite the systemic complexity, the most critical intervention has to come from the waste and resources management sector. In most of the cases, except for accidents and natural disasters such as tsunami, it is sound waste management and resource recovery practices that can prevent used plastics from entering the sea. The most important actions are to identify the sources and transport routes of used plastics into the sea, for the plastics properties (chemical composition, fragmentation/degradation modes, absorption of POPs) and quantities with the potential to cause maximum harm. In particular, it is essential for the waste and resources academics and practitioners to work in close collaboration with all other stakeholders and come up with reliable estimates on the key sources and pathways and how to effectively circumvent them. With this information available, existing intervention action plans can be prioritized according to their effectiveness and new more effective ones can be devised if necessary.

11 Hämer, J., L. Gutow, A. Köhler, R. Saborowski (2014). Fate of Microplastics in the Marine Isopod Idotea emarginata. *Environ. Sci. Technol.* 2014, *48*, (22), p. 13451.

12 Shah, A. A., F. Hasan, A. Hameed and S. Ahmed (2008). "Biological degradation of plastics: A comprehensive review." *Biotechnology Advances* 26(3): 246-265.

13 DG-ENV European Commission (2011)

14 See Section 1.1 on the challenge of sustainable waste management and Topic Sheet 2, the 50 Biggest Dumpsites.

Over the mid- to long-term, the proliferation of sound waste management collection and disposal practices, extended producer responsibility (EPR) that involves manufacturers and retailers, and behaviour change focused on littering individuals are possibly key aspects of any solution, but the details are still missing. At an intermediate stage, innovation will be needed around the litter generation points: upstream, that is, in the production chain, where redesign can potentially be important in reducing generation quantities and in mitigating the inherent risk posed by used plastics in marine environments; and downstream, where there can be innovation in the solid waste and resource management systems regarding collection. Long-term technical solutions for removal and value recovery from the existing used plastics in the world's seas, shores and seabeds are still missing, and some of these places are marine areas beyond national jurisdiction, which could complicate decisions and actions.

There are many important initiatives, from local to worldwide, attempting to contribute in addressing the marine litter challenge. The problem is too big for any one organization or country to deal with alone and requires extensive collaboration. The Global Partnership on Marine Litter (GPML),[15] hosted by the UNEP Global Programme of Action (GPA), was launched in June 2012 at Rio+20. The GPML is a voluntary open-ended partnership for international agencies, governments, businesses, academia, local authorities, nongovernmental organizations and individuals. It provides a platform for increased collaboration and coordination among these groups, promoting a collaborative dialogue to achieve its main goal: to protect human health and the global environment through the reduction and management of marine litter. The resolution on marine plastic debris and microplastics adopted by the United Nations Environment Assembly of the United Nations Environment Programme at its first session on June 2014[16] is also an important step in global collaboration.

Regionally and locally, the Marine Litter Action Network (MLAN)[17] in the UK is an example of a relevant multi-stakeholder initiative, organized by the Marine Conservation Society. The National Marine Debris Monitoring Program (NMDMP) in the U.S. is a five-year program developed by the Ocean Conservancy with support from USEPA to standardize marine debris data collection and assess marine debris sources and trends in the U.S. Ongoing initiatives in Australia include the Coral Triangle Initiative, the Coordinating Body on the Seas of East Asia (COBSEA), and Asia Pacific Economic Cooperation (APEC) has had a working group addressing this issue since 1990. In the EU, the European Environment Agency (EEA) has developed Marine LitterWatch (MLW) for citizen engagement through organized citizen groups, a mobile application and a database.[18]

Targeted interventions through risk-based clean-up initiatives may be useful. The Global Underwater Awareness Association (GUWAA) has conducted clean-up exercises with its S.P.E.E.D. (Special Protection Ecology Environment Diving) Unit that consists of a diving team that aims at cleaning the seabed, rivers, and lakes worldwide.[19] But while clean-up activities are good tools for raising awareness, they are not a long-term solution and are a very costly end-of-pipe approach.

15 http://www.unep.org/gpa/gpml/gpml.asp

16 http://www.unep.org/unea/UNEA_Resolutions.asp

17 https://www.mcsuk.org/what_we_do/Clean+seas+and+beaches/Campaigns+and+policy/Marine+Litter+Action+Network

18 European Environment Agency (2015). European citizens to help tackle marine litter. http://www.eea.europa.eu/themes/coast_sea/marine-litterwatch/at-a-glance/european-citizens-to-help-tackle

19 See Global Underwater Awareness Association (2013). http://whiteflagint.com/guwaa/index.php/guwaa-projects/finished/guwaa-2nd-world-cleanup-2013

TOPIC SHEET 10

DISASTER WASTE[1]

Disasters and conflicts leave behind tremendous amounts of waste, in particular high amounts of debris and construction and demolition waste. Such waste is generated not only through the initial disaster but also during the response and recovery phases. *Figure 1* shows statistics on locations and estimated amounts of disaster waste across the world. Several recent large earthquakes and one war zone have yielded an estimated range of 1-30 million tonnes of disaster waste per incident.

Figure 1 Disaster waste generated at different locations across the world[2, 3, 4]

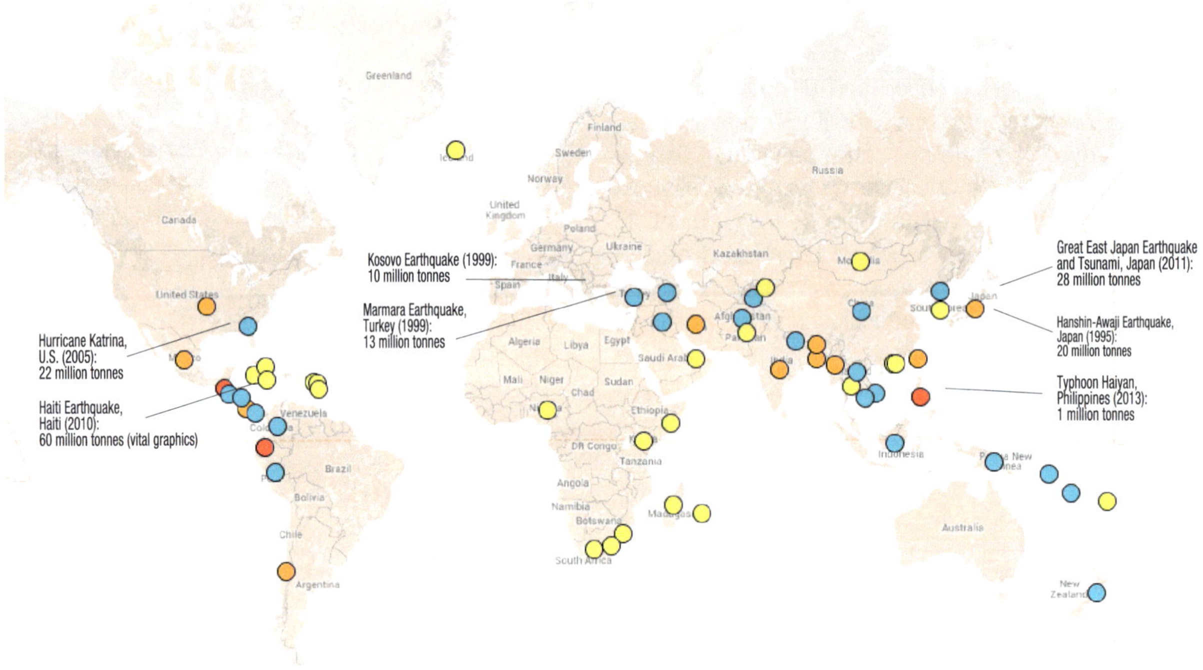

Notes: Color indicates degree of exposure to six major natural hazards. Exposure to 2 hazards – yellow, 3 – blue, 4 – orange, 5 – red

1 Text prepared by Ainhoa Carpintero and Prasad Modak

2 The six major natural hazards are earthquakes, volcanoes, landslides, floods, drought, and cyclones. UNDESA (2012) in Annex A, Chapter 3, Collated Data Sources; and World Bank and Columbia University (2005). *Natural Disaster Hotspots – A Global Analysis*, Tables 1.1a and b. http://www.ldeo.columbia.edu/chrr/pdf/hotspots/maps/table1.pdf

3 Karunasena, G. (2011) Sustainable Post-Disaster Waste Management: Construction and demolition debris. In: Amaratunga, D. & R. Haigh (2011). *Post-Disaster Reconstruction of the Built Environment: Rebuilding for Resilience.* UK, Blackwell Publishing Ltd. https://books.google.co.jp/books?id=yEFliCMKtDEC

4 Ministry of Environment of Japan (2014). Progress on treatment of debris from the Great East Japan Earthquake. See http://www.env.go.jp/en/focus/recycle/eq/ptd20140326.pdf and Environmental Preservation Division, Environment Department, Bureau of Citizens and Cultural Affairs, Hyogo Prefecture, Japan (1997). Treatment of disaster wastes from the Great Hanshin-Awaji earthquake. See http://web.pref.hyogo.jp/wd33/documents/000045493.pdf (only in Japanese)

Debris clearance and waste management are among the highest priorities following a disaster because of the need to provide access, rescue survivors, retrieve dead bodies and address urgent public health and environmental issues.

Management of the disaster waste will differ according to the response phase following a disaster and will also depend on the types of waste and debris generated and the management options available.

- During the emergency phase, the primary aim is removal of debris in order to allow access and to facilitate rescue operations. During this phase, any hazardous waste and human or animal remains should be separated from other streams wherever possible. Temporary, and where possible final, disposal sites need to be identified.
- During the early recovery phase – while still selecting and preparing disposal sites for debris and so on – basic municipal solid waste management and healthcare waste management systems should be reinstated as soon as possible, alongside other essential utility services such as sanitation and clean water.
- In parallel, a disaster waste management programme[5] should be initiated. This should, where possible, include reuse and recycling activities in order to recover large amounts of materials and divert materials from disposal at disposal sites or landfills. Such activities lead to lower disposal costs while facilitating the re-sale of materials such as scrap metal.[6]
- During the recovery phase, the disaster waste management programme will continue to develop, including monitoring and evaluating the situation and implementing specific waste management projects. This programme should also integrate a long-term development vision that enhances local technical capacity, financial self-sufficiency and various other critical elements in order to help ensure the sustainability of the waste management system even after the recovery phase is finished.

Ideally a Disaster Waste Management Contingency Plan should be developed.[7] Having this plan in place before an emergency occurs helps in identifying options for waste collection, recycling and disposal and saves time and resources, critical factors following a disaster. However, if such a plan is not in place, it needs to be developed during the recovery phase, or as a preparedness measure for a future disaster.

5 Disaster waste is much harder to manage than ordinary construction and demolition waste, as the former is very often mixed and contaminated.

6 UNEP/OCHA (2011), listed in Annex A, Chapter 3, Other waste streams, under Topic Sheets, Disaster waste. See also Case Study 2 on disaster waste management, found after this Topic Sheet.

7 The waste management contingency plan can be developed as an initial standalone plan, to be later coordinated with a wider disaster preparedness or emergency response contingency plan, or directly as part of the wider plans.

Japan's experience in Miyagi prefecture following the 2011 earthquake and tsunami

Japan is deemed as one of the most disaster prone countries in the world, with approximately 10% of the world's earthquakes of magnitude 8.0 or greater having occurred there or in its vicinity during the 20th century.[8] Japan is also considered one of the best disaster-prepared countries in the world. Following the 2011 eathquake and tsunami, huge amounts of debris, equivalent to more than 100 years of normal solid waste generation, was impressively managed by the Government of Japan, with concerted efforts by Japan's Task Force on Disaster Debris Management and a later mission team of experts from all over the world, including from UNEP and USEPA. The post-disaster debris challenge, the response by the people of Japan, and key learnings from the event are reflected in the UNEP publication "Managing post-disaster debris: The Japan experience.'[9]

C&D waste management in Umbria and Marche following the 1997 earthquake[10]

The material management strategy for earthquake-affected zones between the regions of Umbria and Marche, Italy has been an interesting case study in managing C&D waste. After the earthquake of September 26, 1997, there were over 70,000 damaged buildings, which led to the issue of more than 1,580 demolition orders in the worst affected municipalities. A directive on the 'Removal of Debris, Demolition of Buildings and Material Recovery' was developed by the region of Umbria in 1998, based on an already existing policy of recycling C&D waste. The directive stated that the local body must utilize 50% of the materials recycled from the debris for reconstruction of public and/or private civil works. The agencies have to adopt a two-fold approach of assessment of the building types in the region, and a subsequent cross-checking with actual demolition activities occurring in the Nocera Umbra region. Based on the estimates of C&D waste generated, a number of recycling initiatives were planned. It led to the recycling of aggregates (189,362 cubic metres/year) and of metals such as steel. This approach was useful when another earthquake struck Italy's Abruzzi region on April 6, 2009.

Disaster waste, Nepal 2015

8 Suganuma (2006).

9 UNEP (2012e).

10 Text taken from Furcas, C. & G. Balletto (2012). Construction and Demolition Debris Management for Sustainable Reconstruction after Disasters. *Journal of Environmental Science and Engineering B.*

CASE STUDY 2

DISASTER WASTE MANAGEMENT FOLLOWING TYPHOON HAIYAN, PHILIPPINES[1]

Context and background

In November 2013, category 5 Typhoon Haiyan, known locally as Yolanda, made landfall in the coastal areas of the eastern Visayas, Philippines. It was among the most powerful storms ever recorded, with rainfall up to 30 mm per hour, winds reaching upwards of 315 km/h and massive storm surges up to 5 to 6 metres high. The typhoon caused massive damage and casualties in nine of the poorest regions of the Philippines, 44 provinces and nearly 600 municipalities,[2] affecting almost 15 million people, including 4 million who were displaced.

Typhoon Haiyan flattened villages, towns and cities, with the islands of Leyte and Samar among the worst affected. Over one million houses, government administrative and social service facilities were totally or partially destroyed, creating massive amounts of debris.

Although this disaster was truly catastrophic, it was not an isolated case, as typhoons and other disasters occur often. Waste management following a disaster goes through different phases, starting with the removal and management of high volumes of debris and construction and demolition waste to improve access, then moving to the rapid reinstatement of both healthcare waste management and municipal solid waste management and finally planning for the long-term sustainability of the system. This case study intends to share the work of, and lessons from, the Waste Management Component (WMC) implemented by UNDP as part of a coordinated international relief programme. The WMC not only undertook the necessary steps for managing the waste but also engaged the affected communities, generated local employment, supported local recyclers and, as a result, boosted local economies.

Overview of the WMC's post-Haiyan initiatives[3]

As part of the humanitarian response activated immediately after the disaster, the UNDP's WMC responded to the debris and waste management needs emerging in the provinces of Eastern Samar, Samar, Leyte, Biliran, Cebu, Palawan, Aklan, Capiz and Iloilo.

It focused on recovery and the disposal of typhoon debris and other waste, and on laying the foundations for long-term sustainable waste management through the interim and emergency rehabilitation of critical waste management infrastructure and strengthening the capacity of Local Government Units (LGUs) in the area of waste management.

Activities undertaken

Approximately 4 or 5 months after the disaster occurred, the WMC's activities transitioned from clearing debris in order to facilitate access for relief work to a broader range of waste management activities.

Debris clearing

The priority task in the first days, weeks and in this case, months, after the disaster was debris and residual debris clearing.[4] Basic public service buildings, including schools, hospitals, markets and local government buildings were prioritized in an attempt to restore a modicum of normality to life. Main streets were cleared

1 Case study prepared by Ainhoa Carpintero from information provided by Aiden Short and inputs from Ana Fonseca, Tim Walsh, and Thorsten Kallnischkies

2 OCHA (2014). *Philippines – Typhoon Haiyan/Yolanda. Environmental Assessment. Final report and Recommendations.* http://www.humanitarianresponse.info/system/files/documents/files/EFA%20Philippines%20Environmental%20Assessment%20June%202014.pdf

3 UNDP (2013). UNDP results: Typhoon Haiyan early recovery. http://www.ph.undp.org/content/dam/philippines/docs/Typhoon%20Haiyan/UNDP%20Results%20Typhoon%20Haiyan%20Early%20Recovery/UNDP%20Results_Typhoon%20Haiyan%20Early%20Recovery_6%20December%202013_final.pdf

4 The debris clearance activity was considered a lifesaving activity as it provided access to other humanitarian actors, enabling them to perform their work. In particular it also gave access to search and recovery teams including a UNDP-sponsored specialized cadaver recovery dog and handler team from the USA.

first. Once those were open to traffic, efforts were broadened to include outlying areas and smaller roads and villages.

Two aspects of the debris clearing efforts are noteworthy. First, the debris clearing adopted a **cash-for-work** system involving labour-intensive debris collection by victims of the typhoon. These workers received both an occupation to take their minds off the disaster and a basic wage to enable them to reduce reliance on food handouts and begin to purchase their own necessities in the local markets, injecting much needed money into the local economies.[5]

The second noteworthy aspect is that it made use of backhoes, loaders and other rented heavy machinery along with dumper trucks that facilitated the transport of debris from city streets to temporary and long-term dumpsites. UNDP prioritized the hiring of equipment from local businesses, themselves heavily impacted by the disaster, rather than exporting the revenue to larger businesses outside the disaster area. The local contractors also proved to be far more motivated to work hard despite long hours, as they were clearing their own city.

Those employed under the cash-for-work arrangements cleared the areas inaccessible to heavy machinery and collected the rubble and debris in temporary dumpsites set up close to the debris generation points or at accessible areas such as the sides of roads, where the debris would be picked up by small trucks and taken to temporary dumpsites, including two main ones within Tacloban City which acted as transfer stations for larger trucks. An effort was made to ensure the material could remain accessible to the affected population for the longest possible time, thereby enabling the temporary dumpsites to function as crude open-air waste transfer stations accessible to everyone. The workers were also encouraged to keep and use any materials they would consider of value. This encouraged and maximized resource recovery and recycling of the debris to lumber, shelter or products such as furniture.

The debris clearing activities finished by March 2014. The quantity of debris cleared across the eastern coast of Leyte was estimated at between 600,000 and 1,000,000 m^3, depending on the estimating technique used.

Temporary dumpsite, Tacloban Abucay

Healthcare waste management

Another key priority was the management of healthcare waste. Hospitals were overwhelmed in the immediate aftermath, as they themselves had been very badly hit and were expected to treat a huge spike in the number of patients. A major concern was getting the minimal public health infrastructure in place. UNDP, together with partners like Australian Aid and the Agency for Technical Cooperation and Development (ACTED) set up an emergency healthcare waste collection programme under guidance from WHO. This programme ultimately covered 14 hospitals in the affected area, including emergency tent hospitals run by such organizations as Médecins Sans Frontières, and ensured regular collection and safe disposal. It was implemented in phases. In the first phase, four trucks were used exclusively for hospitals to clear their backlogs, thereby facilitating the resumption of regular and immediate services. As backlogs were cleared, and pre-Haiyan procedures such as segregation within the wards were reinstated, efficiency savings made it possible for only one truck to service all the hospitals. As the programme was handed over, each LGU took control of the hospitals within its jurisdiction, and from that time onward, they each made their own arrangements.

The methods used for healthcare waste disposal also followed a phased approached based on safe burial practices (because of the lack of landfill in the area and the fact that incineration is banned in the Philippines due to the Clean Air Act). At first, dedicated pits approximately 4 m square and 3 m deep were dug on the dumpsite. These were covered after each load was received and then capped when full. Due to inefficient segregation and the large amount of material, this technique was unsustainable, with each pit lasting only 2 to 3 days. As a result, a separate section of the dumpsite was selected and cordoned off purely for hospital waste. Daily cover was also stockpiled there. This allowed for immediate cover after each load was received.

5 The workers were employed for 15 working days (usually in six-day weeks). They were usually selected by barangay captains (a barangay is the smallest unit of local administration), who were encouraged to assemble a fair breakdown of men and women within the cash-for-work groups.

Healthcare waste cell

Municipal waste management

Municipalities needed urgent assistance with collecting and safely disposing of growing piles of municipal solid waste that residents were dumping in the streets or at temporary dumpsites. Initially, with many civil servants looking after their own families, and many waste collection trucks destroyed or inundated with seawater, there were few resources to deal with household waste and the few accessible roads began to fill with rotting garbage and the like. In the absence of its own resources, Tacloban City hall requested that UNDP temporarily coordinate all actors and agencies in both municipal solid waste and debris clearance. Initial activities in this regard led UNDP to work closely with other local government units to establish routine household waste pickup and to make use of heavy equipment to clear out the temporary disposal sites that had cropped up around the city.[6]

Slowly the establishment, restoration and operationalization of solid waste management facilities got under way at the LGU level.

Municipal waste disposal

One of the biggest impacts of large-scale debris clearance was the impact it had on the operations and the capacity of the local landfill and dumpsites. As well as coordinating waste collection, UNDP was tasked by the Tacloban government with managing the huge number of trucks dumping both debris and municipal waste at the local landfill each day. From its pre-disaster operations volume of a few dozen trucks a day, the landfill was struggling to cope with over 250 loads a day at the peak of the cleanup efforts. UNDP began coordinating the traffic flows and opening new landfill areas. Later UNDP worked to build the capacity of the operators and improve the safety of the facility by upgrading from an open dump to a controlled dumpsite, significantly reducing impacts on the environment. Assistance[7] was also provided to the Ormoc landfill and dumpsites in other districts in Samar and Leyte such as Palo, Tolosa, Basey and Guiuan.

6 See http://www.undp.org/content/undp/en/home/presscenter/pressreleases/2013/11/27/waste-management-for-disease-prevention-a-pressing-need-in-tacloban/ for further information.

7 This assistance included support with heavy equipment and training and guidance for LGU employees working directly with site disposal.

Overview of the active cell without any management – before UNDP WMC intervention, Ormoc landfill

Active cell partially managed: waste compaction, daily cover and vents in place –during UNDP WMC intervention

Demolition of structurally unsafe buildings

After the initial debris removal phase, there were a large number of buildings that were still standing but heavily damaged, often to the point where they constituted a public health and safety hazard. In a region frequently struck by strong winds and earthquakes, many of these buildings were at risk of collapsing and injuring or killing people living in or near them.

UNDP WMC set up two demolition teams, one team of trained long-term cash-for-work employees for manual demolition, including substantial deconstruction, and another team of people operating backhoes. A register of buildings earmarked for demolition was created and then prioritized, after which time the activities were coordinated at the LGU and barangay level. The two teams razed several dozen buildings, both large and small. While the primary objective was hazard control and safety for residents, another objective was to make way for new investments. In several cases the landowners had a reconstruction budget but no demolition budget, thus effectively blocking any progress in reconstruction. Guiuan Municipal Hall, Balangiga Hospital and Tanauan secondary school all faced this budgetary situation.

The demolition of Guiuan Municipal Hall is an especially good example of the value of demolition activities supported by UNDP. Although the Guiuan LGU had a budget of over 900,000 USD for reconstruction, its lack of funds for demolition effectively prevented the project from going forward. The total demolition costs came to less than 20,000 USD, including both the deconstruction of window frames, air-conditioning systems, wiring, lighting, wood etc. done prior to demolition by the cash-for-work employees, and later the demolition itself with a backhoe equipped with a concrete crusher. Once the building was demolished, the LGU was able to invest in construction, thereby boosting the local construction economy and creating employment.

Sewage sludge disposal

Similarly in the initial stages of the emergency and recovery it became apparent that there was nowhere safe to dispose of the sewage sludge that was rapidly accumulating in evacuation centres. In partnership with Oxfam, UNDP, initially with the assistance of equipment provided by the Philippines military, and later with its own heavy equipment, dug sewage sludge ponds at the dumpsite that were lined with an impermeable membrane prior to receiving the cities' surplus sewage sludge.

Development of operational waste management structures

From April 2014, activities started focusing on improvements to or transformation of interim dumpsites, undertaking activities such as waste rearrangement and soil collection to serve as cover material (Tacloban, Palo and Guiuan); construction of infrastructure similar to engineered cells (Basey, Guiuan); management of an existing sanitary landfill (Ormoc); support of MSW collection to make it more efficient; and improvements to healthcare waste collection and disposal.

Resource recovery market

In the immediate aftermath of the disaster, small and medium "junk shops" or "scrap shops" were severely affected. Because they could not operate, opportunities for business and opportunities to recover materials and reduce pressure on final disposal sites were lost.[8] In April 2014 UNDP launched the waste reprocessors programme, a support programme to 26 junk shops across the affected areas to provide them with assets which may have been destroyed or which would allow them to adapt their procedures to the changing waste characteristics. The assets delivered ranged from vehicles to balers and proved to be highly useful for market regeneration.

One lesson learned from this activity is that the injection of cash at an earlier stage than the assets may have

8 It also prevented individuals from selling damaged possessions for much needed cash.

been more cost efficient, due to the liquidity and flexibility it would have provided to reprocessors. A proposal for dealing with future disaster waste situations includes providing junk shops with cash grants or loans if the shops can prove that the funds will be used to restart their business, taking a "cash-for-business" modality. It is anticipated that with these cash injections, junk shops may be able to cope better and conduct material recovery more efficiently.

Capacity building of local government units

Sixteen municipalities in Leyte and Samar received training on the entire waste management chain, on promoting appropriate techniques and technologies applicable in the local context of each target LGU and on improving occupational safety and health. They also received training for heavy equipment operators. In order to identify each municipality's training needs, the trainers first conducted a full-day assessment in each municipality. This process highlighted a need to move away from the LGU- (and even barangay-) centric focus on SWM solutions towards a more regional and inter-LGU coordinated approach. With this in mind, the training was organized geographically to promote regionalized solutions to both common and distinct challenges. The assessments led to draft recommendations to the Government of the Philippines to update the Ecological Solid Waste Management Act RA9003.

UNDP WMC also trained unskilled Haiyan victims in a six-week programme conducted by the Republic of Korea Army and the Philippine Technical Education and Skills Development Authority (TESDA), enabling them to qualify as heavy equipment operators. Forty-eight former workers in the UNDP cash-for-work programme were granted a stipend to receive training in operating backhoes, bulldozers, wheel loaders, forklifts and so on, and 48 Tacloban City Hall staff members were supported with travel costs to be able to participate in the same training. The programme was organized to certify formerly unqualified workers to operate heavy equipment and work in a qualified job. Moreover, the overall capacity of the LGU was improved in the area of heavy equipment operation in the event of a future disaster.

In addition to this training, an asbestos expert conducted capacity building and awareness campaigns for LGUs, the Regional Environmental Management Bureau, NGOs and UN entities in Samar and Leyte. Part of his activities included sampling destroyed and damaged building sites to verify the presence of asbestos, and to assist and train demolition teams managed by UNDP and NGOs in asbestos awareness and general occupational health and safety matters.

Enabling environment

The intervention on debris removal and waste management activities was possible due to the funding made available as part of an early recovery plan funded by the Governments of Japan, the Russian Federation and Saudi Arabia and by the Central Emergency Response Fund and UNDP.[9] The later part of the intervention was funded through a 3.5 million USD UNICEF project running from April 2014 to February 2015. The WMC was mobilized by the SURGE facility of UNDP's Bureau for Crisis Prevention and Recovery (UNDP-BCPR), now the Crisis Response Unit (UNDP-CRU). Additional staff were provided by MSB, the Swedish Civil Contingencies Agency, with funding from SIDA, the Swedish International Development Agency.

The project was also possible due to the close liaison of the UNDP WMC team with its main counterpart, the Environment Management Bureau (EMB), the implementing agency of the Philippines' Department of Environment and Natural Resources (DENR). As decentralization is an important aspect of Philippine governance, the team also worked very closely with each LGU. The team also worked in close coordination with other UN agencies, especially UNICEF, WHO and ILO, as well as local and international NGOs.

Although the waste management programme was scheduled to finish in February 2015, at the time this case study was prepared, a follow-up project in two municipalities was planned for the following 18 months to design and construct two sanitary landfills (SLFs), one in Basey and one in Guiuan. The new project is funded by the Korea International Cooperation Agency (KOICA).

Lessons to carry forward to future disasters

Things that worked well

- The combination of activities undertaken by the UNDP WMC made it possible to clear the debris caused by the typhoon, re-establish and improve waste management services and build the capacity of LGU officials dealing with waste management.
- The **cash-for-work programme** proved to be a great tool for and beyond the debris clearing phase since:
  - It was an effective way to boost the local economy through the injection of cash directly at the community levels, allowing local market regeneration.
  - It developed ownership of the recovery process.
  - It allowed quick engagement for fast debris clearing (one month following the storm, the main streets were open to traffic in affected municipalities). This facilitated the restoration of other priority services.
  - It accelerated the clearing of areas inaccessible to heavy machinery.

9 http://www.undp.org/content/undp/en/home/ourwork/crisispreventionandrecovery/projects_initiatives/Typhoon-Yolanda-Philippines/

- It enhanced on-site safety for workers and heavy equipment on dumpsites by acting as spotters and coordinators on site.

- The **demolition of buildings** proved valuable in providing:
  - Additional safety for residents living near structurally unsafe buildings.
  - Leverage for future investments in construction projects, thus creating income for small local construction enterprises and jobs for residents.
- The **clustering of LGUs** into groups for training had the benefits of:
  - Inter-LGU sharing of experiences and knowledge.
  - increased LGU stakeholder interaction with their peers.
  - Identifying, encouraging and enabling inter-LGU cooperation and asset sharing (e.g. promoting a regionalized approach to landfills which is advocated by the EMB).

Areas for future improvement

Implementation of the asset provision programme began several months after the disaster. Had the programme been in place in the immediate aftermath of the disaster, recovery of materials would have been maximized and even more support for the local market would have been generated. During the cash-for-work programmes in Yolanda, large amounts of recoverable material were being taken to dumpsites as the junk shops could process the material either not at all or not quickly enough.

Trucks that delivered relief goods left the disaster area empty when they could have been used for transporting recyclable materials out to the market in Manila. However, to implement this transport to other markets, the materials would have needed pre-treatment, such as compaction to reduce the waste volume, which the junk shops would not have been able to provide due to damage or loss of their machinery. More importantly, junk shops were unable to trade due to a lack of cash reserves. Small business owners preferred to stockpile cash due to the looming uncertainty of the economic situation (looting, lack of access to markets) rather than invest it in buying scrap. It is believed that an injection of cash to junk shops may have reassured and calmed the market, enabling junk shop owners to keep purchasing material, which in turn would allow members of the public to sell the remnants of their destroyed belongings. A cash-for-business programme may have the potential to address this problem, but the idea has yet to be tested.

Plans for dedicated carpentry workshops to make replacement school furniture, as UNDP had successfully implemented in Aceh, Indonesia, never materialized, in part due to the government desire for immediate removal of all debris from urban areas and in part due to early funding constraints.

UNDP preparing for the interviews with the waste pickers

TOPIC SHEET 11

FOOD WASTE

According to 2013 FAO estimates, approximately 1.3 billion tonnes of edible food – one third of the total produced for human consumption worldwide – is lost or wasted annually.[1] This food, if turned into meals, would be sufficient to provide nutrition and an adequate basis for health and well-being for over 2 billion people – more than double the official FAO estimate of undernourished persons in the world.

Moreover, this loss means that both the resources used and the environmental impacts sustained by the climate, land, water, and biodiversity in order to produce this food have been for nothing. For example, the global GHG emissions related to wasted food were estimated at 3.3 Gtonnes of CO_2 equivalent in 2007.[2] This level of CO_2 emissions would rank the wasting of food as the third largest 'country' in the world, surpassing every nation but the United States and China. The financial loss due to food wastage is about USD 750 billion based on producer prices alone.

It is important to note that these losses occur at different places along the supply chain and for different reasons in developing economies than they do in developed ones, and therefore require different approaches to tackle. In developing economies, almost 80% of the losses are on-farm and in transportation and processing, due to for example the unavailability of selling outlets or inadequate packaging and storage. In developed economies, almost 80% of the losses occur in retail, catering and in the home, due to various reasons including extremely stringent cosmetic standards for fruit and vegetables, relatively low prices for consumers, strict adherence to 'use by' and 'best before' dates despite inadequate understanding of the meaning of such labelling, lack of knowledge on food safety and 'use by' dates set earlier than necessary due to fear of litigation. Not surprisingly, the further along the supply chain the loss occurs, the worse its environmental impact, and indeed the financial losses, as more handling – with greater input of resources – has taken place to get it there.

Preventing this wastage either directly or through redistribution to those in need would undeniably have a major positive effect on global food security.[3] In addition, food waste could be utilised through a variety of applications, including conversion to feed for animals and even industrial chemicals. Food waste reaching the SWM system can be converted into fertiliser, organic matter, fuels and energy, as represented in the Figure below.

Figure 1 Food waste hierarchy

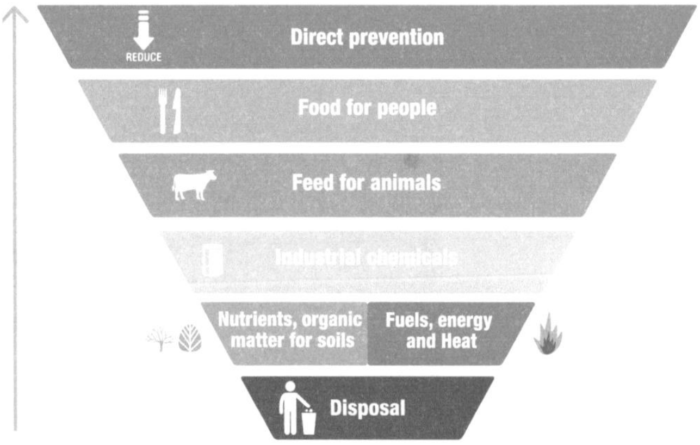

1 FAO (2013), listed in Annex A, Chapter 3, Other waste streams, under Topic Sheets, Food waste.

2 This estimate does not take into account GHG emissions due to land use changes. If land use changes are also incorporated, GHG emissions would be 25-40% higher.

3 This statement is made from the perspective of waste and resource management. Economists will rightly argue that significant prevention of food waste would potentially impact global food markets and thereby food production.

In reflection of its importance, the prevention of food waste is included in the UN Sustainable Development Goal (SDG) 12, 'Ensure sustainable consumption and production patterns' as an explicit sustainable development target 12.3: "By 2030, halve per capita global food waste at the retail and consumer levels and reduce food losses along production and supply chains, including post-harvest losses'.[4] Food wastage is also against social norms of behaviour and traditional cultures. Therefore it is no surprise that currently UN agencies, governments and civil society are taking action to prevent food waste through awareness raising campaigns, regulatory interventions, community action events and exchanges of ideas and experiences around the world. For example, the global Think.Eat.Save campaign of the Save Food initiative is a partnership between UNEP, FAO and Messe Düsseldorf in support of the UN Secretary-General's Zero Hunger Challenge.[7] Love Food Hate Waste is an initiative by the UK government implemented through WRAP.[8] 'Feeding the 5000' is an NGO with global reach that works with partners to organize events where meals are prepared for thousands of people from ingredients that would otherwise be wasted.[9] Large retail companies are also responding to pressure from civic society and gradually getting on board.

The Global FoodBanking Network (GFN)[8]

Dedicated to alleviating hunger and improving nutrition around the world, the not-for-profit organization GFN has been instrumental in establishing food banks in over 25 countries, including Brazil, Canada, Ecuador, India, Mexico, Namibia, Nigeria and Russia. GFN provides training, shares experiences, builds relationships with the food industry and other stakeholders and empowers communities to deploy food banking to feed the hungry. In 2013 the food banks associated with GFN distributed over 400,000 tonnes of food.

Food banks contributing to food and nutrition security in Belo Horizonte, Brazil[9]

The city of Belo Horizonte, Minas Gerais State, Brazil, has been a pioneer in urban governance on food security ever since 1993, through a dedicated agency, the Secretariat for Food Policy and Supply (abbreviated as SMASAN[10] or SMAAB in Portuguese). As one of the six main 'lines of work' within the comprehensive programme, the City Food Bank was established in 2003 to contribute to the prevention of food waste while providing healthy food to those in need. Organized and carried out in collaboration with the Municipal Secretariat for Urban Sanitation, food is collected from partner establishments, including street vegetable markets, green groceries and supermarkets, and then goes through a rigorous process of selection, processing, and storage, with sanitary norms strictly observed. The processed food is distributed to more than 100 institutions registered with the City Food Bank, including day-care centres, homes for the elderly, orphanages and shelters, benefiting about 15,000 people.

Indian weddings feasts feed the poor as well as the guests[11]

Annakshetra Foundation was formed by the NGO Centre for Development Communications (CDC) in Jaipur, Rajasthan State, India, in 2010, with the aim of minimizing food wastage by effectively utilizing excess food leftovers from weddings, parties, restaurants and temples. As of May 2015, the Foundation has developed a network of 1,800 donors including restaurant owners, caterers, sweets makers, temples and individuals called *'Annadoots'*.[12] It conducts awareness campaigns to prevent food wastage, collects excess food from donors, stores and tests the food and then distributes it to those in need, if the food is found fit for human consumption; otherwise it is sent for composting. Some 5,000 tonnes of high quality food have been recovered and approximately 200,000 persons have benefited from the Foundation's activities so far, including manual labourers and slum dwellers. In support of the global Think.Eat.Save campaign and inspired by 'Feeding the 5000', Annakshetra organized a meal in 2013 where almost 10,000 people were served; in April 2015 over 15,000 people were served under one roof using the surplus food collected the previous night. Many organizations are interested in applying the Annakshetra model in their own cities, including organizations in Delhi, Ahmedabad, Allahabad, Hyderabad, Gurgaon, Surat and Mumbai.

4 https://sustainabledevelopment.un.org/sdgsproposal
5 http://www.foodbanking.org
6 Details on the programme can be found in Rocha, C. and I. Lessa (2009). Urban governance for food security: The alternative food system in Belo Horizonte, Brazil. and at http://www.fao.org/fileadmin/templates/FCIT/Meetings/Africites/presentations/WorldFutureCouncil_experience-Belo-Horizonte.pdf and http://worldcongress2012.iclei.org/fileadmin/templates/WC2012/Documents/Presentations/A5-Duffles.pdf
7 http://www.thinkeatsave.org
8 UK Waste & Resources Action Programme. See Case Study 3 on food waste.
9 http://feedbackglobal.org/campaigns/feeding-the-5000/
10 www.pbh.gov.br/segurancaalimentar
11 See http://www.annakshetra.org and Hartley, M. and C. Walker (2012). Battling Hunger with Entrepreneurship at http://www.forbes.com/sites/morganhartley/2012/12/28/battling-hunger-with-entrepreneurship/ and based on input from Dr.Vivek Agrawal, CDC, Jaipur.
12 The word is derived from anna meaning 'food' and doot meaning 'messenger' or 'ambassador'.

CASE STUDY 3

LOVE FOOD HATE WASTE: CHANGING BEHAVIOUR AND REDUCING FOOD WASTE

Online interview with Emma Marsh

Context and Background

Food waste arises during various stages in the life cycle of food production, including agriculture and requires focused attention from both resource management and food security perspectives.[1] A particular issue in high-income countries is waste of food that could have been eaten. A 2006 survey showed that one third of the food bought by households in the UK was thrown away, and at least half of that could have been eaten.[2] In response, 'Love Food Hate Waste'[3] (LFHW) was started in the UK in 2007 as a national campaign led by WRAP[4] to change people's behaviours to tackle this issue.

Food waste has now attracted priority attention around the world, and a target to halve per capita food waste at both the retail and consumer levels by 2030 is included in the Sustainable Development Goals.[5] Variations of LFHW are now being rolled out globally, including through the UNEP and FAO campaign 'Think, Eat, Save – Reduce your Foodprint'.[6] This case study showcases experiences and lessons learned from Love Food Hate Waste over eight years in the UK and how a national initiative scaled up to become global.

1 See Topic Sheet 11 on food waste, found after Chapter 3.
2 http://www.wrap.org.uk/sites/files/wrap/FoodWasteResearchSummary-FINALADP29_3__07.pdf
3 http://www.lovefoodhatewaste.com/
4 www.wrap.org.uk WRAP – the Waste and Resources Action Programme – is a not-for-profit company set up in 2000 by the four UK national governments to facilitate the growth of the UK's MSW recycling rates from less than 10% towards the initial 25% target, which later became 50%. In terms of its roles within the waste management system, WRAP is an example of a 'Change agent' – see Section 4.8, including Figure 4.7.
5 Goal 12.3 as of July 2014.
6 See www.thinkeatsave.org

Some of the key achievements of LFHW include a reduction in avoidable food waste across the UK by 21% and corresponding savings in household expenses, carbon emissions and water consumption. Local authorities are estimated to have saved around 85 million GBP in avoided gate fees[7] and landfill charges in 2012 alone as a result of these reductions. While these are great achievements, progress has slowed since then, and much more still needs to be done. A new strategy was developed in 2013, following the latest comprehensive evidence review, and this is now being rolled out to address this slowdown and increase impact. Was it a smooth ride all these years? How has the campaign evolved? And what are the lessons for replication? These are some of the key inquiries which WRAP team leader Emma Marsh explains in the interview.

How does LFHW work?

When designing the campaign and launching the pilots, LFHW comprehensively included all four 'E's (enable, encourage, engage and exemplify),[8] which are necessary to achieve sustainable behaviour change.

An overarching initiative is our cascade training programme. We have a team of six (three part-time) who go out and deliver free training sessions to any organization – this is a standard training package and we've trained hundreds of organizations ranging from the retailers, to housing associations, to global international IT businesses, to community groups, to policy makers. The training focuses on raising awareness of the issue of food waste, helping people to recognize that this is an issue that affects them personally and help them recognize the value in reducing wasted food. It provides

7 Gate fee or tipping fee refers to the charge per tonne of waste delivered to the waste management facility (gate)
8 For further information on the 4Es, see Section 4.6, including Figure 4.4. The 4Es model was introduced in DEFRA (2005), *Securing the Future: Delivering UK sustainable development strategy*, available at https://www.gov.uk/government/publications/securing-the-future-delivering-uk-sustainable-development-strategy

Know your dates
LOVE FOOD hate waste

Checking dates saves money and lets us enjoy food at its best, but do you know what they mean?
Write down what you think each of the following date labels mean.

USE BY:

BEST BEFORE:

SELL BY:

DISPLAY UNTIL:

What could you do with food that is approaching its 'use by' date?

wrap

insight into what is already being done across the UK from farm to fork to reduce food waste and offers simple ways to take action at home. There are seven interactive activities which can be run depending on the group, ranging from 'know your dates' to 'the freezer is your friend' to 'Fresher for Longer'. We provide attendees with a trainer support pack, and if they are looking to run sessions themselves we provide activities and tools such as measurers, bag clips etc. to facilitate the sessions. We 'encourage' through motivational information and through being part of the LFHW campaign with a certificate of achievement and ongoing on-line support and we 'exemplify' through sharing each others' successes and challenges etc. We send out a monthly electronic newsletter that goes to all attendees and each issue features news and updates from fellow delegates sharing what they have been doing since receiving the training. We also share stories and updates that are sent into us via Facebook, Twitter and the website. On average for every person we train he or she goes on to share that information with 38 others!

In 2014 we developed a new training package, 'Save More', which could be delivered independently of WRAP and Love Food Hate Waste and which is free to download for anyone in the UK.

How does LFHW benefit various stakeholders?

We are helping local government to run local campaigns to help reduce the amount of good food and drink going to waste, saving them money from avoided disposal costs, to grassroots community groups who are looking to address food poverty (helping people to make their budgets go further by making the most of the food they buy), and cookery and food skills which have been lost over the years, to our retail and brand Courtauld[9] partners, who want to help their customers waste less, making their budgets stretch further using a trusted brand such as Love Food Hate Waste.

What have been key challenges and how were they met?

I've been working on the Love Food Hate Waste campaign since 2007 and we've had many challenges along the way, but also a lot of successes. Food waste prevention is incredibly challenging – it's made up of many complex and interrelated behaviours, most of which have little or nothing to do with the act of throwing away. We always ensure that we don't just ask people to 'not waste food', as that's so broad it's actually meaningless. We look at the multitude of barriers, attitudes, habits, motivations and address them while at the same time identifying what key behaviours will have the biggest effect for people. We know that time is one of the biggest barriers and our new call to action #DoOneThingDifferently which is being used as part of the '10 cities' campaign[10] seeks

9 Through the 'Courtauld commitment', stakeholders in the UK food and drink supply chain sign up to voluntary commitments to reduce both packaging and food waste. The first agreement was signed in 2005, and Courtauld 3 runs from 2013-2015. Their progress in meeting the targets is given wide publicity and is important for a company's public image.

10 In 2014 LFHW was taken to 10 UK cities following the local campaign in West London in 2013. The '10 cities' campaign will run until the end of March 2016.

to address this. There are many opportunities during the journey of an item of food from supermarket through to the home when it could be saved – from someone buying only what they need, to freezing any extras before they go past their use-by date, to using up leftovers in another meal – the campaign highlights that you don't have to be great at everything – you just need to find the one thing that works for you and make it a habit. The call to action has been welcomed by all stakeholders as a way to engage people without it being made to sound difficult and 'not for them'. Reducing food waste is not a visible activity – you don't tend to see people 'not' wasting food – and so it is challenging to make it a social norm. We have to make it visible in each of the '10 cities' to make it 'normal' through waste reducing cookery demonstrations, cookery classes, real-life stories and more.

What are the (three) most important tips for planning a new campaign?

1. **Do your research – understand the real issues, barriers and motivations along with external impacts such as economic and social pressures as well as the retail environment.**

In West London we carried out surveys before the campaign was fully developed to really understand what people's particular motivations and barriers were so that we could tailor the campaign activities to address those issues but also evaluate the impact at the end. Those households which had reported to have seen the Love Food Hate Waste campaign and to be doing something different as a result reduced their avoidable food waste by 43% (as measured through waste composition analysis).

2. **Understand where people are, not where you think they should be.**

It was originally suggested that we should use environmentally-focused communication[11] in West London as this was felt to be the right thing to do. However research showed that the environment featured low on a list of motivations for taking action to reduce food waste. Therefore the money saving communication was used to motivate people with the environmental information being shared once they were engaged.

3. **Design an integrated and comprehensive campaign including enabling and encouraging people (awareness and information), direct engagement, and exemplification (being part of something bigger).**

Once we had the right messaging in West London we made sure we had the right mix of activities from advertising, to free training to give people all the information they needed, to a champions network where they could feel part of something bigger, to cookery classes to build skills and confidence in the kitchens to sharing case studies from individuals, businesses and government.

What pitfalls should one be aware of in planning a new campaign?

- Never design a campaign for yourself – think about the audience. In our 10 cities campaign we have tailored the campaign to each city based on insights from partners on the ground and wider research.
- Make sure you can deliver on your promises and prioritize activity – we have had many challenges in delivering the '10 cities campaign', including losing key staff at key points in the project and delays in receiving specifications from partners, but at all times we have ensured that projects are delivered.
- Work closely with partners as the messenger is key – sometimes you won't be the right person/organization to talk to a certain audience but one of your partners might be. We carry out stakeholder mapping before we start any campaign, both internally but also with key partners, to look at who we should be working with and why.

Which of your posters are you most proud of?

Our 'food lovers' faces' posters for the fun and the fact they make people look twice and want to find out more. Also, our 'six top tips' posters for their simplicity and action-based content – each poster provides a specific tip and call to action rather than just awareness raising.

11 WRAP has developed several alternative 'communication themes' that can be adapted to the needs of a local campaign, such as 'environmentally focused', 'money and time saving', 'portion planning, storage and recipe tips'. See the section on featured communication types at https://partners.wrap.org.uk/campaigns/love-food-hate-waste/

What have been major benefits of the campaign?

- From 2007 until 2012 avoidable food waste was reduced by 1.1 m tonnes (from 5.3 m to 4.2 m tonnes). The UK could halve avoidable food waste by 2025 compared to when WRAP started work in 2007. Both avoidable and unavoidable food waste was reduced, with avoidable food waste decreasing by 0.17 kg/household (hh)/week (14%) and unavoidable food waste by 0.27 kg/hh/week (24%).
- This decrease in food waste was accompanied by a shift towards behaviours that are associated with lower levels of food waste, such as planning meals, using leftovers as a meal or as part of another meal, making shopping lists and wrapping items (e.g. cheese) appropriately to optimize their shelf life. Thus, the campaign successfully encouraged people to change key targeted behaviours, which then reduced their food waste.
- Households in West London would have saved millions of pounds by not wasting this good food and drink. There was a reduction of 0.4 kg per household per week – a 15% reduction in food waste – between the pre- and post-campaign analyses. The reduction in food waste overall could have saved around 1.3 million GBP (over a 12-month period) in disposal costs (including gate fees and landfill tax). Thus this campaign saved the local authorities approximately 8 GBP for every 1 GBP spent on the campaign/ behaviour change activities.
- The greenhouse gas emissions associated with the reduction of avoidable food and drink waste in West London is equivalent to 20,000 tonnes of CO_2e prevented.
- Food Waste Champions generated a total of 1,373 volunteer hours across West London, which had an economic value of 21,507 GBP.

What has been the global experience so far with LFHW campaign?

The campaign has been rolled out globally, helping to tackle food waste. We have been working in New South Wales in Australia since 2009-2010, across the EU and with UNEP and UN-FAO for their global campaign under the catchline 'Think, Eat, Save – Reduce your Foodprint'. Love Food Hate Waste is also part of the EU-wide FUSIONS project[12] and delivers across the world under franchise,[13] sharing our knowledge and expertise, building on the brilliant work already being carried out across the UK by dedicated individuals and groups. By working together we can make a real difference globally.

Love Food Hate Waste experiential tour across 10 cities of the UK 2014

12 LFHW worked across the EU as part of its role in the FUSIONS project in 2014. http://www.eu-fusions.org.

13 Metro Vancouver, Canada is preparing to launch Love Food Hate Waste in spring 2015 (carried out under franchise).

CASE STUDY 4

MILAN – THE FIRST METROPOLIS IN EUROPE WITH INTENSIVE SOURCE SEPARATION OF FOOD WASTE[1]

Context and background

In Italy, the most important piece of waste legislation, Legislative Decree 152/2006, not only shaped the national waste management system but introduced separate collection and recycling targets of 65% by 2012. As with EU-wide recycling targets, these high levels are only achievable if food waste is segregated at source and collected separately.

Many smaller towns and cities in Milan Province have been conducting Source Segregation of Organics (SSO) for at least 10 years. However, Milan is the first major city anywhere in Europe to introduce such collections across the whole city, including the 'more difficult' high-rise multi-occupancy buildings and high-density neighbourhoods. To help reach this target, Milan extended source separation and the separate collection of food waste to residential areas. This was previously only available to commercial activities, such as restaurants, schools, supermarkets and hotels. The city has 1.3 million inhabitants, with a population density higher than 7,000 people per square kilometre and more than 80% of its apartments being part of multi-family buildings. The total annual municipal waste generation in the city is 0.66 million tonnes, of which 25% to 30% corresponds to the organic fraction. This case study focuses on intensive source separation and the collection of residential food waste in Milan to divert the food component from landfill to mitigate methane (and other GHG) emissions and recovery via composting and/or anaerobic digestion.

1 Information provided by CIC, the Italian Composting and Biogas Consortium. Rome, with inputs from Marco Ricci-Jürgensen, Italian Composting and Biogas Association. See www.compost.it

About the initiative

In November 2012, Milan started source separation and door-to-door collection of food waste to residential areas. This measure was ordered by the Mayor and supported by a strong public awareness and information campaign. The door-to-door collection and transport are undertaken by AMSA (Azienda Milanese Servizi Ambientali, A2A Group),[2] a publicly owned company. The collection schedules were designed to reduce the impact that waste collection trucks have on city traffic. All pick-up services have been scheduled between 5.50 a.m. and 11.30 a.m., with the highest volume traffic areas in the city centre being served before 8.15 a.m.

Households have been given a ventilated kitchen pail (biobins) to be used in the kitchen, along with a roll of certified compostable bags to last for the first few weeks.[3] Food waste is collected in those bags, or in compostable shopper bags commonly given to the customers in shops, and placed in a kerbside wheelie bin, which is picked up twice per week (See *Table 1*). There is an external monitoring task force set up by the municipality which randomly checks bins shortly before collection.[4] A fining mechanism is in place for biobins (i.e. buildings) that fail to comply with prescribed separate collection rules. Fines are used for the separate collection budget. In 2013, 50,000 fines[5] were collected, each fine amounting to 50 euros. In addition, multiple methods of awareness raising are used, including public

2 AMSA is in charge of all aspects of collection and transport of all types of MSW arising in Milan. Additionally AMSA manages its own energy from waste (EfW) plant, a sorting facility for glass (packaging) and a sorting facility for residues from street sweeping.

3 Households can also use compostable bags that they receive for free when shopping. (Italy has a ban on plastic bags, so only compostable bags or re-usable bags are available.) There is also the option to purchase bags.

4 This task force is made up of representatives of a social cooperative.

5 This is for both improper separation of waste and use of wrong bags.

advertising, a free computing application, newspaper and radio plus TV advertising and a toll-free hot line. Additional awareness activities have been held for those areas where the quality of collection is lower than average.

Table 1 Source separation scheme for food waste in Milan

Kitchen	10 litres vented caddy
Bags	Compostable according to EN 13432 standard
Collection	Wheelie bins 120 or 240 litres in size
Collection frequency	2 per week (households)
	6 per week (for hotels, restaurants and cafeterias)

The initiative that was initially implemented in just one zone of the city in 2012 was expanded gradually to all four of the city's zones by 2014. By December 2013 food waste was being collected in three quarters of the city, reaching 77% of the population, and accounting for nearly 1 million inhabitants. Thus Milan became the first metropolis in Europe to have extensive Source Segregation Organic (SSO) collection. By the end of June 2014, food waste collection had been extended to 100% of the population (see *Figure 1*).

Figure 1 Areas with food waste collection

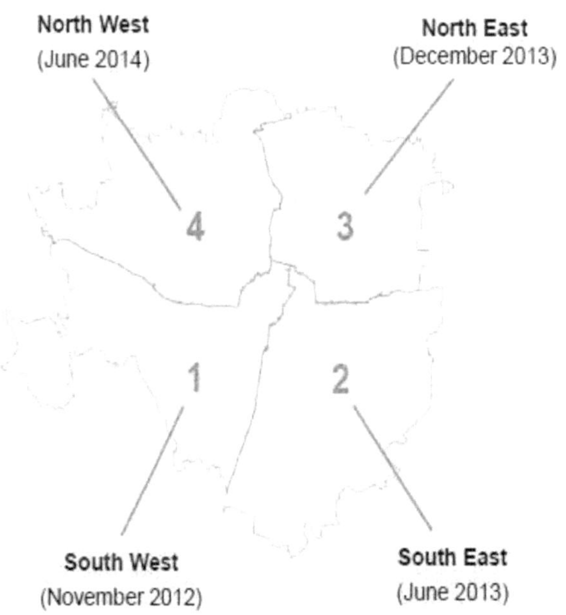

Figure 2 shows the fast growth of collected food waste for each of the three sectors in which separate collection is under way (since November 2012, June 2013 and December 2013, respectively). It clearly shows that collection becomes fully functional in about 4 to 6 weeks and about 1.7 kg is collected per inhabitant per week, with an average annual collection of 91 kg per capita. These amounts also include food waste collected from the hotels, restaurants and the catering sector. A detailed assessment of MSW composition indicates that about 84% of the total food waste produced in Milan is collected by separate collection, while about 16% still remains as residual waste.

Figure 3 appears to suggest that contamination (presence of non-compostable material) is less of a problem in downtown and outer areas once the scheme has been under way a few months, but becomes more of a problem over time in 'popular dwellings' (buildings for low-income families). Popular dwellings are more critical (not only for SSO) and are monitored constantly and the collection with average contamination values (in that area) up to 5-6% is still acceptable at the AD composting plant. Separate collection at high-rise buildings poses challenges everywhere and not only in popular dwellings.

The cost for MSW management service is borne by the Municipality of Milan. Each household and commercial activity has to pay a waste fee, according to the dimension of the apartment and the number of residents. For commercial activities, the fees depend on the type of activity and area used for the activity.[6]

AMSA floated a tender[7] for the transport and recycling of food waste in 2012. A private company (Montello) won the tender and is paid a gate fee of 74 to 80 euros per tonne of waste delivered for treatment to their AD composting facility for the production of biogas and compost. This facility serves 2.5 million inhabitants and is located 60 km from Milan. The produced compost[8] is sold for use in agriculture and fetches modest revenues compared to the gate fee and biogas is transformed into heat and electricity, of which electricity in excess of captive consumption is sold to the Italian National Electricity Grid.

As reported by AMSA and the Municipality of Milan, even after introducing the separate collection of food waste, the total expenses incurred by the MSWM system did not change. In practice the amounts of workers and trucks for residual waste collection were reduced and assigned to the separate collection of food waste. Disposal costs[9] in Milan province vary between 94 to 100 euros per tonne.

6 A 3-person family living in a 90m^2 flat paid about 300 euros per year in 2014. Commercial activities are divided into 30 categories, each with a different cost per square meter.

7 While AMSA is responsible for door-to-door collection and transport, it can tender out or sub-contract some services.

8 The quality of this compost is certification by CIC, the Italian Composting and Biogas Consortium.

9 Gate fee for disposing of waste, not including the collection costs.

Figure 2 Weekly food waste per inhabitant collected in 3 of the 4 waste management zones (Primaticcio, Zama, Olgettina) within Milan City in 2012 and 2013

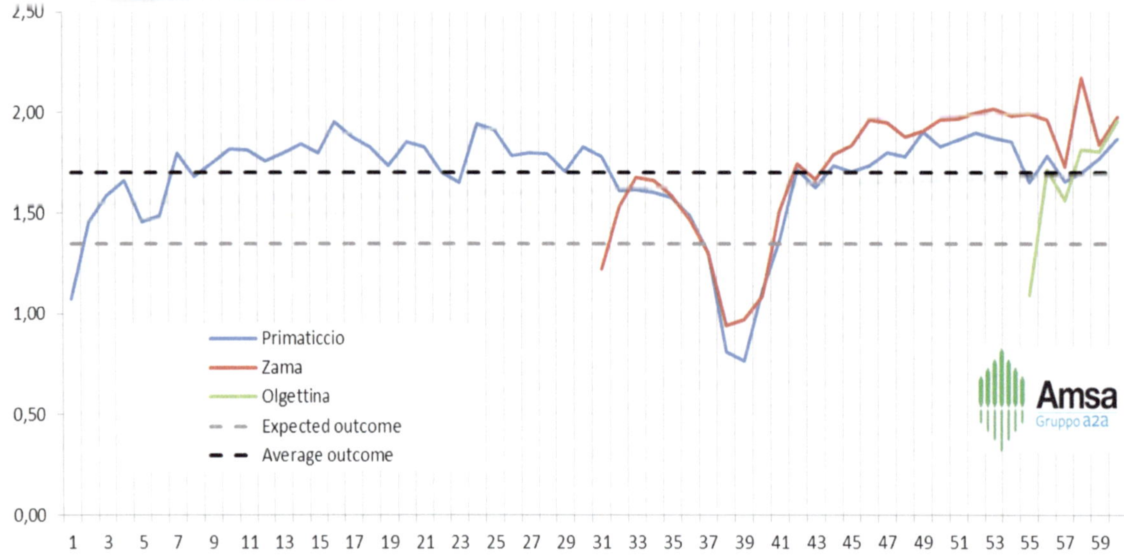

Source: AMSA (Azienda Milanese Servizi Ambientali, A2A Group)

Figure 3 Quality of the food waste collected in Milan. Vertical bars indicate the standard deviation from the mean value of the biowaste collected

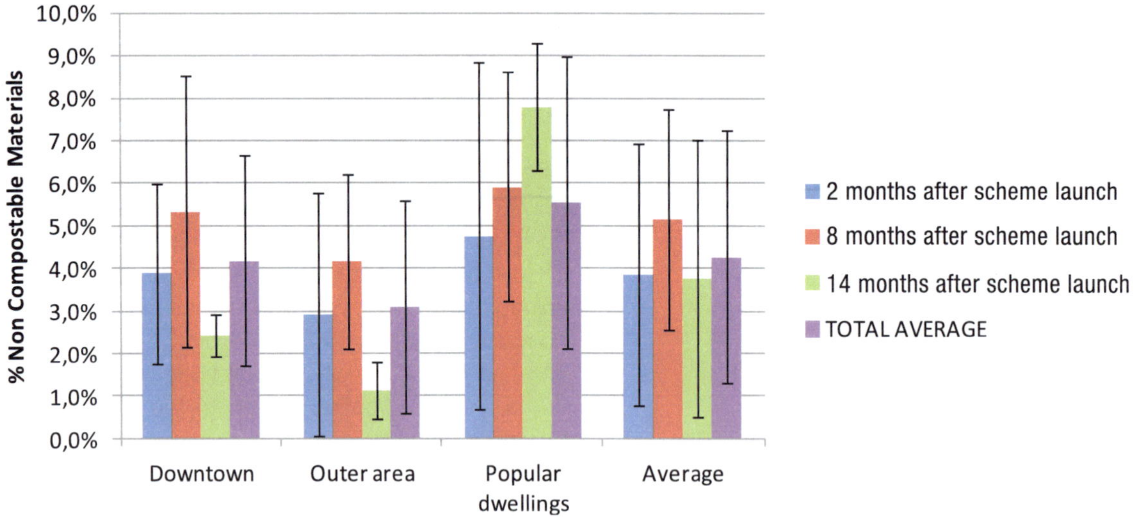

Note: Vertical bars indicate the standard deviation from the mean value of the biowaste collected.

Source: AMSA (Azienda Milanese Servizi Ambientali, A2A Group), waste composition analyses performed by the Italian Composting Association

The importance of good public communications

Initial challenges were the on-time distribution of kitchen caddies and biobins, gaining the involvement of the population and continuously monitoring the quality of the collected food waste. To increase participation and ensure sustained attention to the topic, intensive information activities including repeated press releases were begun one month prior to initiating separate collection. The initiative was rolled out in about 18 months (1 waste area every 4 to 5 months) to allow continuous visibility for about two years, which helped in reinforcing the concept among the residents and other users. While a fully user-friendly scheme, with generous delivery of sorting tools (the vented kitchen bin + biobags) and convenient frequencies for emptying the biobins, acts as an incentive for participation, quality checks at the biobins and fines for defaulters maintain the seriousness of the initiative in the eyes of the public.

Good teamwork between the Municipality of Milan and AMSA has enhanced the extent and visibility of all initiative of communication and PR activities.

Results achieved

The quality of the food waste collected is high. The waste composition analyses performed by the Italian Composting Association CIC aimed at quantifying the amount of non-compostable materials (NCM), show that, on an average, NCM are 4.3%[10] of the collected material, putting Milan in line with good practice. Apparently the amount of NCM does not depend on the collection zone or the time elapsed from the beginning of the new collection scheme (see *Figure 3*). Once contamination starts to rise, control is increased, fines are applied and more information is delivered in critical areas by means of leaflet distribution and advertising in local news media.

Customer satisfaction shows that 90% of citizens are satisfied or sufficiently satisfied by the new collection scheme and participate daily in separate collection.

The new system collects around 91 kg of food waste per capita per year (1.7 kg/capita per week) that would otherwise have gone to landfill. The city's residential and commercial food waste, which are collected separately, have resulted in 120,000 tonnes (about 18% of total generation) of waste per year being diverted from landfill

10 Non-compostable materials are all non-compostable waste delivered together with food waste, like glass, metals, multilayer packaging, etc. Good results in separate collection in Italy, according to the Italian Composting and Biogas Consortium, should stay below 5% NCM, hence Milan's result is quite good. Excessive amounts of NCM (above 15%) can affect the ability to recycle the food waste.

and being treated to generate 5.4 MW[11] of energy, leading to total savings of 8,760 tCO_2 per annum.[12]

Producing compost from the food waste also helps the city to avoid methane emissions and close the carbon and nutrient cycle. About 15,000 tonnes of high quality compost are generated from Milan's food waste and sold for use in agriculture.

Lessons learned

The most important lesson learned is that intensive source separation can be introduced with good results in high-density areas, with population up to 7,000 persons per square kilometre, and that the amounts collected and the quality of the food waste (in terms of contamination by plastics and other non-compostables) is acceptable at industrial composting and AD plants.

Full engagement of local authorities and waste collection companies is necessary to bring about, extend and continue actions and raise public awareness and consciousness about correct methods of separating waste. This can lead to good participation and high satisfaction towards the services provided to customers.

11 The anaerobic digestion facility treats 285,000 tonnes overall, including waste from surrounding areas, and generates 12.8 MW of electricity in total. Almost all of that green energy is used to power a nearby plastic reprocessing plant.

12 Calculated by Italian Composting and Biogas Consortium based on DEFRA UK calculation tool (2011).

CHAPTER

4

WASTE GOVERNANCE

A particular focus of the GWMO is waste governance, starting here by looking at policies, responsibilities and partnerships. The chapter showcases a range of policy instruments that are applied around the world, in relation to the policy goals they seek to achieve. The chapter also highlights the local circumstances under which these instruments are deployed.

Before embarking on an exploration of individual instruments, the Chapter starts with a Summary (4.1) and an introduction on policy, governance and strategic planning (4.2). Subsequently, three main categories of policy instruments – legal, economic and 'social' – and their implementation are presented (4.3 through 4.6). The next two sections discuss stakeholders and their interactions (4.7), and highlight the roles of governmental institutions in particular (4.8). Finally, lessons learned from experiences worldwide are presented (4.9).

© George Clapton

4.1 SUMMARY OF THE CHAPTER – KEY MESSAGES ON WASTE GOVERNANCE

- Waste governance starts from strategic goals and guiding principles. It addresses policy instruments, institutional arrangements and capacities and stakeholder interactions.
  - Formulate explicit and clear goals; plan how to achieve them. Formulate the goals carefully – the goals express the very purpose of the system, and thus determine its elements and the way they interact. Waste prevention is an important goal, not only for developed countries but also and particularly for developing countries.
- Communicate, facilitate involvement, engage in exchange with the actors in the system. Spend time on building citizen and stakeholder engagement into policy-making processes. Make sure that those who are key to the success of the system are on board.
- Good governance requires consistency in decision-making applied to all levels and all dimensions of the system. Make sure that the strategic goals are reflected in choices made throughout the waste system, not only those related to policies and institutions but also, and particularly, technology selection and financing structure.
  - Take charge of the technology selection process at the level of governance, not at the level of technical management. Understand the function and purpose of technological options, rather than just their features, and study their performance and real costs before making a selection.
  - Aim for financial sustainability of the system. As a part of it, aspire to cost recovery in relation to services with directly visible benefits to the users, such as waste collection. In developing countries, the capital needed for the construction of processing or disposal facilities will require other sources of financing such as the central government, while operation and maintenance could be financed locally.
  - In cases where waste collection systems or facilities were developed through donor-financed projects, make sure to establish appropriate cost-recovery mechanisms for their operation for the period after the project – and donor funding – end.
- Experience suggests that an effective waste system calls for a continuous use of three categories of policy instruments in a coherent mix: (a) 'direct regulation', comprising legislation accompanied by its keen enforcement, (b) economic instruments, providing incentives and disincentives for specific waste practices and (c) 'social' instruments, based on communication and interaction with stakeholders.
  - While policy instruments had previously focused on waste generators and the waste sector, they are increasingly focusing on producers, including manufacturers, brand owners and importers, in consideration of the entire life cycle of materials and products. This is part of a broader societal trend toward sustainable consumption and production.
- Direct regulation serves to protect common interests in a society, such as public health and the environment. A combination of legislation (laws and derived regulations) and its credible and consistent enforcement has resulted in the waste industry as we currently know it – otherwise waste would be dumped at the lowest cost.
  - Laws and regulations define basic concepts such as waste and hazardous waste, clearly allocate responsibilities, set standards of environmental performance of facilities and operations, and state sanctions in cases of non-compliance and violation.
  - A 'direct regulation' approach is based on information and monitoring; therefore, a commitment to reporting and inspection in combination with good and continuous data management are essential.
  - In order for businesses – both waste generators and the waste management industry – to plan their operations and investments into the future, they need 'regulatory certainty'. This entails the passing of coherent and clear legislation (laws and derived regulations, including incentive and sanction measures) and fair and consistent enforcement.
- While monitoring, inspection and enforcement tend to be costly in terms of institutional capacities required, 'direct regulation' will remain a preferred policy instrument in situations involving high risks to society and serious consequences of non-compliance, such as the handling of hazardous waste.

- A *relatively* recent addition to waste legislation addresses waste reduction and recovery of resources from waste, in terms of materials, nutrients and energy. These laws and regulations are as much a response to the increasing costs of waste management as to the value of the resources recovered.
- Economic instruments serve to steer stakeholders' behaviours and practices towards strategic goals through market-based incentives and disincentives. For example, a pay-as-you-throw (PAYT) charging system for residual (mixed) waste will reward people for segregating their waste; taxes on landfilling or incineration will discourage opting for these methods; fiscal benefits will encourage private companies' investment in SWM. When in doubt whether or not such instruments are appropriate, look at the system as a whole.
  - If one of the main goals is to get disposal under some level of control, then taxes on disposal are certainly out of place. In contrast, PAYT may turn out to be beneficial to raise awareness and help segregation at source.
  - In developing countries, subsidies may simply not be affordable for the government to finance, in the face of various other claims in the society.
  - Extended producer responsibility (EPR) will hold producers and importers accountable for the products they place on the local market.
- 'Social' instruments rely on communication, awareness raising and interaction between the government institutions and the public and other stakeholders.
  - It takes more than just providing information to change people's attitudes and behaviours. Encouraging people, engaging with communities, and leading by example are *at least as important.*
- Waste governance will also depend on the institutional framework in place and capacities of institutions to prepare legislation and particularly to enforce it, to collaborate among them and with the private sector, and to engage with the public. For the effectiveness and credibility of enforcement, it is better to keep the two roles – that of legislator and that of regulator (enforcer) – separate.
- Incorporate monitoring and evaluation into implementation, to measure performance and gather feedback and lessons from practice. Communicate with the stakeholders throughout – good waste management is in everybody's interest.

4.2 INTRODUCTION

4.2.1 Introduction to waste governance

Current effective waste management systems in countries around the world are the product of a gradual learning process in urban and industrial environmental management that was initiated decades or even centuries ago. Starting from problem recognition, often as the result of some sort of crisis, the systems began to take shape through formulation of policy and its translation into legislation, accompanied by the development of physical infrastructure and facilities to tackle the waste problem. Strategic thinking usually came in later, in accordance with broader national development priorities.

As a domain of high public interest, waste management requires public *policy* to be developed, establishing its general goals, and stipulating guiding principles and decision-making criteria to inform the process of preparing *waste strategies* for achieving the established goals. The general *goals – or driving forces (drivers) behind the policy formulation* – include the protection of public health and the environment, as well as the recovery of resource value from discarded products and waste materials.[1] The *guiding principles* in waste policies in various countries include several of the following: waste prevention, duty of care, polluter pays principle, universal service coverage, inclusivity,[2] subsidiarity principle, precautionary principle, cost recovery, proximity and self-

1 The drivers for waste and resource management are elaborated in Section 2.3.

2 In some Latin American examples, inclusivity is an explicit *goal* rather than 'only' a guiding principle – policies are formulated with the primary goal of inclusion of informal recyclers, in addition to public health and the environment.

sufficiency.[3] The articulated guiding principles may not necessarily be mutually consistent or compatible; hence their translation into instruments may require additional effort to clarify the priorities among them.

The policy and strategies are given 'body' through their translation into legislation and other instruments that are to support their implementation. While starting from the same policy, the instruments can be different in nature and purpose, as illustrated in *Figure 4.1*. Usually they are applied in combination, to ensure their optimal effectiveness and maximum benefit to the system. Which particular instrument is deployed depends on the goal they are envisaged to achieve. For example, in order to ensure installation of (costly) technological measures for environmental protection, these measures should be stipulated as mandatory for a waste facility to get a license. If the goal is to divert waste from disposal to other management options, landfill taxes or bans are likely to be effective. If the goal is to increase waste segregation at source, various communication approaches – as well as suitable infrastructure – are needed to influence people's attitudes and behaviour, based on the understanding of their realities, including their daily routines and concerns.

Figure 4.1 Policy instruments for waste and resource management

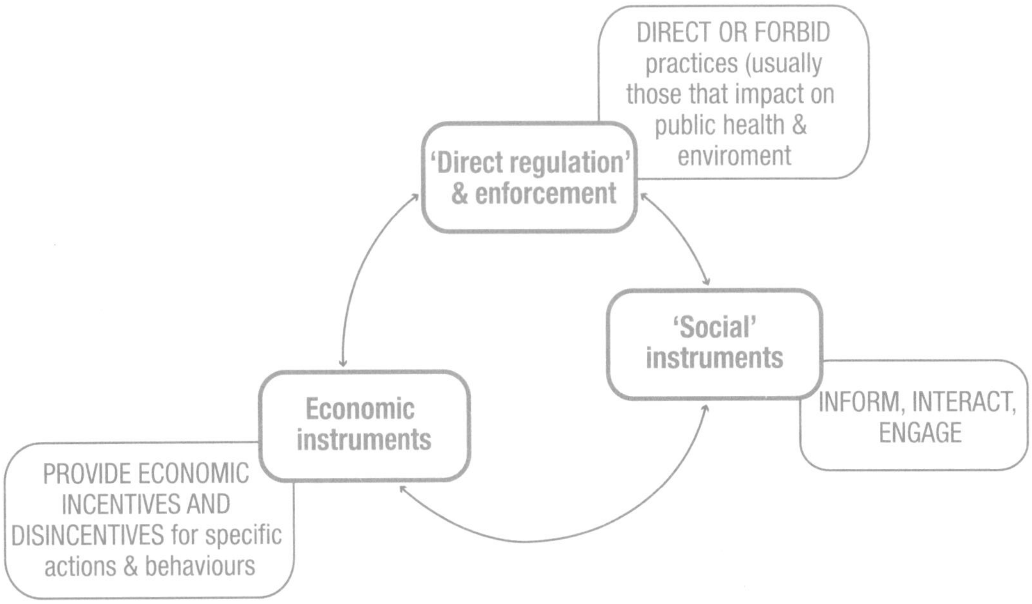

These policy instruments are prepared or implemented by various governmental institutions, agencies or organizations in accordance with their specific mandates and the institutional arrangements defined between them.[4] How the government and its institutions engage in interaction with other stakeholders in the system has to do with a broader issue of governance.

While state power – through its governmental institutions, policies and legislation – is an important, even predominant determining force in developing strategies and taking decisions, it is not the only one. An *engaged* state negotiates policies and seeks agreement to its processes and practices with those who are a party to, or otherwise affected by, its decisions. These include not only a few powerful stakeholders but also others. Hence, the term *governance* is used here to denote institutions and processes, both formal and informal, which provide for the interaction of the state with a range of other actors or stakeholders affected by the activities of government. The concept of governance thus embraces an inclusive consideration of stakeholders, with a particular emphasis on the inclusion of the people in decision-making process, including those in underprivileged and poor communities.[5]

3 These principles are explained in Annex C under Glossary of guiding principles.
4 As they are established for the purpose of serving the system, governmental institutions, agencies or organizations can also be seen as an *instrument*, albeit of a different kind from policy instruments.
5 Sometimes this is referred to as participatory governance.

Aware that this is a normative statement, the GWMO consistently takes the stance that good governance is inclusive. This is in accordance with the Rio Declaration, which promotes "appropriate access to information concerning the environment that is held by public authorities, including … activities in their communities, and the opportunity to participate in decision-making processes."

Accordingly, waste governance is as much about the role of government and policy instruments as it is about the interests and roles of an array of other stakeholders in the system, and how these roles and interests are represented (*Figure 4.2*). Governance is also about *responsibility*, expressed through various legal and financial obligations, but also through a sense of 'ownership' of waste-related issues, which translates into involvement and care about the cleanliness of the open spaces in the community as well as protection of the broader environment and natural resources. The latter means that good waste governance goes beyond street cleaning and waste handling into the realms of production and consumption.

Figure 4.2 Waste governance – a complex interplay of goals, instruments and their implementation

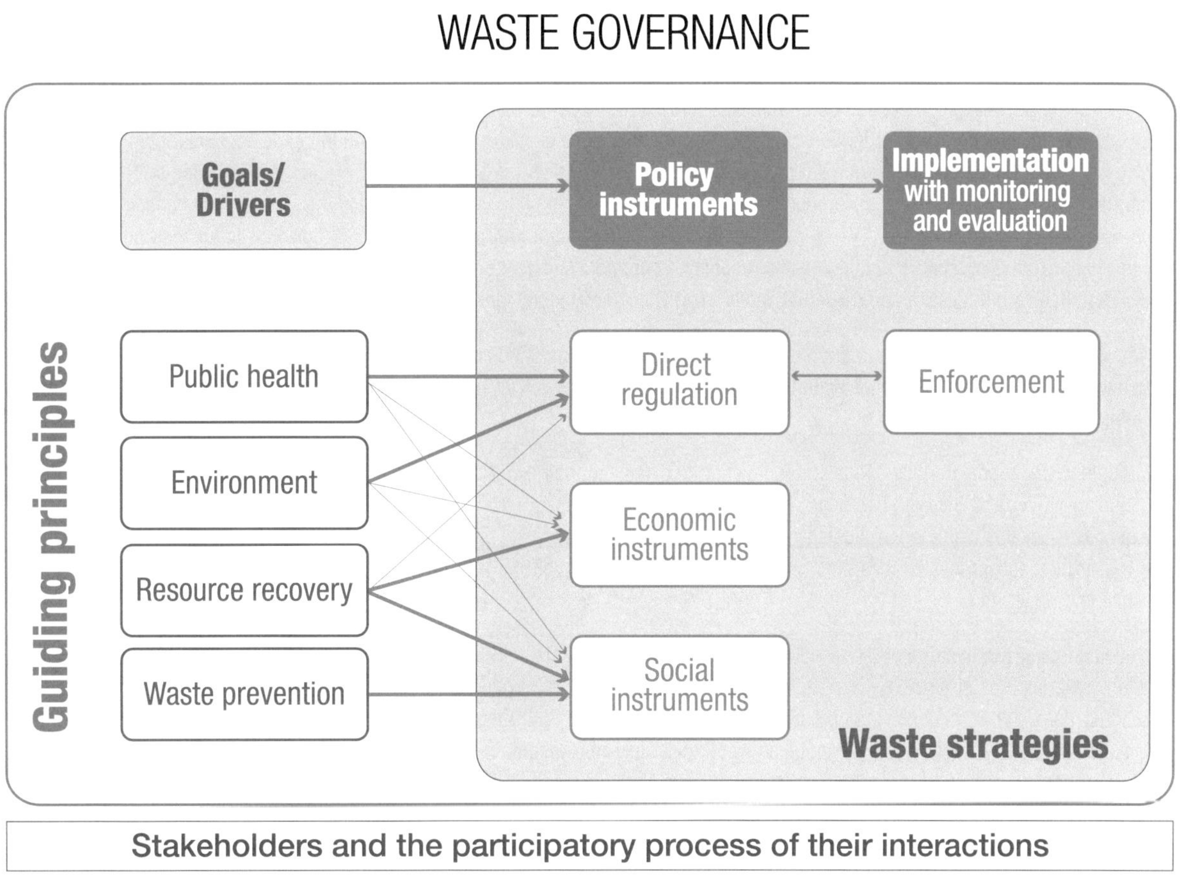

BOX 4.1 TRANSFORMING WASTE MANAGEMENT INTO RESOURCE MANAGEMENT[6]

In accordance with an emerging trend toward what is in various places called a 'circular economy', the shape and the function of SWM are changing, with a growing focus on recycling. Furthermore, the place of SWM at the end of a product's life cycle is increasingly gaining attention from actors placed 'upstream', such as designers and producers. Policy makers are both responding to signals from society and leading the way in exploring new avenues for sustainable development. Examples of national policies include, but are not limited to, Japan's sound material-cycle society,[7] the European Union's recycling society and circular economy[8] and the People's Republic of China's circular economy development strategy.[9]

The circular economy is one of the various concepts proposed for sustainable development, addressing one or more of its three domains – environmental, social and economic. These concepts include planetary boundaries, Cradle to Cradle, The Natural Step, circular economy, performance economy, industrial ecology, resource efficiency, decoupling of resource use from economic growth, dematerialization, zero waste and prosperity without economic growth.[10] A number of indicators have been proposed to measure progress toward the goal of sustainable use of natural resources, such as intensity of material use, material input per unit of service, ecological rucksack and ecological footprint.[11]

The related policies deploy a wide range of instruments, including direct regulation, economic instruments and social instruments. A typical example of direct regulation would be legislation that prescribes the reduction or elimination of hazardous substances in products. Economic instruments such as bans and taxes on landfilling (as a waste management method with limited potential for resource recovery) or tax reduction on products according to the content of secondary raw materials in them (as a way to provide an incentive for the uptake of secondary raw materials, send a price signal to the market and increase consumer confidence) could also contribute to sustainable resource management, particularly if they are accompanied by reuse and recycling targets. Providing information through eco-labels on products can be used to shift people's behaviour toward environmentally conscious purchase choices. Similarly, posters and TV commercials can convey adequate and appealing information that influences people's waste segregation habits at home. In addition, governments can promote and support development of sustainable production through their own purchasing practices, as presented in Topic Sheet 3 on sustainable consumption and production.[12]

As waste prevention assumes a more prominent place as a policy goal indispensable for sustainable development, it is important to make sure that other policy instruments – applied to achieve other SWM policy goals – are compatible and synergistic with the waste prevention goal.

In addition to national policies, cities are connecting to exchange their experiences and best practices. Also civil society members, including NGOs and individual citizens, are taking initiatives to ensure the sustainable development of their communities. For example, 'zero waste' is used as a catchy slogan by various organizations to promote design-for-recycling, reduction and elimination of hazardous substances in consumer products and waste prevention, as well as sound waste management practices including reuse and recycling.[13]

Notwithstading the benefits of these policies, a number of environmental and other concerns are yet to be fully and adequately addressed, such as dissipative material losses during product use, reintroduction of pollutants through the recycling of materials, product design that does not support disassembly and recycling, and others. Also, global issues that await full consideration include heavy reliance and dependence on a few critical materials for some major applications, and, even more broadly, decoupling of economic growth and prosperity from resource use. Various researchers have examined these concerns and have proposed solutions. Some of those are listed in Topic Sheet 3 on sustainable consumption and production, Topic Sheet 4 on waste prevention, Annex C under Concepts of sustainable resource management and in Annex A, under Chapter 4, Sustainability.

Due to the renewed interest in waste as resource, a rather unexpected issue increasingly arises of who gets access to waste and who is the legal owner of waste once the original owner has placed it out for collection.[14]

---

6 "Resource management" is used in this document to refer to the use of materials and energy that can be recovered from waste and to the prevention of use of resources. 'Waste management' implies a focus on what to do with waste once it has been generated, which would include reuse, recycling and recovery. 'Waste and resource management' broadens the focus to include also 'upstream' measures to prevent, minimize or reduce waste prior to the point of generation. See also Figure 2.1 and Section 2.2.2.

7 Japan (2000). Basic Act for Establishing a Sound Material-Cycle Society. Available at http://www.env.go.jp/en/laws/recycle/12.pdf In 2003, this Law was elaborated into a strategic plan with appropriate material flow indicators, as presented in Box 2.1.

8 European Commission (2011), listed in Annex A, Chapter 4, Sustainability

9 PRC (2008). Circular Economy Promotion Law. Available at http://www.fdi.gov.cn/pub/FDI_EN/Laws/GeneralLawsandRegulations/BasicLaws/t20080919_97504.jsp

10 A selection of these concepts is introduced in Annex C under Concepts for sustainable resource management

11 Some of these are discussed in Section 2.5.4 on resource management indicators.

12 Topic Sheet 3 can be found after Chapter 2.

13 Zero Waste Europe and ACR+ are associations of municipalities, as mentioned in Section 4.8.1; Zero Waste Alliance is an international NGO. http://www.zerowaste.org/

14 Box 5.1 in Section 5.3 elaborates on this issue.

4.2.2 Strategic planning

Deciding goals and developing strategies with the guiding principles in mind is at the core of the waste governance process. There is a considerable body of literature that can be consulted to guide the process of preparing national strategic documents, including national waste management plans.[15] This literature is listed in Annex A under Chapter 1.[16] Instead of repeating the literature, this Section provides a few general points.

While waste management services around the world have developed out of necessity, often in response to a crisis, strategic planning enables the decision-makers and practitioners to go beyond the 'firefighting' mode of operation and take a good look not only at the waste system itself but also at the developments in a broader societal context which may impact on that system. This is particularly valuable under circumstances of limited financial and other resources, so as to accomplish allocation to the most beneficial purposes, in terms of particular facilities or activities.[17] Importantly, a long-term vision is needed, as it may take 20 years or more to plan, build and repay the loan for facilities, meaning that waste planning goes well beyond the duration of a typical political cycle.

A national waste management strategic plan has the highest chance of becoming valuable, living guidance for those involved if it is based on profound knowledge and understanding of the local circumstances, including both constraints and the existing strengths (as the latter often tend to be overlooked). In other words, a strategic plan will be substantially better if it acknowledges local realities including waste composition, climate, culture, customs, characteristics of the local commodities market, level of technical expertise, availability of financial resources, and other relevant factors of the societal context. An equally important prerequisite is knowledge of the waste handling activities, as illustrated in Box 4.2.

BOX 4.2 "PREPARATION OF STRATEGIC PLANS IS NOT A JOB THAT CAN BE DONE WELL FROM BEHIND AN OFFICE DESK."[18]

In order to be able to make the right strategic decisions regarding waste facilities, it is essential to make sure you know enough about the details. For that: (a) establish, develop and keep improving a waste information system so as to get reliable information about the waste amounts, the waste composition and the situation and changes in the waste handling sector; (b) when formulating strategies, consult people involved in the day-to-day practice of waste handling in order to become aware of specific situations and the feasibility of strategic objectives; their input can enhance the quality of the strategies and secure their support for the implementation; (c) acquire valuable practical knowledge – go and visit facilities and get a sense of the real situation in the field.

For the strategy to have a higher chance of being accepted and owned by the community, a participatory (inclusive) approach to the planning process is essential – a key point in the World Bank's Strategic Planning Guide.[19] In this manner stakeholders can get involved, contribute their knowledge and ideas, express their concerns, explain their needs and expectations, and raise issues that are important to their community. In such a process, the claims of different interests can be questioned, tested and validated or refuted; differing values and perspectives can be incorporated. For a participatory process to work, both institutional capacity and political will are necessary, accompanied by the consequent allocation of adequate resources.[20]

Developing an effective strategy for a waste management system is not a one-time event; it is a continuous and intensive *learning* process involving various activities – planning, implementation, regular monitoring, evaluation and updating. Valuable information on planning, monitoring and evaluation is provided in the 2011 UNDG Results-Based Management Handbook.[21] The updated strategy can then incorporate new insights that originate from (a) lessons learned through experiences in implementation, (b) information about possible changes in circumstances, (c) newly obtained scientific knowledge and ideas. Such efforts need to be

15 The terms 'strategy' and 'plan' (and 'strategic plan') have been used imprecisely in waste-related literature. While both are used to denote documents prepared to bring about a desired change in the system, a strategy is a more general document that outlines the scope and the main means to be used in order to achieve policy goals; a plan is a more concrete document that elaborates the strategy and stipulates more details.

16 Of particular note are the World Bank's *Strategic planning guide for municipal solid waste management* (Wilson et al., 2001); UNEP's *Guidelines for National Waste Management Strategies* (2013); and ABRELPE/ISWA's *Solid Waste: Guidelines for Successful Planning* (2013). See Annex A, Chapter 1, Strategic planning for SWM. Specific guidance to the EU Member States is available at: http://ec.europa.eu/environment/waste/plans/pdf/2012_guidance_note.pdf.

17 An overview of the countries' waste strategies is available through the UNEP Global Partnership on Waste Management, at http://www.unep.org/gpwm/InformationPlatform/CountryWasteManagementProfiles/tabid/104472/Default.aspx; with some additions in the UNEP's Guidelines document cited above.

18 Based on the experience of the team leader in charge of preparation of the national waste management plan in the Netherlands.

19 Wilson et al. (2001)

20 Inclusivity is elaborated in Section 4.7 on stakeholders.

21 UNDG (2011), listed in Annex A, Chapter 1, Strategic planning for SWM

coordinated with 'adjacent' Ministries that have impact on or are impacted by SWM activities in the country. This is an iterative process of creating 'the best fit' for the local situation.

As change needs time to take hold, it is important to allow time for the implementation and some degree of consolidation of the experiences with the instruments applied, otherwise affected stakeholders may be confused by an array of new requirements, ever-revolving agencies and probably increasing quantities of administration. At the same time, fostering local feedback through a flow of information and exchange among stakeholders could reveal valuable insights and experiences, allow for activities to evolve from this interaction, and strengthen the system from within. This is characteristic of a governance form that embraces an adaptive approach to formulation of strategies and policies, based on complex adaptive systems thinking.

BOX 4.3 COSTLY CONSEQUENCES OF POOR PLANNING – THE CASE OF AN ANAEROBIC DIGESTION PLANT IN LUCKNOW, INDIA

Facing a common problem of lack of waste disposal space, Lucknow, the capital of Uttar Pradesh, India, opted for a solution that would return a 'double dividend' by both diverting organic waste (constituting roughly half of the municipal waste[22]) from disposal and generating much needed energy. The project was approved, investors found, subsidies given, land provided, a proven Austrian technology selected,[23] and the India Biomethanation Plant, a facility with a capacity of 300 tonnes of waste per day, was built in August 2003. The project was given the status of a demonstration project[24] for waste-to-energy facilities by the MNES, the Ministry then in charge of renewable energy.[25] However, while mixed waste was piling up around the premises, the plant had to be closed in 2004, due to lack of 'acceptable waste', at a cost of more than 15 million USD. It appears that, while due attention was paid to the technology and financial feasibility, insufficient consideration was given to securing a supply of suitable organic waste.

The story is in no way unique – several other Indian cities have made the same costly mistake in their planning. Istanbul, Turkey, faced a similar problem in the first years of the operation of a large composting plant.

BOX 4.4 'TOO GOOD' WASTE SEGREGATION AT SOURCE – THE CASE OF STUTTGART, GERMANY

The inhabitants of the city of Stuttgart, Germany, were segregating their organic and other waste so well that in 2011 the city found itself in a costly breach of contract with the Energie Baden-Württemberg (EnBW) incineration plant, as it could not supply the contractually guaranteed annual minimum of 110,000 tonnes of waste. In order to reduce the financial losses, city authorities resorted to having neighbouring municipalities provide waste to the incinerator at a reduced charge, with Stuttgart paying the difference.[26] This situation is repeating itself across Europe and North America. A more adaptive approach to planning and contracting could have prevented it.[27]

Selecting technologies[28]

Once strategic goals are established, in addition to governance, a broad range of technological options is available to provide the system with the necessary infrastructure and facilities to achieve these goals. That entails consideration of not only their technical characteristics but also the financial side of things as well as operational aspects. Concretely, selection of a technology based on glossy brochures or a study tour to state-of-the-art facilities in a highly industrialized country with cold climate may not be wise for a developing country with very different waste composition and climate, and where affordability and institutional capacities for adequate

22 According to the government commissioned research for Indian cities; the results are available at http://www.cpcb.nic.in/wast/municipalwast/Waste_generation_Composition.pdf

23 http://www.entec-biogas.com/5_files/projekte/referenzliste.pdf

24 http://www.pppinindia.com/pdf/ppp_position_paper_solid_waste_mgmt_112k9.pdf

25 India, Ministry of Non-Conventional Energy Sources (MNES), renamed as Ministry of New and Renewable Energy (MNRE) in 2006. http://mnre.gov.in/mission-and-vision-2/mission-and-vision/

26 Since 2011 Stuttgart has been 'short' some 6,000-7,000 tonnes of waste per year. The contract stipulates that Stuttgart pays 141.21 EUR/tonne to the incineration plant. The neighbouring municipalities agreed to deliver their waste to the incineration plant and pay only 95.20 EUR/tonne, with Stuttgart paying the difference. In this way, even though the city has to pay around 300,000 EUR on a yearly basis for waste it does not generate, it still saves twice as much compared to what it would otherwise have to pay due to the contract with the incinerator. http://www.stuttgarter-zeitung.de/inhalt.abfallwirtschaft-in-stuttgart-verkehrte-welt-stadt-kauft-muell-um-geld-zu-sparen.1812add1-6529-4b8d-b2c7-c4debe341312.html

27 Delivery of services by the private sector is elaborated in Section 5.6.3.

28 Some of the main technologies for resource recovery in MSWM are presented in Section 3.5.

operation, maintenance and emissions monitoring are still major issues. Conditions under which facilities operate cannot be packed in a container and shipped across the ocean in the same way as modern technology can.

Furthermore, as facilities and infrastructure are costly, the return on investment period is long and hence likely to bind stakeholders to the chosen solution for a long time, which can cause a multitude of financial problems later on. For example, the Netherlands has invested heavily in waste incineration only to find itself in a position of having far too much incineration capacity for the country's needs, resulting in competition for waste and a substantial lowering of incineration prices, thus jeopardizing the economic viability of the facilities. This is of particular concern to the authorities if they made the investment with public money. In such cases, in their efforts to secure return on investment, authorities may inadvertently slow down broader societal trends towards waste prevention and enhanced environmental practices. Conversely, societal developments can supersede choice for one technology over another and make it inappropriate, as illustrated by the Stuttgart case overviewed in Box 4.4 above. An adaptive approach to planning is of paramount importance, especially now in a time of transition toward more sustainable use of natural resources.

Concerns about the selection of technologies are very different in places that are at the other end of the technological modernization spectrum and still struggling to provide even basic services. A stepwise upgrading of the technology could start by bringing waste disposal under control, accompanied by an expansion of waste collection services to reach the entire population. This would be an excellent strategy with far-reaching benefits to public health and the environment.

Essential considerations of technology selection are translated into direct questions listed in *Box 4.37* at the end of this chapter.

BOX 4.5 APPROPRIATE TECHNOLOGIES FOR SECONDARY CITIES AND SMALL TOWNS IN DEVELOPING COUNTRIES – REGIONAL PROJECT IN SOUTH AND EAST ASIA[29]

While larger cities in developing countries may be able to consider a wide range of technological options for solid waste management, secondary cities and small towns usually lack the economies of scale and the capacities necessary for large centralized solutions. Decentralized and community-based small-scale facilities can provide a viable and affordable alternative. Waste Concern, an NGO in Bangladesh, has developed a model for such facilities, which are called Integrated Resource Recovery Centres (IRRCs). Waste Concern's model uses simple technology based on source separation of organic waste, is low cost and recovers value from waste by converting organic waste into compost and valorizing recyclable materials, while providing livelihood opportunities to the urban poor. With assistance from the United Nations Economic and Social Commission for Asia and the Pacific (UNESCAP), the model has been replicated across secondary cities and towns in several countries. In Matale, Sri Lanka, for example, three neighbourhood-based plants have been installed with a combined capacity of 9 tonnes of organic waste and 3 tonnes of recyclables a day, thus treating a major portion of waste generated by the town and creating employment for 20 urban poor. In parallel, Community Development Officers of the Municipal Public Health Department took care of awareness raising in the community through intensive face-to-face communication with households, resulting in a source separation rate of 60%. The leadership and strong commitment by the mayor and the municipality have played an important role in the success of the initiative.

Worker placing organic waste into composting boxes in a waste-to-resource plant in Matale, Sri Lanka

29 Text provided by the United Nations Economic and Social Commission for Asia and the Pacific (UNESCAP), Bangkok, Thailand. For more information about individual projects, manuals and trainings, see http://waste2resource.org/ and http://www.unescap.org/sites/default/files/Paper_ESCAP%20paper%20on%20IRRC%20ISWA%20Congress.pdf.

BOX 4.6 SOLID WASTE IN TROPICAL ASIA: LESSONS FROM BALI[30]

Facing waste problems common to most of tropical Asia, the island of Bali, Indonesia, has made numerous attempts to address them, with varying success. Bali's intensive tourism-driven development has contributed to these problems but has also prompted and accelerated the search for solutions. The three cases presented here are very different in scale and degree of technological sophistication, offering some important lessons. A fundamental tension remains between high-tech, capital-intensive, large-scale approaches (that, more often than not, fail) and local, low-tech, small-scale ones (that tend to work given favourable circumstances). Yet, the reasons for relative success and failure of various initiatives are primarily socio-cultural and political-economic in nature, rather than technological.

(1) In 1996, a small local company ***Jimbaran Lestari (JL)*** managed to persuade some of the large hotels in the area to hire its waste collection services, with guests' complaints and a smoke problem from burning illegal dumps 'helping' their case. The service is organized separately for dry, wet (organic) and garden wastes. The dry waste fraction, consisting mainly of packaging, is sorted manually and shipped to Java for recycling. The wet (mostly kitchen and food) waste is sold to pig farmers. Garden waste is sorted as well; wood is sold as firewood while lighter materials are composted and the compost is sold back to the hotels. The residual 30% goes to a semi-controlled disposal site. JL has been successful in providing effective services to its clients. Its financial sustainability is supported by its favourable location near major hotels, which are able and willing to pay for collection services, and which provide a waste stream much richer in recyclable materials than the average in Indonesia.

(2) Much larger quantities of waste are generated in the ***capital city of Denpasar*** and surrounding semi-urban, light industrial and tourism zones. Most is dumped in a huge semi-controlled disposal site on coastal land reclaimed from a mangrove foreshore. The only recycling or 'management' at the site is a small army of scavengers (*pemulung*) who pick metals, glass, plastics and paper for sale to recycling plants in Java. The remainder rots in the tropical heat, producing a foul odour and attracting vermin but also releasing methane and leaching contaminants into the sea.

In 2003 the governments of Denpasar and three neighbouring districts, together comprising over 50% of the island's population, joined under the acronym *Sarbagita* and developed an "integrated" plan for waste management, based on a public-private partnership with foreign companies. The basic idea was to convert waste (of which there was too much) into energy (of which there was not enough), combining anaerobic digestion, landfill gas, and pyrolysis, to generate electricity. The economic viability of the project relied on obtaining carbon-credit funding through the UN's Clean Development Mechanism (CDM). While the plan may have made sense, there were technical, environmental, financial and bureaucratic problems from the start and a decade on, after massive investment, both waste processing and electricity production remain minimal. Critics point out that its scale alone is neither efficient nor environmentally safe and it relies on high transport costs, while taking any profits offshore and undermining local community-based approaches including any efforts to reduce waste.

(3) ***Ubud*** is a small town and centre of both traditional Balinese culture and tourism, which has been tackling its SWM problems for 30 years, but which undertook several failed attempts before developing a successful solution. In the early 1980s, a local organization formed by concerned citizens organized waste collection from the main streets and market, in collaboration with the local council (LKMD). Unfortunately, the waste was subsequently dumped into a nearby river gorge. The waste collection service could not keep pace with increasing amounts – waste was lining the streets or was indiscriminately dumped or burnt. Alarmed by the situation, another citizens' organization initiated several projects in 1993, none of which succeeded, for a variety of reasons, mostly to do with local politics and NIMBY attitudes.

In 2004, a pilot project was initiated to sort recyclables from incoming waste at the semi-controlled disposal site for the Gianyar district, located in the village of Temesi, as a result of efforts by the local Rotary Club (which secured international donor funding) in collaboration with the district government, a local NGO and the local village (which provided labour). Waste was sorted from a conveyor belt placed in an open shed. The system worked well, but it was not financially viable because the capacity was too small and waste was mainly organic, with insufficient amounts of recyclables. The solutions were relatively straightforward: enlarge the facility and shift the focus to composting. Through refining their composting technology they were able to produce high-quality compost for sale to the growing market for use in hotel gardens, public parks and a government funded project supplying organic fertilizer to farmers. As the income from the compost still could not cover the costs, the project sought CDM carbon funding. After a complex, expensive and lengthy application process they sold their carbon credits to Kuoni, a large Swiss travel agent that wanted to offset emissions caused by its airline flights, many of them to Bali. While the facility is processing only a small portion of the waste arriving at the disposal site, it provides a model for a district-level multi-stakeholder solution for waste streams high in organic content. It 'removes' methane from the atmosphere and leachates from groundwater and reduces the land area needed for waste disposal.

30 Text is provided by Dr. Graeme MacRae, Massey University, Auckland, New Zealand; for more details see MacRae, G. (2012). Solid waste management in tropical Asia: What can we learn from Bali? *Waste Management and Research* 30(1): 72-79.

4.3 DIRECT REGULATION

4.3.1 Introductory remarks

'Direct regulation' (popularly referred to as a 'command-and-control' approach) is a policy instrument that relies on the legitimacy, authority and sanctioning power of the state for establishing rules of conduct or procedure through legislation, and inducing consent and compliance through implementation, including enforcement.[31] In waste management, the command-and-control approach has been necessary to ensure that public health and the environment are protected, particularly from the impact of waste containing hazardous substances. The application of both criminal and civil sanctions has played a vital role in achieving effective environmental protection. Importantly, the waste industry as we currently know it only exists because of legislation – otherwise waste would be dumped at the lowest cost. Without strong laws and regulations, together with their rigorous enforcement, legitimate businesses would be undermined by criminal operators, causing huge costs to society through health impacts, pollution, clean-up costs and the like.[32]

The command-and-control approach is based on information and monitoring and relies on inspection and enforcement for its effectiveness, which both can be administratively 'heavy' and costly in terms of institutional capacities required. As such this approach is often contrasted to economic instruments, a very different category of policy instruments that rely on market-based incentives and disincentives.[33] In reality, however, the two regulatory models frequently operate in tandem, each providing its own advantages and disadvantages. Countries have different (political) cultures and historical preferences for prevailing instruments, so the balance between the two categories may be different. It should be noted however that the command-and-control approach will remain preferred for situations involving high risks and serious consequences of non-compliance, such as, for example, the use of hazardous substances and the handling of hazardous waste.

Basically, in the field of waste management, direct regulation is used to define what is considered to be waste, and prescribe *who* is obliged or allowed to do *what* with it and *how* (to which standard). *Figure 4.3* illustrates the essential issues to be regulated for an effective waste management system.[34]

Legislation usually comprises two 'layers'. The first of these is so-called framework legislation,[35] which, as the term suggests, defines a 'framework' that sets the stage by establishing definitions, responsibilities and obligations. The second layer of legislation comprises the subsequent subsidiary detailed laws and regulations that serve to transpose the broad goals of the framework legislation into specific measurable obligations. Framework legislation is the necessary first step to give legal status to the strategic goals and create a frame of reference for the subsequent detailed laws and regulations. Furthermore, it is easier to amend and keep subsidiary regulations up-to-date than it is to change a framework law. It is important to recognize this two-step approach, because in some countries the first step of enacting the framework is taken, but then the introduction of detailed regulations gets delayed, which in turn delays the concrete implementation activities.[36]

As introduced in Section 2.3, waste management happens because one or more main driving forces are at play: either the society is concerned about public health and the environment[37] and/or it seeks to promote resource recovery from waste in support of sustainable development. As elaborated below, legislation will have a different focus depending on its goals and the specific driving force behind its development.

---

31 This is accomplished through formulation, enactment and implementation, including enforcement, of legally binding documents such as laws, directives, regulations, decisions, statutes, executive orders and the like. The exact terms for referring to these documents can vary, both for various documents within a country and for similar documents in various countries.

32 See Chapter 5, in particular Section 5.2.2 on the costs of inaction and Section 5.3.6 on the economic dimension of waste crime.

33 Harrington & Morgenstern (2004), listed in Annex A, Chapter 4, Economic instruments.

34 In the process of preparing legislation it is important to hold interdepartmental meetings where possible contradictions, overlaps and gaps can be identified and addressed; also, consultations are needed to ensure synergies between national and municipal rules and regulations and targets. (A step-by-step description of how laws are made and transposed, including descriptions of formal processes and the reasons why each step takes place, is beyond the scope of the GWMO.)

35 If a country has expressed its consent to be bound by an international treaty, such as the Basel Convention, then it has the obligation to reflect the content of that treaty at the national level by adopting adequate legislation.

36 The importance of implementation, including law enforcement, is elaborated in Section 4.6, on social instruments.

37 In recent decades, environmental concerns have also been including global climate change.

Figure 4.3 Essentials to be regulated

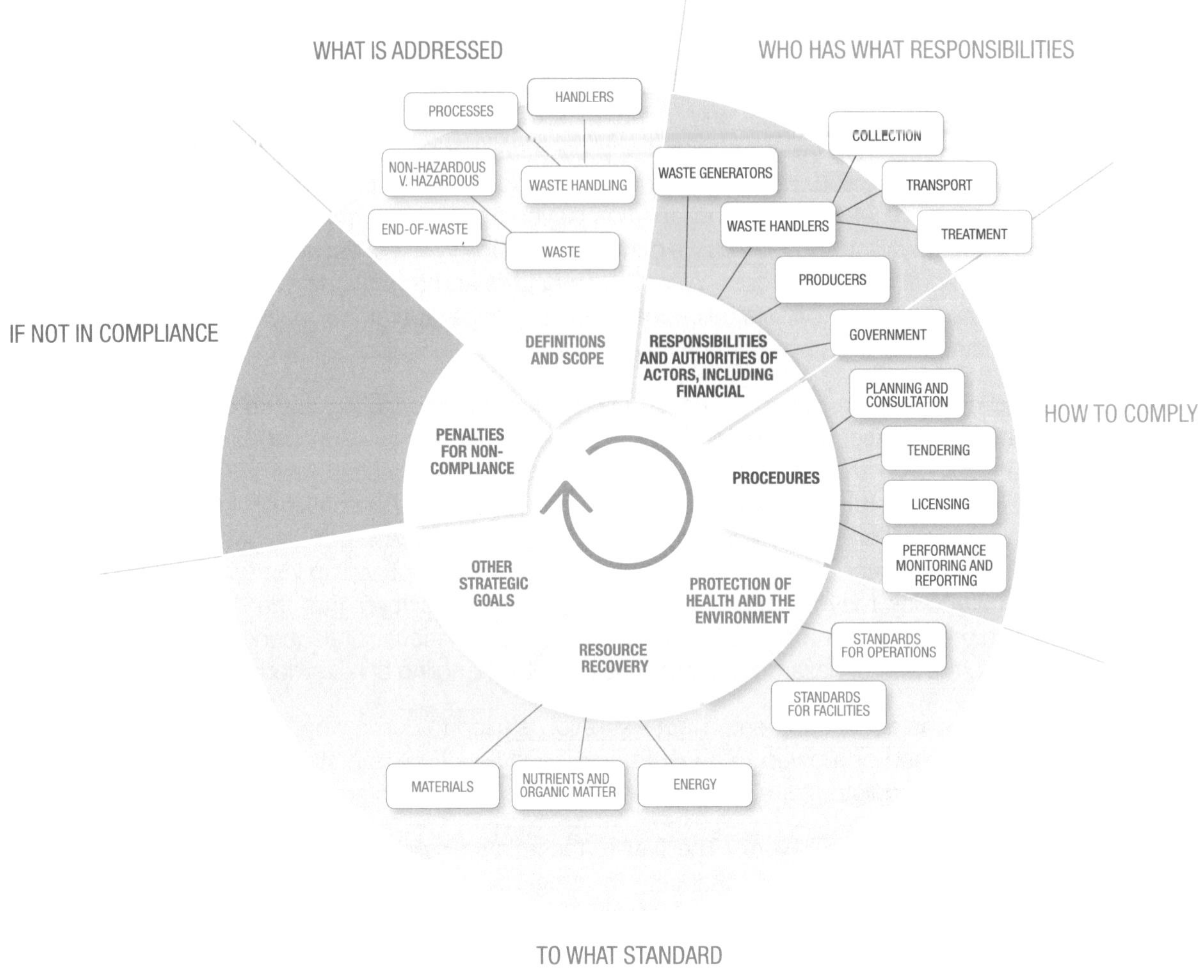

4.3.2 Legal definitions and classifications

Legal definitions and classifications are vital to the correct application of waste legislation. Ideally, they provide clear and unambiguous guidance as to (1) what materials are considered to be waste, what is hazardous waste (in other words, what materials are within and outside the scope of the law), (2) who the actors are to whom the obligations will apply; (3) what constitutes particular waste handling processes (including prevention, recovery and disposal), operations and facilities. If a substance or a material is designated by law as waste, its handling and processing have to comply with pertinent waste legislation. This has direct, and often substantial, financial and other consequences for both waste generators (to be) affected by the legislation and for the regulator in charge of enforcement. The legal definition of waste is also of great consequence for materials recovered from waste, especially in cases of movement of waste from one country to another and regional approaches among countries such as SIDS.[38]

In particular, classification of waste into the categories hazardous versus non-hazardous is of great consequence, as technologies associated with the management of hazardous waste tend to be much more costly than those for non-hazardous waste. This makes legal definitions and classification tremendously important for the parties involved.[39] Establishing these definitions is therefore an intense political process in which diverse interests are manifested, contested and negotiated, so as to collectively arrive at acceptable definitions. In that context, scientific knowledge on innate – hazardous and other – properties of waste substances is only one of the factors that play a role, along with political or economic implications, viability of implementation, and social

38 This is discussed in Section 4.3.5 below, on legislation for resource recovery.
39 For example, a spirited discussion is ongoing on whether e-waste should be classified as hazardous waste.

acceptance. Various industries may lobby their governments to exempt 'their' materials from designation as waste so as to avoid the costs associated with compliance with strict environmental regulations pertaining to waste. The already difficult task is further complicated by the uncertainties inherent in scientific knowledge on risks associated with exposure to (waste) substances.[40]

Finally, particularly relevant for resource recovery from waste is legislation establishing the so-called end-of-waste criteria, which determine under which conditions a waste material ceases to be designated as waste and obtains the status of a product. Such legislation creates legal certainty and reduces administrative burden for the handlers, especially in transport and shipment. Conditions include its suitability for specific purposes, its quality and compliance with pertinent technical standards, as well as the existence of market demand for it.[41]

4.3.3 Public health legislation

Protection of public health is always the foremost policy goal and principle related to waste management. Due to its importance, this goal is translated into laws that clearly define who is responsible for it. Commonly, it is the waste generator. For generators of MSW, municipalities or counties or a similar level of local government as the 'proxy-generator' have a legal responsibility for ensuring that the waste is indeed collected and removed from waste generators' premises as well as from city streets and public spaces in a safe and timely manner. The principle of protection of public health is often combined with the 'polluter pays principle',[42] so that local authorities organize a functional waste system and provide services (or arrange for services to be provided) while (at least the larger) waste generators pay. The municipality's responsibility is usually further elaborated in local by-laws, which stipulate whether the local authorities will engage their waste department to provide the services itself or will instead engage a third party as a service provider.[43] For such a collaboration with others and outsourcing of waste management services to be fruitful and successful, a number of boundary conditions ought to be met; these are discussed in Sections 4.7 on stakeholders and 4.8 on government as a stakeholder. Based on the same principle, commercial and industrial generators are responsible for arranging that their waste is adequately taken care of.

4.3.4 Environmental legislation

In response to environmental concerns, waste-related legislation prescribes measures of environmental protection required for the physical infrastructure, the 'hardware' of a waste management system, thus particularly focusing on facilities, including restrictions on their location. Environmental legislation can also establish legal liability for environmental damage, based on which economic instruments such as mandatory insurance are developed. These are further discussed in Section 4.5.1.

The legislator may choose to formulate environmental legal requirements either as a specific technology or as an environmental performance standard for any technology that may be installed in waste facilities, or it may use a combination of the two approaches. Each of these two formats has advantages and disadvantages, as discussed in *Box 4.7.*

40 The process is even more complicated if it involves multiple countries, such as the 183 Parties to the Basel Convention, http://www.basel.int/Implementation/LegalMatters/LegalClarity/tabid/3621/Default.aspx.

41 Article 6 of the Directive 2008/98/EC on waste (the EU Waste Framework Directive) is an example of such legislation, available at: http://eur-lex.europa.eu/legal-content/EN/TXT/PDF/?uri=CELEX:32008L0098&from=EN. End-of-waste criteria for individual waste streams are available at: http://susproc.jrc.ec.europa.eu/activities/waste/index.html. Topic Sheet 12, found after Section 4.3.5, shows how end-of-waste can be implemented in practice for compost.

42 This and other guiding principles in waste policies are presented in Annex C under Glossary of guiding principles.

43 Outsourcing options are elaborated in Section 5.6, on financial models for delivering MSWM services.

BOX 4.7 CONTRASTING LEGISLATIVE APPROACHES TO CONTROLLING THE ENVIRONMENTAL STANDARDS OF WASTE FACILITIES

OPTION 1. SPECIFICATION (FEATURES)	OPTION 2. FUNCTION (PERFORMANCE)
Environmental legislation prescribes a specific technology type or concrete design features that have to be installed in waste facilities.	Environmental legislation prescribes standards to which any technology installed in a facility has to perform; emission limits for individual pollutants would constitute a typical example.
Taking this approach makes it fairly easy for the regulator to check and establish that required technologies are put in place; this may be useful in situations where capacities and resources needed for frequent routine monitoring (of pollution emissions) are lacking.	Taking this approach enables design of facilities based on their function, which is ultimately protection of the environment (air, oceans, groundwater, soil, rivers,…), and thus leaves space for solutions that achieve the required level of environmental protection in different, possibly better and more cost-effective ways.
However, with such legislation in place, affected stakeholders may not see the incentive to innovate and explore potentially better technological alternatives, and may only install minimum requirements.	However, for this approach to realize its full potential, certain levels of knowledge and expertise are required on the part of stakeholders involved. (That is why smaller firms may prefer prescriptive regulation if they lack the skills or capacity to design their own solutions.)
Installing technological features does necessarily not mean that there is know-how available to operate the facilities as designed.	These policies may fail to achieve the envisaged results due to a lack of qualified staff, technical capacities or financial resources within the regulatory agency to implement and monitor compliance.

Example – Landfill design by function rather than by specification

Instead of prescribing technology to protect groundwater and soil from pollutants present in landfilled waste, the EU Directive on the landfill of waste[44] prescribes the level of protection expressed as its function – attenuation capacity achieved by geological barriers at the bottom and the sides of a landfill. For example, the Directive stipulates that at the class of landfills for non-hazardous waste, these barriers have to provide at least the same level of environmental protection as that provided by a mineral (clay) layer of thickness greater than or equal to 1 m and permeability K less than or equal to 1.0×10^{-9} m/s. This implies that alternative technologies can be developed with at least equal environmental performance while possibly requiring lesser amounts of materials for the liner, instead of clay that would have to be quarried and transported from elsewhere. Even better environmental protection is obtained if a natural site is available with layers of naturally consolidated clay of comparable permeability, meaning that no engineered liners need to be installed at the bottom at all. The focus on the function of technology – rather than on its features – is well understood and has been deployed for several landfills for non-hazardous waste in, for example, Turkey and Nepal.

44 Council Directive 1999/31/EC on the landfill of waste. http://eur-lex.europa.eu/legal-content/EN/TXT/PDF/?uri=CELEX:31999L0031&from=EN For more information on the EU policies on landfilling see http://ec.europa.eu/environment/waste/landfill_index.htm

It is important to note that environmental legislation has a higher chance of serving its purpose and achieving policy goals if it is designed and implemented in stages, with progressively more stringent standards, as this allows time for the actors in the system to develop expertise and raise the necessary financial resources. Copying environmental performance standards from a very advanced, high-income country or going even further to compile a set of the most stringent criteria from several countries – as some countries have done in the process of developing their environmental legislation on, for example, leachate discharge standards to surface water bodies – is actually likely to be counterproductive under the circumstances of limited financial and other resources and will thus fail to deliver envisaged environmental gains.[45] That notwithstanding, the regulations should establish a clear path and stipulate that standards will gradually be raised over time, thus creating regulatory certainty for the parties affected.

4.3.5 Legislation for resource recovery[46]

In order to promote resource recovery from waste, countries may opt to ban disposal (landfilling) of municipal waste or certain waste fractions from which resources such as materials and energy can be recovered, including organic waste, recyclable material streams or combustible waste. For example, Austria, Germany, the Netherlands, Norway and Switzerland have had some form of landfill ban in place since 2005 or earlier.[47] Instead of outright bans, a legislator may prescribe landfill *diversion* targets, which specify the percentages of waste to be processed in a way different from disposal. In order for the diversion targets to actually contribute to resource recovery, they need to be accompanied by explicit requirements to that end and penalties for non-compliance.

Legislation may also explicitly set target recycling rates for specific waste streams, to be reached within a certain timeframe. For example, legislation may stipulate that a (minimum) rate of 50% recycling of metals in packaging waste has to be achieved by the end of 2018. This kind of legislation not only sends a clear signal to local administrations and the public about the importance of recovery of resource value as a policy goal but also provides the legal certainty required for future private sector planning and investments in recycling and energy recovery facilities. It should be noted though that recycling targets could be unrealistic if there are no existing viable markets for recyclables. If this is the case, other measures may be needed to create them.

Recycling waste container, PRC

45 For example, if there are no financial resources or qualified staff to install a low permeability liner or operate leachate treatment at a waste disposal site, there is no benefit to having legislation in place that requires both. Bringing waste disposal under control and gradually upgrading the technology, even if it is not up to the ultimately desirable standard, is still much better in terms of its benefits to public health and the environment, than waste being dumped in rivers and the sea while very advanced legislation sits on the shelf, unused.

46 The term 'resource recovery' refers to the resources that are recovered from waste include both materials and energy sources, and either of them may be addressed by such legislation. Materials include recyclables such as metals, plastics, glass, paper, nutrients and organic matter.

47 For example, in Austria it is prohibited to landfill waste with TOC (total organic carbon) content higher than 5%; in Sweden, landfilling is banned for organic waste (waste containing carbon in organic form, such as food waste and plastics); in Switzerland, landfilling of municipal solid waste is banned.

BOX 4.8 LAW ON FOOD WASTE RECYCLING, JAPAN

Japan's Law for Promotion of Recycling and Related Activities for the Treatment of Cyclical Food Resources (2001; revised 2007) aims to increase recycling rates of waste generated during the manufacture, distribution and consumption of food. The law stipulates the responsibilities of food-related businesses and of consumers and provides specific criteria for reducing the generation of waste, reducing the waste volume and recycling. The competent Ministry is to provide guidance and also facilitate the process by, for example, establishing stable transactions among different parties or linking registered food waste-producing businesses to other registered businesses which can reuse or recycle the food waste.[48]

The recycling is often done by means of processing the waste into animal feed (usually for pigs) under strictly controlled industrial conditions, thereby ensuring safety for consumption. This kind of food waste recycling is a big industry in Japan and the Republic of Korea contributing to resource (nutrient) recovery. Use of food waste for animal feed is forbidden in the EU, due to fear of mala fide processing practices that could spread into the feed any pathogens that may be present in the input materials and thereby infect animals.[49] The EU also bans the feeding of catering waste (swill) to pigs. A UK-based civil society movement 'The Pig Idea' is aiming at lifting the ban and reinstating this time-honoured practice. The movement has gained popularity and support by celebrity chefs, cooks and farmers around the country.[50]

'Ugly' vegetables at reduced price in supermarket, France.

Students gleaning 'too big' vegetables, Netherlands.

In addition to national legislation, or in its absence, smaller administrative entities such as states or (large) cities may enact local legislation, aiming to increase resource recovery. For example, the state of South Australia, and several cities in the U.S., including San Francisco (California), Portland (Oregon) and Seattle (Washington), are examples of good practice of not only enacting such legislation but also – and more importantly – following through and implementing it successfully. In 2005 Buenos Aires City, Argentina, passed a local law (Law 1854/05), commonly known as the 'Zero Waste Law', which requires the waste quantities sent to landfills be reduced by 50% in 2012 and by 75% in 2017 in comparison to 2004 levels; in addition, this law bans waste incineration, with or without energy recovery. While the level of performance is still far below these ambitious targets, the city's waste management system has nevertheless been transformed, with intensified recycling of construction and demolition waste, construction of an MBT plant and the introduction of organized separate collection of dry recyclables by informal recyclers, cartoneros,[51] accompanied by communication campaigns. This leaves 55% of the waste to be directly disposed, which constitutes a 20% reduction compared to 2004 levels.[52]

48 http://www.env.go.jp/en/laws/recycle/10.pdf

49 There have been cases where such practices are believed to have contributed to spread of the foot-and-mouth disease in the UK.

50 http://thepigidea.org

51 In order to emphasise the inclusion of *cartoneros* in MSWM system, they are now officially referred to as *recuperadores urbanos*. Informal waste sector in MSWM is elaborated in Topic Sheet 14, found after Section 4.7.

52 The amount of waste landfilled in 2004 was 1.5 million tonnes. Currently the city generates about 6,000 tonnes per day, of which 3,250 tonnes are directly landfilled.

BOX 4.9 TOWARD ZERO WASTE IN SEATTLE, WASHINGTON, U.S.

Among U.S. cities at the forefront of change, shoulder to shoulder with San Francisco, California, the city of Seattle in King County, Washington State, adopted zero waste as a 'guiding principle' in its Integrated SWM Plan of 1998, which emphasizes managing resources instead of waste and conserving natural resources through waste prevention and recycling. However, its history shows that the policy has developed over almost 40 years.

As early as 1981, Seattle introduced volume-based waste collection charges, supported by measures to tackle the temporary side effect of illegal dumping. In the 1980s, its two landfills closed, leaving the city with an acute waste disposal problem. Although a large incineration plant was considered, the siting process met with considerable public opposition. The mayor decided to put the incineration project on hold and called on the residents to show their commitment and source separate their waste so as to divert it from disposal. A diversion target was set at 60% under the city's 1988 Integrated SWM Plan and reaffirmed in subsequent plans.[53] Similar ambitious targets were set across the entire King County, with home (backyard) composting as an integral part of the plans; the county provided residents with compost bins at subsidized rates as well as with technical assistance. The pre-existing waste collection fee structure helped encourage backyard composting. In parallel, a recycling programme was launched in 1988, to become the cornerstone of the upcoming comprehensive recycling system. Costs were closely monitored and a number of measures were also introduced to cut the costs for the city. Interestingly, each step in the process was supported and informed by invited studies by consultancies and proposals from competing private companies, accompanied by extensive feedback from citizens and other stakeholders.[54]

Zero waste goals, source separation and composting of yard and food waste are embraced by various groups in the city. For example, having started from a 12% waste diversion rate at its stadium in 2005, the Seattle Mariners baseball team approached 85% waste diversion in 2012 through various measures, including compostable food service ware and snack food packaging; it also distributed "Kitchen Catcher" composting kits and other useful items to promote composting to the fans.[55]

Recently Seattle has focused on reducing its greenhouse gas emissions. Based on an extensive 2011 report,[56] the City Council passed resolutions calling for zero net GHG emissions by 2050 in the three sectors that account for the majority of emissions in Seattle and offer the greatest opportunities for the city to influence through direct policies, programmes and investments, namely transportation, buildings and waste. To that end, the 2030 target waste diversion rate was increased to above 70%.[57] In Seattle citizens separate their waste and place it in three different bins located outside of their houses: a green bin to hold yard clipping and food waste, a blue one for recyclables and a black one for residual waste. From January 2015 source separation of organic waste has been made mandatory. Efforts by the city to increase source separation and recycling by the residents have been supported by a system of fines for those who repeatedly demonstrate poor waste separation practices. The system has primarily an educational function – it starts with a friendly reminder through an 'educational ticket' affixed to the waste bins, followed by a symbolic fine. Education of the public is generally seen as important, thus school programmes on waste practices are organized.[58]

The city's waste authorities are looking into waste streams that are currently still difficult to recycle, such as cooking oils, wet or dirty produce bags, cat litter, painted or treated wood etc.

Related to the end-of-waste criteria are the quality protocols and standards for recycled materials and compost produced from waste, as presented in Topic Sheet 12 below.

A number of economic instruments (such as landfill and incineration taxes) as well as information-based instruments have been used or proposed to promote resource recovery; these instruments are discussed in subsequent sections. Obviously, legislation has to be enacted so as to mandate their implementation.

53 Seattle Zero Waste Resolution 30990 of 2007, available at http://www.seattle.gov/parks/docs/zero%20waste%20resolution.pdf

54 USEPA (1994). Waste Prevention, Recycling, and Composting Options: Lessons from 30 Communities. EPA530-R-92-015; Bagby, J. (1999). City of Seattle: Past, Present and Future and the Role of Full Cost Accounting and SWM. Seattle Public Utilities.

55 Farrell Tucker, M. & N. Goldstein (2012). Seattle Mariners Push the Zero Waste Envelope. Biocycle 53(12): 25.

56 Getting to Zero: A Pathway to a Carbon Neutral Seattle, available at http://www.seattle.gov/Documents/Departments/OSE/CN_Seattle_Report_May_2011.pdf

57 http://www.seattle.gov/council/issues/carbon_neutrality.htm

58 Hickman, M. (2014). Seattle residents with too many table scraps in their trash can now be fined. Mother Nature Network, 23 September 2014. Available at http://www.mnn.com/lifestyle/recycling/blogs/seattle-residents-with-too-many-table-scraps-in-their-trash-can-now-be. The explanatory guide for Seattle residents is available at www.seattle.gov/council/bagshaw/attachments/compost%20requirement%20qa.pdf

TOPIC SHEET

12

END-OF-WASTE CRITERIA, ILLUSTRATED FOR COMPOST IN EUROPE[1]

As discussed in Section 4.3.2 and reiterated in 4.3.5, the purpose of legislation establishing end-of-waste criteria is to support resource recovery from waste. That is, end-of-waste criteria provide clarity and legal certainty regarding definitions of waste and the materials recovered from it, which cease to be regarded as waste and become 'products' or 'secondary raw materials'. In the European Union, based on Article 6(1) and (2) of the Waste Framework Directive 2008/98/EC, the end-of-waste criteria are being prepared for priority waste streams.[2] One of those is organic waste, also referred to as biowaste, a generally under-utilized resource whose recovery could result in a broad range of benefits, from improvements to degraded agricultural and other soils to substitution of mineral fertilizers to GHG emission reductions. Composting and anaerobic digestion are two common recycling options for organic waste.[3]

© Natalie Maynor/Flickr/Creative Commons

Quality assurance schemes play an important role in establishing and applying the end-of-waste criteria for compost and digestate. Such schemes have been successfully implemented in the last 20 years in various European Member States, including Austria, Belgium, Germany, Italy, the Netherlands, Sweden and the UK, and contributed to sustainable biowaste recycling. In this way, in many EU countries, quality assurance is connected to legislation and executed by specialized organizations or certification bodies. This legislative push for organic waste recycling is necessary in the early stage of the national implementation of biowaste management. Once there are high quality composts and digestates, which have beneficial applications as soil improvers and organic fertilizers, the market pull will be created in the next stage. Provided with a quality label, those composts and digestates are accepted as products on the market comparable to primary goods, which is the idea behind the end-of-waste criteria and their goal.

On the basis of the positive experiences in countries with running quality assurance schemes and quality assurance organizations in place, the European Compost Network (ECN) developed a European Quality Assurance Scheme (ECN-QAS) for compost and digestate[4] which is used to certify the existing quality assurance organizations, namely in Austria, Belgium and Germany. The ECN-QAS also includes an organization model to support countries just starting out in this area, and this model has already been successfully applied in, for example, Bulgaria and Estonia. The Scheme consists of four pillars: (1) a list of suitable input materials, (2) plant operation quality requirements, (3) criteria for product quality of compost and digestates and (4) recommendations on how to use the recycled organics. The following sections describe the evolution of separate collection of biowaste and the quality assurance of compost in three EU countries.

1 Topic Sheet prepared by Ljiljana Rodic using text and information provided by Josef Barth, INFORMA, Germany, and ECN European Compost Network, EU, Dr Marco Ricci-Jürgensen, Italian Composting and Biogas Association, Italy, and Dr Jane Gilbert, Carbon Clarity, UK.

2 More information on the status of the criteria for individual waste streams and the related scientific research can be found at http://ec.europa.eu/environment/waste/framework/end_of_waste.htm.

3 See Box 3.6 in Chapter 3 on organics recycling and recovery.

4 The European Compost Network is the leading European membership organization promoting sustainable recycling practices in composting, anaerobic digestion and other biological treatment processes of organic resources. More information about the ECN-QAS is available at http://www.ecn-qas.eu

Germany

In Germany, composting started with mixed municipal waste in the 1980s, resulting in what can be described as a disastrous image of compost, due to quality problems related to the presence of contaminants and heavy metals from mixed waste and, thus, no real market for such compost. Consequently, by the end of the decade the waste compost plants were shut down and the application of mixed waste composts was legally banned.

The waste and recycling sector took a different approach – that of producing compost of very high quality through the separate collection of organic waste from kitchens, gardens and parks. This was however not sufficient to convince customers of the good quality of compost or gain their confidence. Therefore the German Compost Quality Assurance Organization (Bundesgütegemeinschaft Kompost e. V., BGK) was founded in 1990 to establish and manage an independent quality assurance system and raise awareness for a better image of compost through an official compost quality label. The BGK's quality assurance scheme is based on independent control by approved sample takers and laboratories following a set of precautionary (e.g. heavy metals thresholds) and use-related (e.g. nutrient content) quality parameters.[5] Only listed input materials can be composted and are analysed countrywide by the same prescribed methods. The analysis results are used to characterize the compost properties and develop product specifications and recommendations for application, which is also managed by BGK. All plants that pass the system can carry the RAL quality label[6] and use it for public relations and advertisements for their high-quality products. So a recognizable and reliable national standard for quality compost was established to which suppliers and customers can refer when meeting each other on the market – a key element to make a waste-derived product marketable on a large scale. These strict quality efforts were instrumental in overcoming the negative image that compost had in the 1980s. In 2000 the BGK scheme was extended to separately collected digestate materials coming from anaerobic digestion.

Due to the proven effectiveness of the BGK quality assurance scheme, its core precautionary elements became part of Germany's legislation on biowastes in 1998 (revised in 2013).[7] This legislation covers the application of biowastes and green wastes on agricultural, forest and horticultural land, as well as suitable raw materials, quality and hygiene requirements and the treatment processes and analyses of such biowastes and mixtures. The treatment plants within the BGK quality assurance programme are widely exempted from the typical monitoring and documentation duties laid down in this ordinance. Germany expects that composts and digestates carrying the RAL quality label will meet the EU end-of-waste criteria on compost when these are finally agreed.

Due to the BGK's consistent strategy of focusing on high quality from the beginning, RAL quality assurance of compost and digestate has established itself as a national brand, well known to all customer groups. Currently the demand for the product exceeds the supply on the market. Agriculture in particular appreciates this organic fertilizer and its soil improving stable organic matter that helps to maintain the humus balance in agricultural soils. What began in 1989 with just a few small composting plants ended up in 2014, after 25 years, as a success story of nearly 700 composting and anaerobic digestion plants for organic waste recycling, treating around 10 million tonnes per annum,[8] which is more than 75% of the separately collected biowaste in Germany. The latest German waste law "Closed Substance Cycle Management Act (KrWG)" of May 2012 implements the EU Waste Framework Directive into national legislation, with a strong focus on high grade recycling and recovery of resources. For biowaste it reflects the idea of recovery of resources such as nutrients (primarily phosphorous), organic matter (humus for the soils) and biogas (a renewable energy source). Section 11 of this Act stipulates the obligation to implement separate collection of biowaste nationwide before 1 January 2015, as much as technically and economically feasible. This is expected to increase the quantities to approximately 12 million tonnes per annum.[9]

Source: http://www.ral-guetezeichen.de/

5 See http://www.kompost.de
6 See http://www.kompost.de

7 See http://www.gesetze-im-internet.de/bundesrecht/bioabfv/gesamt.pdf
8 H & K aktuell (2014), available at http://www.kompost.de/fileadmin/docs/HuK/H_K-12-2014.pdf
9 Schneider, M. (2011). Bioabfallsammlung in Ballungsräumen – Flächendeckung sinnvoll?. *Müll und Abfall* 3/11, 109-113.

Italy

Following the enactment of the general Waste Act (D.Lgs. 22/1997) of 1997, which radically changed the legal framework and vision for MSW management, separate collection of biowaste became a strategic element to reach the recycling targets set out in the national law. In response, separate collection and composting have increased significantly across Italy. Biowaste is usually collected by two complementary schemes for separate collection: a scheme capturing kitchen and food waste, and a scheme intercepting green waste, with lower collection frequencies than the food waste scheme and different collection tools. Separate collection of municipal kitchen, food and garden wastes currently accounts for 40% of all source separation of MSW. Official data for the year 2013 show 5.2 million tonnes per annum of biowaste to be separately collected and recycled in composting or anaerobic digestion (AD) plants.[10]

While in 1993 Italy had only about 10 composting plants, with the spread of separate biowaste collection, this number grew to about 240 operating plants in 2013, of which 150 each have an annual capacity exceeding 10,000 tonnes. In recent years, anaerobic digestion (AD) of biowaste has been steadily increasing. This form of treatment is chosen in almost all newly opening facilities, in combination with aerobic treatment (composting) of the digestate. In 2013 there were 43 AD plants with a total annual throughput exceeding 1 million tonnes.

According to the Italian legislation, compost is a product (rather than waste) if it is obtained from separate collected feedstock, as required by MoE Decree of 1998, and in compliance with the criteria stipulated in the Fertilizers Law, version of 2010, D.Lgs. 75/2010. In 2013 the composting sector produced over 1.5 million tonnes of compost out of separately collected feedstock. About 70% was used as crop fertilizer while 30% was processed into garden and landscape products.

Source: http://www.compost.it Source: http://www.compostabile.com/

In 2003 the Italian Composting and Biogas Association (CIC)[11] started developing a system and a label for compost quality (CQL), with the aim of enhancing the quality of compost produced by its members and making customers aware of the advantages of compost use while ensuring transparency and reliability. Compost samples are taken by qualified sampling operators that are trained and authorized by CIC and are independent from the composting plant. The samples are analysed by independent laboratories, accredited by the Italian Ministry of Agriculture (MIPAAF) for analysis of soil improvers and growing media and for the European Ecolabel. In addition to sample analysis the CQL system requires composting plants to verify the traceability system for feedstock and the final destination of the compost they produce. The trademark "CIC Compostable" logo is issued after audits and tests performed by CIC in cooperation with Certiquality, a market leader in certification standards. As a result, about 50 products successfully comply with Compostabile-CIC certification and over 30 companies are licensed to use the logo to label their products. By 2013 about 30% of the compost sold in the Italian market had obtained the CIC quality label.

Separate collection of biowaste, particularly of kitchen and food waste, has become a standard approach in many waste management districts of Italy. Moreover, the separate collection of biowaste has been showing a long-term trend of increasing by 5% annually, as more municipalities implement it. By 2020 an increase of approximately 50% compared to the 2013 data is expected in the amount of biowaste recycled in Italy, as Southern Regions of the country fulfil the national targets for recycling and extend food waste collection schemes.

In addition, in 2006 CIC started a programme for certification of biodegradable materials and compostable products. The programme ensures the material's ability to be turned into compost during the industrial composting process. The programme applies the European regulation EN:13432 on compostable packaging. A unique characteristic of the CIC's testing procedure is that it tests compostability directly in industrial full-scale composting plants, hence applying real operating conditions to the materials tested.

10 CIC (2014). Report on biowaste management in Italy. http://www.compost.it/english-version-home-english-version-77/861-country-report-on-biowaste-collection-and-recycling-in-italy.html

11 http://www.compost.it/about-us-english-version-79.html

United Kingdom

Composting across the UK has increased significantly in response to EU Directives. As municipalities face ever-higher recycling targets, coupled with restrictions on the landfilling of biodegradable municipal waste posed by the 1999 EU Landfill Directive, composting is seen as a cost-effective and efficient means of transforming biowaste into a valuable product. Recognizing that quality is essential for underpinning the development of sustainable markets, the then UK Composting Association launched a compost quality standard and certification scheme in early 2000. This was developed further a few years later to become the British Standards Institute's Publicly Available Specification 100 (BSI PAS 100).

Despite having in place a national compost standard and certification scheme setting stringent quality criteria coupled with a third-party audit to verify consistency and compliance, compost had remained legally classified as a 'waste', and was therefore subjected to full waste management controls. Needless to say, this posed significant difficulties for composters attempting to market their composts to farmers, landscapers and retailers, and a solution needed to be found. In order to address this problem and create conditions for compost to be removed from the waste regulatory regime, industry, regulators and government bodies came together to review the quality of compost produced to the BSI PAS 100 standard, identify major markets and assess potential risks to the environment or human health. This resulted in the publication of the 'Quality Protocol for the Production and Use of Quality Compost from Source-Segregated Biodegradable Waste' in 2007 and an updated version in 2012.[12] If compost complies with the criteria set by the Quality Protocol, it can be used without the need for waste regulatory controls in its handling, transport and end use.

The Quality Protocol sets a number of criteria, including those concerning, for example, the source of input materials, quality and usage of the product, supplemented by any additional customer specifications, as well requirements for record keeping and proof of compliance with the Quality Protocol through an independent Compost Certification Scheme.[13] Since the launch of the revised Quality Protocol in 2012, over 10 million tonnes of input material has been certified, which is equivalent to approximately three million tonnes a year, processed at over 150 individual composting sites. This represents about half of all composting in England, Wales and Northern Ireland. The UK has been lobbying hard with the European Commission to ensure that the future EU end-of-waste criteria, still under discussion in 2015, are compatible with their already-established Quality Protocol.

Compost
Certification Scheme

Source: http://www.qualitycompost.org.uk/

12 https://www.gov.uk/government/uploads/system/uploads/attachment_data/file/297215/geho0812bwpl-e-e.pdf

13 Further information about the compost certification scheme is available at www.qualitycompost.org.uk; information about other Quality Protocols is available at http://www.wrap.org.uk/content/quality-protocols

4.3.6 Legislation on waste prevention and sustainable consumption and production (SCP)

Going beyond resource recovery within traditional waste management systems (through recycling, composting, anaerobic digestion and thermal treatment with energy recovery), there is an array of possible instruments, either applied in practice or proposed in literature, which can be deployed towards the goal of sustainable consumption and production (SCP), which are 'upstream' from waste management in the product life cycle. Some of them have the form of direct regulation, particularly when it comes to qualitative waste prevention – legislation is put in place to make sure that the designers and manufacturers do indeed opt for non-hazardous materials. This legislation can have various forms, ranging from outright *prohibitions* to *restrictions* on the content of certain hazardous substances used, to mandatory *requirements to obtain and report information* on the substances in products manufactured or imported, so that the information can be shared with other parties in the supply chain. An example of each form is presented in the box below. It should be noted that legislation focusing on production is targeted not only at manufacturers but also at importers.[59]

BOX 4.10 QUALITATIVE WASTE PREVENTION BY LEGISLATION

The European 2006 Directive on batteries[60] *prohibits* the placing on the market of batteries and accumulators that contain more than 0.0005% mercury and more than 0.002% cadmium by weight (with some exemptions). The EU 2011 Directive on RoHS (Restriction of Hazardous Substances)[61] *restricts* maximum concentration values of mercury, lead, hexavalent chromium, and two flame retardants, polybrominated biphenyls and polybrominated diphenyl ethers, to 0.1% by weight, and of cadmium to 0.01% by weight in homogeneous materials in electrical and electronic equipment. The 2006 Regulation concerning REACH (Registration, Evaluation, Authorisation and Restriction of Chemicals)[62] *requires* manufacturers and importers of chemicals to obtain relevant information on their substances, which is shared with other actors in the supply chain so as to manage them safely. Its scope covers all substances, whether manufactured, imported, used in production or placed on the market, with some exemptions. It also establishes the European Chemicals Agency (ECHA) to gather the information and make it available at the EU level.

BOX 4.11 'FREE FROM' DOES NOT YET MEAN SAFE

When the legislator requires certain hazardous substances to be eliminated from a specific product, it is important to include requirements regarding the environmental performance of the substitute material as well. In the past, omissions to do so have led to situations where a product was made 'free from' one hazardous substance only to introduce another one instead. For example, in the case of vehicle brake pads, asbestos was replaced by antimony, which is an officially recognized carcinogen.[63] As a result, abrasion of vehicle brake pads contributes hundreds of kilograms of antimony emissions in an urban area per year.[64]

A ban on plastic carrier bags is a special case of regulatory waste prevention measures put in place to prevent the consequences of their indiscriminate dumping.

59 Other examples of legislation for SCP as well as other instruments for SCP can be found in UNEP (2012b), listed in Annex A, Chapter 1, Precursors of the GWMO. Some instruments are also discussed in Topic Sheet 3 on SCP, found after Chapter 2.

60 Directive 2006/66/EC on batteries and accumulators and waste batteries and accumulators, http://eur-lex.europa.eu/legal-content/EN/TXT/PDF/?uri=CELEX:02006L0066-20131230&rid=1; for more information on EU policies on batteries see http://ec.europa.eu/environment/waste/batteries/

61 Directive 2011/65/EU on the restriction of the use of certain hazardous substances in electrical and electronic equipment, http://eur-lex.europa.eu/legal-content/EN/TXT/PDF/?uri=CELEX:02011L0065-20140129&rid=2; for more information on EU policies on RoHS see: http://ec.europa.eu/environment/waste/rohs_eee/

62 http://eur-lex.europa.eu/legal-content/EN/TXT/PDF/?uri=CELEX:02006R1907-20130701&rid=2; for more information on REACH see: http://ec.europa.eu/enterprise/sectors/chemicals/reach/index_en.htm

63 Uexküll, O. von, S. Skerfving, R. Doyle, M. Braungart (2005). Antimony in brake pads – a carcinogenic component? *Journal of Cleaner Production* 13, 19–31.

64 Månsson, N.S., D.S.T. Hjortenkrans, B.G. Bergbäck, L. Sörme, A.V. Häggerud (2009). Sources of antimony in an urban area, *Environmental Chemistry* 6(2),160-169

BOX 4.12 ALTERNATIVE APPROACHES TO CONTROLLING PLASTIC CARRIER BAGS

Thin plastic carrier bags are strong, lightweight and moisture resistant, thus lending themselves to various beneficial and convenient applications. However, due to these very properties they create serious environmental and health problems in their wider surroundings as they are blown by wind and transported quite some distances. They do not decompose under ambient conditions and thus accummulate and clog drains, aggravating floods. When swallowed by domestic animals, they do not decompose in their digestive tracts, thus leading to bloating, an array of adverse health effects and eventually death by starvation. This often has dire economic consequences for farmers.[65] The problems are aggravated in places with inadequate SWM services, where discarded bags litter streets, watercourses and agricultural fields.[66] Countries around the world have put various instruments in place to reduce the use of thin plastic bags, ranging from outright bans to obligatory charges or taxes or voluntary agreements with retailers.[67] Each of these instruments has its merits and can be successful and effective, depending on the local circumstances. In general, experience shows that public interest is a key factor. In order to create such interest, authorities need to direct concentrated efforts toward organizing public awareness campaigns, providing information and engaging with citizens, in partnership with other stakeholders such as retailers, NGOs and schools.

Bangladesh banned thin plastic bags in 2002 as a response to devastating floods in 1998 when plastic bags clogged street drains and exacerbated the situation. Under similar circumstances of severe monsoon flooding in Mumbai and the entire state of Maharashtra, India, in August 2005, Maharashtra banned the manufacture, sale and use of thin plastic bags.[68] The state of Kerala banned plastic bags under 20µm thickness in 2003 and has been very active in implementation. Kenya, Mauretania, Rwanda and other African countries have banned plastic bags of specified thickness, usually 20-40 µm. The bans have had varying degrees of success, as they are in some cases difficult to enforce on powerful industries or numerous retailers. Even if authorities are committed to strongly enforcing a ban, as is the case in Rwanda, implementation remains problematic as adequate alternative packaging is often lacking, so vendors resort to smuggling bags from abroad and continue to use them. Thin plastic bags have also been banned in the state of California and individual cities in the U.S., in the state of South Australia, and in other countries.

Ireland is an example where a combination of a levy introduced in 2003 and intensive and sustained public awareness campaigns resulted in an impressive 94% reduction in the use of plastic bags over a very short period.[69] In December 2014 the EU approved the proposal for a Directive to reduce consumption of lightweight plastic carrier bags (with thickness under 50 µm), mainly through levies and taxes, with specific targets of a maximum of 90 such bags to be used per capita per year by the end of 2019, reduced to a maximum of 40 bags by the end of 2025. The proposed Directive also highlights the importance of public information and awareness campaigns.

In some cases voluntary agreements have brought about the same results as a ban or a tax. For example, in Luxembourg, the reusable 'eco-bag' project[70] was launched in 2004 as a national packaging waste prevention plan, based on a voluntary agreement signed by the Ministry of Sustainable Development and Infrastructure, Environmental Agency, Luxembourg Trade Confederation (clc) and Valorlux a.s.b.l. (Green Dot Organisation Luxembourg). The agreement was extended for five years in 2008 and again in 2012. Reusable bags were introduced in two sizes, costing 0.70 and 0.50 EUR respectively. In 2007 the distribution of free single-use shopping bags was discontinued and replaced by what was appropriately named *stopgap* shopping bags (sac de dépannage; Behelfstüte), costing 0.03 EUR. As a result of the project, consumers' habits have changed dramatically: nowadays, over 85% of customers use reusable bags for shopping, two-thirds of which are 'eco-bags'. Since the launch, this project has avoided the use of 560 million single-use bags, representing a saving of 3,700 tonnes of plastics. In 2012 the European Commission declared the project as a best practice in waste prevention.

4.3.7 Regulation on waste handlers

In order to make it clear who is responsible for what kind of actions, the law also prescribes responsibilities and duties of the key stakeholders in the system, particularly waste generators, governmental agencies in charge, and service providers in waste collection, transport, treatment and disposal. One feature of the waste sector is its transition towards a process industry with increased pre-treatment of waste prior to the disposal of residues and the co-location of several technologies at a single site, often combining resource recovery with residuals management. This makes the job of a legislator more complex than in a situation in which facilities focus on a single method of waste management.

65 Ramaswamy, V. & H.R. Sharma (2012). Plastic bags – threat to environment and cattle health: Retrospective study from Gondar city in Ethiopia. *IIOAB Journal-India* 2(1): 7-12. Available at http://www.iioab.org/SPI-1(EBT)/Ramaswamy-IIOABJ-2%20(1)-(SP1)-7-12p.pdf

66 Box 1.1 presents an example of flooding in Accra, Ghana; plastic bags are one contributor to marine litter – see Topic Sheet 9, found after Chapter 3.

67 An overview of measures undertaken to reduce use of plastic bags per country can be found at http://www.earth-policy.org/plan_b_updates/2013/update123 An overview of measures in various U.S. states is available at http://plasticbagbanreport.com

68 Wilson, Velis & Rodic (2013), listed in Annex A, Chapter 1, Waste management, and references therein

69 http://www.environ.ie/en/Environment/Waste/PlasticBags/

70 http://www.valorlux.lu/en/operation-eco-bag

A particularly important part of the laws and regulations on waste handlers concerns specific requirements for waste tracking 'from cradle to grave', usually focusing on hazardous waste, in order to ensure that the waste materials do not get 'lost' *en route*. This is not dissimilar to controls on the transport of dangerous goods and chemicals in other stages of their life cycle.[71] Due to the high costs involved with environmentally sound waste treatment and disposal, in combination with ever-increasing waste amounts and public protests against siting waste facilities, the 1980s saw numerous scandals involving the export of hazardous waste from high-income countries, where stringent regulations had just been introduced and enforced, to countries where this was not the case. This was the primary driver leading to several international agreements on the transboundary movement of hazardous wastes (Box 4.13). If a country has expressed its consent to be bound by an international treaty such as the Basel Convention, then it has the obligation to implement the treaty at the national level by adopting and enforcing adequate legislation. There are many examples of good legislation in place to oblige waste generators to record their waste and send it to appropriately licensed facilities. Despite the Basel Convention having 183 Parties (as of 2015), the sheer volume of the international container trade poses practical difficulties with policing it. With the world largest shipping terminals receiving in the order of 50,000 containers every day, it is impossible for the regulator to inspect each container. In some places the problem is exacerbated by the corruption of the custom officers.[72] All this means that the control of waste trafficking is an on-going issue.[73]

BOX 4.13 MULTILATERAL ENVIRONMENTAL AGREEMENTS (MEAs) ON TRANSBOUNDARY MOVEMENT OF HAZARDOUS WASTES AND OTHER WASTES[74]

The Basel Convention on the Control of Transboundary Movements of Hazardous Wastes and their Disposal is the only global treaty specifically focused on hazardous and other wastes. It was adopted in March 1989 and came into force in May 1992.[75] As of August 2015, 182 States and one integration organization are Parties to the Basel Convention. In order to protect human health and the environment against the adverse effects of hazardous wastes and other wastes, the Basel Convention places obligations on the Parties to reduce generation of these wastes; ensure their environmentally sound management, including adequate disposal facilities; control and reduce their transboundary movements as much as possible; and prevent and punish illegal traffic.

Under the umbrella of the Basel Convention, and in collaboration with other international institutions, an array of valuable normative tools have been developed and their use is promoted in order to achieve effective and coherent implementation at the global level, including technical guidelines, manuals, interactive tools, and others.[76]

Capacity building, in particular for developing countries, is an important element facilitating implementation of the Convention through the operations of a network of 14 Basel Convention regional centres and the establishment of public-private partnerships.[77]

Based on the Basel Convention, the Bamako Convention and the Waigani Convention have been elaborated among the countries in their respective regions. Both these conventions contain provisions similar to those of the Basel Convention but go beyond it by *prohibiting* the import of specific waste streams, including radioactive wastes.

As an African regional treaty, the Bamako Convention on the Ban of the Import to Africa and the Control of Transboundary Movement and Management of Hazardous Wastes within Africa came into force in April 1998.[78] According to the latest update, it has 25 Parties. The Bamako Convention prohibits the import of hazardous wastes into Africa from non-Contracting Parties.

71 UNEP (2013b), listed in Annex A, Chapter 1, Precursors of the GWMO.
72 As a result, goods such as household appliances and computers are officially exported for *reuse*, whereas they may actually be heading for environmentally unsound recycling, polluting local environment and impacting health of workers and local residents. Or waste materials that are officially destined for recycling may in fact be sent for cheap dumping overseas.
73 See Section 5.3.4 on waste crime.
74 Text is in part provided by the Secretariat of the Basel, Rotterdam and Stockholm Conventions.
75 http://www.basel.int
76 http://www.basel.int/TheConvention/Publications/TechnicalGuidelines/tabid/2362/Default.aspx
http://www.basel.int/TheConvention/Publications/GuidanceManuals/tabid/2364/Default.aspx
77 http://www.basel.int/Implementation/TechnicalAssistance/Partnerships/tabid/3235/Default.aspx
78 http://www.au.int/en/content/bamako-convention-ban-import-africa-and-control-transboundary-movement-and-management-hazard

As a South Pacific regional treaty, the Waigani Convention to Ban the Importation into Forum Island Countries of Hazardous and Radioactive Wastes and to Control the Transboundary Movement and Management of Hazardous Wastes within the South Pacific Region entered into force in October 2001.[79] According to the latest update, it has 13 Parties. The Waigani Convention prohibits the import of all hazardous and radioactive wastes from outside the South Pacific Forum[80] Island Countries.

The 2014 Minamata Convention on Mercury is a new international treaty, with its Article 11 on mercury wastes.[81] This Article stipulates that mercury wastes should be treated in an environmentally sound manner, taking into account the guidelines developed under the Basel Convention and in accordance with requirements to be adopted by a Conference of the Parties to the Minamata Convention. The Minamata Convention will enter into force when 50 countries have ratified it. As of August 2015, it has 12 Parties.

Rotterdam Convention and Stockholm Convention

Together with the Basel Convention and the Minamata Convention, the Rotterdam Convention on the Prior Informed Consent Procedure for Certain Hazardous Chemicals and Pesticides in International Trade[82] and the Stockholm Convention on Persistent Organic Pollutants (POPs)[83] provide the international legally binding framework governing the environmentally sound management of hazardous substances (including wastes) throughout their life cycles. The Rotterdam Convention was adopted in 1998 to address a dramatic growth in chemicals production and global trade since the 1970s. Among other provisions under the Rotterdam Convention, a list is established of hazardous pesticides and industrial chemicals that are subject to an informed consent procedure prior to their international trade. The Stockholm Convention was adopted in 2001 in response to the urgent need for global action to protect human health and the environment from POPs, as these chemicals are not only highly toxic to both humans and wildlife but also persistent and transported long distances in the environment, bioaccumulative in tissues of living organisms and increasing in concentration along the food chain.

The four Conventions enable and require the effective coordination of local, national, regional and global environmental policies for effective implementation and enforcement of their provisions, to strengthen institutional frameworks and ensure policy coherence among parties, for maximum effectiveness. Under the Basel Convention, import or export restrictions and prohibitions strengthen the ability of Parties to only accept wastes if they have the capacity to manage them in an environmentally sound manner.

Defining illegal traffic in hazardous wastes and other wastes as a crime, as provided for by the Basel Convention, increases the chances of detection of illegal shipments, thereby preventing inadequate disposal of hazardous substances – illegal shipments are those most likely to be disposed of in a manner that can later pose risks to human health and the environment.

Through promotion of environmentally sound management of wastes, including their treatment, recovery, reuse and recycling, the four Conventions contribute to protection and preservation of natural resources and thereby to sustainable development.

Notwithstanding the benefits achieved, challenges for the coming period include further implementation and enforcement of the Conventions as well as active promotion of waste prevention, minimization and resource recovery. The use of cleaner technologies and production methods is particularly important so as to find appropriate alternatives to POPs and reduce use of hazardous substances, taking into account social, technological and economic concerns.

In order to further the implementation and overcome the challenges, current efforts are directed toward further improvement of institutional and technical capabilities, as well as developing or strengthening mechanisms for monitoring and promoting implementation and compliance with the Conventions. This includes the establishing of effective cooperation mechanisms and partnerships between government authorities and key stakeholders, including industry and businesses, NGOs, academic institutions, international and regional organizations, such as regional centres; development of opportunities for technical assistance (enhancing industry support for knowledge-sharing and capacity building); and development and subsequent use of procedures and mechanisms for promoting compliance. The Conventions are also contributing to broader societal efforts to enhance information exchange, education and awareness raising about these important issues.

79 https://www.sprep.org/attachments/Publications/FactSheet/Waigani_Convention.pdf
80 The South Pacific Forum changed its name to the Pacific Islands Forum in 2000.
81 http://www.mercuryconvention.org
82 http://www.pic.int
83 http://chm.pops.int

4.3.8 Voluntary agreements (self-regulation and co-regulation)

In general, a combination of consumer demand and pressure from legislators is increasingly encouraging businesses to pay attention to the environmental performance of the materials used in their products. Accordingly, they may opt for voluntary agreements, which are particularly successful when led by large companies in the sector. Government can initiate, guide and facilitate the process of negotiations in order for the parties to reach agreement. For example, large retail chains may agree to decrease their packaging. Such agreements are often sealed by a formal Memorandum of Understanding. Also, in some countries extended producer responsibility programmes are established based on voluntary agreements, as is the case with, for example, e-waste in Switzerland, packaging in Norway, Singapore and South Africa, and various waste streams in Brazil.[84]

BOX 4.14 A CLEAN LAND, A CLEAN HOME – SINGAPORE'S SWM STRATEGY[85]

In the 1960s, waste in Singapore was dumped in swamps. With the commitment to providing a clean environment and a clean home for the people, the government devised a multi-pronged approach involving land-use planning, construction of infrastructure and facilities, public education, monitoring and enforcement, in developing Singapore's solid waste management system, with the central idea that waste is a resource. Cost-effective systems to collect and treat waste are used to keep waste management costs affordable for citizens.[86]

As one part of the strategy, in order to minimize the use of scarce land, all combustible waste that is not recycled is treated in energy-from-waste (EfW) plants, whereby the waste volume is reduced by 90% and energy is recovered to produce electricity that meets up to 3% of Singapore's total electricity demand. The plants are equipped with state-of-the-art pollution control equipment. Only non-combustible waste, comprising 2% of Singapore's waste, and residues from the EfW plants are sent to the Singapore's only landfill, the offshore Semakau Landfill, which was commissioned in 1999.[87] Covering a total area of 350 hectares, a 7-kilometre perimeter rock bund was built to enclose a part of the sea off Pulau Semakau and Pulau Sekang. The bund was lined with an impermeable membrane and a layer of marine clay to ensure that the leachate would be contained within the landfill body, from which it is removed and treated at the leachate treatment plant. Landfilling operations take place in cells; when a cell is filled up, it is covered with a layer of earth. Vegetation, ranging from grass to trees, forms a green landscape; also, a mangrove area was planted at one end of the landfill, and has since flourished with rich biodiversity. This has become one of the distinguishing features of Semakau Landfill, in that a waste disposal site can co-exist with a vibrant marine eco-system, mangroves, grassland, and shoreline habitats, providing visitors with many possibilities to enjoy various outdoors activities.

Singapore cityscape

84 Examples are presented in Box 4.12 on plastic bags in Luxembourg, Box 4.14 on Singapore, Box 4.19 on EPR in Switzerland and Topic Sheet 13 on EPR initiatives from countries outside the 'usual' OECD countries, found after Section 4.5.

85 Text is provided by the Singapore National Environment Agency (NEA).

86 Households pay a flat monthly fee, which is less than 0.5% of the average monthly household income (2014 levels). The fee is collected via a monthly utility bill that also includes the fees for electricity, water and gas.

87 Semakau Landfill has gained local and international recognition for its engineering excellence and co-existence with nature, such as the Singapore Good Design Mark Gold Award for 2014 and praise as the 'Garbage of Eden' in a *New Scientist* article of 12 April 2007. It also featured in 2012 'Nature Journeys with Zeb Hogan', an educational collaboration of the National Environment Agency, Singapore (NEA) and the National Geographic Channel.

The other major focus of Singapore's waste management strategy concerns recycling. As the recycling rate of domestic waste was very low, the National Environment Agency (NEA) launched the National Recycling Programme (NRP) in 2001 to provide recycling collection services to the residents. It started off with the provision of recycling bags with fortnightly door-to-door collection. Following feedback from residents living in high-rise public flats built by the Housing and Development Board (HDB), where over 80% of Singapore's population lives, regarding space constraints for storage of recyclables, coupled with a demand for more recycling infrastructure, the number of recycling bins and the collection frequency have gradually increased. Since 2011, a 660 liter recycling bin has been provided for every HDB block of flats, with collection service at least three times a week. To increase convenience further, NEA has been promoting the adoption of a dual-chute system for recyclables and residual waste. In light of the encouraging results of the trial projects, all new public high-rise residential developments will be fitted with Centralized Chutes for Recyclables (CCR) from 2014. Recycling for condominiums and high-rise private estates began with voluntary provision of recycling receptacles by the management within their estate grounds. However, response was slow; therefore, following a consultation with stakeholders, it was eventually mandated in November 2008 that estates be required by law to have such receptacles.

As urbanization and technological advancement increase, Singapore is looking into new approaches and improvements to waste management. A facility for recovery of metals from the incineration bottom ash is expected to begin operations in 2015. NEA is also studying the possibilities of ash reuse, so as to further increase resource recovery and prolong the lifespan of Singapore's only landfill.

In parallel to providing the 'hardware' components of the system, NEA and other stakeholders actively promote the 3Rs (reduce, reuse, recycle) to enhance citizens' sense of environmental ownership. For example, the public waste collectors – one of NEA's key partners – provide incentive programmes such as *Cash for Trash,* where cash is paid in exchange for recyclables. NEA collaborates with schools to implement the School Recycling Corner Programme, launched in 2002, accompanied by 3Rs Awards for schools since 2003. Since 2009, all primary and secondary schools, as well as junior colleges, have set up recycling corners in their premises. A Preschool 3Rs Awareness Kit, consisting of a set of pictocards and a teacher's guide, was also developed to assist kindergarten teachers in planning activities to kindle pre-schoolers' interest and educate them on both the importance of practising the 3Rs and what and how to recycle.

Since the implementation of the NRP, households' participation rate has increased from 54% in 2004 to 76% in 2013, with household recycling rates at 20%. The overall recycling rate, for household, industrial and commercial waste together, has risen from 45% in 2002 to 61% in 2013. Looking ahead, the government has set out a vision of becoming a 'Zero Waste Nation' with a target of achieving a national recycling rate of 70% by 2030. The 3Rs will remain an integral part of Singapore's waste management strategy, and NEA will extend its partnership with companies to promote recycling, especially in the areas of food waste and e-waste.

In addition, Singapore has been exploring new avenues toward achieving its ambitious environmental goals. In 2007, NEA signed a voluntary agreement, the Singapore Packaging Agreement (SPA), with industry and other parties to reduce packaging waste, which constitutes about one-third of Singapore's domestic waste by weight. The Agreement aims at creating an opportunity for the industry to assume greater corporate responsibility to reduce their packaging waste and enjoy cost savings through process and product improvements. At the same time, it offers a platform where companies can share their experiences, exchange practical ideas and collaborate to develop cost-effective solutions to reduce waste. Over 160 organizations, including industry associations, companies, and NGOs, have signed this Agreement. Over the past seven years, the signatories have cumulatively reduced about 20,000 tonnes of packaging waste, resulting in cumulative savings of more than 44 million SGD[88] in the material costs of locally consumed products. In addition, Singapore's Minister for the Environment and Water Resources presents the annual 3R Packaging Awards to the SPA signatories for their notable efforts and successful packaging waste initiatives, which include reduction of the thickness, weight or size of packaging, optimization of production processes, elimination of unnecessary packaging and changing the way products are packaged.[89]

88 Approximately 33 million USD
89 More details on the SPA, as well as the work done by the award winners in reducing waste may be found at NEA's website, http://www.nea.gov.sg/SPA.

The enactment of appropriate laws and regulations, such as those discussed in Section 4.3 above, is just the beginning – not the end – of creating an effective legal system on waste management. For laws and regulations to be of value and actually benefit society, they need to be implemented and enforced. Promptly acting, and then following through on the implementation once the legislation is enacted, not only highlights the importance of waste issues but also shows government competence and increases its credibility. The implementation process requires several distinct but interrelated steps:

- **Translation of laws and regulations** into so-called 'interpretative' and 'decisional' documents. As laws and regulations may need clarifications and detailing in order to be properly implemented by those to whom they are addressed, interpretative documents are prepared to support and facilitate their application in practice. Supporting documents offer guidance on the interpretation of definitions; operating procedures; standards for waste handling and controls in place during waste collection, transfer and transport; acceptance criteria for waste destined for treatment or disposal at a facility with a certain level of environmental protection; procedures and standards for recovery and recycling, treatment and final disposal; and reporting requirements. If well devised, such documents can be very helpful in bringing about the envisaged changes laid out in the law. In turn, the input of waste practitioners and other stakeholders during the preparation of these documents can greatly contribute to their practical usefulness and relevance. These documents may take diverse forms, including technical and other guidelines, protocols, codes of practice, communications on best practices, memoranda of professional bodies and trade associations to their members, declarations, and the like. Professional federations such as ISWA also provide valuable input and guidance.

- **Allocation of authority** through the creation or adaptation of institutions. Legislation also establishes duties and authorities of appropriate government agencies and organizations. This includes the regulatory agency in charge of implementing strategies and enforcing pertinent legislation through licensing (permitting) and inspecting waste handling activities as well as controlling financial arrangements with service users. The independence of the regulatory agency is of paramount importance. If this governmental institution also takes on other roles in the system, particularly that of waste operator, then the potential for conflicts of interests and partiality arises due to self-regulation of the inspection and control of its own waste handling activities and financial arrangements. Therefore, it is highly preferable to avoid such a situation.[90]

  The institutional arrangements for such a regulatory agency may be organized via national or lower, regional[91] governments, according to the subsidiarity principle – the level is not the critical factor in its effectiveness, provided that the agency has adequate capacity to perform its duties and the system works across the whole country.[92] Either a separate regulatory agency may be established, or regulatory duties may be assigned to an existing institution. In either case, information about the new authority needs to be communicated well to both the stakeholders affected by the new legislation and the institutions that are to collaborate with it, to enhance its visibility and status.

- Developing/improving **waste collection services**, supported by appropriate operator models and financing structures.

- **Developing waste facilities**.[93] While possibly not that obvious to the legislator – as mistakes made in the early days in European countries testify – this is an essential step for the waste system to function. Ideally, adequate facilities are developed in parallel to the development of legislation and the documents derived from it. This is because if the facilities are not there, waste generators and other handlers will not be able to fulfil their legal obligations. If facilities are built first, before the legislation takes effect, they are likely to remain unused (and fail to generate revenue) until there is some legal pressure on the waste generators to take their waste to the new facilities, rather than to existing facilities operating at (much) lower environmental standards and therefore able to offer a lower price or cost nothing at all. As developing facilities usually requires large or very large investments, this can pose a serious 'implementation conundrum', and various

90 Different roles that governmental institutions can have in a SWM system are addressed in Section 4.8, which discusses government as a stakeholder.

91 Here 'regional' means province (as it is called in e.g. Argentina, Canada, PRC), state (India, U.S.), federal state (Brazil, Germany), region (France, Italy) and other, at a similar level of administrative organization.

92 The issue of institutional coherence among agencies is discussed in Section 4.8.2, on institutional coherence.

93 An interesting background paper on this issue is from SLR Consulting (2005). Delivering key waste management infrastructure: Lessons learned from Europe, available at http://www.twinning-waste-bacau.ro/twinning-3/activities/pollution-prevention/ste-outputs/3-3-2-b-materials/document-britains-waste-the-lessons-we-can-learn-from-europe

financing approaches have been used to resolve it under specific local circumstances, as discussed in Section 5.8. On the same note, as the development of waste facilities as a response to changes in legislation is likely to be costly, legislation should establish a clear path and create 'regulatory certainty' – as opposed to ever-changing strategies and laws with each change in government – for operators to be able to adjust and invest.[94] Paying due attention to the management of the transition period is therefore critical; it is important to allow time for new instruments to be implemented, and for change to take hold; a phased approach to implementation often works best; but equally, all stakeholders need to know when existing, non-compliant facilities will finally be closed down by the regulator.

- **Data collection and reporting**, consisting of developing supporting services, compulsory record keeping and declarations of waste generated by industries and municipalities, and monitoring and reporting on facilities performance. This category includes rules and guidance on how to comply with legal obligations concerning licensing of operations and facilities as well as reporting about operations of a running facility, such as (a) requirements regarding registration and reporting, such as declarations of waste generation and consignments of waste transport including export and import, (b) details of the licensing (permitting) procedures for new operators and facilities, (c) protocols for emission monitoring, sampling, testing and reporting, and (d) reporting of incidents involving hazardous substances.

- **Enforcement**. The law specifies what constitutes illegal activities, accompanied by the specification of penalties for them. Unfortunately, the list of failures to comply with regulations over the years around the world is a long one indeed, and it includes deliberate unauthorized waste dumping, acceptance of non-permitted waste streams at treatment or disposal facilities, inadvertent spills, illegal discharges, falsification of information, illegal waste shipments and acceptance thereof, and other criminal activities. Experience shows that credibility and transparency of enforcement is essential to promote compliance among waste handlers.[95] This entails not only the prosecution of offenders but also application of a significant enough punishment that will have impact and provide effective deterrence to other potential offenders. In some countries, the message that the authorities are taking waste crime seriously is reinforced by broad publicity given to the court cases in the media, as was the case in the Czech Republic after it joined the EU and was in the process of adjusting legislation. The message the Czech authorities sent to the public at the time was unmistakably clear: new rules and regulations (in this case, on environmental protection) are here to stay and we will make sure that they do. An important aspect of enforcement is the institutional capacity to actually monitor and inspect compliance in order to enforce the law. This is often lacking, as further discussed in Section 4.8.3.

No segregation. No collection, Philippines

- **Monitoring and evaluation**. While the implementation stage is important to bring about the envisaged changes in practice, it can also be seen as a test of the suitability, robustness and practicality of the piece of legislation being implemented. This is particularly true for the derived documents. Therefore, after some time passes and experiences accrue with the implementation, taking stock and consulting practitioners and other stakeholders for their feedback is a valuable learning moment for the legislator, after which an update of these documents, and perhaps the basic legislation, may be warranted. This implies that it is important that provisions for review and update are made within the original regulation.

94 The linkage between this need for 'regulatory certainty' and investor confidence is discussed in Section 5.5.2 on raising investment finance and Section 5.8.5 on private sector participation in investment.

95 For example, in some places even street littering cannot be fined due to the problem of corruption, as apprehending offenders may affect the political careers of those in power or the perpetrators may have 'friends in high places'.

BOX 4.15 NATIONAL GUIDANCE DOCUMENT ON WASTE MANAGEMENT PLANNING IN COSTA RICA

In Costa Rica, the Law on Integrated Waste Management No. 8839 (*Ley para la Gestión Integral de Residuos No. 8839 – Ley GIR*) was promulgated as the national framework law by the Ministry of Health in 2010. Among other stipulations, the Law places responsibility on the municipalities for the integrated management of the waste generated in their counties (cantons), including preparation of SWM plans that will then guide their actions. In addition, a more elaborated manual, the 'Guide for the Development of Municipal Plans on Integrated SWM', has been developed as a reference tool to assist municipalities and other local stakeholders in this process and strengthen the capacities needed to implement the Law.[96]

BOX 4.16 ELABORATION OF NATIONAL LAWS INTO MORE DETAILED REGULATIONS – THE EXAMPLE OF HAZARDOUS WASTE IN ARGENTINA[97]

Argentina is a federal republic of twenty-three provinces and one autonomous city, Buenos Aires. While each province has its own environmental legislation with National Laws as the framework, implementation arrangements are different for non-hazardous and hazardous waste. Non-hazardous waste facilities and operations are regulated and authorized by provincial and municipal authorities, in accordance with National Law No. 25.916 on the management of household waste. Hazardous waste, however, is addressed at the national level to ensure traceability from its generation to its final destination, with the Secretariat of Environment and Sustainable Development of Argentina (Secretaría de Ambiente y Desarrollo Sustentable, SAyDS) as the national agency in charge. This is based on National Law No. 24.051, promulgated in 1992, which regulates management (generation, handling, transportation, treatment and disposal) of hazardous waste generated by industrial activities and services, concerning especially: 1) hazardous waste that is generated in one province but transported to another for treatment and/or disposal; 2) export of waste (hazardous and non-hazardous) or import of non-hazardous waste. (The import of hazardous waste is prohibited in Article 41 of the National Constitution.[98])

As some hazardous waste is generated by households, and it is necessary to control its destiny once discarded, this waste is addressed by specific additional regulations. For example, Resolución (Decision) No. 1729/2007 and its amended form Res. 204/2010[99] address operations of businesses for repair and maintenance of toner cartridges (waste stream Y12 Annex I of the Basel Convention; Annex I of Law No. 24.051). The Resolución requires that these businesses, once the cartridges cannot be refilled, repaired and reused any more, must dispose of this waste (the cartridges and the waste resulting from operations) according to the same Law No. 24.051 that applies to industries. In order to implement this Resolución, every subject that performs such activities is registered as a Generator in the National Register of Hazardous Wastes,[100] which was established under the same Law. Furthermore, the SAyDS may inspect their premises at least once a year. Municipal and provincial authorities have the right to control the management of this waste as well, under their own norms and jurisdiction.

96 http://ley8839.go.cr/blog/seleccion-documentos/guia-elaboracion-planes-municipales-girs/; http://www.ifam.go.cr
97 Text is provided by Alberto Santos Capra and Irina Talamoni of the Autonomous City of Buenos Aires, Argentina.
98 Ley 24.051 de Residuos Peligrosos, http://www.ambiente.gob.ar/?aplicacion=normativa&IdNorma=147&IdSeccion=22
99 Manejo de Residuos de Cartuchos de Tonner, http://www.ambiente.gob.ar/?aplicacion=normativa&IdNorma=933&IdSeccion=22
100 Resolución Nº 204/2010 Registro Nacional de Generadores y Operadores de Residuos Peligrosos, http://www.ambiente.gob.ar/?aplicacion=normativa&IdNorma=1217&IdSeccion=22

4.5.1 Economic (market-based) instruments

In combination with 'command-and-control' policy instruments, economic instruments can be used to achieve waste policy goals.[101] Economic instruments are a typical example of a 'carrot and stick' approach, combining financial gains ('the carrot') for waste generators (both citizens and businesses) and other stakeholders that can be derived from adoption of a specific practice or behaviour, with financial losses ('the stick') associated with another practice or behaviour. As such they usually aim at changing the behaviour of waste generators, influencing practices of product manufacturers, or attracting the interest of the private sector to invest in the development of waste facilities or services.

As briefly introduced in Section 4.3.1, economic instruments rely on market-based incentives and disincentives.[102] Accordingly, some of the main benefits include lower implementation costs and greater flexibility in industries' responses to the policy as well as encouragement for technological innovation. However, the potential benefits strongly depend on and are determined by the political context in which the economic instruments are applied.[103]

An array of economic instruments has been devised to support waste management. Looking from a public administrator's perspective, they are commonly divided into revenue-generating, revenue-providing, and non-revenue instruments, as presented in *Box 4.17*. Alternatively, these economic instruments could be divided according to the policy goals that they aim to achieve, for example, improving resource use, preventing pollution, redistributing responsibilities among stakeholders, and creating jobs.

BOX 4.17 COMMON ECONOMIC (MARKED-BASED) INSTRUMENTS IN SWM[104]

REVENUE-GENERATING INSTRUMENTS	REVENUE-PROVIDING INSTRUMENTS	NON-REVENUE INSTRUMENTS[105]
– User charges (including PAYT) and gate fees[106] – Taxes on waste management options (e.g., landfilling, incineration) – Green taxes (eco-taxes) on consumption and production (e.g. taxes on plastic carrying bags, packaging, or the use of hazardous substances in products)	– Subsidies – Tax credits (fiscal instruments) for private companies – Development rights and property rights (e.g. for land reclaimed from disposal sites) – Host community compensations for facility siting – Grants (e.g. for research) – Funds for environmental improvements (e.g., Superfund)	– Liability for environmental damage – Public procurement requirements (e.g. price preference[107] for goods with specified percentage of recycled materials) – Tradable pollution rights – Deposit-refund systems (e.g. for beverage containers) – Extended producer responsibility (EPR)

A few policy issues in each category are highlighted below.

Charges and fees as revenue-generating instruments are based on the premise that, in accordance with the 'polluter pays principle', people are supposed to pay for the services that take care of their waste. Cost recovery from waste fees has been broadly advocated, but in fact cities can be clean and have their waste

101 It should be noted that economic instruments are also introduced and given legitimacy through legislation.
102 Harrington & Morgenstern (2004), listed in Annex A, Chapter 4, Economic instruments.
103 UNEP (2004); Helm (2005)
104 Adapted from the IDB report by Cointreau & Hornig (2003).
105 In this category, the IDB report also lists various legal instruments (such as bans on landfilling specific waste streams and procurement laws that create a level playing field for the private sector) as well as social instruments (such as name-and-shame lists of polluters, cleanliness competitions among cities, eco-labeling, and information disclosure).
106 See Section 5.7.3 on direct charging.
107 A percentage by which offered prices for recycled products are reduced for purposes of bid evaluation.

properly treated also if the financing comes from other sources, such as the central government budget. For example, in a number of countries, charging for municipal SWM services is contested by the service users, who tend to see these kinds of communal services purely as a 'duty' of the government for which citizens should not have to pay. How governments deal with this particular issue, and by extension, their attitudes toward cost recovery, vary across a broad range. In practice, solutions borne out in the local context tend to result in viable and effective services, and cannot readily be copied from one place to another. Charging for services can be organized in various ways, as elaborated in Section 5.7.

Pricing is often used as an economic instrument to encourage and enhance desirable behaviour by the citizens. A promising approach concerns a waste collection charging system for residual mixed, unsegregated waste (based on receptacle volume, actual weight, or frequency of collection), popularly known as 'pay-as-you-throw' (PAYT). This system has been applied in a number of municipalities in, for example, Austria, Belgium, Denmark, Germany, the Netherlands, and the U.S., as a financial incentive to encourage households to either reuse or segregate their waste. It has however been expressly prohibited in the UK, after a political battle and public opposition, based on the media framing this charging system as 'bin tax' and issues related to invasion of privacy.[108]

This way of charging for services has been effective in bringing down the amounts of residual mixed waste by 20% on average.[109] Despite doubts that this difference might be illegally dumped or burnt, evidence from some communities where this is easy to measure (because they have only one service provider for all waste streams) shows a corresponding increase in the amount of segregated waste.[110] It must be said that enforcement had a prominent role during the transition. For example, when differentiated tariffs for different waste streams were introduced in Flanders, Belgium, household waste that was illegally dumped was examined for clues to ascertain who had owned the waste (from letter envelopes and other post with an address on them) and offenders were heavily fined. This example also shows the importance of a mix of instruments.

A tax or surcharge (levy) is often used as a disincentive so as to deter waste handlers from opting for certain methods of waste management, usually disposal (landfilling) of organic or combustible waste, or incineration. Such a tax is imposed by the governmental authority (usually at the national level) on the top of the gate fee charged by the operator of the waste facility, thereby increasing the total charge to be paid. Denmark introduced a landfill tax as early as 1987 (in parallel to an incineration tax), due to a lack of disposal space in combination with the country's recycling ambitions. A landfill tax was introduced in the Netherlands in 1995 and increased to 85 EUR in 2000, which was more than double the real price of landfill operations and the highest in Europe. At 107.5 EUR per tonne of combustible waste, the Dutch landfill tax remained the highest in Europe in 2011. In combination with a ban on the landfilling of more than 60 waste streams, the tax proved very effective in redirecting waste from landfills to incineration plants in the Netherlands, but also to landfills in neighbouring Germany, which did not have such a tax at the time.[111] In the UK, landfill tax in the amount of 8 GBP per tonne of landfilled waste (in addition to the gate fee charged by the landfill operator, typically in the range of 10-30 GBP at the time) was introduced in 1996. The tax had been gradually increased to 80 GBP per tonne[112] by 2014 so as to prompt councils and companies towards recycling by making landfill more costly and thus less attractive, in expectation that it will help the UK achieve the diversion goals (for organic waste) of the EU Landfill Directive. Several other EU countries have a substantial landfill tax for organic, combustible or any waste, including Austria, Belgium, Finland, Ireland and Sweden. South Australia also has such a tax (a waste depot levy) in place. If sufficiently high, such a tax has proven to be very effective in diverting waste from disposal in these countries.[113]

Incineration taxes have been introduced in Austria and in Scandinavian countries to boost recycling. For that purpose, Denmark introduced incineration tax at the same time and in the same amount as its landfill tax. Sweden and Norway had incineration taxes but abolished them in 2010 to steer the waste market in the region. The revenue generated by landfill and incineration taxes is used for various purposes; for example, in Austria, it raises funds for the remediation of contaminated sites, in South Australia (SA) it partly finances the activities of Zero Waste SA – a specialized government agency promoting the 3Rs.

108 For more details see the 2014 report at http://www.serco.com/Images/Serco%20Eunomia%20Incentives%20Full%20Report_tcm3-44276.pdf

109 Goorhuis, M. et al. (2012). New developments in waste management in the Netherlands. *Waste Management and Research* 30(9 SUPPL.1): 67-77.

110 Finnveden et al. (2012), listed in Annex A, Chapter 4, Policy combinations

111 Such disparities in taxes between neighbouring countries usually prompt waste handlers to transport waste to the country with the lower tax. Accordingly, after a landfill ban was introduced in Germany in June 2005, the export of waste from the Netherlands to Germany for disposal ceased.

112 Approximately 124 USD (as of August 2015).

113 Watkins et al. (2012)

BOX 4.18 POLICY INSTRUMENTS AND THEIR EFFECTS IN THE NETHERLANDS[114]

In addition to the strategic decision of opting for incineration, two policy instruments have played a major role in shaping the SWM system in the Netherlands: landfill bans and a very high landfill tax, both introduced in mid-1990s. As a result, 60 landfills have been closed and, currently, out of some 60 million tonnes of waste generated per year, only 1.5 to 2 million tonnes are landfilled. As of 2014, there are 20 landfills in operation, with a combined remaining capacity of some 50 million m^3. However, due to low quantities of incoming waste, landfill operations are run at a financial loss. Waste cannot be imported from abroad as the EU legislation does not allow transboundary shipment of waste if it is destined for disposal. In fierce competition for waste from local clients, operators have decreased prices below the actual cost level – the gate fee is less than 20 EUR per tonne, while actual costs amount to 35-40 EUR per tonne. But by staying in business at a loss, landfill owners are eating away at their obligatory financial provisions for closure and aftercare. In 85% of the cases public entities are the owners; as such they are also responsible for landfill aftercare. They cannot find a buyer in the current situation, while the excess landfill capacity and possible bankruptcy of operators represent a huge cost to society. In such a way, they find themselves 'between a rock and a hard place' – either they incur losses as a shareholder now, while the landfills are still in operation, or later on, after the landfills are closed and need regular maintenance.

In contrast, revenue-*providing* instruments are devised as incentives to motivate actors in the system to engage in specific practices and behaviours. Typically, various fiscal instruments can be used to encourage private companies to invest in waste management services and facilities, through tax breaks on equipment, reduced import duties, low-interest loans and the like. Such instruments are often combined with contractual obligations that reduce the financial risk for private companies, such as guaranteed access to sufficient waste, as well as project support such as help in permitting and siting processes or even providing land for facilities. While attracting private companies, it is important to balance their business interests with the affordability of the introduced services as the key to ensuring the financial sustainability of the system.

In support of local authorities as they implement policies at the local level, central governments can provide (part of) the necessary funding.[115]

Based on the legal liability for environmental damage caused by accidents and emissions from waste facilities, an array of economic instruments has been developed to secure the finances necessary for waste generators and waste companies to pay for the clean-up and restoration of contaminated sites, as well as for compensation, penalties and fines. For example, in Argentina, environmental insurance is mandatory for prescribed industrial operations that pose risks to the environment.[116] Major insurance companies now have divisions that provide financial services to private companies handling hazardous substances and waste. In the case of waste landfills, operation costs include funds set aside for post-closure maintenance of landfills. Instead of relying on financial insurance, companies are often willing to apply more stringent measures of environmental protection than the law requires or participate in voluntary certification schemes so as to avoid negative publicity associated with litigation and pollution while also protecting and enhancing their public image and market share.[117]

Non-revenue instruments can also be used to create a market in rights (expressed, for example, through permits) to use a resource or generate pollution. Starting from established quotas on emissions of, for example, greenhouse gases (usually expressed as CO_2 equivalent) or amounts of waste that are allowed to be landfilled, tradable rights have been introduced to enable waste handlers to buy their way into compliance. The rationale behind this instrument is that the total effect is maintained – it is just the relative contribution of the individual waste handlers that changes. Also, if the rights are bought from a developing country, it is argued that the financial gain could be channelled towards other beneficial activities.[118] At the same time, these tradable rights have been criticized as inconsistent with the broader global and national efforts towards sustainability and shared burden, particularly among those who create most of the emissions and waste, but who can afford to buy additional rights to pollute.

114 More details and other aspects can be found in Scharff, H. (2014). Landfill reduction experience in the Netherlands. *Waste Management* 34(11): 2218-2224.

115 See Case Study 8 on Flanders, found after Chapter 5.

116 Article 22 of Argentina's National Environmental Law (Ley Nacional 25.675 – Ley General del Ambiente de la República Argentina); http://www2.medioambiente.gov.ar/mlegal/marco/ley25675.htm. This serves to avoid situations in which companies go out of business and taxpayers pay the bill for the remediation of polluted sites, as in the case of the U.S. Superfund, which was established to finance the clean-up of the messes resulting from uncontrolled hazardous waste disposal, as mentioned in Chapters 1 and 3.

117 UNEP (2011). An Analysis of Economic Instruments in Sound Management of Chemicals. http://www.unep.org/chemicalsandwaste/Portals/9/Mainstreaming/LIRA-Country%20Workshop/DevEconInstruments/Economic%20Instruments%20_Survey%20analysis__Final_JM120127b.pdf

118 This is discussed as a revenue source in Section 5.7.6. Waste and climate are addressed in Topic Sheet 1, found after Chapter 1.

4.5.2 Extended producer responsibility (EPR)

Breaking out of the traditional paradigm of post-consumer waste management as government responsibility, extended producer responsibility (EPR) is an innovative policy approach focusing on products instead of waste,[119] which (1) introduces consideration of the entire life cycle of a product, thus also the product's design and production, instead of only its end-of-use stage and (2) transfers, fully or partly, physical and/or financial responsibility for discarded products and the costs involved with their collection and recycling from government authorities (and thereby the taxpayer) onto producers (brand owners, first importers and manufacturers). Essentially, the EPR economic instruments enable creation of the necessary cash flow to organize collection and recycling whereby they internalize the part of the environmental costs of a product involved with the waste stage into the product price. Producers in the same industry join forces and establish producer responsibility organizations (PROs) to represent them in the system and make it work in practice. While the producer is, fully or partly, responsible for the financing, EPR is inherently a system in which responsibilities are shared among actors in the supply chain, as distributors, retailers and consumers also play important roles. Local authorities are often involved in the organization of the physical waste collection system (for which they get reimbursed by the producers); retailers often have an obligation to take products back or even collect them from the consumer in the case of household appliances; and consumers' participation is crucial as they deliver their discarded products or packaging to a collection point.[120]

Container for separate collection of packaging and paper, Scotland

Financing for EPR schemes is implemented through a number of mechanisms applied at the interface between individual actors. For example, PROs are financed through fees based on the weight or volume of the specific materials they collect, the fee per unit of product put on the market by their members, or the membership and registration fees the companies pay based on their turnover. Ultimately, the costs are passed on to the consumer of course, as they are added to the price of the product one way or another. An advance recovery fee charged at the time of purchase is a typical example.[121] If the fee is explicitly stated as a separate item on the bill, it serves to inform consumers and raise their awareness – this would be a successful mix of instruments, combining an economic instrument for the producers with a social one for the consumers, so as to establish a new social norm of behaviour.

Through the development of suitable waste collection systems in the form of various take-back schemes, EPR aims to reduce waste and increase recycling rates. In addition, EPR is envisaged to encourage environmentally superior product design, which would entail a reduction in resource use (the conservation of resources used to make the product) and selection of materials with fewer and lesser amounts of hazardous substances in them, hence, less hazardous waste to manage at the end of the product's use period.[122] Besides environmental improvements, EPR also seeks to achieve economic benefits in terms of job creation in waste collection and the recycling industry.

While the terminology used may differ from country to country, many countries have enacted legislation that places the responsibility for waste on producers,[123] most notably for packaging and e-waste, but also batteries, end-of-life vehicles, tyres, and other products. In some countries EPR is mandatory and in some it is a result of voluntary covenant agreements among the central government, producers and local authorities, usually based on an MoU signed by the parties involved. Industries often opt for voluntary EPR agreements, as these commonly leave space for the industries to influence the conditions under which they will operate, so as to pre-empt legislation that would make EPR mandatory and potentially create unfavourable conditions.

---

119 The degree to which producers and other actors are responsible differs from country to country. In North America, 'product stewardship' is a more prevalent term, which typically places less financial responsibility on the producer but still adopts a sharing of responsibilities for a product throughout its life cycle, to include the producer, distributor, retailer, consumer and recycler as well as the government. More information can be found at http://www.productstewardship.us

120 References to some widely cited and recent documents on EPR are listed in Annex A, Chapter 4, Extended producer responsibility.

121 Some countries already had deposit-refund schemes in place for some types of packaging before EPR was introduced, and have chosen to keep both. In cases where the two separate collection systems exist in parallel, there will be a degree of 'duplication' and some stakeholders will complain about 'unnecessary costs'.

122 These are, in effect, measures of quantitative and qualitative waste prevention respectively.

123 These include Argentina, Australia, Brazil, Chile, the EU Member States, Japan, the Republic of Korea, Norway, South Africa, Switzerland and Thailand, as well as multiple Canadian provinces and U.S. states.

While producers generally seek to avoid this responsibility and the ensuing costs, participation in an EPR scheme may actually be beneficial for a company's *green* image, in line with advanced corporate social responsibility, and enhance its access to markets and consequently increase sales, as illustrated in *Box 4.19* by the success of a voluntary system established in Switzerland in the absence of specific binding legislation. Governments can apply social instruments (discussed in the next section) to make producers aware of these potential benefits. Also, if producers identify their own interest in voluntary EPR, government can save resources as voluntary schemes tend to be more flexible and cost less than government-imposed ones. In the case of voluntary EPR schemes in place, some retailers or waste handlers may charge the consumers but ultimately engage in illegal dumping rather than proper processing of the e-waste. In such cases, intensified control by the government and pressure from *bona fide* industries have helped.

BOX 4.19 THE SWISS EXPERIENCE WITH VOLUNTARY EPR

When large Swiss institutional customers of IT equipment, who often have multiple suppliers (manufacturers and importers), required a comprehensive solution for their discarded equipment, rather than brand-specific ones, their suppliers readily agreed as they saw the potential for a new marketing strategy – taking care of the obsolete IT equipment would be an incentive to the clients to buy new equipment, particularly in light of ongoing technological advancement and ever-shorter use periods of electronic equipment.

EPR policy instruments have been implemented for over 20 years, mainly in industrialized countries. They have been highly successful in shifting (part of) the economic responsibility for post-consumer waste management from local governments to producers, establishing various take-back systems and involving consumers. They have resulted in diversion of waste from disposal,[124] removal of specific wastes of high environmental concern from the waste stream, dramatic increases in waste recycling rates through the recovery of materials with high market value, and reductions in the administrative and management costs of handling these waste streams for the municipal authorities. In addition, regarding the envisaged contribution of EPR to the changes in design for sustainability, the amount of packaging put on the market has strongly decreased in a number of countries, as industries revised packaging of their products to save money.[125] Notwithstanding such achievements, there has been strong opposition to EPR legislation by some companies, and this seems to have been successful in preventing its adoption, particularly since the financial crisis of 2008-2009. For example, in the U.S., while industries' claims about regulations 'killing jobs' and the need for reducing regulations to stimulate the economy were not successful before the economic crisis, such arguments seem to have become effective since that time, as the pace of adopting EPR laws has slowed down across the country and the EPR legislation that is getting adopted is weaker than earlier legislation was.[126]

Due to the increase in commodity prices over the last decade, some end-of-life products such as e-waste have been attracting private recyclers into the system, resulting in the producers and their organizations that run EPR systems losing out on the value that can be recovered from the waste. EPR legislation has not adjusted to this development.[127]

124 EPR policy instruments are not the only ones applied to divert waste from disposal. Landfill taxes and outright bans may have been the main reasons for the results in this respect.

125 As for other products targeted by the EPR, for example, household appliances, it cannot be stated with any degree of certainty that the changes that have taken place in design and material use in this period are directly linked to EPR policies.

126 Prindiville, M. (2014) Extended Producer Responsibility: Moving forward in hard times. Presented at NAHMMA (North American Hazardous Materials Management Association) National Conference, Orlando, Florida, U.S., 18-21 August 2014.

127 Annex A, Chapter 4 in the section 'Extended producer responsibility' includes a number of documents from 2014 to provide a representative overview of the situation worldwide.

TOPIC SHEET 13

EPR INITIATIVES FROM COUNTRIES OUTSIDE THE 'USUAL' OECD COUNTRIES[1]

Extended producer responsibility (EPR) schemes have been established differently across countries. The potential for EPR to share financial and other responsibilities along the supply chain (and provide municipalities with additional income to cope with the rising quantities of end-of-life products entering municipal solid waste) has attracted interest from countries around the world. Most experience has been in the industrialized countries. Four initiatives from countries from different regions are presented here.

Sharing responsibility in Brazil

Brazil hosted an early voluntary initiative in the form of CEMPRE, an association of major producers of consumer goods established in 1992 to promote recycling of packaging. Legislation for the handling of specific waste streams, including packaging that constitutes hazardous waste (such as packaging for pesticides[2] and lubricant oil), tyres, and batteries, and subsequent management systems have been gradually introduced since 2000. However, it was the National Solid Waste Policy Bill No. 12.305/10 *(Política Nacional de Residuos Sólidos, PNRS)*[3] that established 'shared responsibility' among relevant stakeholders over a product's life cycle, including industry, governments, consumers and municipal SWM services. As the first-ever SWM legislation at the federal (central government) level in Brazil, the law came into force at the end of 2010 following 20 years of discussion, to provide a unified set of policies and requirements for SWM within the broader context of Brazil's sustainable development.

As the key instrument for implementing shared responsibility, the Law introduces reverse logistics – a 'take-back' system for recovery of discarded products, packaging and other wastes from consumers and their return to the production process, including the management of related information flows.[4] Importantly, financial responsibility lies with the private sector. Also, if a public entity responsible for SWM services conducts any activity that the law allocates to the private sector, that entity must be adequately remunerated by the private sector. In order to make this happen, the law provides several options: (a) voluntary 'sectorial agreements' to be made among the stakeholders for the sectors specified by the law, (b) 'commitment terms', which are a semi-voluntary solution in which the competent public agency creates the basis and the private sector commits to it, or (c) by a decree, which is a legally binding solution prescribed by the competent public agency.

© Ainhoa Carpintero

1 Topic Sheet prepared by Ljiljana Rodic, with inputs from Victoria Romero, student at the Imperial College London, UK, and Carlos Silva Filho, ABRELPE, Brazil.

2 As Brazil is the world's largest consumer of pesticides, accounting for some 19% of the world's consumption, this waste stream is an environmental priority. Law 7.802/89 regulates the handling of pesticides.

3 The original in Portuguese available at: http://www.planalto.gov.br/ccivil_03/_ato2007-2010/2010/lei/l12305.htm. An unofficial English translation by the Ministry of the Environment of Brazil available at: http://www.mma.gov.br/estruturas/253/_arquivos/125_publicacao17052011041349_253.pdf

4 Reviews of scientific literature on reverse logistics, its structure and processes include Meade, L.M., J. Sarkis, and A. Presley (2007). The theory and practice to reverse logistics. *International Journal of Logistics Systems and Management* 3(1): 56-84, and Pokharel, S. and A. Mutha (2009). Perspectives in reverse logistics: A review. *Resources, Conservation and Recycling* 53(4): 175-182.

Furthermore, the law identifies informal waste pickers (*catadores*) of reusable and recyclable materials as actors in the system, and requires integration of their cooperatives into the reverse logistics operations, except for hazardous waste, where their inclusion is not allowed. *Catadores* conduct separate collection, based on waste segregation at source. The Ministry of the Environment is developing qualification courses and 'minimum standards' guidance for cooperatives.

© L. Rodic

Soon after the PNRS, Decree no. 7404 prompted the creation of the Steering Committee for the Implementation of Reverse Logistics (CORI) involving five Ministries, with the Ministry of the Environment (MMA) and the Secretary of Water Resources and Urban Environment (SRHU) appointed as the coordinating authorities. Assisted by the Technical Advisory Group (GTA), CORI has developed five Technical Working Groups (WGs), addressing waste electrical and electronic equipment (WEEE), packaging, waste oils and their packaging, lamps, and medical waste. Their purpose is to design their legal sectorial agreements (subject to the approval by CORI, after a public consultation) and establish conditions, targets and obligations for the parties bringing the products on the Brazilian market for the first time – manufacturers, importers, distributors and retailers, related to the implementation of reverse logistics strategies.[5] As of April 2015, despite a delay caused by prolonged negotiations with industry, the WGs have finalized their work, which has been concluded with a Technical and Economic Assessment Study for each specific waste flow under consideration. Furthermore, after completion of all the steps, the WGs for lamps and for waste oils have signed a Sectorial Agreement. The WG for packaging waste received over 1,000 submissions during its public consultation and the GTA is consolidating a new text based on those contributions. The other two WGs, for WEEE and for medicines, are preparing the first draft of their respective sectorial agreements to be sent for public consultation.

A number of challenges will need to be addressed when implementing this ambitious programme. First, there are very large differences in development across Brazilian States, compounded by the country's geographical vastness. Unlike the populous, industrialized south-eastern region with its high concentration of industries, relatively short transportation distances to recycling facilities, and markets for recycled materials, other States face prohibitively high transportation costs to facilities that are both fewer in number and farther apart. Therefore, these capacities will need to be developed regionally.[6] A lack of public awareness and waste segregation habits is another barrier, which results in the need for the costly sorting of waste at facilities.[7] Furthermore, facilities often lack the expertise needed for proper treatment of complex products such as electronic equipment. Last but not least, consistent and accurate data across the country, including local legislation, would help stakeholders fulfil their obligations and track progress towards waste targets.

Tackling the 'mountain' of e-waste in the People's Republic of China

As a major supplier of electrical and electronic equipment (EEE) to the global market, in 2012 the People's Republic of China (PRC) produced in total more than 1 billion refrigerators, washing machines, air conditioners, colour TVs and personal computers and computer equipment, and 1.18 billion mobile phones. While a large portion of this is exported, the domestic market has also grown enormously. For example, while there were 30.8 air-conditioners per 100 urban households in 2000, this had risen to 126.8 in 2012. Ownership of computers has increased almost tenfold in the same period, from 9.7 to 87.0 per 100 urban households.[8] This means

5 Consonni, S. (2013). Brazil's incoming e-waste recycling regulations explained. *Waste Management World.* http://www.waste-management-world.com/articles/2013/03/brazil-e-waste-recycling-regulations-pnrs-explained.html

6 Two examples: Quariguasi Frota Neto, J. and L. N. Van Wassenhove (2013) Original Equipment Manufacturers' participation in take-back initiatives in Brazil: An analysis of engagement levels and obstacles. *Journal of Industrial Ecology* 17(2): 238-248; Silva, D.A.L. et al. (2013). Comparison of disposable and returnable packaging: A case study of reverse logistics in Brazil. *Journal of Cleaner Production* 47: 377-387.

7 Silva Filho, C. (2011) The New Waste Law and the Recycling Sector in Brazil. Available at http://www.beacon2011.com.ar/presentaciones/Carlos-Da-Silva.pdf

8 *China Statistical Yearbook 2013,* Table 11-9 'Ownership of major durable consumer goods per 100 urban households', Table 14-12 'Output of industrial products', at http://www.stats.gov.cn/tjsj/ndsj/2013/indexeh.htm.

literately hundreds of millions of EEE items that will eventually end up as waste. While the PRC has a very active repair, refurbishment, reuse and recycling market for discarded EEE, until recently it was almost entirely in the informal sector, with no rules for operations in terms of occupational safety or environmental protection measures, no quality standards, and inadequate inspection. Such enterprises have also attracted large quantities of illegal e-waste imports from high-income countries.

The EPR approach for WEEE was first proposed in the PRC in the early 2000s. Its foundations were laid by the Solid Waste Pollution Prevention and Control Law of 1996, amended in 2004, and the Cleaner Production Promotion Law of 2003. The new course was reiterated by the PRC's milestone legislation, the Circular Economy Promotion Law (CEPL) that came into force in January 2009. At the heart of CEPL are the life-cycle approach and producer responsibility for the products listed in the newly established 'catalogue of articles subject to compulsory recycling'. As the flagship WEEE legislation for the PRC, the Ordinance for Administration of Collection and Treatment of Waste Electrical and Electronic Products, promulgated in February 2009, provides details about each party's responsibility in e-waste management, including funding.[9] While the legislation was being developed, four national pilot projects were launched in 2004 at different administrative levels involving government authorities, large manufacturers, retailers and consumers. The projects explored the entire range of operations, from a collection network in the city and financial reimbursement to consumers for their e-waste, to development of technologies and equipment for disassembly and recycling, to construction and operation of facilities.[10] Among other findings, the pilot projects revealed people's clear preference to sell their goods to informal recyclers rather than to official recycling companies, due to higher reimbursement. This changed completely during the nationwide old-for-new Home Appliances Replacement Scheme (HARS), which operated in 28 cities and provinces between 2009 and 2011 within a package of measures to stimulate domestic consumption in the face of the economic crisis. HARS provided a 10% discount on new electrical or electronic goods if consumers submitted a voucher proving they had sold their old appliances to a certified recycler.[11]

To support the development of a legitimate recycling industry, a WEEE funding scheme for TVs, computers, refrigerators, washing machines and air conditioners was established in 2012. This funding scheme had its basis in the 2012 legislation entitled the Measures for the Collection and Administration of the Funds for the Recovery and Disposal of Waste Electrical and Electronic Products. Government agencies operate the scheme. Tax and customs authorities collect charges per unit product from producers and importers respectively. The funds are subsequently used to provide subsidies to recyclers certified by the Ministry of Environmental Protection (MEP). In over a year, MEP certified 64 companies for WEEE recycling, mostly situated in the more developed eastern and central provinces. The Ministry of Finance is responsible for the coordination of collection, utilization and administration of the funds.[12]

Notwithstanding the achievements so far, the development of a formal e-waste recycling industry is proving difficult as some major, interrelated challenges remain. First, certified recycling companies are at a financial disadvantage compared to informal recyclers, who do not invest in the health and safety or environmental protection measures needed to comply with legal requirements, and thus can offer higher prices for used equipment. Consequently, formal recyclers often struggle to collect sufficient e-waste with market value for their

9 The implementation was postponed until January 2011, due to the financial crisis. Zeng, X., Li, J. et al. (2013) Perspective of electronic waste management in China based on a legislation comparison between China and the EU. *Journal of Cleaner Production* 51: 80-87.

10 Pilots included the city of Qingdao (Tsingtao) in Shandong Province, the entire Zheijang Province, and the large cities Beijing and Tianjin. Yu, J., E. Williams et al. (2010). Managing e-waste in China: Policies, pilot projects and alternative approaches. *Resources, Conservation and Recycling* 54(11): 991-999; Zhou, L. and Z. Xu (2012). Response to waste electrical and electronic equipment in China: Legislation, recycling system, and advanced integrated process. *Environmental Science and Technology* 46(9): 4713-4724.

11 Tong, X. and L. Yan (2013). From Legal Transplants to Sustainable Transition: Extended Producer Responsibility in Chinese Waste Electrical and Electronic Equipment Management. *Journal of Industrial Ecology* 17(2): 199-212.

12 OECD (2014). How Does the Chinese E-waste Disposal Fund Scheme Work http://www.oecd.org/environment/waste/China%20case%20study%20final.pdf

operations.[13] Second, official recyclers are yet to define or occupy a clear position in the market for second-hand EEE products. Third, organizing an effective but costly administrative system, organizing reverse logistics operations, marketing, creating successful business models and finding adequate funding are common problems in other countries that have introduced EPR, and these are even more challenging in country as large as the PRC.

However, the PRC may have a considerable advantage in the form of available low-cost labour for manual dismantling and sorting.[14] This advantage could be used for the benefit of the entire system, in accordance with the strategic direction of the circular economy. Moreover, the concentration of people in megacities offers possibilities for economies of scale. Building upon the positive aspects of current practices, including those of the informal sector, would be a good starting point for future action.[15] In order to facilitate administration, more responsibility may be transferred to producers through the development of PROs.

Valuable experiences in South Africa

South Africa has accrued considerable experience with EPR, both voluntary and mandatory, with different degrees of success. While voluntary programmes have been successful and a mandatory programme failed to deliver on its envisaged goals, the difference in their legal status is arguably not the main reason why. A mandatory EPR programme for plastic bags failed probably because of the low recycling potential and low market value of plastic bags. In contrast, a voluntary EPR programme for metal packaging was initiated out of producers' interest in a product (packaging) with high market value, supported by their technological expertise and

13 For example, mainly old TV sets with cathode ray tubes (CRTs) are collected, as they have negative value. Obviously, this is funded from taxes paid by producers and importers of all five product groups, which creates an imbalance among them that is hardly acceptable.

14 Wang, Y., Y. Ru et al. (2010). Recent developments in waste electrical and electronics equipment legislation in China. *International Journal of Advanced Manufacturing Technology* 47(5-8): 437-448.

15 This is, in general, one of the key messages of this entire Chapter and indeed the GWMO. Topic Sheet 14 addresses the informal waste sector in greater detail.

cost-effective operations. In the case of PET and glass packaging, voluntary EPR programmes were undertaken by the relevant industries while still involving government, based on MOUs signed among parties. This approach was taken in order to avoid legislation imposed by the government and unfavourable conditions that such legislation could create, as had been the case with plastic bag industry. This is a good example of a policy situation where just 'threatening' with legislation was sufficient to precipitate pre-emptive action by industry and create a system that works.

Historically, as early as 1993 Collect-a-Can was established as a joint venture between ArcelorMittal South Africa and Nampak, to recover and recycle beverage cans across southern Africa. Buying beverage cans from local collectors for cash and processing them in their own plants, Collect-a-Can has been successful in markedly increasing the recovery of beverage cans to the current rate of 72%, which is among the highest in the world.[16]

The ubiquitous presence of discarded plastic bags in the late 1990s is thought to be one of the reasons why South African EPR regulatory efforts initially focused on packaging waste. The 2003 plastic bag regulation imposed requirements on bag thickness and a mandatory levy on manufacturers and importers, which was passed on to consumers. The intent of the legislation was to support recycling and reduce demand for plastic bags. The latter was temporarily achieved, until the manufacturers decreased the price to such a low level that it was significantly below the former (pre-regulation) price. Recycling activities never developed, due to a complicated financing scheme and lack of adequate business plans for viable recycling.[17]

In parallel with this mandatory EPR, other industries have engaged in voluntary EPR activities to avoid similar legislation being imposed upon them as well. In 2004 PETCO was established to enable the PET industry to coordinate its recycling activities, with the government as one of the partners based on an MOU. Similarly, in 2006 the Glass Recycling Company (GRC) was established through a partnership involving government, glass manufacturers, fillers (companies using glass packaging for their products) and recyclers, with the aim of facilitating recovery of glass packaging for recycling. Unlike Collect-a-Can operations, neither PETCO nor GRC processes the materials recovered.[18] The PET system is financed by voluntary levies and annual grants by the members, whereas the system for glass is financed by a levy paid by fillers per tonne of glass bottles they buy from the manufacturers. Both organizations use the revenues to support the recyclers through guaranteed prices, ensuring reliable supply from collectors and support for their livelihoods.[19]

Concurrent with these changes, EPR legislation has also progressed. The 2008 National Environmental Management: Waste Act[20] addresses EPR explicitly by mandating the Ministry, in consultation with producers and other pertinent parties, to identify a list of products to which EPR is to be applied, define specific EPR measures and financial arrangements and set targets for waste reduction. In doing so, the Waste Act opens the door for future EPR legislation to target specific products. This will be supported by a pricing strategy currently under preparation, as required by the Waste Act amendments of 2014.[21] This is in a broader context of a promised 'radical transformation' of the country's approach to environmental management, where environmental considerations take a prominent place in development policies.[22]

Mandatory EPR programmes may be necessary for specific products, such as those involving hazardous constituents that may pose risks at the end of use, those large in size that are difficult to handle and carry to a collection point, and those composed of multiple materials in combinations that make them difficult to recycle, as stated in the country's National Waste Management Strategy of 2011.[23]

Steadily progressing in Tunisia

Tunisia has the oldest EPR legislation in its region, originating in the first National SWM Programme launched in 1993 (Programme National de Gestion des Déchets Solides, PRONAGDES), the keystone of the country's national solid waste policy. Guided by the 'polluter pays' and EPR principles, PROGNADES provided a strategic framework for the development of SWM in Tunisia in the period 1995-2006 and prompted preparation of a number of waste-streams specific EPR decrees, starting with one on plastic packaging waste. In order to facilitate its implementation, Eco-Lef was established in 1997 under Decree 97-1102 as a public system for the recovery and recycling of plastic packaging. Eco-Lef's operations include collection infrastructure and recycling activities, as well as encouragement of industry and community involvement in and education of the public about sound waste management. Financing has been secured by shifting the financial burden onto producers. Namely, the law requires producers and importers to pay a levy of 5% of the value of the plastic goods and resins.

16 http://www.collectacan.co.za/

17 Nahman (2010). Extended producer responsibility for packaging waste in South Africa: Current approaches and lessons learned. *Resources, Conservation and Recycling* 54(3), 155-162.

18 http://www.petco.co.za; http://www.theglassrecyclingcompany.co.za

19 Nahman (2010). Extended producer responsibility for packaging waste in South Africa: Current approaches and lessons learned. *Resources, Conservation and Recycling 54* (3), 155-162.

20 Act No. 59 of 2008, http://www.gov.za/sites/www.gov.za/files/32000_278.pdf

21 http://www.gov.za/sites/www.gov.za/files/37714_2-6_Act26of2014Nat-EnvironManWaste_aa.pdf

22 http://www.engineeringnews.co.za/article/molewa-to-radically-transform-sa-approach-the-enviro-management-2014-07-17

23 http://www.gov.za/sites/www.gov.za/files/nationalwaste_management_strategy_0.pdf

© Verele de Vreede

Informal sector workers search through collected waste for plastics. Sorted materials are transported to one of the licensed collection points. The success of Eco-Lef has resulted in over 100,000 tonnes[24] of plastic waste collected since 2001, across 372 collection points and 110 recycling facilities.[25]

Other EPR decrees include the 2002 Eco-Zit Decree aimed at lubricants and used oils and the 2005 Eco-Batteries and Eco-Pile Decrees aimed at accumulators and used batteries. Those for tyres and WEEE are currently under development, with the financial arrangements under review. EPR law enforcement was strengthened through the establishment of the National Waste Management Agency (ANGed) under the Ministry of the Environment and Sustainable Development in 2005. It is positioned as the government agency in charge of waste management activities, through the monitoring, control and implementation of the national strategy, including technical assistance to municipalities.[26] While ANGed holds these responsibilities, it is private operators that provide the services and treatment. As a part of its institutional capacity strengthening, ANGed is involved in international development co-operation with the German GIZ. Tunisia is also a partner in SWEEP-Net,[27] a regional network for exchanging experiences and expertise in Mashreq and Maghreb countries. Financial assistance to support the development of major facilities is provided by several foreign and international funding agencies.[28]

Over the last two decades, significant progress has been achieved, including the success of Eco-Lef with plastic packaging, a 65% recovery rate for lubricating oils, and very encouraging initial results regarding e-waste.

EPR funds are earmarked and used in part to subsidize the system, which is artificial and based on licensing criteria that have little to do with actual performance, while it has competition from a thriving private chain. Part of the EPR funds is used to subsidize landfilling as an appropriate first step, in parallel with recycling, to eliminate uncontrolled disposal and thus ensure environmental protection. Tunisia is well on the way to close or rehabilitate its estimated 400 uncontrolled waste dumps and construct engineered landfills. However, waste collection services have deteriorated since the revolution, apparently due to vandalism toward containers and vehicles. They are in need of attention that they are not currently receiving. This is because EPR and landfilling are under the remit of the Ministry of Environment and Sustainable Development, while waste collection is the responsibility of municipalities, which are under the Ministry of Interior and Local Development. Again here, looking at the goals of the waste management system as a whole would provide much needed coherence to help foster the maximum benefits for society.

24 Of this amount of collected plastics, 70-90% has actually been recycled, depending on the type of polymer.

25 This large number of facilities is at the same time a weakness, as many of them are too small, lack working capital and technical know-how, have difficult working conditions, and compete with strong Asian traders. See http://www.sweep-net.org/sites/default/files/4.%20Ilyes%20Abdeljaouad_EPR%20-%20Eco-lef.pdf

26 http://www.anged.nat.tn

27 http://www.sweep-net.org/country/tunisia

28 http://www.sweep-net.org/sites/default/files/TUNISIE%20RA%20ANG_0.pdf

4.6.1 Social instruments for behavioural change

Social instruments are applicable to situations in which government aims to raise people's awareness and influence, or indeed *change*, people's attitudes and behaviour, and where 'command and control' and economic instruments are deemed either inadequate on their own, or even inappropriate or undesirable. Social instruments are based on interaction and communication among stakeholders, with or without the direct participation of the government. While there are many possibilities for intervention through social instruments, it is difficult to ensure that the chosen instruments will indeed be effective in achieving behavioural change. This is because people are triggered and motivated by any number of very diverse factors and circumstances, including, for example, social status and aspirations, accepted norms of behaviour, the broader societal context including the media, a sense of agency (the belief that our actions matter) and daily habits.[128] In other words, what people actually do with their waste depends on a complex interplay of factors and circumstances, which may not necessarily be within the reach or the sphere of influence of the waste management department.

A useful framework for designing initiatives that aim to change people's behaviour consists of four complementary elements, as shown in *Figure 4.4*; to be successful, an initiative needs to combine actions from all four elements.

Figure 4.4 The '4Es' framework for designing initiatives to influence (environmental) behaviour[129]

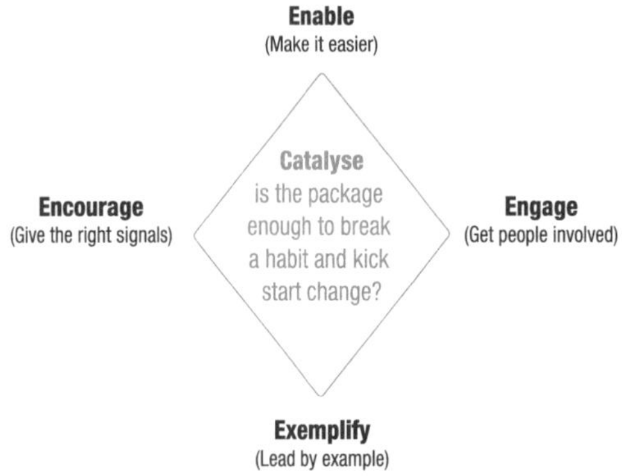

In order to capture people's attention effectively and raise their interest, it is essential for social instruments to take as a starting point the 'sense making' of the citizens and their realities – their lives and surroundings. Furthermore, as many officers in national and local authorities have experienced and research evidence consistently shows, information *alone* is seldom sufficient to bring about the desired and lasting change in people's behaviour. While information and instruction are necessary, it takes engaging people at a level deeper than intellectual understanding to prompt them into action.[130] Also, encouraging people to adopt new patterns of behaviour *while not invading their freedoms* is a challenging task, with possible political implications.

In addition to actions of reaching out *towards* the public, government agencies and organizations have at their disposal a 'powerful, symbolic and highly visible signal of changing norms' in society if they exemplify the desired behaviour and thereby lead by example. For the credibility of any policy, there is no substitute for 'walking the talk', as stated in the 2006 UK Government report of a telling title: '*I will if you will. Towards sustainable consumption*'.[131] Waste segregation at source and, particularly, public procurements towards sustainable, green choices, are some typical examples. In the latter case, governments can include requirements of design for sustainability into their public procurement rules and thereby directly deploy their purchasing power to encourage and support manufacturers (and importers) who adopt such approaches.[132]

128 For critical overviews of the models and theories of change, see for example, Jackson (2005), Darnton (2008) and Shove (2010), listed in Annex A, Chapter 4, Social instruments.

129 DEFRA (2005). Securing the future – delivering UK sustainable development strategy. https://www.gov.uk/government/uploads/system/uploads/attachment_data/file/69412/pb10589-securing-the-future-050307.pdf Examples of using the 4Es model can be found in Case Study 3 on food waste, after Chapter 3, and in Table 6.5. Various other approaches exist to the design of interaction and communication with citizens and other stakeholders; see, for example, Windahl, S., B. Signitzer & J.T. Olson (2009). *Using Communication Theory: An Introduction to Planned Communication*, 2nd ed. Sage Publications, California, U.S.

130 Heath & Heath (2011)

131 Sustainable Consumption Roundtable (2006)

132 See Topic Sheet 3 on sustainable consumption and production, found after Chapter 2.

Design of concrete activities from the basket of social instruments requires specialist knowledge of *social marketing*[133] – something that is probably not readily available in a waste management department. This is an area where engineers and public health officials working in the field of waste management would greatly benefit from contributions from sociology, social psychology, evolutionary psychology, cognitive neurosciences, and other related fields that *corporate marketing* commonly utilizes for its purposes. Admittedly, marketing budgets of governments are incomparably lower than those of corporations; yet governments have not fully exploited the possibilities that collaboration with civil society members could create in this area. For example, in African cities, governments often work with NGOs and community-based organizations to carry out campaigns incorporating a variety of tools – informal meetings with the community and its leaders, informative posters, music and dramatic plays on the radio. In Bologna Province, Italy, once the infrastructure was installed, door-to-door visits by the municipality's advisory teams have worked particularly well to increase waste segregation at source. In Austria, street theatre has been used to raise environmental awareness. In many cases, gatherings where people engage in a dialogue within their communities (rather than reflecting on their environmental habits and behaviour individually, on their own) have proven to be very useful, as they tend to enhance people's sense of community and their common interests. Announcements of public events and other information are increasingly being shared through social media; this trend is very likely to continue and gain pace. Also, governments can collaborate with schools and universities in projects soliciting students' opinions about sustainability and what it takes for them to engage in, for example, reuse of products or segregation of their waste for recycling. In any case, the form of any campaign needs to be suitable and fit in with the local culture and customs.

Educational Material for Awareness Raising "Let us get together to create a waste free environment", Sri Lanka.

Educating women on composting, Nepal.

4.6.2 Information-based instruments

Targeted information provision including non-binding guidance[134]

The prerequisite for any effort to change behaviour related to a certain subject is being aware of the subject and its implications — simply put, knowing about it in the first place. This is how environmental movements started in the past; in the 1960s and 1970s these movements began with learning about oil spills, the effects of dichlorodiphenyltrichloroethane (DDT) on wildlife, urban air pollution, ozone layer depletion by chlorofluorocarbons (CFCs), atmospheric nuclear testing, ocean dumping of toxic and radioactive waste and other human activities imparting adverse effects on the natural environment. Similarly, awareness of the impacts of uncollected waste on health, or awareness of the finite nature of material resources, or awareness of the limits of the assimilative capacity of Earth's ecosystems constitutes the first step towards any kind of action related to waste management. Around the world, a wide range of activities has been organized to provide information to citizens and other waste generators, on, for example, waste segregation at source or similar

133 This term is used in one of the models of behavioural change.

134 This instrument is very different from reporting requirements that legally oblige waste generators and handlers to keep records and report to the government.

environmental behaviour, with the idea: if people are well informed, they will follow through and adopt the behaviour promoted by the information activity.[135] Well, in fact, they will not — or rather, most of them will not, due to an array of factors and circumstances interfering with the person's internal pathway from awareness and knowledge to actual behaviour, as said above.

For information-based instruments to have any chance of success, the waste-related information has to bear some relevance to people's daily realities and concerns. Otherwise, it will lose out among diverse competing claims on citizens' attention and limited (free) time. For those members of the public who do engage in the process and pay attention to the information provided, if they are already aware of areas of ambiguity and public debate taking place in the media, they will not simply accept information that is presented as definitive and simplistic. Experience also teaches that the chance of success generally tends to be higher if people engage within their communities, rather than on their own.

In addition to education and awareness raising on the importance and benefits of reuse, reduction and recycling, information is also needed in the form of instruction on how exactly to segregate waste – which products and materials belong to recyclables and which do not. Information can be provided on the products themselves, so that citizens know that the product is suitable for recycling and ought to be segregated from other waste. *Box 4.20* provides some examples of illustrations promoting 3Rs.

BOX 4.20 3RS LOGOS AND MESSAGES[136]

Source: https://vienna-wv.com/portal/city-information/recycling/

Source: https://commons.wikimedia.org/wiki/File:Recycle001-perc.svg

Source: Paper Recycling Promotion Center, Japan; http://www.prpc.or.jp/menu03/cont09.html

Source: https://commons.wikimedia.org/wiki/File:Tidyman-glass-recycling.svg

Source: https://en.wikipedia.org/wiki/Waste_Electrical_and_Electronic_Equipment_Directive

Source: http://www.kronenbourg.com/1er-brasseur-France/actualites/responsabilite-solidaire/Pages/Kronenbourg-soutient-une-politique-de-traitement-durable-des-emballages.aspx#.Vc2e94v_9-U

Source: e-Stewards Certification Program, USA, http://e-stewards.org; http://www.ban.org

Source: Novelis South America, Brasil, http://www.novelis.com/en-us/pages/novelis-recycling-capabilities.aspx

Source: http://www.cleanup-sa.co.za

135 This was often done based on the rational choice model of human decision-making, which has been strongly criticized and largely replaced by other, more empirically confirmed models. In addition, rational choice model often neglects the historical and contextual (irrational) factors that influence decision-making.

136 http://www.keepcalmandposters.com/poster/keep-calm-and-throw-garbage-in-the-dustbin

Also, citizens can be informed about the producer's involvement in an official recovery scheme, such as the 'Green Dot' scheme for packaging, which originated in Germany but is now active in 28 countries (*Box 4.21*).

BOX 4.21 RECOGNIZABLE SYMBOL OF AN EPR SCHEME

The 'Green Dot' logo on packaging means that the producer has paid a financial contribution to the national organization for packaging recovery that has been set up and authorized in accordance with the European Packaging and Packaging Waste Directive 94/62 and the respective national law. As of 2015, organizations in 28 countries are using the Green Dot as a financing symbol to facilitate organization of a system for collection, sorting and recovery of discarded (mainly household) packaging.[137]

Various information-based instruments have been used to, for example, give visibility to best- or worst-performing companies; such instruments are popularly known as 'name and fame', or 'name and shame', respectively.[138]

BOX 4.22 JAPAN'S 'TOP RUNNER' PROGRAMME[139]

This programme sets energy efficiency standards for products in 21 categories (for example, vending machines, air conditioners, TVs) in Japan. On a regular basis, officials test all the products available on the Japanese market in a given category, determine the most efficient model, and make that model's level of efficiency the new baseline. Manufacturers have the obligation to make efforts to achieve the new baseline within four to eight years. Products that meet the energy efficiency standard receive a Top Runner label. If a manufacturer does not meet the target or fails to make a good faith effort, this fact gets publicized.

4.7 INCLUDING STAKEHOLDERS

4.7.1 The range of stakeholders and their roles

Waste management involves a multitude of stakeholders (or actors), directly and indirectly. The stakeholders play different roles in the system as they are interested in waste from different perspectives and for different reasons. Four *main* categories of stakeholders' roles can be distinguished, as presented in *Figure 4.5*.

137 More information about the 'Green Dot' scheme and individual countries can be found at http://www.pro-e.org/Green-Dot-General-Remarks.html.

138 WasteWise is the U.S. EPA award for outstanding organizations and companies. More information available at http://www.epa.gov/epawaste/conserve/smm/wastewise/index.htm

139 More information can be found at http://www.asiaeec-col.eccj.or.jp/top_runner/index.html

Figure 4.5 Stakeholders and their roles in the waste management system

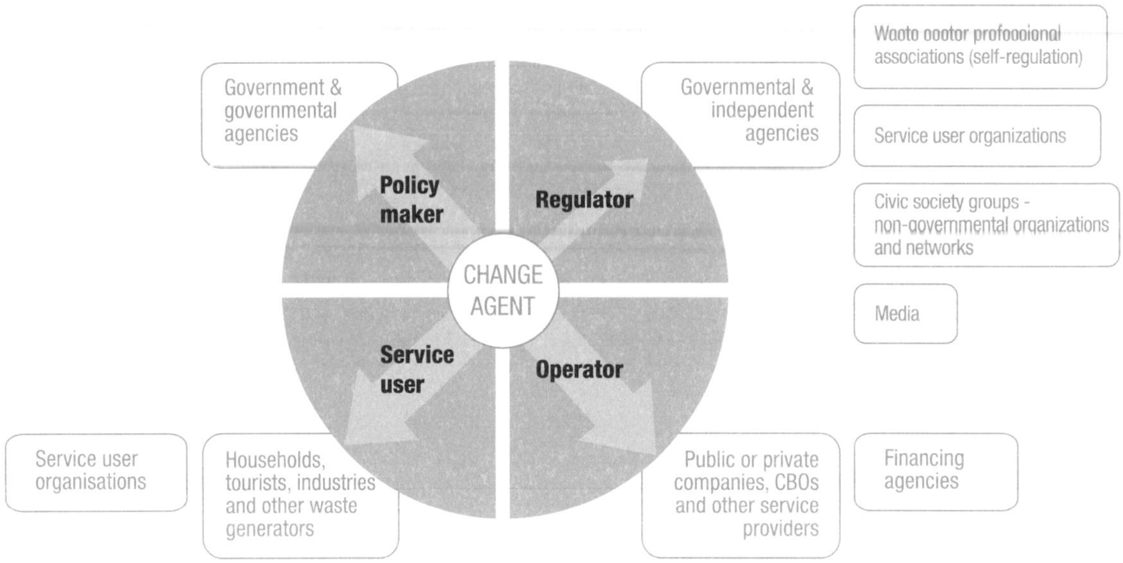

Note: Any stakeholder can step up and be a change agent, taking the system to the next level. While this is not a governmental function, governments are likely to contribute to the initial funding of a 'dedicated' change agent in the system, to help make it happen. Typical examples of such an agent include WRAP in the UK and Zero Waste South Australia.[140]

Waste management starts with waste generators, which includes basically everyone, either as a person and consumer, a resident or a tourist, or as a business owner or an employee of a company or an organization. Waste generators are also users of waste management services and thus have a dual role in the system, which pertains to two different stages in a product's life cycle – the consumption stage and the end-of-use stage. Hence these stages probably require different approaches to involve citizens and influence their behaviour.[141] As a stakeholder, citizens may be represented by community-based organizations, advocacy NGOs and community leaders. Depending on the culture, community leaders may have an authoritative position and their stance may determine that of the community, which is important to bear in mind when designing governance instruments.

Another important category of stakeholders includes government and government agencies, as they create the political setting, initiate and guide the process of strategic planning, prepare a supporting legal framework within which waste is addressed, and define and select various other instruments that will be applied to support the implementation.

At least equally important for the waste management system's performance, albeit possibly in a position of lesser power, are various service providers, from waste collection to transport to operators of facilities. These can be municipal departments, publicly-owned companies, privately-owned companies (varying in size across a very wide range, from multinational corporations to self-employed individuals), community-based organizations, informal workers or their cooperatives, or any combination of these. Looking beyond providers of waste collection services, other actors include those engaging in the repair or reuse of products, the recycling of materials or the agricultural value chain, as well as other actors that recover resources from waste. In this category, international networks of practitioners in waste management, e.g. ISWA,[142] have been instrumental in sharing knowledge and best practices and have contributed to the professionalization of waste management worldwide.

140 http://www.wrap.org.uk; http://www.zerowaste.sa.gov.au

141 In the consumption stage, where waste can be prevented through, for example, purchasing choices or repair and reuse, citizens are consumers and the instruments should be chosen accordingly; in the end-of-use stage, where for example fly-tipping and littering may be a problem, citizens are (unwilling) service users and different instruments are called for. Another example of citizens' behaviour in the end-of-use stage concerns waste segregation for recycling. It is noteworthy that even though it may seem that the same instruments can be applied to encourage citizens to segregate their waste for recycling and to prevent waste in the first place through, for example, reuse, scientific evidence from industrialized countries suggests that people's motivations to engage in one or the other differ; this implies that appropriate policy instruments will acknowledge this difference. Namely, people who engage in waste prevention tend to be more concerned about the environment and are prepared to adopt more frugal lifestyles; people who segregate their waste for recycling often do so because they consider it as the norm for socially expected or desirable behaviour. Moreover, they may not be willing to engage in prevention. The motivations in developing countries may be very different – people who engage in reuse tend to be the poorest ones; their motivations are borne out of economic necessity.

142 http://www.iswa.org/

Finally, the role of a regulator is indispensable for an effective waste management system to monitor the quality and price of services as well as compliance with permits and legislation.[143] The regulator function is backed by various other organizations that provide monitoring and analysis of the environmental performance of waste facilities, including laboratory and other supporting services.

In line with these four direct roles, an array of stakeholders has keen professional or advocacy interest in waste management. For example, they may do research and provide support to policy makers,[144] or they may advocate a particular policy direction or support the position of informal service providers or underprivileged service users. These actors include advocacy and operational NGOs, the media, education and research institutions and engineering and other consultants. For example, international networks such as Zero Waste Alliance,[145] GAIA,[146] RED LACRE[147] the Latin American Network of Recyclers, WIEGO[148] and CWG[149] advocate various ideas on sustainable or inclusive waste management and provide a platform for the exchange of experiences, including those of grass-roots practitioners. Funding agencies such the World Bank, IDB, KfW and others may also have a prominent role, as their interest and input often accelerate activities to improve services and upgrade facilities.

In addition to their *roles* in the system, stakeholders may be in positions with varying degrees of power to *influence* the system and varying degrees of *interest* in waste management topics. These two dimensions, influence and interest, provide a useful analytical framework for stakeholder analysis with the aim of identifying their positions in, for example, the process of strategic planning.[150] Accordingly, different approaches are appropriate for their participation and inclusion:

- For stakeholders of *high influence and high interest* (often referred to as 'key players') it is advisable to ensure their close involvement throughout the decision-making process and subsequent implementation. Examples include large cities (which may have more power than a ministry), federal states (which may have more power than the central government), and large industries.
- For stakeholders with *high influence but low interest*, it may be sufficient to keep them informed and acknowledge their views in order to avoid unnecessary disruption. These could include politicians, governmental agencies and financing agencies that have waste low on their agenda.
- Stakeholders in the unfavourable position of *high interest but low influence* require special attention, to make sure that their needs and interests are adequately addressed. Typically, these include citizens, particularly those in underprivileged neighbourhoods, as well as community-based organizations and informal service providers.
- Finally, stakeholders with *low influence and low interest* are unlikely to be involved and require no particular participation strategies.

Forming multi-stakeholder partnerships is a successful model to provide a structure for result-oriented problem solving. An example is the Partnership for Action on Computing Equipment (PACE),[151] which was developed as a multi-stakeholder public-private partnership to provide a forum to tackle environmentally sound refurbishment, repair, material recovery, recycling and disposal of used and end-of-life computing equipment. Through the Partnership, a network of expertise from different perspectives has been created, where manufacturers, users, refurbishers, recyclers, academia, NGOs, etc. can work together in a strictly neutral structure; PACE has achieved results which have a very high degree of acceptance by all its stakeholder groups.

143 For a discussion of this, see Section 4.4 on implementation and enforcement.

144 Academia and other research institutions, including NGOs, expand the body of knowledge on specific aspects of waste and resource management. In response to legal requirements, researchers have been developing various technologies including engineered controls for environmental protection and technologies for resource recovery. Researchers also investigate various waste management options by comparing their environmental benefits, for example, through LCA and MFA, and work on optimization of logistics networks for recyclables. In the domain of social sciences, governments have sought scientific explanations about human motivations for behavioural change so as to inform their design of corresponding instruments. The scientific community has frequently raised issues of resources use and provided the frameworks and metrics to address them.

145 http://zwia.org

146 http://www.no-burn.org

147 http://www.redrecicladores.net/es/

148 http://wiego.org

149 http://www.cwgnet.net

150 The interest-influence matrix is a popular and widely used method of stakeholder analysis. For a useful recent overview of other methods see Reed et al. (2009).

151 See Topic Sheet 8 on e-waste, found after Chapter 3.

4.7.2 User inclusivity

In general, the need for a basic waste collection service is essentially undisputed. If a service is not provided, or is (perceived to be) unaffordable, then people will take steps to deal with their waste on their own – often by illegal dumping, burying or burning. When a SWM system is being designed and developed, either for the very first time or in a renewed form, waste generators – householders, businesses and others alike – usually have a lot to say about their preferences and concerns. For example, people living in densely populated neighbourhoods, in houses without yards, will be less likely to segregate their waste, whether in Lyon, France or Kumasi, Ghana.[152] Therefore, instead of insisting on one or the other technical 'solution', experience shows that new or revised service provisions have a higher chance of success if they are discussed, negotiated and agreed with those whose needs they are to address. This means that citizens participate in decision-making on the SWM system if they wish to do so.[153] The instruments include an array of possibilities, from various platforms for public consultation in planning the SWM system and siting the facilities,[154] to feedback mechanisms through which the service provider can learn about system performance.

BOX 4.23 WASTE MANAGEMENT PLATFORM IN MOSHI, TANZANIA[155]

Moshi is a municipality of 185,000 inhabitants at the foot of Mount Kilimanjaro in north-east Tanzania. The streets are spotless and lined with trees. There is a clear focus on cleanliness, driven by concerns over public health and the culture of the local Chaga and Pare tribes that both hold cleanliness in high esteem regardless of income level, and are outspoken if someone litters on the street. Accordingly, the municipality is known for its strict environmental and sanitation by-laws. In addition, heavy penalties are imposed on the spot. As a result, Moshi won the official title of 'cleanest city in Tanzania' for several years in a row.

A stakeholder platform on solid waste has been active since 1999, making strategic and action plans that are subsequently implemented. Pilot projects have been used to test new models of service delivery, involving both the local private sector and community-based organizations (CBOs) that provide primary collection in unplanned settlements. This is a result of a broader commitment of the Council and the citizens to urban infrastructure and governance issues, as demonstrated by their active participation in various countrywide initiatives, such as the Sustainable Cities Programme and the Urban Sector Rehabilitation Programme. The Moshi Municipal Council is also a partner in IPLA,[156] a member of ICLEI[157] and maintains collaboration with sister cities in developed countries.

In the special case of siting new waste facilities, if timely and appropriate communication and transparency from the authorities are lacking, members of the local community are very likely to respond with NIMBY attitudes,[158] which may precipitate fierce protests and opposition. This is particularly likely to result if people have been 'burned' by bad experiences with poor waste management practices in the past, such as indiscriminate dumping of hazardous waste that affected their residential areas and water wells or soot from early incineration plants with inadequate emission controls. Reassurances that experts will address possible risks, or explanations that modern engineering practices are much better, will not suffice; people will be reluctant to believe that the governance factors required to make that happen will actually be delivered. For example, this was typical of the dire situation with facility siting in Europe in the 1980s, where in some cases waste had to be temporarily stored until a permanent disposal solution was found. Under such circumstances it is important to engage in dialogue with the community on the possible sites that are selected based on sound environmental and technical criteria, as demonstrated by a high-quality environmental impact assessment (EIA), rather than settling for an unsuitable site simply because there are no objections there from the local community. Such processes may involve negotiations of compensation to the community as well as more fundamental discussions about the need for the waste facility in the first place.

152 Owusu et al. (2013), listed in Annex A, Chapter 4, Inclusivity.

153 In many instances however the public may appear uninterested, indifferent or even apathetic. Such attitudes may have to do with previous experiences in which their contribution had no tangible impact on the outcomes, or with a more general lack of trust in the consultation process. Moreover, people may not be interested to discuss topics that they perceive as 'resolved', such as waste services, which is often the case in industrialized countries. Or simply, waste-related topics will not make it into their 'window of attention' due to more pressing issues in their realities. In some cases, while most of the general public may not be interested, a small but highly visible group will engage in strong opposition and protests.

154 If people feel inferior due to their possibly limited eloquence or lack of knowledge in the technical fields involved, they may feel empowered if they are represented and supported by community groups or advocacy NGOs; in other situations they may prefer to voice their concerns themselves.

155 Information is provided by Alodia Ishengoma.

156 IPLA – International Partnership for Expanding Waste Management Services of Local Authorities, http://www.iplaportal.org

157 ICLEI – Local Governments for Sustainability, http://www.iclei.org

158 A number of abbreviations have been coined to describe attitudes of public opposition to siting industrial and waste facilities, including NIMBY (Not In My Back Yard), BANANA (Build Absolutely Nothing Anywhere Near Anything), NOPE (Not On Planet Earth). In order to avoid confrontation, officials in charge may, in turn, respond with NIMTO (Not In My Term of Office).

BOX 4.24 NEW MINING SITES AND PUBLIC SCRUTINY – THE CASE OF ROSIA MONTANA[159]

Romania has experienced many mining accidents in its history, with the one in 2000 in Baia Mare being the most notorious. Considered the worst industrial accident since Chernobyl in 1986, the collapse of the Aurul tailings dam caused spillage of cyanide that polluted the major rivers in that part of Europe, including the Danube. Even though there were no human casualties, the wave of cyanide and heavy metals killed thousands of tonnes of fish in Romania, Hungary and the former Yugoslavia. The costs of the emergency response alone totalled 1.17 million EUR.[160] The Baia Mare accident precipitated changes in laws and regulations, including the revision of the Seveso Directive on major accidents involving hazardous substances, to cover risks arising from storage and processing activities in mining.[161]

Social media played a significant role in raising awareness on the Baia Mare accident after the news spread about a new mining project that was about to be developed in Rosia Montana in the Apuseni Mountains.[162] While politicians including the president initially supported the project, civil society organized massive protests against it in 2013, receiving extensive media coverage.

Rosia Montana Gold Corporation (RMGC), owned by the Canadian mining company Gabriel Resources, aimed at developing a massive gold exploitation in the area, using cyanide in the industrial process of extracting the precious metal from the ore.

Arguments on both sides were strong. The strong marketing campaign of RMGC focused mainly on the positive economic and social impacts of the project, with some 2,300 jobs to be created, including 880 in mining, and 4 billion USD revenue for the state (thus winning the support of the government). While the technology proposed by RMGC constituted a Best Available Technique (BAT), that message failed to reach or convince the public. Civil society and NGOs fiercely campaigned against the project through online social networks as well as large gatherings in major cities, causing what is considered the largest civic movement in Romania since the revolution in 1989. The arguments against the project included the high quantity of cyanide to be used (13 times higher than the average yearly use in Europe), the impact the project would have on the landscape and historical sites (scarring four mountains, relocating three villages and destroying ancient Roman mining galleries in the area) and insufficient economic benefit. The opponents deemed groundless the state's need for the economic benefits, as they claimed that the state had failed to absorb fully a much higher amount of 26.8 billion EUR that had been made available to Romania through EU grants between 2007 and 2013. The main argument against the project however was the enormous tailings pond that would be left behind after the exploitation, containing 215 million m^3 of cyanide-laden tailings.

With the 2000 Baia Mare accident fresh in collective memory, and given the fact that the RMGC project had been full of controversy and lacked transparency throughout the approval process, combined with the generally low level of confidence people have in the government and authorities, the fear of another accident was understandable. The protests in Romania and throughout Europe pushed for a broader consultation process, which ultimately resulted in the project being rejected.

In an operational system, the transparent sharing of information and the existence of an effective complaint (grievances) mechanism are means of 'downward accountability' toward service users. Complaints provide direct and valuable feedback about the service performance and the quality of services provided. Furthermore, if the complaints are timely and adequately addressed, this helps develop trust and goodwill with customers, which is a good starting point for collaboration (which can be built upon) concerning, for example, good habits of placing appropriate waste at appropriate places at appropriate times, and waste segregation at source.

While participatory processes hold the promise of broader societal support for the policies or legislation at hand, in order for a constructive dialogue to take place, a genuine interest in each other's views is key, guided by clear and agreed goals, and supported by the right setting.[163] Otherwise, there is a danger that the exchange becomes an exercise of 'going through the motions' without any real substance to it, resulting in 'participation fatigue' and a deepened divide between the authorities and the citizens.

While a participatory process may not guarantee collaboration, consensus and agreement, i.e. while it may not result in *pragmatic* benefits for the decision-making process at hand, it has *normative* benefits for the society in a broader sense. Hence, stakeholder participation is generally promoted as a process of social learning and a means of not only enhancing procedural fairness but also challenging what is in the public interest in the first place; it also contributes to the integration of social values into technical decisions, and quality assurance into expert-centred decision-making. Moreover, it helps increase institutional legitimacy and contributes to public

159 Text is provided by Diana Gheorghiu and Flaviu Petean of Green Partners, Cluj-Napoca, Romania.

160 Prommer, M. & K. Skwarek (2000). Report on the Economic and Social Impacts of the Cyanide Spill and Heavy Metal Pollution on River Tisza, Budapest, Hungary: Kornyezettudomanyi Kozpont – Centre for Environmental Studies

161 The Seveso Directive states that companies that store or process significant amounts of hazardous substances have to prepare plans to demonstrate the knowledge and ability to manage the risks from their activities, so as to keep the risks within an acceptable range.

162 A traditional mining area, with mining activities dating back to the Roman period

163 For a useful critical overview of benefits and best practices in stakeholder participations see Reed (2008).

trust and confidence in decisions and decision-makers, and ultimately helps advance democracy. Any one of these enhancements would of itself be a valuable outcome in its own right.[164]

BOX 4.25 TAKING A DIFFERENT PERSPECTIVE ON CUSTOMERS IN BRAZILIAN FAVELAS

Sometimes it is necessary to take an altogether different, novel perspective on a particular stakeholder to make the system work. Perhaps a lesson can be learnt from some of the public service (water supply and electricity) companies in Brazil's *favelas* (slum areas). The companies decided to change the way they treat *gatos* – those who unlawfully tap water and electricity – as thieves; instead they started looking at them as people trying to become customers. Working with the communities, sometimes through NGOs, they are now gradually improving their infrastructure and services and building a new customer base.[165]

4.7.3 Provider inclusivity

Generally, municipal (or other comparable local level) authorities have a legal *responsibility* to ensure that an adequate waste service is provided to citizens. The law may prescribe or allow various operator models. In many places, the law obliges the public entity (either a municipal department or publicly owned waste company) to actually provide the service in the city; in others, the public service provider is obliged – and is the only service provider allowed – to serve households, while commercial waste generators may choose whether they contract with them or with a private service provider; in yet others, contracted private companies may be engaged throughout the city. In addition or instead, the service provider is a local community-based organization or entreprise, or informal sector workers. (This service typically pertains to primary collection, i.e. collection of waste from a residential house or commercial premises and transport to a collection point in the neighbourhood. In such cases, the formal service provider then takes over and transports the waste to the final destination.) Inclusivity of service providers represents the degree to which service providers from both municipal and non-municipal (including the formal private, community or informal sectors) are included in the planning and implementation of solid waste and recycling services and activities. The topic of how to select the appropriate 'operator model' – in order to deliver effective and affordable SWM services in a particular situation – is discussed in more detail in Chapter 5, beginning with Section 5.4.

In some cases, where local authorities, particularly in megacities, are overwhelmed by the daunting task of addressing ever-growing amounts of waste, citizens may help both in the preparation of strategic plans, like in Bengaluru, India, and in the organization of waste collection services, like in Dhaka, Bangladesh, as illustrated in *Box 4.26* below. Such situations, where the system shows a remarkable capacity of self-organization, require an attitude of openness and willingness to collaborate on the part of institutions. This also concerns interest in and support for grass-roots initiatives.

BOX 4.26 GRASS ROOT INITIATIVES IN DHAKA, BANGLADESH

Frustrated by the garbage accumulating on the streets of their neighbourhood in Dhaka, civil society activists started community-based primary collection to clean up their surroundings. The operational success of some initiatives, notably, Parichana Kalabagan in Dhaka and a few others in Chittagong and Khulna, has received considerable attention from policy makers, city government leaders, and international development agencies. Dhaka City Corporation has extended its support to these initiatives by providing earmarked containers for the waste collected by the community-based organizations (CBOs) and also transport assistance. The practices have been replicated around the country by non-government and community-based organizations. Similarly, the local operational CBO WasteConcern has initiated door-to-door waste collection from households and vegetable markets, in which organic waste is subsequently taken to a community-based composting plant where it is turned into valuable product. In order to ensure utilization of the fertilizer and sustain the system, WasteConcern assists communities in marketing the product by contacting and negotiating with fertilizer companies. For the *bastee* (slum) settlements, barrel composting has multiple benefits. It can help in achieving behavioural change by minimizing littering in the slums, improve the environment and create a source of income for slum dwellers. In addition, in cooperation with European private investment partners, WasteConcern was also successful in securing support from the Clean Development Mechanism (CDM) of the Kyoto Protocol for a larger-scale composting plant.[166] These successful experiences are now being replicated elsewhere in Asia.[167]

164 Petts (2004)

165 Neuwirth, R. (2004). *Shadow Cities: A Billion Squatters, A New Urban World.* Routledge.

166 Sinha, A.H.M.M. & I. Enayetullah (2010). *Waste Concern and worldwide recycling: financing Dhaka market composting with public-private partnerships and carbon credits.* See Key Sheet 9, in Scheinberg et al. (2010), listed in Annex A, Chapter 1, Waste management, pp. 117-123.

167 Such cases are presented in Box 4.5.

BOX 4.27 ROLE OF NGOs AND CBOs IN SWM IN KAMPALA, UGANDA

Due to a combination of liberalization reforms in the 1990s and the insufficient capacity of local authorities to provide adequate SWM services to the nation's growing urban population, the Government of Uganda privatized urban sanitation and solid waste services. In 1997 the Kampala City Council prepared a Strategic Framework for Reform (SFR), which promoted the diversification of service providers, including private companies as well as NGOs and community-based organizations (CBOs). The Kampala Capital City Authority (KCCA) – the new governing body that superseded the Kampala City Council in accordance with the Kampala Capital City Act of 2010 – remains committed to promoting public-private partnerships as a means of harnessing the financing, expertise and efficiencies that the private sector can bring to the delivery of public services. However, it is difficult for CBOs to fulfill the requirements and thus to bid directly for the work. In practice, large private companies that bid and win waste management operations in specific zones of the city subcontract CBOs and other micro and small enterprises to provide services in areas that are inaccessible to their vehicles. Importantly, KCCA has divided the city into seven zones based on the equity principle, which is to ensure that informal settlements and slums also receive adequate services, while being cross-subsidized by affluent areas.

Some 15 years after the policy was first implemented, Kampala now has over 40 international and local NGOs and local CBOs actively involved – through partnerships – in the development and implementation of sanitation and solid waste activities. Local branches of international NGOs, which tend to have more financial resources, better knowledge and information, and better access to politicians, focus mainly on community sensitization and mobilization, aiming to change people's behaviour regarding proper waste management and disposal and their attitudes regarding payment of the collection service fees; their activities include advocacy, capacity building, advisory work and monitoring.

In contrast, the domestic local NGOs and CBOs are more involved in physical operations such as waste recycling, drainage cleaning and waste collection, especially in poorer neighbourhoods. The involvement of local NGOs and CBOs has mainly been hampered by a lack of resources, donor dependency, central policies that favour large private companies and a lack of government recognition, as well as their lack of trust and confidence in partnerships or collaboration. Nevertheless, local NGOs and CBOs are fully involved in sanitation and SWM in the city, moving beyond just implementing marginal projects in poor neighbourhoods.[168]

Staff and owner from Wisdom and Insight Investments Ltd

To take one example, Living Earth Foundation, a UK-based NGO, secured funding from the European Union in 2010 for the 'Waste to Wealth' programme to improve living conditions in the slums of Cameroon, Nigeria and Uganda, by capitalizing on the economic opportunities in waste management.[169] A local, community-based waste management company *Wisdom and Insight Investments Ltd.* started in Kampala in 2010. It operated informally, collecting garbage with wheelbarrows and bicycles and disposing of it at dumping sites. In 2012, they became one of 120 enterprises to undertake Living Earth's 'African Urban Enterprise Development Programme'. The Programme was designed to build the capacity of micro and small enterprises in their business development, combining business theory with practical application and follow-up mentoring in the workplace.

Since completion of this training, Wisdom and Insight has registered as a business and, as a result, has been able to secure a formal contract with the local District Council to collect waste in under-served areas. The company has been able to expand its customer base and increase its turnover by 300%. They have pioneered a model of household waste collection in a new area of town and the payment rate for this new service is currently at 75%. While more work needs to be done to address the remaining 25% of householders (who either do not pay, or who require persistent reminding in order to pay), this represents a significant increase in willingness to pay, which is due to both the quality of the service delivered, and the proximity of the company to the community in which it operates.

168 Tukahirwa, J.T., A.P.J. Mol & P. Oosterveer (2013). Comparing urban sanitation and solid waste management in East African metropolises: The role of civil society organizations. *Cities* 30: 204-211.

169 http://www.livingearth.org.uk/projects/waste-to-wealth-uganda/

INFORMAL WASTE SECTOR[1]

In many cities around the world there is a considerable presence of the informal sector in waste management, particularly in cities where there is no formal separate collection system for recyclable materials.[2]

Informal activities tend to intensify in times of economic crises when employment is difficult to find and in cases where imported raw materials are relatively expensive due to inflation or currency depreciation.[3] While political and professional opinions differ about the desirability of informal activities, and the scholarly debate is ongoing about the place, role and contribution of informal service providers, informal economic activities appear to constitute a very significant portion of the economy, accounting for as much as 60% of jobs among the urban population in some places.[4] It is important to realize that the informal sector is very heterogeneous and complex; it operates at various scales, from individuals and small family businesses to large and mighty enterprises.[5]

In recognition of its presence, the informal waste sector is examined here along the three sustainability domains – environment, economy and society. As is often the case, whichever sustainability domain is considered, there are two sides to every aspect.

© Asmare Junh

With regard to the environment and resources, the benefits are evident in many cities. In some places informal-sector service providers are responsible for a significant percentage of waste collection. For example, in Lusaka, Zambia, this is more than 30%. In many cities, the informal sector recovers, reuses or recycles valuable materials from waste and thereby contributes to sustainable resource management. On the other hand, due to the ubiquitous lack of both capital and knowledge at small recyclers, and without any agreements or cooperation with the authorities, working conditions are generally unsafe and unhealthy, and processing residues are often just crudely dumped, causing serious pollution, particularly in the case of e-waste recycling.

Regarding the economic aspect, in the absence of a formal recycling system, informal recyclers benefit cities directly, as they remove materials from the city and thereby save money that would otherwise need to be spent on waste collection, transport, and disposal,

1 Topic Sheet prepared by Ljiljana Rodic.

2 While informal activities are more prevalent in cities without separate collection for recyclables, they also take place in situations where such services exist, particularly in times of economic downturn. The effects of informal activities are quite different for the two situations. Where the informal recycling sector is referred to in the GWMO, the focus is primarily on those developing countries where there is currently no formal, separate collection system for recyclable materials.

3 One example is the informal activities increasing after the currency depreciation in Argentina in the early 2000s. See Parizeau, K. (2013). Formalization beckons: A baseline of informal recycling work in Buenos Aires, 2007-2011. *Environment and Urbanization* 25(2): 501-521; Sternberg, C.A. (2013). From "cartoneros" to "recolectores urbanos": The changing rhetoric and urban waste management policies in neoliberal Buenos Aires. *Geoforum* 48: 187-195.

4 UN Habitat (2003). The challenge of slums: Global report on human settlements. London: Earthscan.

5 Under such circumstances, realities often even defy a clear distinction between formal and informal service providers, as discussed in the case of water supply in Maputo, Mozambique, by Schwartz, K. et al. (2015). (In)formality: The meshwork of water service provisioning. *WIREs Water* 2(1): 31-36. doi: 10.1002/wat2.1056 (open access)

and in effect extend the use period of disposal facilities. Diversion of 10% or more of a city's waste is quite common, even according to official estimates. Examples can be found in Buenos Aires, Argentina, and several major Indian cities.[6] Also, as recyclers generate income for their families, their role helps alleviate poverty. In some instances, they generate substantial income, well above subsistence level. However, in places with a formal recycling system, where public or private waste companies have made investments and placed containers and other infrastructure for separate collection of recyclables, informal activities are often seen as unfair competition or outright theft, because they diminish the revenues of the formal companies without bearing any of the costs of the infrastructure. For example, due to a rekindled market interest in secondary materials, in European cities, individuals take materials of value from collection containers and sell them for their own financial gain. The access to waste and, by extension, the competition for waste materials with market value are issues that directly impact on waste service providers. In Latin America, in 2009 the Constitutional Court of Colombia ruled in favour of waste recyclers, guaranteeing their customary rights to access, sort and recycle waste and their legitimacy to compete in the waste recycling business.[7]

© KKPKP

The social aspect of informal work has received considerable attention from advocacy NGOs and social scientists, particularly through action research. In some cases, the informal waste sector comprises ethnic minorities, such as the Zabaleen in Egypt, or other separate social groups; in other cases, like in Nigeria, they are a mixed group. Women are often in a vulnerable position[8] and waste recovery practices often involve child labour. With a few exceptions, authorities' attitudes are typically characterized as negative, ranging from disinterested and indifferent to harassing and outright hostile.[9] One such exception is the Nansana Town Council in Kampala, Uganda, which has championed the promotion of MOUs between micro enterprises and local government as a means of achieving cost-effective waste management service delivery in Nansana District. This attitude may also depend on the circumstances: waste pickers may be tacitly tolerated at a simple waste transfer site or a simple disposal site, but they get expelled when the site is upgraded into an engineered facility. A common complaint is that informal waste pickers tear bags or empty waste bins on the street in search of recyclable materials and leave a mess behind them. On the other hand, there are also examples in Brazil and the Balkans in which waste pickers contribute to the cleanliness of urban open spaces, if mutually beneficial agreements are made. In some countries, it has been reported that the public feels a certain degree of social connection with itinerant waste collectors taking old clothes and other reusable items or recyclables at their door, whereas they feel aversion towards waste pickers taking 'their' recyclables from street containers. This aversion can be very strong and waste pickers are often viewed, particularly by more affluent citizens, as 'dirty and suspicious' in countries as different as Argentina, Malaysia, Slovakia and South Africa.

There is a considerable body of scientific and activist literature that highlights the benefits and advocates the rights of these workers throughout the world, from Brazil to Nigeria to India. Some NGOs like Chintan Environmental Research and Action Group, India, have received international praise for their activities empowering underprivileged women[10] and their work is regularly reported in the national and international press. Despite the positive side, informal activities also

6 A systematic study for GIZ produced estimates of the economic contribution of the informal sector often far in excess of 10%. See Scheinberg, A., M. Simpson, Y. Gupt, et al. (2010), listed in Annex A, Chapter 5, Service delivery and private sector participation in SWM.

7 The issue of ownership of waste is elaborated in Box 5.1 'Who owns municipal solid waste?'. The Colombian example is explored further in Case Study 5 on the inclusion of recyclers in waste collection, found after Chapter 4.

8 See Topic Sheet 15 on gender and waste management.

9 Medina (2007), listed in Annex A, Chapter 4, Inclusivity.

10 Chintan received the Innovation Award for the Empowerment of Women and Girls from the U.S. Secretary of State, Hillary Clinton in March 2012. See http://www.state.gov/r/pa/prs/ps/2012/03/185349.htm

include some egregious examples of exploitation and aggression.

The inclusion of the informal sector in waste management may remain a problematic and contested issue. However, starting from the premise that the informal sector is a significant stakeholder in waste management in many places, especially those where the formal SWM system struggles to provide good SWM services to all their citizens, it is more valuable to work with the existing favourable practices and use them for the benefit of the system than to put effort, already strained capacities and limited resources into dismantling them and starting from scratch to develop new systems. The position and the role of informal sector is particularly important to address when a city is establishing a formal separate collection scheme, either through public municipal or private business initiatives. The same holds for the preparation of long-term goals and strategic plans, including selection of technology for a city's SWM system.

The challenge for authorities is to support and promote the entrepreneurship, flexibility and productivity that characterize the informal sector, while striving to reduce the sector's negative aspects. Rather than criminalizing informality, such an approach would imply giving a mandate to the informal service providers, integrating them to work alongside the formal sector and thus incorporating them in the system. In some cases where public authorities have made attempts to involve the informal sector, they have faced resistance, for various reasons, one of them being that the economic situation may be better when operating informally, as the case of the city of Douala, Cameroon, illustrates. The government introduced legislation in 2014 that bans informal recycling and waste-picking activities and requires all waste pickers to register and pay taxes, which would increase government revenue from this growing economic sector. However, the result has been a decrease in waste collection and recycling in some neighborhoods.

Opening a dialogue in order to better understand each other's roles and to negotiate responsibilities holds the potential for identifying common goals and interests as well as creating partnerships for the benefit of the entire community. For example, it is important to distinguish between waste pickers who work in the streets, waste pickers who work on dumpsites, and itinerant waste buyers who go door-to-door, actually paying householders (or household servants) for clean, source-segregated waste materials. Accordingly, the introduction of waste segregation at source would not only be likely to result in more and better materials but would also be an opportunity to develop an integrated system[11] that successfully combines formal and informal actors, as experiences in Bangladesh, India, the Philippines and many other countries confirm. Latin American countries are particularly advanced in such efforts, including Argentina, Brazil, Colombia and Ecuador. *Figure 1* shows a useful framework that can serve as guidance in designing initiatives for integration of the informal sector into mainstream waste management.

11 The importance of waste segregation at source is discussed in Section 3.5.2.

Figure 1 Framework for integration of informal sector

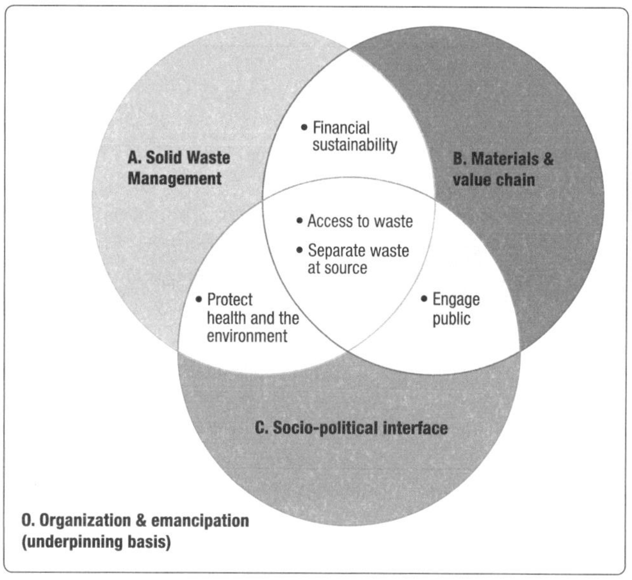

Note: Adapted from Velis et al. (2012).

Organizing waste pickers into associations and cooperatives in Brazil[12]

As recycling companies favour suppliers who can consistently deliver large amounts of clean and compacted materials, Brazilian informal waste pickers, working individually, were dependent on middlemen for amassing adequate volumes and for access to recycling companies, which was making them vulnerable to exploitation. Dissatisfied with their position, they started organizing and creating their own cooperative businesses in the 1990s, with support from local NGOs and international development agencies and foundations. In addition to organizing state meetings and then national meetings, they created the Brazilian National Movement of Recyclable Waste Pickers (Movimento Nacional dos Catadores de Materiais Recicláveis, MNCR) in 1999. Currently representing hundreds of such cooperatives and associations, MNCR has more power in negotiations with other stakeholders and was instrumental in the establishment of an inter-ministerial committee in 2003 to coordinate public policies towards the social and economic integration of waste pickers. As a result of these dedicated and concerted efforts, the national government has launched financing lines through the Brazilian Development Bank and the Bank of Brazil that aim at job creation through waste picker cooperatives. However, both the environmental and economic benefits of material recovery from waste could be much larger, as less than 10% of Brazil's 5560 municipalities have systems for separate collection of recyclables, and out of those only 150 partner with waste picker associations and cooperatives.

12 Fergutz et al. (2011)

BOX 4.28 PIONEERING SOURCE-SEPARATED COLLECTION IN BOLIVIA[170]

© Imanol Zabaleta of Sandec/EAWAG

With 630,000 inhabitants, Cochabamba is one of the biggest cities in Bolivia, and generates 500 tonnes/day of domestic waste, of which 61% are organics and 18% recyclables.[171] In the early 2000s, Cochabamba had a SWM situation typical of other Bolivian urban centres, where only mixed (unsegregated) waste collection was in place, with incomplete coverage of the city. The collected waste used to end up at the Kara-Kara dumpsite. The only resource recovery was carried out by informal recyclers who picked recyclables from containers, waste piles in the streets and at the dumpsite itself. It is relevant to note that these informal recyclers had been beggars and started waste picking as a last resort to make a living.

The Bolivian branch of the Swiss Foundation for Technical Cooperation (Swisscontact), a non-profit-organization created in 1959, has three main offices in the country, with predominantly Bolivian staff. The organization's mission is to provide technical and managerial support to city governments on how to tackle issues of waste, sustainable tourism, small and medium business initiatives and other environment-related topics.

In 2007 the first try-outs of separate collection schemes took place in the Amanecer and La Camana Periodista neighbourhoods of Cochabamba. The schemes were operated by informal recyclers and supervised by a member of the neighbourhood council. The fact that several other neighbourhoods in the city joined the initiative motivated Swisscontact to launch the first phase of the Ecovecindarios (Eco-neighborhood) project in 2009, with a budgeted cost of 2 million CHF (Swiss francs).[172] The project aimed at valorizing the economic potential of solid waste by establishing new structures for collection, treatment and recycling in Bolivian cities, at various administrative levels.[173]

Based on the positive experiences at the neighbourhood level, Swisscontact, in alliance with EMSA (Empresa Municipal de Saneamiento Ambiental), the authorized municipal operator for the solid waste management in the city, implemented source-separated collection at the district level, a first for Bolivia. Distrito 3 was the front-runner and achieved a 50% reduction of mixed waste. This motivated EMSA and the municipality of Cochabamba to enact a regulation (By-law 4588/2013)[174] that included source separation in its integrated solid waste management plan, thus becoming in 2013 the first municipality in Bolivia with such legislation.

In 2010 the Ecorecolectores self-employment programme[175] was established, to provide some visibility and acknowledgement to the informal recyclers. Swisscontact coordinated recyclers willing to participate, by assigning them a collection route in which households would voluntarily source separate recyclables and give them to the recyclers, now called Ecorecolectores. By selling the recyclables collected, they could generate income of approximately 1,200 BOB (Bolivian bolvianos)[176] per month, while also contributing to higher recycling rates.

The Ecorecolectores programme is now incorporated into the integrated solid waste management system of the municipality. Consequently, EMSA is now in charge of registering the waste pickers in a database and providing them with credentials as well as collection routes and collection gear (gloves, bags, trolley, etc.).

After these successful experiences in Cochabamba, the Ecovecindarios project spread to the three other biggest cities of El Alto, La Paz and Santa Cruz, and then to other municipalities. At the moment, securing financial sustainability is a key issue – as municipal SWM services are heavily subsidized in Bolivia (60% in Cochabamba and 100% in small municipalities) citizens are not accustomed to paying and are also unwilling to pay for deficient services (resulting from the interrelated problems of lack of resources, other more pressing issues and other political priorities). Despite this shortcoming, the first phase of the project (2009-2012) saw many achievements, such as 397 Ecovecindarios working in 14 municipalities, the provision of information on source separation of waste to 475,000 households, the creation of 443 jobs (46% held by women), 29,000 tonnes of solid waste collected and treated and 4.5 million USD invested on behalf of municipal governments.

170 Text is provided by Imanol Zabaleta Altuna of Sandec/EAWAG, Switzerland, and Carola Ortuño, Bolivia. More information on this case is available at http://swisscontact.bo/swisscontact.php

171 According to a 2009 waste characterization study commissioned by the Ministry of Water and Environment of Bolivia, carried out by the National Solid Waste Board

172 Approximately 2 million USD (as of August 2015).

173 Administratively, Bolivian cities are divided into neighborhoods, which are clustered into districts (several districts constitute a *comuna*). It is important to mention that the promulgation of law No. 1551 (1994) (Ley de Participación Popular) gives legal status to neighborhood councils (locally known as OTB, organización territorial de base), whereby they can implement their own projects using their own public resources. This was a necessary precondition for the implementation of the Ecovecindarios project.

174 Ordenanza Municipal 4588/2013 correspondiente al "reglamento municipal para la gestión integral de residuos sólidos domiciliarios y asimilables".

175 Programa de Autoempleo Ecorecolectores

176 Approximately 175 USD (as of August 2015).

4.8.1 Possible roles of governmental institutions

Governmental institutions are established and given a specific mandate by law, as one of the *instruments* through which a nation's policy goals are defined and reached. For an effective waste management system, a strong, robust and transparent institutional framework is essential, within which institutions take various roles at different levels of government, from national to local.

At the national level, the main role of the government and its institutions is that of **policy maker and legislator**, which includes strategic planning and preparation of policies, and their translation into legislation. Institutions deploy various participatory governance processes so as to benefit from the interaction with other societal actors in which they can contribute their knowledge, insights, queries and concerns, as discussed in Section 4.7 on stakeholders. In order to integrate policies across a broader thematic scope, the process usually entails exchange and discussions on the topics from various adjacent areas, including air and water pollution, spatial planning, local governance and others. While governments usually take the lead in formulating strategies, policies and legislation, they may be prompted to do so through pressure from the public, interest groups or even individuals.[177]

Figure 4.6: Possible roles of government institutions in a waste management system

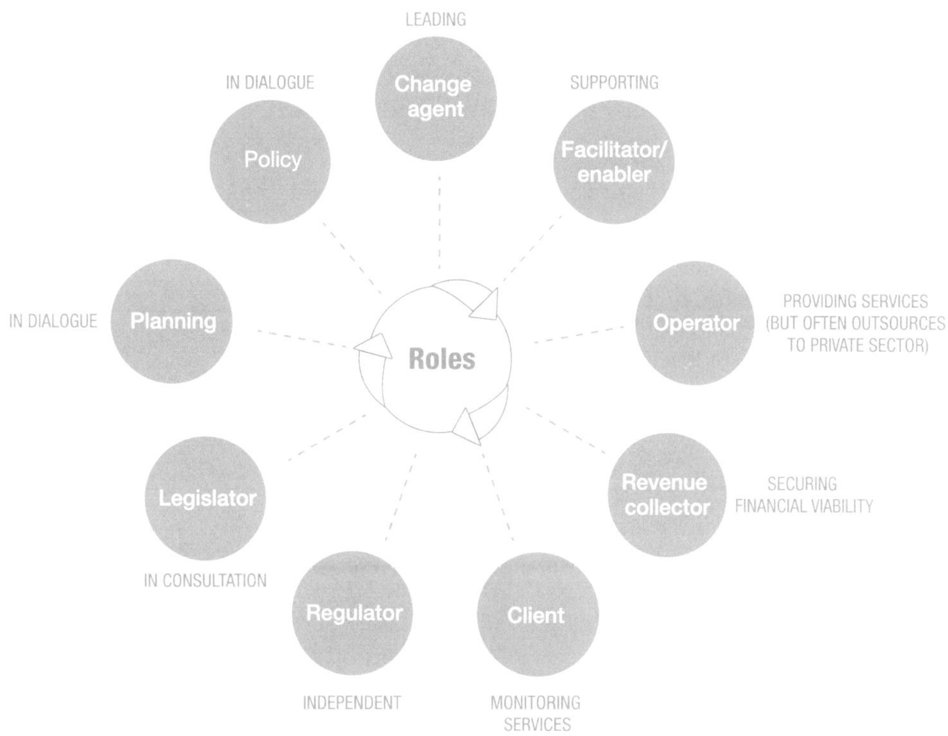

Complementary to the policy maker/legislator role, governmental institutions have a prominent role as a **regulator** in charge of implementing strategies and enforcing the legislation. As discussed in Section 4.4 on implementation, the regulator role can be undertaken at different administrative levels, as long as the institution in charge has adequate capacity to perform its duties and the system works across the whole country. Another key point of discussion is the *independence* of such a regulatory agency, as it needs to enforce standards

177 For example, the discovery of hazardous waste dumped near residential areas in several European countries in the 1970s resulted in public outrage accompanied by broad media coverage; in response, pertinent environmental legislation was prepared and enacted right away. The UK Deposit of Poisonous Waste Act 1972 – the first ever legislation in the UK to control hazardous waste – was drafted in just 10 days and passed by Parliament within a month. In India, it was an individual action – the case of Almitra H. Patel v. Union of India, Writ Petn. (C) No. 888 of 1996, D/- 15-2-2000 (Municipal Waste Case), that accelerated the preparation of the landmark environmental legislation, the Municipal Solid Wastes (Management and Handling) Rules of 2000. Finally, in a number of Latin American countries, sustained efforts of advocacy NGOs have been instrumental in bringing about legislation that establishes the rights of the informal sector in waste management.

equally on both private and public sector operators. Therefore, the accountability and transparency of the regulator are essential.

While it is relatively common for countries to put comprehensive environmental legislation in place (albeit after a long period of preparation), it is less common to have the institutional capacity, resources and commitment to effectively enforce the newly enacted legislation. Therefore, the role of regulator is arguably more critical for a well-functioning system. It is in this role that governmental institutions are often failing, due to problems of inadequate resources and capacities for the task and, equally importantly, due to problems with conflicting political interests, political interference, and corruption all playing a part.

At the local (municipality, county or similar) level, the role of government institutions is more about taking on the 'client' function, as it has responsibility as the 'proxy-generator' of municipal solid waste. The client function includes making decisions as to who delivers the service, how the revenues are collected and how the necessary finance for investment is raised.[178] In many countries, municipalities find the necessary efficiency of scale for implementing the SWM system through inter-municipal cooperation.[179]

While it is important that the authorities show that they 'mean business' and ensure that the rule of law is upheld, the government also has a role in working with other actors in the system to increase their knowledge and understanding of the newly established situation, for example industries handling hazardous substances in their products or their production processes. In this role, the governmental institutions are not just law enforcers, but also educators and persuaders.[180] Finally, as reliance on regulation alone will not produce the envisaged results, governmental institutions have an important role in communicating with other actors in the system and facilitating their involvement, in order to get people to 'own' the issue of waste and sustainable resource management. This means that the role of institutions is to provide information, instructions, enable, encourage, and last but not least, lead by example, with the goal of influencing people's attitude and ultimately their behaviour, as discussed in Section 4.6 on social instruments. It is important to note however that the roles of governmental institutions in their interaction with stakeholders evolve and change over time, as the policy regime and physical waste management system gradually mature and a culture of compliance takes root among stakeholders.

In addition to institutions in these diverse roles, political commitment and keen leadership are indispensable in producing tangible results. In some cases the development is driven by the mayor of the capital city who takes waste management to the next level within a comprehensive vision for the city; sometimes it is a minister of the environment or another politician or government official. Or it may be strong central planning for development and determined implementation – in which local affordability is not an issue as the costs are covered by central financial resources – that result in reliable, robust and modern waste services, such as in many cities in the PRC.

Discussing plans.

178 These four aspects of a financing model for MSWM are discussed in Sections 5.4 and 5.6 through 5.8.

179 See Section 5.6.4 on achieving economies of scale. Examples of inter-municipal cooperation are given from Asia and Latin America in Box 5.7, and from Europe in Case Study 6 on SYSAV and Case Study 8 on Flanders, both found after Chapter 5.

180 However, while this description is very typical of the situation in Europe in the 1980s – when government would be leading the way in environmental management – at this point in time, in the 2010s, it is often frontrunner businesses that take the lead in innovation for sustainable production. The role of governmental institutions in such a regime is that of enabler, creating conditions under which businesses will be comfortable and trusting enough to make the necessary investments in research and development for environmental sustainability.

BOX 4.29 WORKING THROUGH PARTNERSHIPS IN CEBU CITY, PHILIPPINES[181]

The passage in the Philippines in 2000 of the Ecological Solid Waste Management Act, widely known as the Republic Act No. 9003 (RA 9003), provides the required policy framework, institutional mechanisms and mandate to Local Government Units to establish sound SWM practices. The main policy targets include the elimination of open dumping and a 25% reduction in waste disposal through a focus on material recovery from waste, including the introduction of separate collection and facilities at the barangay level (the smallest unit of local government).

When implementing this law, the local government in Cebu City, the country's second-largest city, considers SWM to be a multi-stakeholder engagement involving innovative strategies and partnerships. Authorities promulgated local legislation mandating a wide range of instruments, so as to address multiple issues in synergy. Under Ordinance No. 2017, a newly established SWM Board, headed by the mayor, developed a long-term vision for SWM in the city and translated it into a 10-year plan for solid waste reduction, drafted in 2005 with technical assistance from its sister city, Fort Collins in Colorado, U.S. and in consultation with various stakeholders in the community. Due attention was given to collaboration with the private sector and local NGOs, so as to promote and ensure the viability and the effective implementation of the plan.

Local Ordinance No. 2031 was passed in 2004 to introduce waste segregation at source, establish penalties for violations, and facilitate the creation of a special fund for incentives. Since April 2011 the city has introduced separate collection of three waste categories: biodegradable, recyclable and residual, under the slogan 'No segregation, no collection'. One of the innovative approaches is the creation of the Cebu Environmental Sanitation and Enforcement Team (CESET), whereby not only officials but also designated residents and civil society groups can apprehend offenders and are entitled to 20% of the fine. Effective enforcement has resulted in not only a substantial amount of money for the city from penalties but also a steep decrease in the number of violations. From these funds, the city supports barangays in establishing material recovery activities through an annual grant of 400 USD, awarded upon request.

Recognizing that implementation of the SWM programme depends on people's level of environmental awareness and involvement, the municipality introduced information and education campaigns in which volunteers served as community leaders. These campaigns were based on communication and exchanges with and among householders, who had previously been uninterested in SWM.

Pilot projects involving simple, cost-effective composting technologies were initiated in parallel to develop the evidence base to convince both the people and the barangay officials of the merits of composting. Various stakeholders such as companies and shopping malls are involved in resource recovery from waste as part of their corporate social responsibility efforts. Economic value of waste materials is highlighted through a 'Cash for Trash' programme, focusing on the recovery of recyclables, led by women's organizations and other local NGOs. Particularly in the area of capacity development, the city authorities actively collaborate with international entities, including the city of Kitakyushu (Japan), the Institute for Global Environmental Strategies (IGES, Japan), UNESCAP and its sister city of Haarlemmermeer (the Netherlands).

The city's concerted efforts have borne fruit: its target of reducing waste by 30% by 2012 has been achieved; treating organic waste within neighbourhoods has led to lower transportation costs and a longer use period of the landfill; and a couple of hundred jobs have been created, mainly for the poor. Cebu City has shown a successful mix of instruments in its SWM system: a clear vision supported by strong political commitment and collaborative interaction with local stakeholders, ranging from businesses to NGOs to barangay councils to householders, coupled with strict law enforcement and economic incentives. As emphasized by the authorities in Cebu City, solid waste can exist only when people lose sight of its value.

One of the very important – albeit not that obvious – roles of government is to facilitate communication and interaction among stakeholders in the system, for sharing and exchanging ideas and experiences. This can be done through various platforms for knowledge diffusion and through demonstration projects.

181 More details can be found in Ancog, R.C., N.D. Archival & C.M. Rebancos (2012). Institutional arrangements for solid waste management in Cebu City, Philippines, *Journal of Environmental Science and Management* 15(2): 74-82; Premakumara, D.G.J., A.M.L. Canete, M. Nagaishi & T.A. Kurniawan (2014). Policy implementation of the Republic Act (RA) No. 9003 in the Philippines: A case study of Cebu City, *Waste Management* 34(6): 971-979.

BOX 4.30 PEARL PEER EXPERIENCE AND REFLECTIVE LEARNING IN INDIA

Jawaharlal Nehru National Urban Renewal Mission (JNNURM) is a flagship project launched by India's Ministry of Urban Development (MoUD) in 65 'mission cities' for planned development and urban reforms, with a focus on efficiency in urban infrastructure and service delivery. Capacity building and knowledge sharing are important elements of the project. Under the name PEARL – Peer Experience and Reflective Learning – a network was launched in 2007 for cross-learning and exchanging experiences. Groups were formed among mission cities having similar socio-economic profiles, complexities of urban problems and issues, size and urban growth patterns, along with their affinity in pairing them. Besides national and group workshops, PEARL also includes a website to update and document best practices in projects, reforms and innovations.[182]

Experimentation is essential for innovation and for changes to existing ways of thinking and doing. The flip side of the experimentation coin is that experimentation does not necessarily pay off. Therefore, as ample evidence shows, companies and organizations interested in innovative waste management practices (and, more broadly, sustainable consumption and production) like to see examples of best practices already in action and learn from them. With this in mind, various government-funded programmes have been established to facilitate exchanges among companies and thus support businesses as they transition toward sustainability. For example, among a diverse range of such schemes funded by the UK government within their Best Practice programme since the early 1990s, so-called *waste minimization clubs* or *resource efficiency clubs* have been particularly successful; the best-run among them generates tenfold 'returns on investment' in terms of cost savings realized by the participating companies compared to the budget allocated by the government to run such a club. A similar approach to the role of government in this field is shared by the Dutch, whose government agencies facilitate gatherings of a diverse range of actors and enable the creation of communities engaged in practice regarding specific topics, such as zero waste regions and the prevention of food waste.

BOX 4.31 EXPERIENCE EXCHANGE IN THE NETHERLANDS

In their efforts to respond to a societal transition toward better recycling, several publicly owned waste companies have experimented with various novel waste collection systems so as to encourage householders to segregate more of their waste more effectively.[183] For example, in the Netherlands, the ROVA public waste company introduced a 'pay-as-you-throw' (PAYT) charging scheme linking payments directly to quantities of residual mixed waste presented for collection; this resulted in a 20% decrease in costs and a 25% decrease in residual mixed waste. After the positive effects had stabilized, ROVA introduced a so-called 'reverse' collection system, whereby only source-*segregated* waste material fractions are collected door-to-door, while householders have to bring their *residual mixed* waste themselves to designated collection points, which is exactly the opposite of what is commonly done in the Netherlands.[184] The results are very positive, both in terms of the amounts of segregated waste and the cost savings. This and other experiments are shared through Royal NVRD[185], the largest national waste association, as their peers are curious about the positive results achieved and the financial implications. This attitude of willingness to 'compare notes' and learn in a network of peers is quite typical of the Dutch public waste sector in general. There are two voluntary benchmark systems in place, where participating public companies compare their own performance with that of their peers.

Government also has a role as a facilitator or a participant in network building, participation platforms and joint learning. As mentioned in Section 4.7 on stakeholders, international networks have a significant role in expanding knowledge and sharing experiences. Concerning governmental institutions, the UNEP Global Partnership on Waste Management assembles national agencies in charge of waste matters;[186] IPLA is the International Partnership for Expanding Waste Management Services of Local Authorities;[187] the U.S. National

182 See http://jnnurm.nic.in/ and http://www.indiaurbanportal.in

183 These innovative waste collection schemes are described in Goorhuis, M. et al. (2012). New developments in waste management in the Netherlands. *Waste Management and Research* 30(9 SUPPL.1): 67-77.

184 Interestingly, this system is *de facto* present in many places around the world, where itinerant buyers visit households to obtain their valuable waste materials, whereas householders walk to collection points to bring their other waste.

185 NVRD – Nederlandse Vereniging van Reinigingsdiensten

186 http://www.unep.org/gpwm/InformationPlatform/CountryWasteManagementProfiles/tabid/104472/Default.aspx

187 http://www.iplaportal.org

Recycling Coalition (NRC) has formed the basis for knowledge-sharing during the rapid modernization period in the U.S. and Canada since the 1980s;[188] ACR+ is an association of cities and regions for waste prevention, reuse, recycling and sustainable resource management;[189] and Zero Waste Europe advocates zero waste strategies and lifestyles and supports communities, businesses and individuals in implementing them.[190]

4.8.2 Institutional coherence

Depending on the academic and ideological tradition, opinions differ on the desirable degree of institutional autonomy needed for effective functioning of the system. Public administration scientists spend their entire careers attempting to strike a balance between on the one hand institutional autonomy, which entails decisions about, for example, management positions and budgets, and on the other hand accountability for the ways in which that autonomy is used in practice, which entails some degree of central control. Reality defies being captured in one definitive theory and description of best practice – there are successes and failures on both sides. Nevertheless, valuable experiences have accrued around the world and some are shared here.

A dedicated agency may be created at the national level as a part of the implementation of a newly enacted law. Alternatively, this can be decentralized to the regional level governments. Experience teaches that, in the latter case, it is still beneficial to have some form of a central agency so as to (1) establish the same interpretation of national laws across regional entities, (2) have some overarching level of control to ensure consistency of approach throughout the country, (3) facilitate better cooperation and use of facilities in the country, and (4) facilitate communication and exchanges of experiences and enable collective learning.

Questions to ask include: Is there such a single institution (governmental agency) at the national level that is in charge of implementing, or coordinating the implementation of, the national waste management policy or strategy, with clear delineation of authority and responsibilities? Is this institution adequately and professionally staffed?[191] Is it separate from the environmental regulator so that it does not get entangled in conflicts of interest by having to license and inspect public waste operations (which is to say, its 'own' activities)? If the institution is newly established, is information about this entity fully communicated to all the actors in the system? If there is no such institution, are there institutional mechanisms in place for coordinating strategy implementation by the relevant ministries or by other institutions or organizations?

A more general point to be made here is that current institutional arrangements are usually based on a sectorial approach and tend to ignore interconnections among sectors. Thereby, in decision-making processes, regulatory and institutional fragmentation encourages a focus on single issues and a limited set of criteria. This is at odds with public perception of the problems at hand – people do not think in terms of ministries and jurisdictions; they think in terms of real issues that affect their daily realities. For example, in Kolkata, India and many other urban agglomerations around the world, the city is governed by a multitude of administrative agencies whose jurisdictions do not coincide. Therefore it is difficult to adequately address issues related to, say, the utilization of organic waste in aquaculture in peri-urban areas.[192] Furthermore, solid waste institutions may facilitate effective performance in other fields, including water supply, sanitation, agriculture and others. While waste topics are spread over an assortment of ministries (the ministries of public health, the environment, local government, natural resources and industry, as some examples[193]), from a holistic perspective, there is a need for some formalized concerted effort so that air, water and waste are considered in an integrated manner. Whether this is in the form of a common ministry, governmental agency or some other institutional form will probably depend on the local historical development of institutions.[194]

Coherence of institutions at the local level is of paramount importance for clean cities. One definition of organizational coherence in this context is the degree to which all waste management responsibilities are concentrated into a single organization or department that can be held accountable for the performance of the SWM system as a whole; or, if there are multiple organizations, the presence of a significant concentration of responsibilities within one named agency. The basic concept is that the local officials responsible for SWM should be able to exercise that responsibility reasonably autonomously without having to seek multiple approvals for each and every small decision from a multitude of ministries or other agencies, at either the local

188 http://nrcrecycles.org
189 http://www.acrplus.org/index.php/en/
190 http://www.zerowasteeurope.eu
191 The issue of institutional capacity development is further discussed in Section 4.8.3 below.
192 Hofmann, P. (2013). Wasted waste – Disappearing reuse at the peri-urban interface. Environmental Science and Policy 31: 13-22.
193 Yet others include the ministries of finance, spatial planning, infrastructure or even energy or agriculture.
194 The need for a holistic approach is discussed in Section 2.2.1.

or national level. This has proven to be advantageous for timely responses to the situation in the field while encouraging a proactive attitude by the local officials in charge of waste. Again the concept of leadership is vital, as is the role of civil society organizations in calling attention to crises.

4.8.3 Institutional capacity development

For the institutions to perform their tasks, it is essential that there is adequate capacity at three levels – the so-called 'enabling environment', the institutional level, and the individual level. Institutions in charge of SWM operate within a given 'environment' that directly impacts on their performance, not only through policies, rules, regulations and procedures to which the institution must adhere, but also through the relative importance of various policy topics and general political culture. At the institutional level, the internal organizational structure, including responsibility and accountability mechanisms, will also determine how the institution performs its tasks. Is there a detailed organizational chart of the agency, with a clear structure, tasks and internal accountability mechanisms? Is the procedure that governs how people are to be hired clearly stipulated? Is this procedure adhered to, or are people appointed based on their political affiliation or some other criterion? Ultimately, performance depends on the staff, whether they are suitably qualified, adequately educated and trained, and their knowledge regularly refreshed and expanded, both in the classroom and in the field. Has a needs assessment been performed so that programmes for capacity development can be designed and prepared to fill any gaps in knowledge, skills or other competencies? Is there structured career progression and motivation for staff to remain with the institution?

Figure 4.7 Levels of capacity development[195]

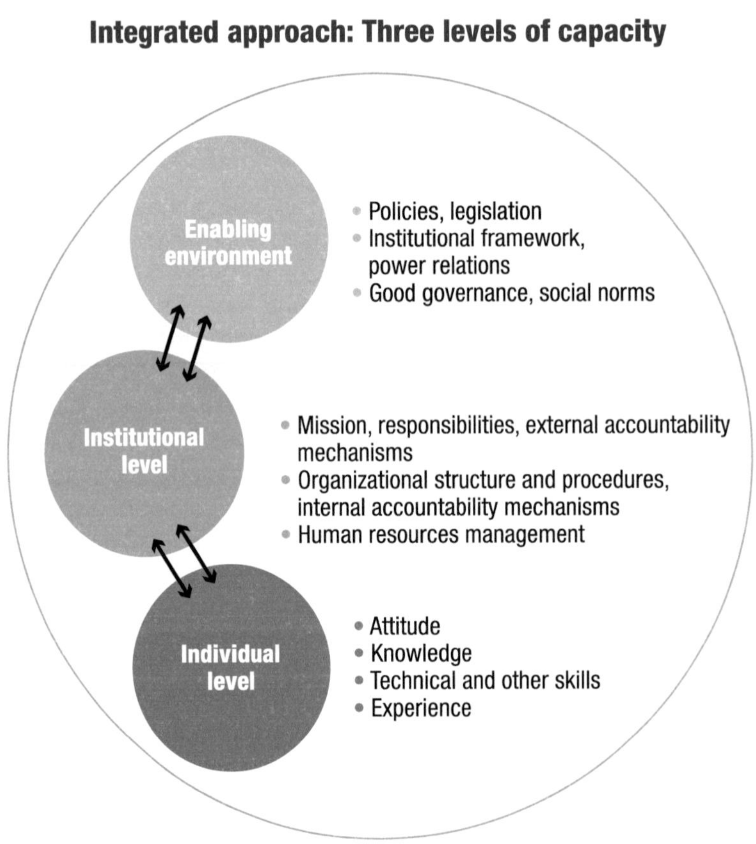

195 Based on UNDP (2008) Capacity Development. Practice Note. http://www.undp.org/content/undp/en/home/librarypage/capacity-building/capacity-development-practice-note.html and UNDP (2009) Capacity Development: A UNDP Primer. http://www.undp.org/content/dam/aplaws/publication/en/publications/capacity-development/capacity-development-a-undp-primer/CDG_PrimerReport_final_web.pdf

There is little point in establishing new laws and regulations with consequent additional requirements for monitoring, inspection and control if the existing institutional, organizational and administrative capacities are inadequate to accommodate additional data processing, reporting and compliance control. An effective waste management system, particularly for hazardous waste, relies on the availability and competence of a number of support services, including laboratory analytical services as well as various information systems, such as information systems on waste generators, waste movement tracking systems, performance assessment systems, and the like.

BOX 4.32 CAPACITY FOR PUBLIC-PRIVATE PARTNERSHIPS[196]

More often than not, governmental institutions lack the competencies required to exercise their 'client function' with respect to municipal SWM, in particular when it comes to interacting with the private sector through diverse models of public-private partnerships (PPPs). The skills and experience needed to successfully initiate, develop, negotiate,[197] award and manage PPPs in the areas of public services and (urban) infrastructure projects are quite specialized in terms of risk allocation and contractual frameworks, and therefore quite different from traditional public sector procurement models. Moreover, developing such competencies may not be a priority for (local) governments, particularly in developing countries, which often function in 'firefighting' mode to meet the needs of their citizens, and therefore only limited time and resources remain for longer-term planning for staff development. In addition to staff training programmes, special purpose agencies such as Partnerships UK (PUK) in the UK, Parpublica in Portugal, and Infrastructure Investment Facilitation Company in Bangladesh,[198, 199] have been established to assist local governments and bridge the gap between existing competencies and needed competencies, with varying success.

In order to develop staff capacities and competencies, a good starting point is for governmental institutions to collaborate with local universities, NGOs and consultants. Strengthening formal education on waste management topics, both technology- and governance-related, is an investment in a steady supply of qualified staff for both government and industry. In parallel with this, a central lead agency can play a prominent role in connecting local governments and facilitate an exchange of experiences for collective learning, as illustrated by the example of PEARL in India.[200] This is particularly valuable for smaller towns, for which the securing of funds and experts to provide training is a real challenge in the face of strained financial resources and increasing decentralization of public services. As a direct, short-term measure, some local governments such as that of Cebu, the Philippines, have established contacts and joined forces with their colleagues from waste departments in industrialized countries through their consular representations located in Cebu or through "twinning" or "sister city" programmes that already exist.[201]

The discussion of capacity development also touches upon the challenge of retaining staff in governmental institutions while they may have better employment options outside the public sector. Especially in developing countries, the staff, once trained, may seek better opportunities elsewhere, and the investment made through the training will then be lost. While research on human resources management in the public sector in developing countries is scarce, the findings from industrialized countries indicate that four variables determine the rate of retaining staff: salary, secondary terms and conditions of employment (which usually include ample training opportunities), motivation, and the direct manager. In a number of developing countries the salaries may be too low to live on, which makes those positions vulnerable. On the other hand, secondary terms and conditions of employment can be quite good. If this is combined with training opportunities, a 'good boss' and a motivating working environment, circumstances can be created that work in favour of staff retention.

---

196 Delivery of services through PPPs is discussed in Section 5.6.3, and revenue and investment finance options in Sections 5.7 and 5.8.

197 Some experiences show that the best improvements in performance are obtained through direct negotiation with private sector service providers rather than decisions made solely through tendering procedures. Tendering does not allow for development of a relationship between the partners before the decisions are made. Since negotiation processes take a long time, the parties get to know each other and develop trust. Thus a mix of traditional contracting (as a safeguard) and relational contracting developed over time through negotiation is seen as a positive option.

198 Farrugia, C., T. Reynolds & R.J. Orr (2008). Public–Private Partnership Agencies: A Global Perspective. Working Paper #39. Collaboratory for Research on Global Projects, Stanford University. https://gpc.stanford.edu/sites/default/files/wp039_0.pdf

199 A very informative analysis of several waste-related cases in India is provided in Devkar, G.A., A. Mahalingam & S.N. Kalidindi (2013). Competencies and urban Public Private Partnership projects in India: A case study analysis, *Policy and Society* 32(2): 125-142.

200 See Box 4.30 in Section 4.8.1 above.

201 See Box 4.29 in Section 4.8.1 above. Similar exchanges provided a starting point in the city of Bo, Sierra Leone. See Case Study 7 on Bo, Sierra Leone, found after Chapter 5.

BOX 4.33 FEDERAL GOVERNMENT STEPPING IN TO IMPROVE MSWM IN MALAYSIA[202]

What started as a privatization tender for the integrated management of municipal solid waste at the national level in 1994, turned out rather differently when it was finally implemented in September 2011, after 17 years of deliberations and interim operations. Following a conventional privatization model, the initial intention was to award MSWM to four concession companies according to geographical regions on a Build-Operate-Transfer (BOT) basis. The concession companies would fund and manage the entire integrated MSWM system from the supply of bins for storage, to transport, transfer, treatment and final disposal, as well as public awareness campaigns. The financial model would ensure financial viability and sustainability. The government was in favour of direct billing to households, but this was considered to be very difficult to implement, as public resistance and potentially high bad debts made the project non-bankable, based on previous experiences with the privatization of wastewater services.

In parallel, new legislation had to be enacted to transfer the powers over MSWM and public cleansing to the Federal Government for them to be able to enter into concession agreements with companies. Although this process took more than 12 years to complete due to the low political priority of the topic, the new SWM and Public Cleansing Act (Act 672) was finally enacted in July 2007.

While the details of the concession agreement were being sorted out, the concession companies were directed to take over the SWM and public cleansing services from the local authorities in their respective regions on an interim basis in 1997, a situation which was supposed to last no longer than two to three years. Existing assets as well as the workforce and equipment were transferred in their entirety from the local authorities to the concession companies under an interim management agreement, which was renewed annually. The local authorities would pay a monthly fee based on the actual operating expenditures incurred during the preceding two years, projected forward to the current year. However, the interim management agreement was not bankable due to its short, annual renewal period[203] and concession companies had to inject substantial capital to sustain operations for the following two to three years while the concession agreement was being prepared.

While the takeover of the staff was subject to the Privatization Guidelines provided by the Department of Civil Service, many resented the takeover and a great deal of persuasion had to be exercised to change their mindset and convince the employees of their future in the new organization. Public response was positive, as their complaints were resolved more quickly than before. Local authorities could now focus on other municipal services, which contributed to improved performance in other areas.

However after three years of interim operations, the state governments,[204] local authorities and the public became impatient as the concession companies were unable to improve services further. As these higher expectations were not met, some local authorities began to decrease payment. Also, there was substantial disparity in revenue among the local authorities and this was reflected in their ability to pay the companies. Smaller local authorities had little revenue apart from a property tax, which, moreover, had not been adjusted for inflation for the previous 15 years, partly due to political reasons.

This undetermined situation went on for another 12 years without significant improvements in the performance of the SWM system, until the government decided to privatize only the waste collection and public cleansing operations to the concession companies while waste treatment and disposal facilities would be tendered separately. With the enforcement of the 2007 SWM and Public Cleansing Act (Act 672) in September 2011 came a major change in the system, as the Act enabled the Federal Government to take over the authority and responsibilities for SWM and public cleansing from the local governments in the country, through its newly formed National Solid Waste Department, which is the policy maker, and the Solid Waste (SW) Corporation, which is the agency in charge of implementation, monitoring and enforcement.[205] Consequently, the concession companies would deal directly with the SW Corporation, after the local authorities entered into a tripartite agreement with the Federal Government and state government.[206] A long-term Concession Agreement was finally made between the SW Corporation and the concession companies in September 2011, whereby the concessionaires would provide a standard service level for waste collection and public cleansing services for the next 22 years, based on a set of key performance indicators.

Due to the political demography of the country, only six state governments out of 11 in West Malaysia plus the Federal Territory of Kuala Lumpur and Putrajaya entered into the tripartite agreement. More state governments are expected to sign the agreements upon seeing significant improvements in service levels and public acceptance in the areas managed by the concession companies since September 2011. Some of the major improvements include the distribution of standard HDPE bins for households and commercial premises and larger bins for apartment buildings, condominiums, institutions and public places; good waste collection services; and extensive cleansing services (including drain cleaning, street sweeping, grass cutting, beach cleaning and litter picking). With the new service levels and modern fleet of collection vehicles, waste is collected regularly, which prevents spillage of leachate that used to leave a stench behind – previously a common problem. A 24-hour customer hotline for public complaints, combined with quick responses, has added to the public acceptance of the privatization.

---

202 Information is provided by Ho De Leong, Chairman of the Waste Management Association of Malaysia

203 Banks would not approve loans for investments, as they were concerned that the loans would not be repaid should the agreement not be renewed.

204 Malaysia is a federation of 13 states and three federal territories. The 13 states are governed by their state governments headed by a Chief Minister, who is appointed by the King under the recommendation of the Prime Minister. Each state is administratively divided into districts, under which there are local (municipal) authorities; these are addressed by the Local Government Act. The function of SWM and public cleansing services appears in both the Federal Law and the Local Government Act.

205 The SW Corporation was established in June 2008, under a related piece of legislation, Act 673 of 2007. More information can be found at http://www.ppsppa.gov.my/index.php/en/pengenalan

206 Based on such an agreement, the local authorities shall pay an amount equivalent to their waste management expenditure prior to privatization. The Federal Government supplements this with any additional funds required to pay the concession companies via the SW Corporation, in accordance with the terms of the Concession Agreement.

The implementation and enforcement of the pertinent legislation, the establishment and actual empowerment of the central, federal-level agencies in charge of SWM and the long-term contracts with service providers have been very successful in raising the standard of SWM services in Malaysia. Through cross-subsidies, the federal model has relieved the burden especially from the smaller local authorities that are constrained by lower levels of revenue and a lack of technical expertise necessary to provide adequate SWM services. It has also seen extensive economic activities in related industries such as the manufacture and supply of bins and waste management equipment, including vehicles.

The Federal Government is now focusing on the establishment of treatment and disposal facilities as well as promoting the 3Rs through public awareness campaigns, which are expected to further contribute to employment opportunities in the waste sector.

4.9 SELECTING AN APPROPRIATE SET OF POLICY INSTRUMENTS

4.9.1 Knowing where you stand versus where you want to be – performance measurement of national and local waste management governance

As stated in Section 4.2.2 on strategic planning, for both policy formulation and preparation of strategic plans to be meaningful, they need to be informed by a profound understanding of the current situation. Ascertaining the current state of affairs – identifying the essential elements of the system, who is involved and how, how they interact, what drives the system, what works well and what are the limiting factors, and particularly where the leverage points for change lie – does not have to involve years-long studies of details, but rather a well thought-out and concerted effort to understand the key features of the system at hand. To that end, the right indicators need to be devised; 'right' meaning that indicators are informative and relevant to their purpose, representative of the phenomenon they are to assess, fairly easy to measure and interpret with a sufficient degree of reliability, and, when combined, provide a comprehensive and complete indication of the state of the existing system.[207] Assessment based on such indicators provides a basis for articulating the goals of the desired state of the system and assists in 'charting the waters' towards them.

BOX 4.34 TO MEASURE IS TO KNOW[208]

Statutory notification of hazardous waste movement was for the first time introduced in the UK in 1972, in accordance with the Deposit of Poisonous Waste Act, the first-ever piece of legislation addressing hazardous waste in the UK. As no specific detailed mechanism for the consistent delivery of the required information was in place yet, the then Department of the Environment prepared and printed 50,000 forms as an example for generators and handlers of hazarous waste to use in the country in one year. This number of notifications was exceeded in the City of Birmingham alone in just nine months. While there was certainly some level of over-notification by producers for fear of getting it wrong, the introduction of this legal obligation revealed the extent of hazardous waste movement in the country. Before that time, hazardous waste had always been largely invisible until it turned up in the wrong place and caused a problem. With statutory notification and tracking in place, a new situation was created that enabled authorities to measure the number and quantities of waste movements, get a better understanding of the day-to-day practices and address them appropriately.

Usually, progress towards envisaged goals is measured quantitatively, using indicators such as recycling rates or percentage of disposal in controlled facilities. As important as such numbers are, they do not provide complete information about a system's performance. Experiences in cities around the world have shown over and over again that technology alone will not do the job without qualified and professional staff and without political

207 The abbreviation SMART has been widely used to describe attributes of good indicators in general: Specific, Measurable, Achievable, Relevant and Time-bound, which is derived from Doran (1981), who actually referred to them as criteria for management's goals and objectives. The meanings of the individual letters has been adapted and expanded since.

208 Information is provided by George Clapton, who was a Pollution Control Inspector with the City of Birmingham at the time that the Deposit of Poisonous Waste Act of 1972 was implemented in the UK.

commitment to environmental protection and cleanliness – factors that are not easily captured by numbers. The available indicator sets address this to varying degrees. Recent performance indicators on the 3Rs and resource efficiency, prepared by experts of the Asia Resource Circulation Policy Research Group,[209] include indicators that reflect preparation of specific policies such as extended producer responsibility, standards of environmentally sound management of e-waste or green procurement policies. However, once the policies are functional, the expectation is that the policy goals and corresponding indicators are quantifiable, and that the quantifiable variables are sufficient to represent system performance. Other indicator sets, such as the Performance Assessment System (PAS) for SWM used in India, include only the existence of legislation (albeit at various levels of government) as a non-technical indicator. As elaborated in Section 2.5.3, the WasteAware indicators give a more balanced overview of both physical components and governance aspects, both of which have proved to be essential for system performance.

In evaluations and discussions of performance, numbers are given a place of special importance; consequently, there is a tendency to give more importance to what can be measured than to what cannot. But some things that are very important just cannot be measured unambiguously and may even defy clear-cut definition. But that does not mean that we should therefore just ignore them as if they did not exist. To the contrary, they should also be looked at, given attention, discussed and addressed.

BOX 4.35 PAYING ATTENTION TO WHAT IS IMPORTANT INSTEAD OF WHAT IS (EASILY) MEASURABLE

Late in the evening, a man was looking for something on the ground around a street lamp. A policeman came along and asked what he was looking for; the man said: I am looking for my keys. They searched together for a while. Then the policeman asked: I don't see your keys here – are you sure you dropped them here under the lamp? The man answered: 'Oh, no, I dropped them there in the bushes but there it is too dark to look.'

4.9.2 How to select an appropriate set of policy instruments that will be most effective in a particular situation

Selection of an appropriate set of policy instruments starts with knowledge about the system. As highlighted above, knowing about the stakeholders – waste generators, staff in institutions, service providers and other actors in the system – is as critical as knowing about waste and facilities. In parallel with the gathering of information about the current system, the guiding principles and goals for the future are to be discussed. Once these are established, instruments can be selected. While this may sound like a neat and structured procedure, most existing policy arrangements or regimes have developed incrementally over a relatively long period of time through piecemeal adjustments and often in an ad hoc fashion and thus contain a wide mix of policy instruments. More often than not, these regimes are the result of a process termed 'policy layering' in which policy instruments and programmes are stacked on top of each other without abandoning or dismantling the outdated ones.[210] The results are arrangements of policy instruments that lack a unifying overall logic and often contain counterproductive instrument mixes and are thus both complex and costly to administer and may ultimately be ineffective in achieving the policy goals. For example, Argentinians have an expression 'legislative pollution' to denote countless rules that are not interconnected, operational or practical, and only complicate environmental management rather than serve it.

209 Hotta (2014), listed in Annex A, Chapter 2, Waste data and indicators.

210 'Policy layering' means adding new goals and instruments without abandoning previous ones, most often leading to both incoherence among the goals and inconsistency with respect to the instruments. Other detrimental processes include 'policy drift' and 'policy conversion'. 'Policy drift' is a process that allows policy goals to change without formal revision and thus without altering the corresponding instruments. 'Policy conversion' happens when a mix of instruments is redirected from a policy domain where it was successful in tackling another policy domain. In addition, a lack of flexibility could result in regulatory instruments ceasing to be suitably targeted and their effectiveness diminishing as a result. The necessity for change may arise as a consequence of social, economic or environmental change, or as a result of new information becoming available. See Howlett & Rayner (2007), listed in Annex A, Chapter 4, Policy combinations.

For these reasons, governments have become increasingly interested in crafting and adopting more carefully designed arrangements of instrument mixes (sometimes referred to as 'new governance arrangements') when creating or reconstructing a policy domain.

BOX 4.36 'RULES OF THUMB' FOR DESIGN OF POLICY MIXES[211]

- Define coherent policy goals and guiding principles
- In order to get there, engage in dialogue with stakeholders
- Mix the instruments carefully so as to create a *consistent set* of instruments
  - Make sure that the instruments explicitly target and address specific aspects of the goals
  - Choose policy instruments in such a way that they support (rather than undermine) one another in the pursuit of the goals
- Double-check that the instruments are appropriate to the local circumstances regarding both the waste sector and societal context
- Consider the full range of policy instruments available
  - In the context of continuing pressures on governments to do more with less, be selective with 'direct regulation' to cover areas with high risks, such as those involving hazardous chemicals and hazardous waste
  - Consider a wide range of market-based instruments, various forms of self-regulation by industry, and policies that can employ civil society organizations to achieve the goals
  - Don't overlook 'social' information-based policy instruments and the government's role as a facilitator of networks and joint learning

While designing policy from scratch is by itself a complex task that requires a lot of effort and expertise, developing capacity at all levels is at least as important – it is the people who carry the system, and their level of awareness, expertise and involvement that eventually determine the level of performance.

Similarly, keeping institutional arrangements simple and clear is usually a good idea; it makes life easier for the legislator and the regulator as well as waste generators and waste service providers.

On the political side, there is a need for policy, legislative and regulatory stability, without changes in direction at every election. Last but not least, political conflict that exploits waste issues for its own agenda often has a debilitating effect both on the services and on the credibility of the government.

Good governance requires consistency in decision-making applied to all levels and all dimensions of the system. This means that the strategic goals lead to choices throughout the waste system, not only those related to policies and institutions but also, and particularly, technology selection and financing model. Therefore it is essential to take charge of the technology selection process as a matter of governance, not only as a matter of technical management. For the long-term benefit to the system, it is important to understand the function and purpose of technological options, rather than just their features, and study their performance and real costs before making a selection. *Box 4.37* provides a 'checklist' of questions that can be useful in guiding the technology selection process.[212]

211 Adapted from Howlett & Rayner (2007).

212 Some of the main technologies for resource recovery in MSWM are presented in Section 3.5.3.

BOX 4.37 SELECTING APPROPRIATE TECHNOLOGIES[213]

The checklist below provides some questions to ask yourself (and any salesman that may come along and offer a specific technology) in order to help you evaluate and select technology that is appropriate for your situation. This checklist is intended for use during the selection of waste treatment and disposal and resource recovery technologies; choice of technology for waste collection is discussed elsewhere.[214]

- Do you know your waste in terms of its composition? For example, how much organic waste is in it, and how much plastic? Is the heating value of your waste high enough to burn without support fuel? Are there significant seasonal variations?[215]
- How much waste is collected? How much is that as a percentage of waste generated? What is the waste collection coverage?
- What are the current trends – are your waste volumes growing, stagnating or decreasing? Do you know the factors or causes underlying those trends? Are new waste streams emerging that require specific treatment, for example e-waste?
- Has the technology that you are considering been proven elsewhere? Are the conditions similar to yours, in terms of waste composition, climate, people's habits, and affordability? If yes, what documentation is there for you to see? If a salesman is suggesting a particular technology to you, how readily can you get this documentation? How complete is this documentation in terms of performance data? How difficult is it to arrange a visit to an existing facility?[216]
- What is the designed use period (lifetime)? What does the financing scheme look like? Have sources of funding been identified? What budget will be required for operation during the designed use period? Can the necessary cost recovery mechanisms be put in place?
- Are the costs both realistic and affordable for local service users? Are local markets available for the products from the facility (heat, gas, compost, recyclables)? If yes, how do you know this? If not, are there plans to develop such markets? Who will finance the development of those markets?
- Can the technology be run and maintained locally, using local labour and local spare parts?
- Does your country have the institutional capacity to regulate facility operations, including licensing, inspection and compliance control?
- Does the choice of a particular technology make the system more robust and resilient? In other words, does it allow for flexibility if your situation changes significantly in the future, in terms of the characteristics mentioned above – amount of waste, composition of waste, people's habits, level of income, or even climate?
- Is there sufficient flexibility regarding contractual conditions with the operator?
- Have you sought independent advice? Have you – or your staff in charge – read this book to the end, including Chapter 6?

Simple but effective windrow composting, India.

Baling packaging, Germany.

213 Adapted from Scheinberg et al. (2010), prepared on behalf of UN-HABITAT, listed in Annex A under Chapter 1, Waste management.

214 Coffey, M. & A. Coad (2010), listed in Annex A, Chapter 3, MSWM – trends, composition, collection, disposal.

215 Oteng-Ababio, M., J.E. Melara Arguello, O. Gabbay (2013) Solid waste management in African cities: Sorting the facts from the fads in Accra, Ghana. *Habitat International* 39: 96-104, and references therein.

216 Even if it may seem impolite, it is particularly important to ask pertinent questions about the existing facilities, so as to detect fraudulent proposals and inappropriate solutions.

4.9.3 Lessons learned from around the world

As Winston Churchill famously said: 'However beautiful the strategy, one should occasionally look at the results'. Here are some lessons that have been learned by looking at the results of the application of various instruments for waste governance around the world.

- The envisaged goals and articulated guiding principles may not necessarily be coherent or mutually compatible; therefore, their translation into instruments may require additional effort to clarify the priorities among them.
- Instruments are only as good as the degree to which they are consistent with the goals they seek to achieve.
- Multiple categories of instruments are available and needed: 'direct regulation', economic instruments, and 'social' instruments. They tend to be most effective when applied in (consistent) combination.
- Instruments are only as good as they match the local circumstances under which they are (to be) applied, including the expertise and financial resources of the affected stakeholders. Also, instruments need to correspond to the capacities of the government as legislator and, particularly, as regulator in charge of implementation and enforcement.
  - If there are deficiencies in stakeholders' expertise, financial resources or capacities, go for implementation gradually. For example, introduce legislation in stages, with progressively more stringent standards, to allow the actors in the system time to develop expertise and raise necessary financial resources. Meet people where they are.
  - Take time to implement new instruments, allow time for change to take hold, for a system to emerge in a different form and for stakeholders to adjust to the new situation.
- When translating your goals into concrete policy instruments, go for the 'best fit' instead of just copying a 'best practice' that has worked elsewhere. Knowledge and information are essential but also need to be tested under the local circumstances. Familiarity with global 'best practices' is very useful because one does not have to make all those discoveries and mistakes oneself. But best practices should be used to inform, and perhaps to inspire, the process of developing local solutions, not replace it. In this sense, going for the 'best fit' means that the system evolves in and for the local situation, gets rooted, supported and 'owned' under the local circumstances, and thus becomes well-embedded in the local context. This is a long-haul learning process, which takes time, effort and commitment, and it requires developing capacities locally, but, in return, it leads to a more viable and robust system.
- In one form or other, some sort of waste management exists in most places already; starting from scratch – even with the best of designs – may not be the most constructive approach to change the situation for better. Learning about the current system and identifying what already works and building on it has a higher chance of success. Build on what already works in your system – even the best masters build on previous efforts.
- Technological solutions are at their best when used to support goals. An array of highly effective technologies has been developed for waste management and resource recovery. If deployed in isolation however, due to their attractive features or their environmental performance somewhere else, rather than match with the local situation, needs and capacities, the technology may result in a costly disappointment. Also here knowledge is power, so learning about the function and purpose of various technologies is an important input to the decision-making process and one of the key strategies to improve waste management performance.
- Whatever the envisaged improvements of the system, they need to correspond to the local financial circumstances. While people are usually willing to pay for good services that meet their needs, affordability can be a serious issue in low-income countries. Working with local partners is essential. Also, the current policy trend of placing financial responsibility on producers (e.g. through extended producer responsibility) can be very beneficial in alleviating the burden that has traditionally been on the shoulders of local authorities.

- The reasons for the relative success of some initiatives are primarily socio-cultural and political economic in nature, rather than technological.
  - The buy-in of stakeholders is an essential ingredient in all success stories.
  - Political commitment and keen leadership are indispensable for tangible results.
- Direct regulation – accompanied by credible and consistent enforcement – is essential for the very existence of the waste sector.
  - When preparing legislation, hold inter-departmental meetings to identify and address possible contradictions, overlaps and gaps. You don't want affected stakeholders to get discouraged because of conflicting or endless repetitive requirements.
  - Legislation consisting of various laws, rules and regulations is more likely to be effective if it makes sense to those for whom it is intended. They may not necessarily agree with it, but they should be able to understand and agree with its logic.
  - Think long-term! While providing regulatory certainty to encourage investment, legislation also needs to be flexible enough to accommodate future changes.
- While cost recovery from waste fees is desirable, cities can be clean and have their waste properly treated also if the financing comes from other sources. More often than not, people are willing to pay for good primary waste collection, getting waste out from under foot. Central government sources may be needed to build facilities. The cost of *running* the facilities is also essential to bear in mind.
- Social instruments that rely *solely* on providing information are insufficient to change people's attitudes and behaviour. Information is necessary but not sufficient to prompt people to change. Engaging with people, particularly in the community setting, works better. Visibly leading by example is necessary, too. While the Internet and social media certainly have their place, also go and meet people, interact with communities, answer their phone calls.
- Embrace adaptive governance. No policy comes with a guarantee of success; therefore, for a robust and resilient system, rather than a focus on the 'right' policy, adopt policy processes that stimulate experimentation, adaptation and learning. The evidence from monitoring and evaluation can provide valuable insights about the system's performance and lessons grounded in practice. This will help to further develop the policy and inform future policies in the same and related areas.

TOPIC SHEET 15

GENDER AND WASTE MANAGEMENT[1]

The concept of waste is not a neutral, objective concept but rather determined by such factors as lifestyle, social structure, gender, class and ethnicity. These factors also influence the social organization of the work related to waste and the political landscape in its management.[2]

Several studies have highlighted the impact of gender issues[3] on the perception, handling and management of waste in both the formal and informal and private and public sectors. Recognizing this and mainstreaming gender by ensuring that waste management policies, programmes and systems take into account their implications and impacts for women and girls facilitates the design of more effective, efficient and inclusive waste management systems while also providing more equitable access to livelihoods and enhancing women's roles in economic and political spheres.

In many societies, women are responsible for managing the waste within their household and globally they are the primary users of waste management services. Therefore, it is important to understand their needs and preferences, which may differ from those of men. For example, in a study carried out in South Africa,[4] men preferred a drop-off system while women preferred door-to-door collection, due to the time constraints women face as a result of their multiple roles and due to the mobility limitations found within certain cultures.[5]

Women from a cooperative showing their products, Philippines

In order to maximize the quality and efficiency of SWM services, it is important to know the challenges women face when using the system. For example, conflicts between the waste collection schedule and household responsibilities could result in waste being put out for collection at the wrong times.[6] Therefore to enable the system to meet expectations and improve its degree of success, women need to be consulted and included in decision-making processes. Unfortunately their degree of participation is still low. This becomes even more pronounced when the formality of the meeting increases, when consultations are organized under a highly masculine structure, or when consultations are scheduled at times that conflict with women's household and family care duties.

1 Topic Sheet prepared by Ainhoa Carpintero, with input from Lios McGilchrist of Living Earth Foundation.

2 Fredericks, R. (2008).

3 To understand many of these issues, it must be recognized that in most societies, women take on triple roles (a reproductive/caretaker role, a productive role, and a community management role) whereas men are assigned a single role (a productive role, combined on occasion with community political activities), through which they are considered the providers who sustain the family.

4 Poswa, T. (2004).

5 This preference might be linked with the willingness to pay for the system. Often men prefer cheaper systems, which frequently means a central collection point within the community.

6 In some countries the citizens are required to bring the waste to the collection truck at specific times. Waste bags left in the streets will attract flies, mosquitoes, rats, dogs and so on that can act as disease transmitters and pose potential health hazards.

People working with waste around the world. Most of the workers are men, although there are also many women.
Photo compilation prepared by Jeroen IJgosse

Gender differences may also be reflected in attitudes towards public health and community cleanliness. As women and children are the ones typically responsible for dealing with household waste, it is they who need to make alternative arrangements in the absence of a regular and reliable collection service. They may have direct interaction with local uncontrolled dumpsites or carry out hazardous disposal practices such as open burning of waste. The related health risks make women more aware of the need for appropriate management practices and services and this influences their willingness to pay for such services. This willingness becomes stronger as the accessibility, quality and efficiency of the primary collection services increase. Their direct involvement with waste handling, exposure to health risks and willingness to pay to have waste removed from their community may also affect their preferences among policies, technologies or general approaches.[7]

In many communities, in the absence of adequate waste management services women engage in voluntary community activities such as cleaning, sweeping the streets and providing primary collection of waste.[8] However, when these cleaning services or waste collection services are contracted to private enterprises or institutionalized, it is often men who get the jobs – there is a general perception in many places that formalized (and therefore paid) work in SWM is a man's job.[9] This denies women the opportunity to earn a living or improve their position. Some women might be hired, but typically they are hired for low-level jobs related to the domestic environment such as street sweeping, rather than for other stages of waste management such as transportation, treatment or disposal. A study carried out by Living Earth Foundation and ANPEZ in Port Harcourt, Nigeria in 2013[10] showed a low level of female participation in the formal waste management sector. The same survey showed women running only 7% of the waste management companies having waste management service contracts with local governments.

Nevertheless, women workers can bring benefits to the system in all stages of the waste management chain, since they are the ones that might have organized waste-related services in the first place and possibly have different views of the local situation, which can result in significant improvements or different approaches. In the same way, having more women in leadership roles can incorporate or bring emphasis to specific values within waste management systems that complement and enhance those provided by men. For example, according to a study carried out by GIZ on gender issues

Women dismantling electronic equipment, Japan

in municipal SWM in the state of Mexico,[11] municipalities headed by women had stronger environmental concerns and placed more relevance to recycling and source separation than those headed by men.[12]

Gender inequalities have also been identified in the informal sector. Access to waste items is conditioned by power relations established within the group working at a particular location. It is common to find that men take control over waste items with a higher recycle value and leave the items of lesser value for women. It is also common to see a clear division of labour within families working in the waste-related informal sector: men mainly work out in the streets collecting waste and selling the materials, while women segregate the collected waste materials at home and are responsible for disposing of those with no value. For example, the Port Harcourt study showed that only 4% of waste pickers working at the city's sites were women. Such a division of tasks makes women's work invisible, which leads to the under-reporting of female participation,[13] a lack of recognition and inclusion in public

Informal sector, Japan

7 Muller, M. and A. Scheinberg (1999).

8 Gonzenbach, B. and A. Coad (2007).

9 Most waste or recycling business owners are men. However this is related to the fact that generally women face greater constraints to access to credit or might not have any assets in their name and therefore cannot afford the initial capital investment required to establish a business.

10 Lawal, N. and S. Di Vicenz (2013). Gender Analysis of waste management structures in Port Harcourt, Living Earth Foundation.

11 GTZ (2006).

12 Chapter 4, Section K of the Beijing Platform for Action, focusing on women and the environment, recognizes such facts and states: "Women have often played leadership roles or taken the lead in promoting an environmental ethic, reducing resource use, and reusing and recycling resources to minimize waste and excessive consumption. Women can have a particularly powerful role in influencing sustainable consumption decisions."

13 Espino, A. and N. Bidegain (2011).

policies, and a reinforcement of the belief that men are the ones sustaining the family by failing to acknowledge the economic contributions of the activities carried out by women. As a result, when waste-picking activities are formalized, in many occasions women are excluded or do not have the same opportunities as men.

When waste value streams are derived at the household level, these appear to be much more accessible to women, such as clothing repair and reuse. Another example comes from a Living Earth project in Uganda, where micro- and small enterprises fabricate fuel briquettes from organic waste. This small-scale processing requires little capital investment and can be done at home; perhaps for these reasons, the briquetting sector has proven to be much more accessible to women. However, constraints on women in terms of opportunity for business expansion and investment mean that such activities often remain confined to the cottage industry level.

Additionally, the inclusion and formalization processes should recognize the constraints, such as rigid schedules or the lack of suitable care services (e.g. for children), which women often face when attempting to combine work with family or household care activities. Recognizing such gender issues within the informal sector and issues of women's empowerment will provide women with better opportunities to earn a living while helping to alleviate poverty.

An increased awareness of the potential of gender perspectives and experiences as well as related specific needs within policies, plans and projects will have a greater social, political and economic impact and benefit society as a whole.

Women Initiative the Gambia (WIG)[22]

The Njau Recycling and Income-Generating Group (NRIGG) began in 199[23] in Njau, the Gambia, with the aim of finding a creative solution to the substantial local problems of inadequate waste management, littering of fields by plastic bags[24] and low income levels of women and youths in rural Gambia. NRIGG has focused on the separation of organics, metals, paper and plastics from waste and has created markets for specific end-products that they produce: organics are composted, plastics are up-cycled into purses, mats and bags;[25] rubber is used to make necklaces and metals are sold to traders. In addition, training in home composting and recycling has been carried out in communities, where women act as recycling champions within their communities, advising other residents on handling waste and earning income by selling value added products. Women have also received literacy education and health training, and those who do not have a bank account can use a savings account system managed by the organization. The savings system requires apprentices to re-invest an agreed amount of their profits into their business in accordance with a business plan they must develop.

What started as an initiative by a five-member team has now grown into a community-based organization (CBO) under the official name of Women Initiative the Gambia (WIG)[26] with 12 volunteer staff, as well as a Director and two advisors. They have trained over 290 women and youths in skills for recycling plastic waste and empowered communities by educating them on environmental protection through better waste management. In 2013-2014, WIG trained eight communities on recycling plastic bags, bicycle tyres and paper, on business management, leadership, value added products and the importance of networking in development. The project is also focusing on other, non-waste-related activities which are generally regarded as being gender specific (crocheting and knitting as a women's activity, for example) but could provide opportunities to both genders.[27]

WIG is currently supported by the UK-based livelihood charity Concern Universal as part of the Women Empowerment Project funded by the European Union.[28] WIG is also part of the global so-called waste-to-wealth movement[29] and in 2015 WIG expects to receive technical support from the newly established charity WasteAid,[30] which seeks to connect poor and vulnerable communities to waste management resources and expertise.

14 Text provided by Lakshmi Narayan from KKPKP.
15 See http://www.wastepickerscollective.org/
16 Waste-pickers retrieve paper, plastic, metal and glass scrap from waste bins or receptacles and from landfill sites where the collected waste is transported and dumped. Itinerant buyers purchase small quantities of scrap from households, offices, shops and other small commercial establishments.
17 Officially named SWaCH SevaSahakariSansthaMaryadit, Pune. See http://www.swachcoop.com/index.html
18 These waste pickers have bought shares in the cooperative and are thus both stakeholders and owners of the cooperative.
19 KKPKP has been fighting for the rights of the waste pickers and since its inception has addressed social issues such as violence against women, child labour, school enrolment, child marriage, domestic violence and gender inequality.
20 As of 2011, 25 women drivers were operating under an agreement with Pimpri-Chinchwad Municipal Corporation (PCMC) (see "We SWaCH" – http://www.youtube.com/watch?v=bMvU5bOHpTU). Currently, even though SWaCH is not working with PCMC, the new contractor has appointed the same women who were trained by SWaCH. The number is around 30.
21 As of early 2015 SWaCH was not operating at full staff strength (awaiting renewal of a MoU with PMC). However, management and supervision positions are largely filled by women – over 70%.
22 Text prepared from information provided by Isatou Ceesay (Director of WIG), and from http://www.theguardian.com/lifeandstyle/2014/may/01/small-african-recycling-project-tackling-mountainous-rubbish-problem and http://www.theecologist.org/News/news_analysis/2507701/gambia_recycling_for_womens_wealth_and_independence.html
23 In conjunction with a Peace Corps initiative.
24 See Box 4.10 regarding plastic bags and problems they cause for people, cattle and the environment.
25 For a demonstration of how NRIGG women make a purse out of plastic bags, see https://www.youtube.com/watch?v=1zZw7yrl22M
26 The group registered as community based organization (CBO) in 2012.
27 Other non-related waste management activities include tie dye and beadwork.
28 Concern Universal Gambia-Senegal Annual Report 2013-14.
29 http://www.wastetowealth.co.za
30 https://wasteaid.wordpress.com

Breaking through gender barriers[14]

Kagad Kach Patra Kashtakari Panchayat (KKPKP)[15] is a membership-based trade union established in 1993 by waste pickers and itinerant waste buyers[16] in Pune and Pimpri Chinchwad, India, to advocate recognition of the waste pickers' role in SWM in the city, their status as workers, their contribution to the environment and their integration in the waste collection and disposal system. In 2008, KKPKP established SWaCH (Solid Waste Collection and Handling),[17] the India's first wholly-owned[18] cooperative of self-employed waste pickers and waste collectors and other urban poor who provide primary and secondary collection to the citizens in Pune.

As of February 2015 KKPKP has around 9,500 members, of whom 80% are women. Since its inception, KKPKP has been working to address gender inequalities reflected in the waste collection sector.[19] KKPKP has resolved that, in the spirit of democracy and participation, the Union's leadership and decision-making positions should reflect the composition of the sector, which has always had a higher percentage of women. The first gender-related debate was over the KKPKP logo – a woman carrying a sack. While itinerant buyers (who are generally men) argued for a 'male' logo, the majority of union members found the logo justified as a result of the predominantly female composition of the trade union. Some 200 male members quit, only to return after two years, recognizing the real gains for workers through the union.

Other gender-related concerns are regularly debated in the monthly training programmes, covering issues such as legitimacy of leadership, true representation (e.g. can a male be a representative leader of a largely female group), the quality of earnings of men as compared to those of women (e.g. percentage of net income contributed towards family expenditures), expenditure patterns, differences in men's and women's priorities with respect to expenditures, etc.

In accordance with the ideology and the history of KKPKP, the internal rules of SWaCH mandate a 75% representation of women on the Executive Committee. Additionally, women waste pickers of SWaCH have forestalled concerns that upgraded livelihoods (through the transition from waste picking to waste collection) will be usurped by males; they have also broken through stereotypes by stepping into typically male-dominated roles, such as garbage truck drivers,[20] as well as managers and supervisors within SWaCH.[21] This has been achieved through systematic training programmes, discussions convincing women (mothers and daughters) to take on such roles and the provision of formal training in driving skills.

© KKPKP

In addition, efforts are made to fill all administrative positions in both KKPKP and SWaCH by the daughters and sons of waste pickers who have completed formal education.

CASE STUDY 5

BOGOTÁ – A GROUND-BREAKING SCENARIO OF THE INCLUSION OF RECYCLERS IN WASTE COLLECTION AND RECYCLING[1]

Context and background

The Colombian capital city of Bogotá with a population of 7.5 million generates over 7,500 tonnes of municipal solid waste per day. The waste collection and transportation system that was in place between 1994 and 2012 was operated exclusively by private cleaning companies. Then, from December 2012, the public operator Aguas de Bogotá started service provision for some parts of the city. As measured by the Special Administrative Unit of Public Services (UAESP), as of December 2014, the public operator took care of cleaning in 63.15% of the city. Private operators had been appointed for the remaining areas of the city.

In parallel to the formal system, for more than 50 years the city´s waste pickers have been collecting and selling specific waste materials, including glass, paper and cardboard, metals and plastics. According to the *Registro* Único de Recicladores (Unique Register of Recyclers), as of December 2014, there are 20,643 registered recyclers in the city, with an estimated additional 8,000 unregistered ones. Their work allows for the operation of over 1,500 recycling warehouses that reintroduce recyclable materials into the recycling value chain and provide feedstock to industries. While they have been the *de facto* recycling system in the city of Bogotá, their work remained unrecognized for a long time. This case study highlights the process of organization and inclusion of waste recyclers in the 'formal' waste management system in Colombia and implementation of the landmark Order 275 of 2011 by the Constitutional Court in Bogotá, which recognized recyclers' organizations as legal bidders for formal waste management systems.

1 Text elaborated with inputs from Mariel Vilella, Zero Waste Europe/GAIA, with the support of: Federico Parra, Coordinator of the program of recyclers in Latin America, WIEGO. For further information see: Parra, F. (2015). Reciclaje sí, pero con Recicladores. WIEGO Policy Briefing (Urban Policy) No 9.

Process of organization and inclusion

The history of informal recyclers in Colombia starts in the 1950s, when a wave of violence forced farmers to migrate to the city, where through waste they found a way to survive. The first successful efforts to organize themselves date back to early 1960s. The process of organizing intensified in the 1980s, culminating in the creation of the National Association of Recyclers in Colombia (ANR) in 1990. The Association of Recyclers of Bogotá (ARB) was formed in 1992 and ultimately became one of the strongest of these organizations, leading the process for the recognition of informal recyclers as public service providers in the country.[2]

Most of the legislation related to waste management was framed within the National Law for Public Services 147/1994 and National Decree 17/13 that regulates public cleaning services, which were established in 2002. This legislation was in favour of awarding contracts to private companies and was detrimental of the inclusion of recyclers in the formal system.

In parallel, the recyclers led the campaign to get a judgement from the Constitutional Court in recognition of their rights. The results of this representation of the informal sector through strong and dedicated leadership by individual recyclers, with the advice and support from NGOs and professionals, mobilization and raising-awareness campaigns in the media, as well as undertaking legal action at the national level have been remarkable – leading to seven rulings and orders by the Constitutional Court of Colombia since 2002 protecting the rights of recyclers. One of the major achievements was the recognition of recyclers' organizations as legal bidders within formal waste management systems in

2 In 1994, several recycler organizations such as ARB and Rescatar, with the support of advocacy NGO 'Fundación Social', participated in the process held by the municipality to award waste management services. From October 1994 to June 1995 the Cooperative Association of Recyclers of Bogotá (ARB) was hired to help with health emergencies when local dumpsites were not used. This was an emergency measure and it did not result in granting the ARB the right to bid for the services. In fact, the National Law for Public Services (Law 142 de 1994) made public bidding very inaccesible for participation by the ARB.

2011. This reinforced a 2010 court ruling that had asked private companies that had won the waste disposal tender to include and provide job opportunities for informal recyclers. Noteworthy also was Order 275 of 2011 by the Constitutional Court, as it combined all the Court's previous rulings and orders, delineating guidelines for a recycling operating model with inclusion of the informal sector and requiring, among other things, that waste pickers be remunerated for the services of collection, transportation and recycling of waste materials. Official recognition of waste pickers as service providers came as a paradigm shift in the outlook of both the national and the city level governments towards the informal sector, as it constituted a formal acknowledgment of the benefits of informal recycling efforts.

As a follow-up, the Constitutional Court ordered the government to amend National Decree 17/13 so that it was consistent with the rulings from the Constitutional Court and provides a level playing field for the recyclers. In December 2013, the government approved National Decree 2981, which is the prevailing legislation.

Implementation in Bogotá

The application of the 2011 Constitutional Court rulings in Bogotá involved significant changes in how the waste management system was organized in the city. The municipality created a Register of Recyclers that identified the areas where they worked and how they operated, in order to map, define and characterize them as beneficiaries of the affirmative actions prescribed by the Court decisions, and thereby prevent misuse by opportunist groups. Furthermore, the municipality committed to promote source separation and highlight the role of waste pickers. It also provided resources for substituting animal-pulled vehicles with motorized vehicles, in order to improve recyclers' working conditions.[3] To make the system operational, in December 2012, the municipality established a network of 250 weighing centres (*bodegas*) for recyclers to take the collected recyclable materials.

The major innovation of the system was the introduction of a bimonthly payment to recyclers who would commit to bringing their recyclable materials to the *bodegas* on a daily basis, at an equal rate per tonne as the rate used to pay private companies for collected and transported waste.[4] Every two months, the 8,250 Bogotá recyclers enrolled in the Bogotá Zero Waste Programme (which ran until December 2014) received a payment equivalent to 87,000 Colombian pesos[5] per tonne collected and transported. The payment was conditional on the recycler being registered in the official Register and having a bank account. This system is a significant departure from the previous situation where sales of the materials were the only source of revenue for informal recyclers. Of all the recyclers participating in this initiative, 1,500 are associated with ARB.[6]

Animal/handpulled carts used by waste recyclers

As these have been the first experiences with including waste pickers at this scale and level, the stakeholders involved are exploring new territory together and searching for workable solutions. For example, efforts are ongoing by the city council and ARB to establish routes and optimize the use of motorized vehicles for transport to *bodegas*. Herein, the right balance has to be found between, on the one hand, the tendency of the authorities to design a system from scratch focusing on efficiency and optimization, and the existing realities and pickers' established practices on the other. Also, WIEGO has provided funds to improve the health and safety conditions of recyclers and from June 2015 onwards recyclers will have access to specialized equipment, professional gloves, masks, boots, identity cards etc., along with capacity building training to educate them in the use of this equipment.

More broadly, the city authorities have initiated a Zero Waste Programme and increasing recycling rates is a part of the programme. Within this, the inclusion of informal recyclers has a clear place. Several areas need attention in order to take the recycling model in Bogotá to the next level. In order to facilitate door-to-door collection of recyclables, source separation habits need to be developed further, perhaps with support through appropriate incentives. A specific problem concerns

3 In 2012, the municipality agreed that each registered and verified recycler may receive, upon surrender of his cart and horse, compensation equivalent to 21 million Colombian pesos (approximately 7,200 USD as of August 2015), in the form of motor vehicle, or in the form of "seed plan" to their business plan. In December 2014, 2,800 animal-drawn vehicles had been replaced.

4 These *bodegas* act as "points for authorized weighing" where recyclers bring their recyclable materials recovered daily and record the weight on an official form. The type of material does not affect the rate of payment. After weighing, the recycler can sell the recyclables to the intermediary by type, following the price set by the market for recyclables, as traditionally done.

5 Approximately 30 USD (as of August 2015).

6 The ARB currently has 2,500 associated recyclers organized in 17 base organizations, represents recyclers, provides training for capacity building, improves the organization of the value chain and makes sustained efforts to mainstream the informal sector.

organic waste, as currently there is no formal system for it in the city (some of it is collected by the recyclers and used as animal feed). Ultimately, any waste that is not recycled by the recyclers, including organic waste, ends in sanitary landfills, which still represents a high percentage of Bogotá's waste.

Weigthing operations at Bodega

Lessons learned and the way forward

The case of Bogotá is significant for two main reasons. First, the Goldman Prize awarded in 2013 to Nohra Padilla, one of the ARB's leaders, was a major victory in gaining global recognition and visibility.[7] Moreover, the orders of the Constitutional Court have been included in public policy at the national level. The Ministry of Housing and Urban Planning issued Decree 2981 in 2013 on public services[8] and Resolution 0754 in 2014 on solid waste management planning,[9] requiring municipal authorities across Colombia to shift their waste management plans to inclusive models that incorporate informal recyclers. The same system is now under consideration to be developed and implemented in several other cities in Colombia.

Second, through the inclusion of some 8,250 recyclers into the waste management system, their role and their work have been formally recognized and remunerated according to rates stipulated by the competent authorities. The investment of the council towards the recyclers' salary has doubled the recyclers' earnings, which has significantly improved their quality of life, enabled them to send their children to school, and contributed to the well-being of the community.[10] Finally, through their activities, recyclers divert an estimated 1,200 tonnes of recyclable materials per day from the landfill, thus having a major contribution to the city's 16% recycling rate.[11]

The initiative has just begun and the new system is making a significant contribution to recycling and to solid waste management overall. The way forward is to build on what has been achieved, to increase the recycling rate to levels beyond the national average and to make this a truly win-win solution for both the recyclers and the city.

7 http://www.goldmanprize.org/recipient/nohra-padilla

8 Decree No. 2981 of 2013, by which the provision of public services is regulated http://diario-oficial.vlex.com.co/vid/decreto-2013-reglamenta-servicio-aseo-482847738?_ga=1.20099835.239573512.1425225743

9 Resolution No. 0754 of 2014, for which the methodology for the formulation, implementation, evaluation, monitoring, control and updating of Integral Management Plans for Solid Waste was adopted. http://diario-oficial.vlex.com.co/vid/resolucion-numero-0754-2014-547338526

10 While this case shows the improvement in the working conditions of the recyclers and the waste management system in the city, there remains room for improvement. According to the National Survey of Recycling 85% of the recycling population in the country belongs to socioeconomic strata 1 or 2, living in precarious housing conditions with low social coverage and low schooling. Aluna Consultores Limitada (2010).

11 Aluna Consultores Limitada (2011), as quoted in (Parra, 2015). Aluna Consultores Ltda, 2011. Estudio Nacional de Reciclaje y los Recicladores. CEMPRE. The same report gives the national average recycling rate as 19%, compared to the 16% in Bogotá. There is however no data available yet on any changes in the recycling rate in Bogotá following the formalization programme.

CHAPTER 5

WASTE MANAGEMENT FINANCING

Money matters in waste management, so a particular focus of the GWMO is waste management financing. After providing a summary of the key messages (5.1), this chapter seeks to understand the costs and benefits, including the difficulty of quantifying the costs of inaction (5.2) and the less tangible benefits to society of good waste and resource management. After introducing waste and resource management as both a public service and a business (5.3), a generic financing model is introduced (5.4) and applied to business-to-business waste services (5.5). The next three sections then apply the financing model to municipal solid waste management, looking in turn at the available options for service delivery (5.6), revenue collection (5.7) and investment finance (5.8). Finally, a discussion on how to decide on the appropriate financing model in a particular situation provides a 'toolkit' to serve as a basis for action (5.9).

© Jayavilal Fernando

5.1 SUMMARY OF THE CHAPTER – KEY MESSAGES ON WASTE FINANCING

- Understanding both the financial and economic costs of waste management remains a global challenge. Sustainable waste management solutions need a reliable economic evidence base, and much more work needs to be done, both globally and in particular in developing countries.
- The economic costs of not addressing waste management problems in developing countries are difficult to quantify, but the available evidence suggests that they (greatly) exceed the financial costs of environmentally sound waste management. So action on waste management is an urgent political priority – waiting for better evidence is no excuse.
- The evidence base on the impact of inaction is strongest in the case of SIDS, reflecting a special case of scarce land, expensive solutions for transport, treatment and disposal, and the strong economic value of coastal aesthetics and tourism.
- Financing models for waste management can be analysed in terms of four components: the relationship between the 'client' and the 'operator' and the sources of 'revenue' and 'investment finance'. Different combinations give almost an infinite number of permutations; There are myriad combinations, with no right or wrong answer. Each local situation requires a tailor-made solution.
- There is no evidence to show that either private or public service provision or financing for MSWM is more frequent or is more efficient or beneficial than the other.
- It takes technical knowledge and management skills to contract out waste management services. Using performance-based contracting, capping fees, building up funds for maintenance and replacement and tightening policy obligations are all good practices.
- Cost recovery is possible where there is a demand for a service (e.g. primary waste collection) or a product, and is increasingly more difficult if activities are policy-driven (such as environmentally sound treatment and disposal). Full cost recovery is more affordable as income levels increase, even though the absolute costs also increase.
- Low-income and lower-middle income countries can barely afford current collection costs, so even the first steps of extending collection coverage and eliminating uncontrolled disposal will raise affordability issues.
- Collecting revenues indirectly with other taxes or directly as user charges can both work, provided that the selected system is a transparent one which fits with the local custom and tradition, and that the service fits customer needs.
- It is possible to increase payment rates and cost recovery by smart enforcement mechanisms and by providing support for those who cannot afford to pay.
- There is a huge need for investment finance in the waste sector. Active announced projects currently total about 300 billion USD, of which 85 billion USD is for MSW projects.[1] Raising investor confidence is a challenge, even in high-income countries.
- Commercial financing to cover the investment need is often not possible because of the poor banking service coverage in the region or the high level of indebtedness of the cities of the developing world. The World Bank estimates that out of the 500 largest cities in the developing countries, only 4% are deemed credit worthy in international capital markets and 20% on the local markets.
- International development financing in SWM represents just 0.3% of the total. Of the available funding over the last 10 years, two-thirds has gone to just 10 middle-income countries. Making the necessary investment finance available to those who need it most, in the low-income countries, is an urgent priority and a particular challenge.
- Environmentally sound treatment and disposal facilities have a higher capital cost and operate at higher gate fees, so are more affordable in upper-middle or high-income cities or countries.

1 These figures need to be treated with caution, insofar as they indicate the total of projects 'active' over a two year period, and include many projects that will never be approved, financed or built. The vast majority of this proposed investment activity is in the high-income countries, with the UK particularly active.

- Such 'higher tech' facilities are more often provided by the private sector under some form of public-private partnership (PPP) than directly by the public sector. They may also involve long-term, inflexible contracts for guaranteed waste amounts.
- Facilitating access to micro-financing for MSE, CBO and informal sector waste services is a relatively simple intervention that can help communities to help themselves and improve the livelihoods of the urban poor and of marginalized people.
- Innovative financing instruments can enhance the efficiency of funds directed to waste management. For example output-based financing has shown positive results.
- Economic instruments such as EPR and product charges can be a very strong incentive to provide for the necessary cash flow allowing investment in collection and recycling facilities.

5.2 UNDERSTANDING COSTS AND BENEFITS

Traditional financial analysis, looking at costs and revenues, is the analysis that any private company will carry out when deciding between two financial projects or project portfolios. The investment performing better on the financial indicators, such as internal rate of return and payback period, will be the one chosen. In the case of spending public funds, one also needs to look at the wider societal costs and benefits of any investment project. In many cases, waste management investments are competing with investments in health, education or other infrastructure, so weighing the relative benefits to society becomes crucial. The economic analysis is analysing costs and benefits, where possible attaching a monetary value to different positive and negative environmental, economic, social and health impacts of the investment.

Waste management activities are a net financial cost to the public, so justifying investments in waste management need to use an economic cost benefit analysis, factoring in costs for environmental protection and for safeguarding public health.

5.2.1 Financial and economic costs and benefits

The basic definition of waste in Section 2.3 is *unwanted or discarded materials rejected as useless, unneeded or excess to requirements*. If waste is perceived by the generator as having little or no value, it will tend to be managed either at the lowest possible cost to themselves, or at minimum cost in compliance with legal requirements. The costs normally considered are the *financial* costs of waste management that occurred for the collection, treatment and disposal of waste. These costs may be offset in part by direct revenues from selling recovered materials or energy or the benefits of action to the society as a whole, such as resource efficiency, green job creation and a healthy and clean living environment. The waste management industry depends on the consistent implementation and enforcement of strong public health and environmental legislation to create a level playing field for the 'waste management market'.

Why spend this money on waste management? This question can be answered at many levels: because it is the 'right' thing to do; because it is a legal requirement; or because it makes economic sense, as the costs to society of the indiscriminate accumulation, littering, burning and dumping of waste likely exceed the financial costs of environmentally sound waste management.

Even the best available waste management technologies have environmental impacts of their own and are not able to eliminate waste without a trace, so there are environmental costs of waste generation that the waste management costs do not account for, but which the environmental economist may attempt to quantify. However, in the absence of waste management, or where waste management services are failing, environmental and social negative impacts of waste are much greater, so the focus here is on attempting to estimate the economic costs of inaction, rather than the somewhat lower remaining externalities (economic costs) of environmentally sound waste management.

There are benefits to society and the wider economy of environmentally sound waste management beyond those which are accounted for as the waste industry's contributions to either GNI or GDP. These benefits include avoiding economic costs, the revenues from the sale of recovered products and energy, and also include the wider economic benefits of resource efficiency and waste prevention such as increasing resource and food security and creating green jobs.

5.2.2 The financial costs of taking action

Financial costs and revenues are those that are recorded (or should be recorded) by accountants, which show up in the financial accounts and are used in financial analysis. In waste management, it is important to distinguish between direct investment, operation costs and direct revenues.

Investment costs

Investment costs generally receive more attention and tend to be better understood largely because of the number of projects focusing on modernizing waste management infrastructure. Investment costs are all those related to developing and constructing a project, among others: project preparation, including planning, siting, feasibility studies, permitting and the associated public involvement and consultation; detailed design; land costs, especially in the case of landfills; equipment, facilities and construction.

Sanitary landfill under construction

In principle, investment costs are relatively easy to benchmark, as equipment and technology suppliers know the cost of their products. So the costs of a compactor truck, a container, a sorting line, or a landfill liner are estimated based on price quotations from equipment and technology providers, in accordance with the technical design requirements. Necessary building and foundation works are estimated by engineers and professionals based on cost standards for the country where the investment is to be carried out.

However, in practice, various complications may appear: it is difficult to obtain reliable comparative costs for the likely capital costs of alternative types of treatment and disposal facilities; waste properties are not uniform, so different or especially tailored technologies may be necessary; some component costs, especially those locally produced, vary between countries and between sites; environmental protection standards vary; many proprietary technologies, being relatively new, have limited experience at full-scale operation under a wide variety of local conditions; and commercial confidentiality may restrict the public availability of data. *Table 5.1* below provides order of magnitude comparative costs for some alternative treatment and disposal technologies for MSWM.

Operation costs

The main components of operation costs are the costs of labour, fuel, energy, maintenance and repair, emission control and monitoring, revenue collection, public communication and management and administration. Some important costs are often overlooked or not properly budgeted for, such as the costs of awareness raising campaigns, customer care, environmental auditors, and training and capacity building, and, in case of private sector participation, the client's costs related to tendering, contract negotiation, supervision, inspection, insurance and control of activities. Though in some cases tax exemptions may apply, operators are normally subject to sales taxes (VAT). Profit and dividend taxes are often not relevant for public operators but are relevant for private commercial operators.

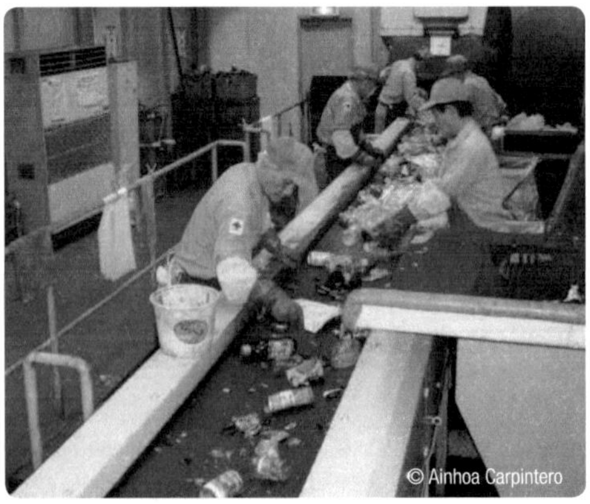

Material recovery facility, Japan

Some waste management costs arise because of requirements set out in legislation. Environmental emission control costs tend to rise for the operator as standards become more stringent, while the environmental regulator will also incur higher associated costs through enforcing those standards. Entirely new cost 'centres' may also occur, such as the after-care costs of a landfill site no longer actively receiving waste; the cost of introducing a user pay system; and costs related to new policy priorities such as incorporating re-use and prevention in the waste management programme.

Overall, 60-70% of the total costs of advanced waste management are operation costs.[2] An activity-based cost accounting makes it possible to understand the costs associated with collection, different treatments and disposal and is helpful for financial management and for identifying inefficiency in the waste management system. Understanding these costs also helps with investments and upgrades in the system, focusing investment on cost-efficient technologies with due care in designing facilities at the correct size.

Operation costs are recorded in the bookkeeping of cities worldwide and are used to estimate annual budgets. Nevertheless, operation costs of SWM services are still often insufficiently known and it remains an enigma why benchmarking for example collection costs or landfilling costs is so difficult even in cities where conditions are similar. This is due in part to differences in the accounting systems. Cities often allocate only a part of the operation costs to the activity of waste management. For example waste management may be organized in a department together with other public utilities and allocation of costs to a particular utility service such as SWM is not common practice, as was the case with public utilities in former communist countries. In many Indian cities the cleansing department used to be in control of waste management, but the vehicles were supplied by the mechanical engineering department and the workers by the 'zones' within the city, while the costs were recorded by each department. Things are further complicated because of the different aggregation methods used for costs.

Other constraints are the frequent absence of information and sensitivities related to sharing cost and revenue data. The city staff are often very helpful in sharing all sorts of technical, organizational data, even inventories or the investment costs of facilities and equipment that are purchased, but when it comes to operation costs, these are often regarded as 'confidential' and cannot be made available. Therefore looking at data across cities is challenging, both in terms of obtaining any useful data, and in further comparing 'apples and oranges'.[3]

An initiative to benchmark operation costs could start by collecting activity-based cost information in a standardized way. This methodology collects only the costs that are related strictly to a distinct activity in waste management, such as the collection of mixed waste, collection of source-separated waste or final disposal. Collecting of information can increasingly be organized through an interactive web-based interface that instructs the user and has built in checks on input information for errors.

Revenues from resource recovery

Resource recovery activities generate revenue streams from the sale of recovered and recyclable materials, compost and energy.[4] The quality of the products and the dynamics of the different markets for secondary raw materials and energy determine the price of sale and thus the revenue potential. Some of these markets are global and are volatile and dependent on price fluctuations on global markets for secondary and primary raw materials. The market for compost is more local but usually underdeveloped and needs effort to become sufficiently developed based on local potential and market requirements. Both of these markets have very specific requirements for quality.

Baled aluminium cans

While it is very difficult to generalize, one 'rule of thumb' is that revenues from the sale of compost are unlikely to cover more than 40% of the costs of separate collection and processing of organic waste.[5] For recyclables recovered from MSWM, this ratio is quite different for different types of materials: metals, high-grade paper, and in some places PET 'pay for themselves' while recycling of lower-grade paper, other plastics, glass, wood and textiles often represent a net cost, but are financially worthwhile at a system level as a cheaper 'sink' when the price of gate fees for disposal (of mixed waste) rises above a level of perhaps 40 USD per tonne.[6]

2 Pfaff-Simoneit (2013), listed in Annex A, Chapter 5, General reading.

3 There have been attempts to collect primary data based on templates that would facilitate collection of comparable data, including applied research work carried out for example for the UN-Habitat Solid Waste Management in the World's Cities publication (20 cities profiled). See Scheinberg et al. (2010) and Wilson et al. (2012), as listed in Annex A, Chapter 1, Waste management. See also a recent SWEEP-Net project to collect cost information for countries in North Africa and the Near East at SWEEP-Net (2014), listed in Annex A, Chapter 5, Estimating economic costs and benefits. The latter information has been consolidated in the Waste Atlas, which aims to collate similar data from around the world – http://www.atlas.d-waste.com/

4 See Boxes 3.6 and 3.7 in Section 3.5.3.

5 The maximum achieved within the EU is less than 20%.

6 See also Section 2.3 on drivers for waste and resource management.

For energy, there is generally a demand, and a long term trend of increasing real prices,[7] but the extent to which energy generated from waste can substitute traditional sources depends on the composition of the waste, the local cost of energy and access to the appropriate energy grid. Various policies, taxes and other economic instruments have a high positive or negative impact on these revenue streams. Subsidies for renewable and/or biogenic energy and heat and the strict regulation on waste disposal encourage energy from waste (EfW)[8] in Europe and increase the earning potential from these facilities. Recycling is similarly impacted by the policy regime. For example the implementation of an EPR system increases recycling and tends to diversify the recycled waste streams, similarly a high gate fee at the landfill, or the imposition of a landfill tax, tends to encourage recycling.

Estimating revenues from resource recovery is again subject to much uncertainty. Markets vary locally and in terms of product quality, while prices of internationally traded materials are subject to the fluctuations of the global markets. Waste management services come at a net cost that needs to be supported in one way or another by society. Every tonne of waste that 'pays for its own' recovery due to its intrinsic value is one less tonne of waste to be treated or disposed of and paid for by a fee, user charge, tax or subsidy, decreasing the net cost of waste management services.

'Typical' net costs for MSWM

For all of these reasons, it is extremely difficult to present 'standard' costs of waste management. *Table 5.1* brings together two published sources, which each look at the typical costs of implementing different unit operations for MSW management, in countries at different income levels. The unit costs increase with income level, due to the higher costs of both personnel and of compliance with more stringent environmental regulations.

This table compiles estimates from the literature to show variations in the combined net costs for different unit operations, taking into account investment and operating costs and resource recovery revenues. Parts A and B show net cost per tonne for each unit operation.[9] Part C estimates what total cost per tonne for MSWM could potentially be affordable in each income band, based on a 'rule of thumb' upper limit on affordability of 1% of the GDP/GNI per capita[10] and the MSW generation per capita given in Part A. The additional column in Part B also shows a 'typical' estimated investment cost for each type of technology, shown in million USD for a facility with a capacity of 100,000 tonnes per year (approximately 300 tonnes per day).

Maishima EfW Plant, Osaka, Japan

Biogas Plant at Gampaha, Sri Lanka

7 The real prices for energy tend to fluctuate over time, depending on among other things global trends of recession or growth, embargoes on gas and oil related to conflicts or discovery of new reserves and energy sources; a notable price slump occurred at the time of writing in 2014-15.

8 In this context, EfW includes a wide variety of thermal energy from waste facilities (e.g. combustion, gasification and pyrolysis); anaerobic digestion; and landfill gas recovery. See also Section 3.5.3 and Box 3.7.

9 Note that the national income bands, as specified in the first row are different in the two Parts.

10 Scheinberg et al. (2010); Wilson et al. (2012), listed in,Annex A under Chapter 1, Waste management.

11 Hoornweg & Badha-Tata (2012), listed in Annex A, Chapter 3, Collated data sources.

12 This is just one data source. See Section 3.4.1 for a detailed discussion of collection coverage.

13 Pfaff-Simoneit (2013).

14 The term 'RDF' (refuse-derived fuel) is used in this table to designate all processed fuel outputs. The properties of the RDF will vary widely depending both on the exact technology used, on the input waste, on the extent of any segregation prior to the process, and on the specification of the end-user of the fuel. Processed fuels can vary widely in calorific value, handling properties/size and size distribution, ash content, contaminants content etc. It is common to differentiate what is termed here as 'RDF' into two categories, with 'RDF' being used for lower grade and 'SRF' (secondary recovered fuel) for higher grade fuel products. See also Box 3.7.

15 Estimate based on a site with a capacity of 3 million m³, which might provide around a 30-year lifetime for a city generating 100,000 tonnes per annum, depending on the density achieved.

16 Calculated on the basis of the income ranges and waste generation data shown in Part B.

17 Spending 1% of GNI on MSWM is quite a high figure – some authors have suggested using rather figures between 0.3% – 0.6% as the upper limit on affordability, which would extend affordability constraints also to upper-middle income countries.

Table 5.1 Comparative cost data for different unit operations on MSWM

Data columns 1-4 show estimated total cost per tonne (net of operation and investment costs, less revenues from resource recovery). Column 5 of Part B shows the 'typical' investment cost for a capacity of 100,000 tonnes per year.

DISCLAIMER: All costs are estimates for comparative purposes only; they are NOT indicative of actual costs in any particular local situation. Equipment costs may be similar worldwide but civil engineering, building and land costs vary widely, as do labour costs. In Part B, The costs of different technologies cannot be directly compared because these mostly have different scopes of application and different outputs.

PART A: WORLD BANK PROJECT DATA (NOMINAL DATE 2006)[11]	LOW INCOME COUNTRIES	LOWER MIDDLE INCOME	UPPER MIDDLE INCOME	HIGH INCOME COUNTRIES
Income (GNI/capita) 2006	<876 USD	876-3 465 USD	3 466-10 725 USD	>10 725 USD
Waste generation (kg/cap/yr)	220	290	420	780
Collection coverage[12] (percent of households served)	43%	68%	85%	98%
	Cost of Collection and Disposal (USD/tonne)			
Collection	20-50	30-75	40-90	85-250
Sanitary landfill	10-30	15-40	25-65	40-100
Open dumping	2-8	3-10	NA	NA
Composting	5-30	10-40	20-75	35-90
Waste-to-energy incineration	NA	40-100	60-150	70-200
Anaerobic digestion	NA	20-80	50-100	65-150

PART B: RESEARCH STUDY COMPARING ALTERNATIVE TECHNOLOGIES (2012 DATA)[13]	LOW INCOME COUNTRIES	LOWER MIDDLE INCOME	UPPER MIDDLE INCOME	HIGH INCOME COUNTRIES	'TYPICAL' INVESTMENT COST
GDP [USD/capita/year]	<2 700	2 700-5 400	5 400-8 100	34 000-41 000	USD million for 100 000 tonnes per year capacity
Waste processing technology	[USD/t]	[USD/t]	[USD/t]	[USD/t]	
Material recovery facility (MRF) for separately collected dry recyclables	25-40	35-50	45-60	80-95	8-10
Sorting of high-calorific value fractions + preparation of refuse derived fuel (RDF)[14]	20-35	25-40	35-50	65-80	13-20
Windrow composting of separately collected bio-waste	25-40	25-40	25-40	50-70	13-20
In-vessel composting/anaerobic digestion (AD) of separately collected bio-waste	65-80	65-80	65-80	95-120	25-50
Simple mechanical biological treatment (MBT) of mixed waste	20-35	25-40	25-40	50-70	7-13
MBT of mixed waste + aerobic/anaerobic treatment prior to landfill disposal	50-70	50-70	60-75	100-120	40-60
MBT/biodrying (partial stabilization) to produce RDF	50-70	50-70	60-75	90-110	35-45
Energy from waste (EfW) using RDF	80-110	80-110	90-115	120-160	80-100
Energy from waste using mixed waste	95-120	95-120	100-130	150-190	80-100
Sanitary landfill	15-30	15-30	20-35	50-80	12-25[15]

PART C: CALCULATED FOR GWMO[16]	UPPER LIMIT ON AFFORDABILITY CALCULATED AT 1% OF GNI[17] (USD/TONNE)			
Affordability limit for total cost of solid waste management	<40	40-120	120-255	>255

Collection costs make up more than 90% of total costs in the lowest income countries where waste is disposed in open dumps after being collected. Early steps to improve MSWM focus on both extending collection coverage and phasing out uncontrolled disposal, while subsequent steps gradually increase the environmental standards of treatment and disposal. This inevitably increases the total costs. The proportion of the total represented by waste collection remains high, but that due to treatment and disposal gradually rises to 30% or even higher in modern systems.

Part C of Table 5.1 shows for comparison an affordability limit for the total cost per tonne of MSWM for each income band, calculated on the basis of a 'rule of thumb' upper limit of 1% of per capita income. For other essential services such as water and wastewater, an upper limit of 3 to 4% of per capita income is quoted by the UNDP and OECD, whereas the 1% figure used here is sometimes seen as 'too high'. In the case of citizens in the low quintile of developing countries, these affordability limits are often grossly surpassed. This shows that low-income and lower-middle income countries can barely afford current collection costs, so even the first steps of extending collection coverage and eliminating uncontrolled disposal will raise affordability issues. This makes the essential first steps needed to protect human health and the environment, of extending collection services to all and eliminating uncontrolled disposal and open burning, even more of a challenge.[18] As income levels rise, more sophisticated technologies generally become more affordable, in spite of the significant increase in the absolute specific costs of the different waste management activities.

5.2.3 The cost of inaction

Environmental pollution can be considered in economics as a 'market failure', as the market does not price the limited capacity of the three environmental receiving media – air, water and land – for absorbing emissions, discharges and waste. The costs of environmental damage – the negative externalities of pollution and waste – are borne by society and the economy as a whole, rather than being recorded by accountants as a cost of production.

Photo courtesy: Raphael Odeniyi

Estimating such economic costs has been a major focus of environmental economics over the last 50 years. A large number of methodologies have been developed for attaching monetary values to different forms of environmental impact, which often provide inconsistent or even conflicting results. Each case study carried out needs to select a particular methodology appropriate to its specific aims and objectives. So, for example, in estimating the economic costs of waste management, some studies focus on negative impacts on people living near waste facilities or waste workers; on the entire population of a city or region; on an economic sector, such as tourism or fishing; on the value of assets; on the quality of life; or on the wasted resources. The economic valuation methods used are various and include among others abatement cost (what it takes to clean up pollution); willingness to pay (asking people what they would be willing to pay for a cleaner environment or saving a particular landscape or ecosystem); and market price (looking up the value of a service or good in places where it does exist, or comparing property prices based on distance from a waste facility). Most studies are selective in the impacts which they try to quantify in monetary terms.

Table 5.2 collates the available evidence on the costs of inaction, focusing on the economic costs of an existing, largely uncontrolled situation (in a lower-income country) where waste is dumped on land or watercourses, or burned in the open air.[19] The Table is divided into two parts:

A. **Public health impacts**, including diarrhoea and gastroenteritis from direct contact; respiratory diseases and dioxin poisoning linked to open air burning of waste; infectious outbreaks and spread of vector-borne diseases when drains are blocked by waste; flooding; risks to animals feeding and hazardous substances entering the food chain; and health impacts from uncontrolled hazardous waste disposal.

18 This is discussed again in Section 6.2.

19 It should be noted that there is also a considerable body of environmental economics literature focused on estimating the residual external costs of emissions from modern, controlled waste treatment and disposal facilities.

B. **Environmental pollution**, including surface, groundwater and marine contamination, greenhouse gas emissions, impacts on fisheries and agriculture, loss of biodiversity and amenity losses to residents and impacts on tourism.

The final column in *Table 5.2* attempts to express the estimates of environmental damage in USD per capita per year. In many cases this is just not feasible; but for some case studies such an estimate has been possible. The range of these estimates is very wide, from 1 USD to more than 100 USD per capita, with a significant number in the range of 10-50 USD. The different rows in the table do overlap when identifying the extent or monetary value of the same type of negative impact and adding the same type of impact would lead to double counting. But equally it is clear that the total cost of inaction requires rows identifying different impacts such as public health, environmental damage or impact to resources and local economy to be added together, including some of those for which it has not been possible to estimate a cost.

To try to put these cost estimates into context, consideration is given to a 'typical' city in a country on the borderline between low-income and lower-middle income (with a GNI/capita of 1,000 USD per year) which is currently collecting waste from 50% of its residents, and delivering the collected waste to a municipal dumpsite (where hospital and industrial wastes are also disposed) at a cost equivalent to 3-5 USD per capita per year. The city is aiming to provide an affordable and sustainable MSWM, collecting waste from more than 95% of its citizens and eliminating uncontrolled disposal. When the previous rule of thumb for an upper limit of affordability for the SWM service of 1% GNI/capita is applied, then the maximum affordable financial cost of the improved service would be 10 USD per capita per year, which is an increase of just 5-7 USD per capita over the current costs of providing an inadequate service.

The available evidence base as summarized in *Table 5.2* is very weak but is sufficient already to demonstrate the point that action is cheaper than inaction. On the basis of the best estimates here, comparing the incremental costs of proper waste management of 5-7 USD with a likely cost of inaction of perhaps 20-50 USD per capita or even more,[20] **it is reasonable to conclude that it is much cheaper for society as a whole to manage its waste now in an environmentally sound manner than to carry on dumping.**

Disposal site, Freetown

20 It should be pointed out that much of the available evidence is for small island developing states (SIDS), where estimation is easier, but where the impacts may be more pronounced due to scarce land and high dependence on certain resources, and the small population may result in a higher cost per capita.

Table 5.2 Collating available evidence on the cost of inaction on solid waste management (SWM)

No.	EXAMPLE OF IMPACT	EVIDENCE FOR ECONOMIC COST	SOURCE	ESTIMATE OF USD/ CAPITA/ YEAR
A Public health impacts in the absence of proper SWM				
1	Health impact on children in households where waste is dumped or burned in the yard	Diarrhoea rates twice and acute respiratory infections six times as high as in areas where waste is collected regularly.	UN Demographic and Health surveys, see Ref. 1, p.88	N/A
2	Health impacts on those living near open dumps[21]	Half the children and adolescents living near a Kenyan dumpsite had respiratory ailments and blood lead levels exceeding the international threshold.	UNEP (2007)	N/A
		In Nigeria, a study indicated an increase in malaria cases in a residential area in the vicinity of a dumpsite.	Nkwocha et al. (2011)	
		Residents nearby and far away from the Granville Brook dumpsite in Freetown Sierra Leona, suffered from malaria, chest pain, cholera and diarrhoea. Low birth weight incidence has also been reported among those living in the vicinity of dumpsites.	Sankoh et al. (2013)	
3	Health and illness costs due to solid waste-related pollution – includes both direct health costs avoided and indirect loss of productivity due to illness. Focus in particular on leptospirosis, dengue fever and gastroenteritis	Total solid waste-related health cost for Palau (population: 19,000) estimated at 700,000 USD per year, based on pharmaceutical costs, time in hospital and lost labour productivity.	GEF/UNDP/ SPREP (2006)	36
4		Avoided public health damage/risks in Saint Lucia (pop: 176,000) estimated in the first year of new SWM system at 3 million USD.	Phillips & Thorne (2011)	16
5		Avoided public health damage/risks in Trinidad and Tobago (pop: 1,328,019) estimated in the first year of new SWM system at 23 million USD.	Phillips & Thorne (2011)	17
6	Health risks to animals feeding	Both direct impacts on the animals (e.g. holy cows in India) and human impacts via the food chain.	No economic estimates found	N/A
7	Waste blocking drains, increasing the risk of major floods	Waste (and in particular plastic bags) blocking drains is cited as a major contributor to major annual flooding in South Asia, East and West Africa.	No economic estimates found	N/A
8	Burden of disease to those living near waste sites receiving hazardous waste	Estimate made for 363 sites in India, Indonesia and Philippines. Main risk factors are lead and hexavalent chromium. Disease burden estimated at 0.1-0.2 disability adjusted life-year (DALY) per exposed person, comparable to outdoor air pollution or malaria.	Chatham-Stephens et al. (2013)	N/A
9	Infections caused by the reuse of contaminated disposable syringes	In 2000, the World Health Organization estimated 21 million hepatitis B virus (HBV) infections, two million hepatitis C virus infections and 260,000 HIV infections worldwide due to such infections.	WHO[22]	N/A
10	Groundwater contamination from, and health effects on those living near, uncontrolled disposal sites which have received hazardous waste	USEPA data for Superfund sites. Clean-up is expected to take 50+ years, at a cost of somewhere between 1bn-5bn USD per year. The problem was largely caused by industry between 1950 (pop 150m) and 1970 (pop 200m), while current pop (2014) is around 317 m. So if a mean pop of 250m is used (correct around 1990), then per capita figure is 4 USD per 1bn USD of expenditure.	USEPA	4-20
11	Health impacts due to lead poisoning from mine waste	There have been numerous cases of lead poisoning in Jamaica. 61 children tested in 1996 all had blood lead levels over the threshold (10 µg/dL). Attributed to lead mine waste from the Old Hope Mine.	Planning Institute of Jamaica (2007)	N/A

21 A new report for ISWA, due to be published in September 2015, reviews the health consequences of large-scale dumpsites. ISWA (2015b) listed in Annex a, Chapter 5, Estimating economic costs and benefits of waste and resource management and Waste and Health. Available at www.iswa.org

22 http://www.who.int/mediacentre/factsheets/fs253/en

No.	EXAMPLE OF IMPACT	EVIDENCE FOR ECONOMIC COST	SOURCE	ESTIMATE OF USD/ CAPITA/ YEAR
12	Health impacts on waste workers	Uncontrolled waste collection, recycling, treatment and disposal is likely to have health impacts on the workers: e.g. fatalities and serious injuries have been reported from ship breaking conducted on beaches in the Indian sub-continent. The health impact on informal sector recyclers has been studied. e.g. In Nigeria, 7.8%, 10.1% and 5.1% of the waste pickers suffered from typhoid, dysentery and cholera respectively. Such research constitutes a huge field in its own right.	Afon (2012)	N/A
B	**Environmental pollution in the absence of proper SWM**			
13	Water contamination from inappropriate solid waste disposal and health impacts	A World Bank report puts the environment cost of water contamination from improper waste disposal at 10 billion Nigerian Naira each year and the lives of about 40 million Nigerians as being at risk.	Reported in Ref. 1, p.89	1.4
14	Loss of near-shore fish catch from water pollution due to solid waste dumping	Palau's near shore fisheries resources include reef fish, lobsters and crabs, which are consumed on a subsistence basis and also marketed. The total value of fish resources lost due to land-sourced pollutants is estimated at 88,000 USD per year.	Hajkowicz et al. (2006)	4.5
15	Pollution of beaches by solid waste and marine litter	Estimates of loss of tourist income: (i) Tangiers estimate of 23m USD per year, converted using current population of 850,000. (ii) Estimate for Palau 960,000 USD per year. The cost of litter control, including cleaning marine litter from beaches, is orders of magnitude greater per tonne of waste than proper waste management (prevention of litter) in the first place.	(i) Reported in Ref. 1, p.89 (ii) GEF/UNDP/ SPREP (2006)	26 50
16	Impact on residents and tourism from loss of aesthetic values	Based on willingness to pay for preservation of the environment. Estimated at: (i) 27m USD for St Lucia. (ii) 3m USD for Trinidad and Tobago.	Phillips & Thorne (2011)	(i) 156 (ii) 2
17	Degradation from waste affects primarily the natural resources	The cost assessment of solid waste degradation in Beirut is high (66.5 million USD) and represents 0.2% of the national GDP in 2012.	SWEEPNet (2014)	20
18	Threat to the terrestrial (especially groundwater contamination) and marine environments	Groundwater contamination in Jamaica has led to the closure of about 25 per cent of groundwater sources. The loss in groundwater caused by contamination could be made worse by frequent and longer periods of drought. The continued expansion of the tourist sector also hinges on the availability of clean water.	Planning Institute of Jamaica (2007)	N/A
19	Fish and soil contamination by heavy metals – impact on agricultural exports	Both the fish around Jamaica and the nation's agricultural soil are contaminated with heavy metals, particularly cadmium.	Planning Institute of Jamaica (2007)	N/A
20	Marine litter gyres, garbage patch	Plastics and other solid waste from land sources and ships have accumulated and formed what is known as marine litter gyres in the world's oceans, which is causing major environmental problems (see row 14 above).	See Topic Sheet 9, found after Chapter 3	N/A
21	Environmental impacts associated with open burning of accumulated waste	Open burning of accumulated waste is quite widely practised to reduce waste amounts when waste collection service is inexistent or failing. The emissions associated with these include dioxins, polyaromatic hydrocarbons (PAHs) and black carbon (BC), which are highly toxic, carcinogenic and powerful short-lived climate pollutants (SLCPs)[23] respectively. The UNEP Dioxin Toolkit addresses this category of emission source and found that for developing countries often more than 60% of the total dioxin emissions are attributable to this source.	UNEP/DTIE Chemicals and Waste Branch (2010) UNEP (2005)	N/A

Details of the sources cited in Table 5.2 can be found in Annex A, Chapter 5, Estimating economic costs and benefits. Ref. 1 in row 1 refers to Scheinberg, Wilson & Rodic (2010). *Solid Waste Management in the World's Cities.* UN-Habitat. Listed in Annex A under Chapter 1, Waste Management.

23 See also Topic Sheet 1 on Waste and climate, found after Chapter 1.

5.2.4 Benefits to society and economy

The waste and resource management industries bring many benefits to society through their contribution to GNI, which has already been reflected through the financial costs and revenues of waste management. In addition, environmentally sound waste management improves public health, provides for a cleaner environment and potentially increases income from tourism as well as investment decisions by visitors who enjoy the clean city: these 'benefits of action' are the inverse of the 'costs of inaction' discussed above (this is included as the first row in *Table 5.3*).

The main focus in this section is rather on a broader range of benefits, including the business benefits of resource efficiency and waste prevention, increased resource security, job creation, greenhouse gas mitigation, food security and facilitating air and water pollution control. These are all real and tangible benefits, but often fall outside of an accountant's financial analysis. They are generally calculated comparing two alternative waste management pathways. Thus, even when the individual benefits are quantified, it is impossible to aggregate them because each is calculated against a different baseline and hardly ever against an 'inaction' scenario. *Table 5.3* briefly elaborates on some of the benefits in a qualitative manner.

Moshi, sitting below Mount Kilimanjaro, voted for several years as the cleanest city in the United Republic of Tanzania

Table 5.3 Wider benefits to society and the economy from sustainable waste & resource management

No.	CATEGORY OF BENEFIT	EXPLANATORY COMMENTS
1	Broader benefits of a clean city	Effective and environmentally sound waste management contributes to a clean city and a pleasant and healthy living environment, which is attractive to residents, to tourists and visitors, and to businesses and inward investors.
2	Social and political consensus and community cohesion	Good waste management is a visible sign of good governance.[24] Where local authorities have failed to tackle waste management they often are not re-elected, while authorities that do tackle waste issues have a greater chance of being re-elected. Similarly, cities and communities that work together to segregate waste and reduce littering tend to also reduce crime, vandalism and social deprivation and enhance community cohesion.[25] Good waste and resource management promotes a sense of community and security, of belonging and well-being.
3	Business benefits of resource efficiency and waste prevention	These have been quantified in a number of authoritative recent reports as in excess of 1 trillion USD per annum worldwide.[26] Waste prevention avoids end-of-pipe waste management costs, but also saves much larger raw material, energy and labour costs embedded in wasted products.
4	Public benefits of resource efficiency and waste prevention	Lower waste generation and higher resource efficiency reduces the municipal cost (and therefore the per-citizen cost) of providing municipal waste management services. Reducing waste saves municipalities anywhere between 35 to 400 USD per tonne of waste depending on where the prevention occurs and what sort of technologies would be used for handling waste.
5	Increased resource security	After a century of steady decline, resource prices in real terms doubled between 2000 and 2010.[27] So despite continued price volatility, developing indigenous supplies of raw materials from recycling makes good sense, particularly in rapidly industrializing countries. E-waste comprises a richer 'ore' for many scarce and critical metals than the natural ores mined for the virgin raw materials.[28]
6	Green jobs	Environmentally sound waste management, the recycling of dry materials, organic materials recycling, and energy recovery from waste all represent 'new' green industrial sectors with the potential for substantial job creation. For example, UK employment in the sector increased by 50% between 1993 and 2013. EU employment doubled between 2000 and 2010 to 2 million,[29] alongside an increase in municipal solid waste recycling rates from 20% to 35%.[30] Box 4.3 provides an estimate of 4.2 jobs per tonne per day of material that is recycled. The wider 'circular economy' holds further promise. The McKinsey report cited under Row 3 above estimates the potential to create between 9 and 25 million new jobs worldwide.
7	Improved livelihoods and cleaner working conditions for the informal sector	The transformation of a city's informal sector to a more formalized part of the mainstream waste and resource management system is a win-win situation. The recyclers can work under cleaner conditions, earn a better livelihood and educate their children. City recycling rates can potentially increase and in addition, the transition can facilitate environmental control, reducing littering and dumping of residual waste as well as bringing the 'informal' sector inside the legal and tax systems.[31] The result can be more and better jobs[32] and a reduced burden on the city's already stretched waste management budget.
8	Reduction in greenhouse gas (GHG) emissions from waste disposal	The Intergovernmental Panel for Climate Change (IPCC) reported that in 2010 MSW accounted for around 3% of total worldwide GHG emissions, mainly as methane from landfills. However, this is a significant underestimate of the potential of better waste management to mitigate GHG emissions. Around half of total worldwide waste arisings in 2010 came from high-income countries, which had been taking steps to control methane emissions since the 1970s and to divert biodegradable municipal waste from landfill since the 1990s. For example, Germany estimated that improved waste management had saved 5% of its total 1990 GHG emissions by 2010, and it had already made significant progress prior to 1990.[33]
9	Reduction in GHG emissions from recycling & waste prevention	Using a life cycle approach, it has been estimated that a 10-15% reduction in global GHG emissions could be achieved through improved SWM including landfill diversion, energy from waste and recycling.[34] Including waste prevention could further increase this to 15-20% – these savings of GHGs embedded in raw materials and products would be achieved across a broad range of sectors of the economy.
10	Reducing food waste – improving food security	The direct economic cost of food waste is estimated at 750 billion USD. It is estimated that more edible food is wasted than what is needed to feed all of the malnourished people in the world.[35]
11	Reduction in air and water pollution by transfer of contaminants to solid waste for proper management	The huge progress made over the last 50 years in cleaning up urban air and water pollution around the world has concentrated contaminants into air pollution control residues and wastewater treatment sludge, which is now managed as solid waste. So environmentally sound waste management underpins clean air and clean water.
12	Energy recovery by using waste to generate energy	Energy is recovered through conventional and advanced energy-from-waste and co-processing technologies and anaerobic digestion. Energy recovered from the biogenic fraction of waste is considered to be renewable energy.

24 See Section 1.3 on waste management as an entry point for sustainable development.
25 One example is Qena in Egypt. See Mourdzhev & El Shebriny (2013), listed in Annex A, Chapter 5, Service delivery.
26 McKinsey & Company (2011).
27 World Bank Commodity Price Data ('Pink Sheet'). See http://data.worldbank.org/data-catalog/commodity-price-data.
28 See Table 3.14 in Section 3.5.2.
29 EEA (2011). EEA Report No. 8/2011. http://www.eea.europa.eu/publications/earnings-jobs-and-innovation-the
30 Average recycling rates across the EU calculated from Eurostat data, as compiled in the EMC Master database prepared for the GWMO. (See Annex B).
31 See Topic Sheet 14 on the informal waste sector in Section 4.7.
32 Globally, between 15 and 20 million people are estimated to be employed in the informal waste sector at present. One estimate suggests that recycling one tonne of materials per day in Cairo supports around 7 jobs. See Linzner and Lange (2013), listed in Annex A, Chapter 4, Inclusivity; and Wilson et al. (2012), listed under Chapter 1, Waste management.
33 For full references, see Topic Sheet 1 on waste and climate, found after Chapter 1.
34 Dehoust et al. (2013), listed in Annex A, Chapter 1, Waste and climate. Also see Topic Sheet 1 on waste and climate, found after Chapter 1.
35 See Topic Sheet 11 on food waste, found after Chapter 3.

5.3 UNDERSTANDING WASTE MANAGEMENT AS A PUBLIC SERVICE AND AS A BUSINESS

The previous section has shown that the costs of environmentally sound waste management are likely to be (much) less than the indirect societal costs of uncontrolled waste accumulation and dumping, and that there are substantial additional economic benefits of 'doing it right'. So it follows that it makes both political and economic sense to implement the 'polluter pays principle': MSWM is provided as a public utility service and paid for by the public, and (larger) private waste generators manage their own waste or pay for its proper management.

Enforcing environmental performance standards, which must be complied with by municipalities and by industry, has created the opportunity for the emergence of the waste management industry, which sells its services both to municipalities to deliver the public (MSWM) service, and to businesses to collect and treat commercial and industrial waste. Selling recovered products is also a substantial industrial sector in its own right. The dependence of the waste industry on legislation to create the market has also created a 'shadow business' which cuts costs by ignoring the law. So this section looks in turn at these different aspects of the 'waste management industry' as a public service, as a business delivering waste services and as a resource recovery business, while also examining the shadow business of waste crime.

5.3.1 Waste management as a public service (and a 'public good')

Concerns over public health in the 19th century led to city authorities being given legal responsibility for the safe collection and disposal of 'municipal solid waste'. One rationale for this is the 'public good' nature of MSWM. It is difficult to control access to MSWM services,[36] and if some citizens escape from their responsibilities for MSWM, they themselves may escape some costs but will cause harm to others and to society as a whole.[37] Unlike other public or utility services that run on separate meters such as energy or water, waste management is a utility service that does not allow for disconnecting users who do not pay without impacting on others. Users of the service are paying for collection and a clean environment and accumulation of waste in the non-paying neighbour's yard impacts the users who regularly pay for the service.

When looking at the different elements of MSWM, the degree to which each component can be considered as a 'public good' varies. For example, street sweeping and safe disposal are public goods, since consumption of the clean environment cannot be restricted; the consumption of door-to-door collection is individual but the use is again common, so this is mainly a public good; the extraction, trading and processing of recyclables are examples of private goods. So in the case of waste management, the boundaries between public goods and private goods are blurred. This is sometimes referred to as an 'impure public good'[38]. These concepts are explored further in *Box 5.1*.

The responsibility for ensuring an adequate service for MSWM has generally been allocated to municipalities. An interesting issue is when the extraction of value from materials within the waste becomes a profitable activity in its own right, so that the question arises as to 'who owns municipal solid waste?' Increasingly there is a competition for access to valuable waste streams (*Box 5.1*). Until recently, relatively low and unstable prices for secondary raw materials have meant that in developed countries, the costs of separate collection of recyclable materials are only partly covered by the revenues derived from selling the materials. The cost is justified when compared to alternative means of treatment or disposal, or in meeting recycling targets set by policy. Informal recyclers in developing countries do manage to make a living simply from the revenues from selling the recovered materials.[39] But with increasing resource scarcity and rising prices, situations also arise in which people can make a living by removing recyclables from the 'formal' systems in developed countries, or where it becomes in the interest of large national manufacturing companies in developing countries to 'compete' with both the city and the informal sector in offering MSWM services, in order to secure access to raw materials (see *Box 5.1*).

36 For example, if a householder does not pay for her waste to be collected, she could still access the service by putting her waste in a neighbour's bin, or in a street litter bin, or by taking the waste in to her place of work.

37 For example, if a householder dumps or burns his waste, he may avoid paying for the waste service, but the public health and environmental impacts affect everyone in the neighbourhood.

38 Cavé (2014), listed in Annex A, Chapter 5, General reading; Cointreau (1994), listed in Annex A, Chapter 5, Service Delivery and private sector participation in SWM.

39 See Topic Sheet 14 on the informal waste sector in Section 4.7.

When the recovery value of the materials starts to play a role in the deciding who is allowed to collect the waste, the quality of the services may be at stake. A private collector can have higher profits by only focusing on densely populated regions where recoverable waste quantities can be collected within a small area, leading to regional 'cherry-picking'. In such cases, only the non-profitable regions are left to be serviced by public authorities. Equally, if prices suddenly fall (as does happen, see *Figure 5.1* below), then the private operator may (temporarily or permanently) withdraw, leaving the public authority to 'pick up the pieces'. The different drivers for the different activities often create tension between public service providers and businesses that may compete for the same waste stream. If ownership rights over materials and responsibilities for cleanliness and service provision are not clearly defined and delineated the tension may lead to vulnerable or under-performing systems.

BOX 5.1 WHO OWNS MUNICIPAL SOLID WASTE? APPROPRIATION CONFLICTS IN EMERGING COUNTRIES[40]

In Coimbatore (Tamil Nadu, India), the implementation of a new municipal SWM scheme in 2008 revealed the coexistence of several parallel mechanisms for delivering SWM services in the city.

- Municipal authorities deliver a solid waste collection public service from street bins to transfer stations. Treatment and landfilling services are delegated to a private Indian consortium (UPL) through a 20-year public-private partnership. The private operator gets its revenues from gate fees, compost sales and the separation and sale of dry recyclable materials.
- Itinerant waste buyers conduct informal recovery by purchasing dry recyclable materials from householders or maids.
- Wastepickers conduct informal recovery of dry recyclable materials by collecting them from street bins and dumpsites.
- Junk shops buy dry recyclable materials from the itinerant waste buyers, wastepickers or directly from the residents themselves.
- The Indian Tobacco Company (ITC), a huge Indian industrial group, has started to implement an innovative scheme called 'Wealth Out of Waste' (WOW). ITC considers solid waste as a reliable supply of recycled paper for its factories, and it claims to be addressing a public problem while reducing its secondary paper imports from overseas[41] The WOW agents distribute large bags to households and frequently pass door-to-door to buy the dry recyclable items.

Such a coexistence of heterogeneous waste management mechanisms results in **appropriation conflicts**. UPL claims that the informal recovery agents are 'creaming off' its solid waste flow, thereby decreasing its revenues. Informal agents complain that the WOW program is damaging their source of livelihood. And municipal authorities struggle to introduce separate collection when private agents are already buying recyclable materials from the residents.

On the one hand, the solid waste management sector is **driven by a tax** and inhabitants have to pay so that their household waste is taken away by the municipal service. On the other hand, there is a proper **business sector**, both formal and informal, through which inhabitants can benefit by selling their dry waste items either to itinerant buyers, local junk shops or to the WOW program. In addition, there are wastepickers who **have free access** to waste.

What then is the economic status of waste activities? Is solid waste a private commodity? Or is it a type of null or negative value good, for which a removal service should be paid? In order to address this issue, two main aspects need to be taken into account:

- First, **waste is what has been abandoned**, i.e. *res derelicta:* a thing on which its former owner has renounced his property right. This is the reason why clashes arise: there are appropriation conflicts in a sector in which property rights are not clearly defined. Municipal authorities have the responsibility to address waste, but they do not *own* the waste.
- Second, **the nature of waste is not intrinsic**: a good becomes waste if it is abandoned, and becomes a private good again if it is re-appropriated. Considering a specific item to be 'waste' or a 'resource' depends on local practices and on the existing recovery mechanisms. It may vary over time and because of space constraints.

If one considers municipal solid waste as a whole, it is neither a private good (a resource) nor a public 'bad' (a burden) – it always has elements of both. It always contains both valuable items and residual waste that requires (expensive) management to protect the public. In economic terms, it is **an impure public good**.

40 Text written with the assistance of Jérémie Cavé, drawing heavily of on Cavé (2014). Additional references: Lupton (2011), Lane (2011), Chaturvedi & Gidwani (2010). All listed in Annex A, Chapter 5, General.

41 See Section 3.6.5.

5.3.2 Waste management service delivery as a business

Waste generated by industry or by large waste generators in general should not be the responsibility of the municipality but rather that of the waste generator directly. Commercial, institutional, and C&D waste may or may not be the responsibility of the municipality. In practice, in lower-income countries, where environmental legislation may be weaker, these tend to fall under municipal responsibility.

However in middle- and higher-income countries where legislation tends to be more stringent and ensures that a certain standard is met in not only operations but also environmental protection, the municipality is no longer responsible for these waste streams.

Thus, large waste generators, whether private or public, are generally expected to make their own private arrangements in compliance with the legislation and their environmental permit. They can either manage the waste themselves or contract a private sector waste management company, or they may sometimes be offered to pay to 'opt-in' to the city's municipal waste management service and generally are expected to pay a higher price for the service so as to not further burden the municipality that delivers the service, and instead to improve cost-recovery by cross-subsidizing the service.[42] An example of the latter could be ship-generated waste, e.g. from visiting cruise ships, that is generally high-value, mostly recyclable waste and can be turned into a source of revenue but is often collected for a fee by the municipal service.

Starting in the 1980s, there was a push to privatize municipal waste management services alongside other public services, stemming from the neo-liberal economic idea that a market-based solution is the best solution for any problem. A combination of waste management legislation and the competitive tendering of public services that requires all operators to meet the same criteria has generally ensured a level playing field and has enabled the emergence of private sector waste management companies around the world, serving both public (municipal) and private (large waste generating) clients.[43]

The industrial sector delivering solid waste management services to both private and public sector clients is large and global in nature. There are some very large international service providers but also major national companies and a multiplicity of local small- and medium-sized enterprises, in both developed and developing countries. Global turnover of the industry in 2011 was around 640 billion USD, of which 95% was in Asia, Europe and the Americas. Of this, collection accounted for 44% and recycling and landfill for 25% each.[44] Another source gives an approximate 50:50 split between municipal solid waste and industrial ('business-to-business' [B2B]) waste services[45].

BOX 5.2 DAR ES SALAAM MICRO-PRIVATIZATION APPROACH[46]

Micro privatization has become an approach to leveraging service delivery in capital cities in East Africa. It was promoted by the International Labour Organization (ILO) as a job creation strategy in the 1990s. The city of Dar es Salaam, Tanzania organizes waste collection via more than 55 micro-, small and community-based enterprises that tender for micro-zones, some with less than 500 households. Since then, the ILO has been working with cities all over East Africa to replicate Dar es Salaam's success. The approach has become a primary service provision approach in the region.

42 More on the division of responsibilities for waste management can be found in Section 2.2 on the scope and coverage of the GWMO.

43 More on the policies ensuring a level-playing field can be found in Section 4.7.3, and on the different models for delivery of municipal solid waste management services in Section 5.6.3.

44 De Angelis (2013), listed in Annex A, Chapter 3, Global secondary materials industry.

45 Chalmin & Gaillochet (2009), listed in Annex A, Chapter 3, Collated data sources.

46 Prepared by Alodia Ishengoma, ILO Office Dar es Salaam (2010) for the UN Habitat *Waste in the World Cities* publication. See Scheinberg et al. (2010), listed in Annex A, Chapter 1, Waste management

5.3.3 The resource recovery business

The resource value of waste is a driver for a series of different businesses, including extraction and recycling of various dry waste streams and composting green and wet waste streams. Certain energy from waste activities are also categorized as resource recovery, including biogas extraction, digestion, thermal treatment to generate electricity and/or heat and the processing of refuse-derived fuels (RDF) or secondary recovered fuels (SRF) for combustion in industrial boilers or cement kilns.[47]

Resource recovery of the various waste streams depends on how much demand there is for that secondary resource and how feasible it is to meet the quality and quantity requirements of the buyers. Economic and related policy instruments, such as extended producer responsibility, pricing disposal and introducing a landfill tax, banning the landfilling of certain material streams and providing subsidies, have contributed especially in the developed world to increase demand and diversify the material streams being recovered.

The secondary materials industry has been important since the Industrial Revolution. Earlier in the 20th century, industry relied mainly on relatively clean industrial waste, but the quantities of materials separated from municipal solid waste have increased again since the 1980s.[48] The economics of recycling is changing over time. For example, the cost of recycling electronic equipment in the US is approaching the point of producing a profit without individuals or businesses having to pay a fee to recycle their equipment.

Recycling has become a truly globalized business since 2000,[49] with a current turnover of more than 200 million USD.[50] Any price fluctuations are typically quickly communicated and have an immediate impact on markets globally. *Figure 5.1* shows that secondary resource prices closely follow fluctuations in their corresponding virgin material. The economic downturn in 2008/09 that affected most industries caused both primary and secondary resource prices to plummet, as has the oil price slump at the time of writing in early 2015.

47 See Boxes 3.6 and 3.7 in Section 3.5 on resource recovery.
48 See Section 2.3 on drivers for waste and resource management.
49 Chalmin & Gaillochet (2009)
50 De Angelis (2013)

Figure 5.1 Comparing global prices of primary and secondary resources[51]

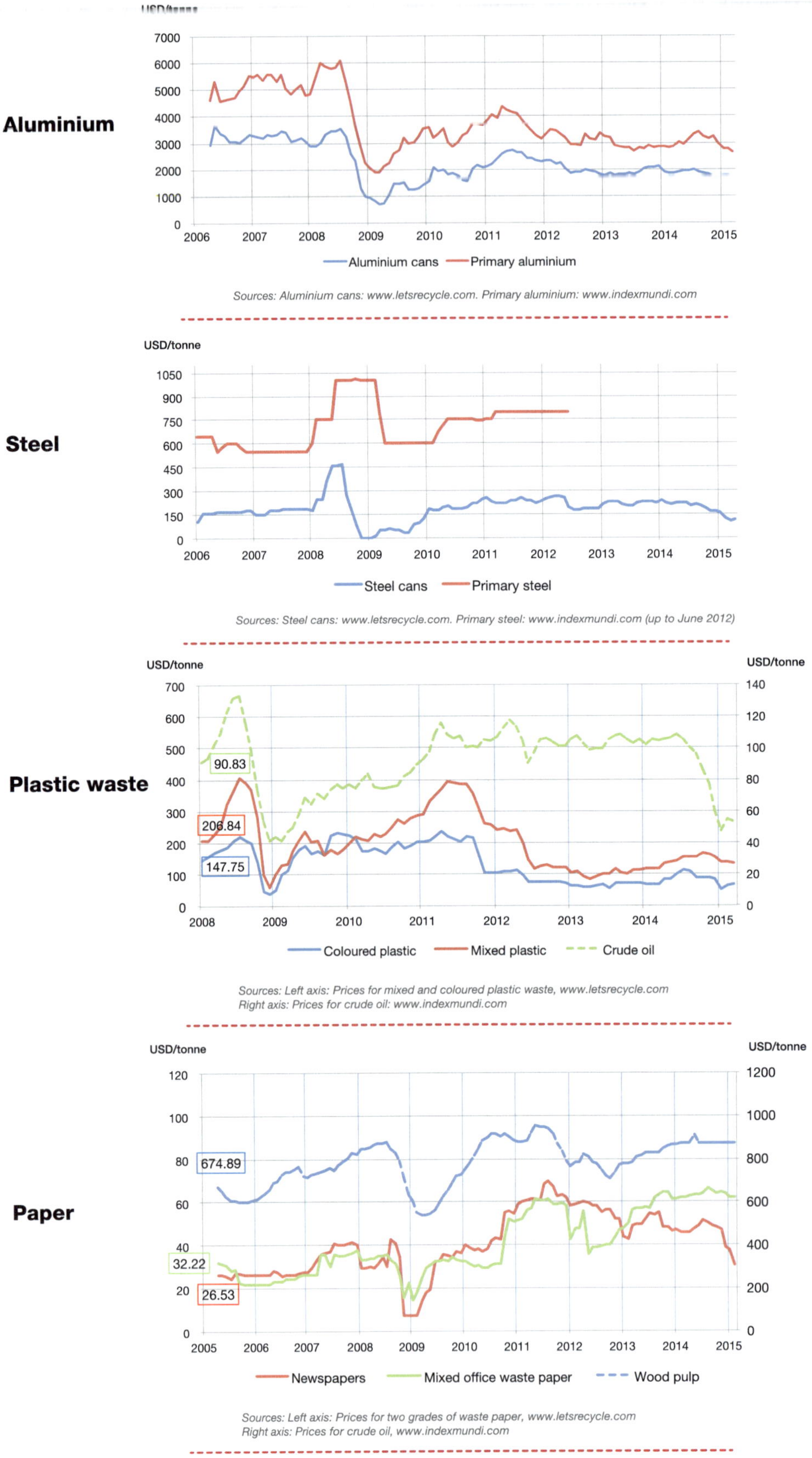

Note: All prices sourced in GBP. Prices converted to USD using monthly average rates.[52] Price data for primary steel was only available on publicly accessible websites up to July 2012.

51 Figures prepared by Flaviu Pop of Green Partners, Cluj-Napoca, Romania.

52 http://www.ozforex.com.au/forex-tools/historical-rate-tools/monthly-average-rates

The valorization value chains, especially in middle-income countries of Asia and Latin America, are often complex, involving a lot of actors from individual waste pickers to scrap yards and processing plants. Research has shown that the involvement of waste pickers and the informal sector can reduce waste management costs of the municipality by diverting waste away from municipal waste collection and landfilling.[53]

Because money is being exchanged each time when the material changes owners, and the financial viability of each deal depends on the feasibility of the logistics, the amount of material being traded is very important. The availability of working capital to pay for transactions is a key factor for sustaining and developing businesses and even small price fluctuations affect the businesses involved in the value chain (see *Figure 5.1*).

BOX 5.3 IFC RECYCLING LINKAGES PROJECT IN THE EASTERN BALKANS, 2005-2008[54]

The goal of the project was strengthening recycling value chains through technical assistance and financing in the Eastern Balkans, which has a total population of 17 million people with an average per capita GDP of 2700 EUR. The information collected in the Eastern Balkans – Serbia, Albania, Bosnia and Herzegovina and Macedonia – during the implementation of the project revealed the existence of a dynamic and labour-intensive recycling sector. The sector recycles 0.11 tonnes of waste per capita annually of the 0.36 tonnes of waste per capita generated, representing a 30% recycling rate.

The size of the business is 400 million EUR annual turnover by recycling 1.8 million tonnes of waste (as of 2008) with the participation of over 400 SMEs, 20 mills, and 20,000 individuals. The investment need and working capital gap in the sector was estimated at 200 million EUR in the baseline year, 2005. Over the project period, IFC disbursed about 180,000 EUR worth of micro-loans to over 100 entrepreneurs and waste pickers working in the sector and facilitated the investment of about 30 million EUR in equity and loans in the sector. The companies that benefited from investment have grown on average 20% in turnover following the investments. On average, the recycling of one tonne of waste per day in the Balkans provides employment for 4.2 people.

5.3.4 Waste crime

The waste industry exists because of strong regulation, consistently enforced. The same conditions that create the legitimate waste market also create opportunities to make money by circumventing the system. Naturally, there are cases in which waste producers or waste transporters break the law unintentionally, for example by producing incorrect transport documents or misinterpreting existing law. However criminals and criminal organizations, and even mafia groups or paramilitary organizations in some countries, are able to take advantage of the complex regulations and the producers' lack of knowledge or lack of concern or some producers' desire to substantially reduce the cost of having their waste disposed of.

A large fraction of the organized dumping of waste in high-income countries, and of illegal exports of waste, is suspected to be the result of deliberate illegal actions. The overall worldwide illegal trade in waste is estimated at a value of between 10 and 12 billion USD annually and generates very high revenues to the criminal actors involved in the trade.[55] Focusing in particular on international waste trafficking, the modus operandi includes, among other means, deliberate misclassification of waste types, false declarations of waste products or unchecked items as re-usable products and other kinds of fraudulent shipment documents.

European enforcement of waste transhipment regulations between 2007 and 2009, targeting waste movements within the EU and to countries outside it, showed that of the waste shipments inspected, 15–18% infringed EU regulations. In spring 2009, a similar but larger operation steered by the World Customs Organization (WCO) led to the seizure of more than 45,600 tonnes and 1,800 pieces of illegal hazardous waste (scrap metal, household waste, e-waste, used vehicle parts). Most seizures were made in Europe, especially in the Netherlands, Belgium and Italy, which have mainland Europe's main ports.[56]

The gigantic global volume of waste and the number of containers moving around the planet make systematic monitoring and control at the container level impossible. In 2010, around 24 million standard-size containers passed through the port of Hong Kong, over 11 million through Rotterdam and around 2.8 million through Gioia

53 See Topic Sheet 14 on the informal waste sector in Section 4.7.
54 Information taken from Popovska (2010). http://www.transwaste.eu/file/001311.pdf
55 Mavropoulos et al. (2014), listed in Annex A, Chapter 1, Waste management.
56 Secretariat of the Basel Convention (2012). Vital Waste Graphics 3, p. 36-39. http://www.grida.no/publications/vg/waste3/

Tauro, Italy's (and the Mediterranean's) biggest harbour. Preventing and punishing illegal traffic in waste needs good governance – and effective crime detection.[57] Many international agencies are actively working together to meet the challenge of tackling organized waste trafficking.

5.4 WASTE MANAGEMENT FINANCING MODELS

A financing model in waste management has four component parts, as shown in *Figure 5.2*. The *client* is either the waste generator, or the municipality which has assumed responsibility for municipal SWM. The *operator* delivers the waste management service 'on the ground'. *Revenue* needs to be raised to pay for the costs of the service; *investment finance* is required to pay for the capital costs if any new infrastructure is required.

Figure 5.2 Components of a waste management financing model

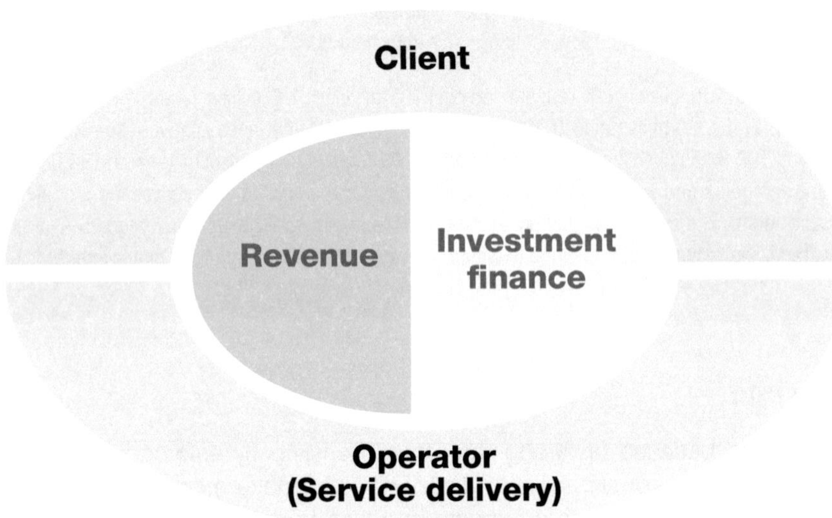

The way these elements are chosen and organized to deliver, manage, finance and re-finance waste management defines the financing model. There are lots of options in each of the four components, so there is no single financing model for waste management that will be 'best' in all circumstances. Essentially this is a space or matrix with four major variables, which has presented considerable challenges in writing this chapter, as there is no single, obvious sequence to present the material as a logical and linear flow.

The next section, Section 5.5, looks at Business-to-Business (B2B) financing models, in which a private waste generator contracts directly with a private-sector waste company to provide a service. B2B models are appropriate in particular for larger waste generators[58] who act directly as the client, and who generally contract directly with a private sector waste company as the operator. These models are relatively simple, so Section 5.5 will consider all four components of the financial models.

For MSWM, the public authority is the client. Sections 5.6 through 5.8 look in turn at the other three components of the MSWM financing model. Major differences between MSWM financing models stem from the choices of (i) whether to engage a public or private operator, (ii) how revenues are collected, (iii) sources of investment capital, (iv) the level of government; and (v) the extent of integration of waste management services. Some options are a better fit for a waste management activity driven by resource value. Others are a better fit for waste management activity driven by the need to remove the waste.

57 See also Section 4.3.7 on regulation on waste handlers. Also Secretariat of the Basel Convention (2012). Vital Waste Graphics 3.

58 See Section 2.2 on the scope and coverage of the GWMO.

5.5.1 Polluter pays

Business-to-business (B2B) models are apparently simple – the (large) waste generator is the client. It contracts directly with one or more private waste management companies as operators, and pays an appropriate service fee to the provider. The revenue from the service fee covers the cost and the profit margin of the provider. This market is a 'construct' of waste regulation – the legitimate waste management industry relies on strong regulations which are robustly enforced to create a 'level playing field' – and to protect them against the 'shadow market' of those undercutting their gate fees by bypassing the required environmental and health and safety standards and specific collection and sorting obligations.

In the case of large single-point producers such as industrial or commercial enterprises, volume or weight-based charges are commonplace. This has the advantage of linking the costs and revenues of waste management to the actual volume of services provided, thus encouraging waste reduction. As noted below, such charges are relatively uncommon for MSW. On the other hand many enterprises still have a more 'service based' fee, in which a container will be emptied a number of times, regardless of the weight or volume of waste inside; such contracts give no incentive to reduce waste generation.

Affordability of charges may be an issue for companies as well as households, as services meeting 'proper' standards may be relatively expensive. This is most likely to be common among smaller waste generators, which is why many countries allow small businesses to use the MSWM service. Large waste generators, especially multinational companies, have shown that they can comply with environmental 'polluter pays' legislation, but will often campaign to avoid new legislative requirements such as extended producer responsibility which they perceive as increasing costs, particularly if competitors in other countries are not subject to the same controls. This interface between industrial/commercial/'private' waste generators on the one hand and the 'public service' of MSWM on the other complicates the discussion and the selection of financing models. This interface and the associated challenges will be discussed in the following sections.

5.5.2 Raising investment finance

A further complication of B2B financing models comes when one considers the fourth component shown in *Figure 5.2*, investment finance. For B2B waste, there is generally a free and very competitive market with many alternative service suppliers. The existence of this market depends on strong regulation. If those regulations in effect require the waste to be sent to a relatively high-technology and expensive facility, then until the new compliant facilities become available, the regulator cannot enforce the regulations, because the generators have no means to comply, as discussed in Section 4.2.2 on strategic planning. At the same time, the waste industry will have problems raising commercial finance to build, say, a 100 million USD high-technology facility (see *Table 5.1*) while cheaper non-compliant facilities still exist. This has been termed the 'implementation conundrum'. For B2B models, the problem is compounded as most individual customers are relatively small waste generators compared to the capacity needed for reaping the benefits of the economies of scale for a modern facility.

Financing in this model is privately sourced and recovery of these costs is fully factored into the service fee. So in effect the waste industry will have trouble securing long term contracts to cover most of the waste inputs and provide reassurance to their banks and other investors. The problem of access to financing was exacerbated by the 2008-09 financial crises, and is still a major constraint on new facility development in many developed countries. One country where data is available is the UK. Data reported in Section 3.5.4 show that some 45 billion USD of waste investments are currently at some stage of development (with a split of 25 billion USD for B2B projects involving a variety of waste types and 20 billion USD for MSW projects)[59] – but access to finance is a major limiting factor, so that the infrastructure actually built will be much less than these figures.

59 AcuComm's Waste Business Finder database http://acucomm.net/

5.5.3 Integrated service (resource management) providers

Waste and resource management policy, especially in some high-income countries, is driving towards waste prevention and the circular economy.[60] This means from a business perspective that a new market is arising that considers production and waste management from a life-cycle point of view and strives to integrate resource and waste management services into a single vertically integrated service package.

A new business model is emerging for this market, called 'resource management contracting' or 'product-service systems' that incentivizes innovation in the fundamental redesign of the product and service combination of a business and its suppliers in order to reduce life cycle impacts. The contractor is paid for a service package rather than a tariff per tonne of waste handled, which provides a strong incentive to minimize resource usage, extend product lifetimes, regenerate and reuse products and minimize wastage. This approach is suitable for manufacturing facilities, institutions, school districts and property managers.[61]

BOX 5.4 EXAMPLE OF A PRODUCT-SERVICE SYSTEM OR RESOURCE MANAGEMENT CONTRACTING

General Motors (GM) outsourced its chemical management services, including procurement of chemicals, their use, monitoring, data tracking, recycling, treatment and disposal. After contract implementation, General Motors reported a 20% overall reduction in waste generation, a 65% increase in recycling and a 60% decrease in disposal across 50 plants in North America, resulting in considerable cost savings.

GM generated USD 2.5 billion in revenue between 2007 and 2010 through various recycling initiatives, including resource management contracting. It now approximates its annual by-product recycling and reuse revenue at about USD 1 billion a year, made possible through using a holistic GM by-products management system combining the environmental and financial benefits of all plant materials. One example seen at Pontiac Metal Center in Michigan resulted in the generation of 7.5 million USD in recycling revenue, including metals, in 2011 alone.

5.6 MSWM FINANCING MODEL – DELIVERING SERVICES

Referring to *Figure 5.2*, this section focuses on alternative models for the delivery of MSWM services. Much of the evidence base comes from a recent GIZ study on 'operator models' as part of a sector project on Concepts in Sustainable Solid Waste Management.[62]

5.6.1 Options for delivering MSWM services

Figure 5.3 shows that there is a continuum of options in choice of models between purely public and purely private service delivery. The research revealed that there are a variety of models in use in low- and middle-income country cities and that neither private nor public service provision is dominant. Rather, the results suggest that some form of small-scale service provision, by micro- or small enterprises (MSEs), community-based organizations (CBOs), non-governmental organizations (NGOs) or the informal sector, is the most common among the cases considered. On average there are 2.5 different service delivery models per city and no city chooses a single homogeneous model for all its waste management activities. This is because in any particular local situation, different models may be more suitable for particular activities, so, for example, different models may be chosen for street sweeping, primary waste collection, secondary collection, composting and landfilling.

60 See Box 4.1 in Section 4.1, as well as Sections 1.2, 2.2 and 4.3.6,. See also Topic Sheets 3 on SCP and 4 on Waste Prevention, both found after Chapter 2.
61 Wilson, Parker, et al. (2012) – listed under Chapter 2, Waste prevention.
62 See also Section 4.8.1. The GIZ work is reported by Soos et al. (2013a, 2013b), listed in Annex A, Chapter 5, Service delivery and private sector participation in SWM.

Figure 5.3 Continuum of options between public and private service delivery models in cities in low- and middle-income countries

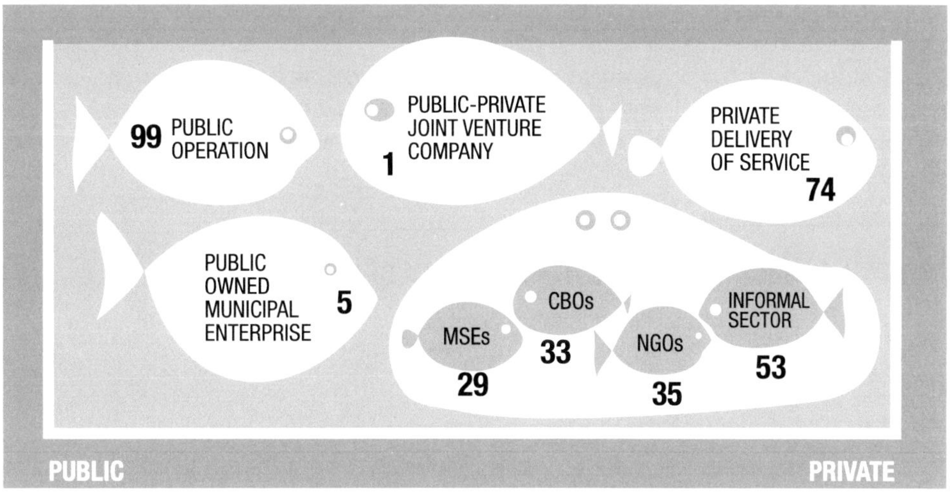

Note: Numbers indicate the frequency of different models in 134 case studies identified in the literature in 2011. The case studies were chosen primarily from low- and middle- income countries: 40 case studies are from Africa, 50 from Asia, 20 from Eastern Europe and the former NIS countries, 13 from Middle East, 10 from Central/Latin America and 1 from Australia. The average number of service delivery models per city was 2.5.[63]

Source: The graphic on which the numbers have been superimposed was sourced from Soos et al. (2013a). Copyright GIZ. Reproduced with permission.

The following sub-sections review various service delivery models for delivering MSWM services, looking at the sustainability of each. The decision as to which model or combination of models will be most appropriate depends on the particular local situation and will vary widely among different cities and countries.

5.6.2 Public models

In public models, the municipality or public authority is both the client and the operator. The operation of the service may be done by a municipal department or a municipal utility company (publically owned municipal enterprise).

This is a traditional model and most cities have had experience at some point in service delivery. The evidence shows that this is a model that works well. For example in Qena, Egypt, there was a decision at the level of the Governorate in 2002 to implement a city cleaning and beautification plan. The decision was to keep waste management a public service and attempt to address at least partially the city's unemployment problem by intentionally keeping the service labour-intensive. The governor's project to beautify the city of Qena was a success for a decade and as long as the budget allocations from the national government were sufficient to sustain the service.

However, public service is traditionally run as a cost centre, not as a business that needs to care for balancing costs and revenues and ensure liquidity. In this case, funds come from a yearly approved municipal budget, the sources of which are a combination of local taxes, transfers from the national budget and, to a lesser extent, from municipal revenues related to subletting assets or facilities, fines and penalties and the like. Thus, public models are vulnerable to political factors and national economic problems. Continuing the example above of Qena, due to the economic crisis budgets became more constrained, and after the Arab Spring in 2011 payment defaults on utility charges increased, and this combination of problems caused a decline in the quality of the service.[64]

Similarly, throughout Tunisia, historically there had been an effective public waste management system — an essential for a country relying largely on the tourism sector as income. The system was financed partially by a local tax collected and redistributed by the central government. After the 2011 revolution, the tax collection rates dropped, investments into maintenance and replacement of equipment stalled, and the quality of the public service declined.

63 Wilson, Kanjogera et al. (2013), listed under Chapter 5, Service delivery and private sector participation in SWM.
64 Mourdzhev and El Shebriny (2013).

5.6.3 Private delivery of services

Private waste management service providers may deliver services to a municipality based on a service delegation contract, franchise agreement or concession contract with a municipality or public authority. A wide variety of options are available for public-private partnerships (PPP) or private sector participation (PSP), some of which are illustrated in *Box 5.5*. It takes knowledge and experience on the part of the authorities to select the best service delivery model and write a suitable tender for private sector participation. Private to private arrangements sometimes emerge without the involvement of the municipality in entire cities or districts where municipal service is not provided.

BOX 5.5 OPTIONS FOR PRIVATE SECTOR DELIVERY OF SERVICES

There are a variety of options for public-private partnerships (PPP) or private sector participation (PSP) in the delivery of MSWM services. These may be based on a short-term or long-term service contract, a concession, a lease, a franchise, a joint venture or contracting out a thin slice of the service such as interim management.[65] The methods of PPP most common in SWM are contracting, concession, lease, franchise and open competition each with its own particularities.

- **Contracting:** The government awards a finite-term contract to a private firm for the delivery of municipal solid waste collection service, street sweeping service, the collection of recyclables, transfer station operation, disposal site operation, or fleet maintenance. The contract award is made after a competitive procurement process. The private firm is paid for service delivery by the government under the terms of contract.
- **Concession:** The government awards a concession to a private firm to set up a facility that utilizes government-owned resources. This concession may require the private firm to recycle materials from waste and/or to transfer or dispose of waste. The concession is in the form of a long-term contractual agreement, whereby the private firm builds and operates the facility. In some cases, the private firm may maintain ownership indefinitely; in others, the private firm may transfer ownership of the facility to the government after a specified period.
- **Lease:** The government owns assets and invests in assets and leases their use to a private operator. The private operator is responsible for maintenance and repair as well as collecting revenues from users. Profits are shared with the public authority who is also responsible for making new investments in upgrading the system.
- **Franchise:** The government awards a finite term zonal monopoly (a franchise) to a private firm for the delivery of solid waste collection services. The franchise awarding is made after a competitive qualification process. The private firm deposits a performance bond with the government and pays a license fee to cover the government's costs for monitoring. The private firm recovers its costs and profit through direct charges to the households and establishments that are served. Government provides control over the tariff charged to the consumer through the development of adequate competition and control of price collusion or through price regulation.
- **Open competition:** The government freely allows qualified private firms to compete for refuse collection, recycling or disposal services. In open competition, individual households and establishments make private arrangements with individual firms for refuse collection and/or recycling. No firm holds a zonal monopoly, and any number of firms may compete within the same zone. Similarly, in open competition, the government grants a licence to qualified individual firms for the private provision of disposal services. One city may be served by several disposal sites competing for business from the area's local governments and private haulers, as well as for business from remote governments and haulers. The government's role in open competition is to license, monitor and, as needed, sanction private firms. Under open competition, costs are directly billed by the private firm to their customers.

Longer term service contracts may be further differentiated from each other depending on which combination of components are included in the contract from Design (D), Finance (F), Build (B), Own (O), Operate (O) and Transfer (T) components. Note that inclusion of the 'Finance' component means that this Box is also relevant to Section 5.8.4 below. The main types of PPP contracts are included below, but it should be noted that this list is not exhaustive:[66]

- **Design, Build and Operate (DBO)** – The private contractor is responsible for the design, construction and operation of the SWM facility.
- **Design, Build, Finance and Operate (DBFO)** – The private partner is responsible for the design, construction, financing and operating of the SWM facility. It is the most complex contractual relationship between an authority and a private investor.
- **Build, Operate and Own (BOO)** – The private partner builds a facility based on a defined design and owns and operates it.
- **Build, Operate, Own and Transfer (BOOT)** – Same as BOO with an additional clause for transfer of the asset to the public partner at the end of the contract.
- **Rehabilitate, Operate and Transfer (ROT)** – The public good created is transferred to the private investor. The investor has the obligation of financing, rehabilitation and operating the public good for a certain period of time.
- **Build, Operate and Renew (BOR)** – The private investor assumes the financing, building, and operational costs and the costs of renewing of the public good for a certain period of time.

65 Based on Cointreau et al. (2000) and Coad (2005), listed in Annex A, under Chapter 5, Service deleivery and private sector participation in SWM

66 Soos, R. and C. Ardelean (2010). Modernizing Landfills through PPPs: A comparison of Romania, Bulgaria and Serbia, Presented at the ISWA Beacon Conference, Novi Sad, Serbia.

There is little up-to-date information on the extent of private sector involvement in municipal waste management services worldwide. In 2003 in the US private companies provided about 80% of all urban waste collection services. In aggregate, some 80% of the treatment market and 35% of the collection market was located within the private sector in selected European countries at about the same time.[67]

The market share of private-sector involvement varies over time and among countries. There is an ongoing debate regarding the net benefits of PPP in municipal service provision in terms of cost efficiency and technological efficiency.

One word of caution is that private-sector involvement does not remove the possibility for political abuse and/or corruption; rather, it provides a different set of opportunities to abuse the system than when services are delivered by the public sector. For example, a case study in the southern region of Brazil, Rio Grande de Sol, showed that, due to price and tariff control through the forming of cartels, the municipalities decided to switch back to publicly operated systems.[68]

Development partners have often argued that private-sector involvement is beneficial and brings efficiency gains and have thus supported PPP in many developing countries. However, recent global research comparing the efficiency of publicly and privately delivered waste management services in 28 case study cities stated: **'We have not been able to conclude that any particular ISWM[69] operator model is inherently better than any other.** The research [...] leads us to the conclusion that stable political and financial backing, and the existence of professional client organizations, working transparently in a non-corrupt environment, matters far more than whether the ISWM service is operated by either the public or private sector'.[70] Over the years, many cities around the world have implemented public-private partnerships, some more successfully than others. When such contracts come to the end of their duration, most cities renew or retender their contracts, but some have chosen to switch back to public models.[71]

The concept of *partnership* is crucial to the success of private sector participation. Both sides should have rights upheld by the courts and duties backed up by the threat of sanctions. As illustrated in *Figure 5.4*, such an equal partnership is far more likely to result in effective, economical and long-lasting services. The public sector sometimes dominates, with little concern for the rights of the private sector, which can mean corporate bankruptcy or reluctance to bid on future work. Equally, an influential company can achieve a monopoly which may leave the public partner obliged to accept high prices and/or poor service.

Figure 5.4 Need for a balanced partnership between the client and private contractor

Note: Cartoon prepared by Dorsi Germann, published in Coad (2005) and reproduced by permission of CWG.

67 Hall et al. (2003), listed in Annex A, Chapter 5, Service delivery.

68 See Box 5.7 below on examples of inter-municipal cooperation for waste management.

69 See Section 2.4 and Figure 2.3 for details on Integrated Sustainable Waste Management (ISWM)

70 Soos et al. (2013a)

71 Examples where municipalities have switched back include several in France, including Paris. See Hall (2010), listed in Annex A, Chapter 5, Service delivery and private sector participation in SWM.

A strong and balanced contract is key to ensure beneficial public-private partnership. Thus, for example it is more beneficial to choose output and performance-based agreements; include 'cash-out' activities[72] such as landfill closure in the contractual obligations; include caps on tariffs and tariff increases; include proportionate penalties for non-performance; build up funds for replacement and maintenance; and avoid guaranteeing large waste amounts that may be difficult to deliver over a long period as waste recovery and prevention targets are gradually imposed. The last point is illustrated for example by the cases of Stuttgart[73] and of the municipally-owned company SYSAV in Sweden.[74]

Collection services tend to benefit from the use of performance based contracts, with contractors being paid based on the quality of service and the cleanliness of the area they collect from; disposal and treatment facilities are usually paid by tonnes of waste treated or disposed. If the contracts are relatively small and restricted and the authorities are looking to encourage the participation of local companies, they will look for ways to simplify the tendering procedures and adjust the criteria to make sure that small bidders are eligible and capable of participating in the tender.[75]

BOX 5.6 WASTE COLLECTION IN MAPUTO, MOZAMBIQUE

Encouraging the participation of small-scale operators in service delivery[76]

The Municipality of Maputo is divided into five mainland administrative districts and two other small settlements. The municipality provides various waste collection services across the city. Inner-city waste collection and secondary collection in the suburban area is covered by private large-scale operators with public containers and compactor trucks. Primary collection in the suburban areas is provided by micro-enterprises with manual door-to-door collection and handcarts. The City Council's own operation covers the residential areas in the inner city with door-to-door collection of plastic bags and a pick-up service in rural areas from several collection points using tractors. Services to large waste generators which are not allowed to use the public collection system are based on the free market, with private operators licensed by the City Council.

Based on various experiences from NGOs, the City Council implemented a primary collection system for the suburban neighbourhoods to encourage small community-based companies to participate. The model is based on small-scale operators contracted by the City Council to provide primary collection using handcarts and to transfer the waste to secondary collection points spread throughout the city. The city's ambition was to extend primary collection to areas that would be difficult to reach with modern collection equipment. The procurement procedure for the primary waste collection was adjusted and simplified to cater for the limited capacity of the small-scale service providers.

This model is particularly interesting because the costs of the collection service are covered in their entirety by the fees collected, and also because the fees increase according to the social-economic level of the users and the commercial sector cross-subsidizes household waste collection in poorer areas. This is a remarkable achievement in one of the poorest cities of the world, where 50% of the people live below the poverty level (1 USD per day). The approach was trialled in a pilot and then successfully extended to large districts of Maputo. Currently 35 micro-enterprises or associations contracted by the Maputo City Council are servicing 900,000 people living in suburban areas with narrow and sandy roads that had no access to collection services as recently as 2006. Service roll-out was implemented earlier than planned due to popular demand from residents, after seeing the benefits in the pilot areas. Political pressure then rose at the national level, as it had become a success for the government and a potential vote-winner.

There can sometimes be a mismatch between the local specific conditions, such as affordability, willingness to pay, tradition of waste collection system, and the company winning an open tender. This was the case for example in Cairo, Egypt, where in 2002 Cairo had decided to privatize the collection service that had traditionally been carried out door to door by the local community of informal waste workers, the *Zabaleen*. The new service providers entered into a competition with the locally already-established informal sector service providers, who found strategies to keep their market. The international contractors were then forced to subcontract the traditional collectors. However, some of them – also due to other difficulties – abandoned the market. In 2014, Cairo initiated a process to invite the traditional service providers to bid, in an effort to formalize the decades-long service into an agreement that would be a win-win situation, benefiting client (public authorities), service users and providers.

72 'Cash-out' activities require the contractor to spend money that is either not covered or not completely covered by direct revenues, either from gate fees for incoming waste or revenues from recovered products.
73 See Box 4.4 in Section 4.2.2.
74 See Case Study 6, found after Chapter 5.
75 See Section 4.7.3 on Provider inclusivity.
76 Stretz (2013), listed in Annex A, Chapter 5, Service delivery and private sector participation in SWM.

Operation of uncontrolled or semi-controlled disposal sites more commonly stays with the local authorities. The private sector is more likely to be involved in the operation of sanitary landfills or treatment plants that require the operation of more sophisticated technology and therefore command higher fees.[77] Waste treatment and resource recovery is a business driven by the intrinsic value of the materials and further strengthened by policy and economic instruments in more developed countries, such as pricing of disposal, introducing EPR and the like. In resource recovery the private sector has a significant role in technology development, investment and operation and is often more experienced and better prepared to handle resource, market and financial risks in the sector. The municipality as a client needs to be prepared to evaluate different proposals and select proven technologies that are affordable for the communities. A more detailed discussion on private sector participation in investment can be found in Section 5.8.5 later in this chapter.

When developing service delivery contracts, it is important to be aware of certain environmental obligations that are 'cash-out' activities, but which should be either contracted at the same time with the rest of the services or managed through a different arrangement. These may include, for example, closure and aftercare of both old (existing) landfills and of new landfills at the end of the operational life; and reaching designated recycling targets (for both dry materials and organic waste) where the revenues may not cover the full costs.

Some cities focus on encouraging source separation and formalizing their cooperation with waste pickers and informal sector workers that enables them to provide primary collection service for a small fee combined with a right over the recyclables. Depending on the cultural setting, the income level of the users, the payment capacity of the municipality and the recyclable content of the waste collected, the waste picker may get paid for the collection service, may do the service for free in order to access the recyclables or may need to offer a small fee in exchange for the recyclables. Where householders separate recyclables at source, they may receive a small payment from the informal collector in exchange for the recyclables (who in this case is often termed an itinerant waste buyer).

5.6.4 Achieving economies of scale

In the developing world, 'waste' is still largely an urban phenomenon.[78] As more people move into towns and cities globally, the by-products of human consumption and its consequences become more concentrated. Urbanization also removes the spaces that could be used to manage this waste, so transport to more distant sites becomes essential. Urban waste therefore must often be dealt with in places far from its origin, under a different local administration, requiring inter-municipal cooperation. This offers economies of scale and upgrades of performance indicators, so inter-municipal arrangements may be a suitable model even without joint treatment facilities. This applies both where there are numbers of relatively small authorities in a well-developed region, as for example in Flanders in Belgium or in southern Sweden,[79] and also for municipalities with vast surface area and low population density, like those in the Brazilian Amazon. Some countries, usually fairly small ones, go one step further and have a national SWM organization in place to operate (or contract out) local services, which may leave waste collection organized at the local level and separated from treatment and disposal. In small island states, in light of their small territory and low number of municipalities, consolidating waste management at the national level may make sense.

These inter-municipal models, or regional models, may be either publicly or privately implemented through the cooperation of municipalities, with different levels of governance chosen for administrative or economic reasons. Usually the collection stays as a responsibility at the municipal level, while transfer, treatment and disposal facilities become regionally owned and operated. Therefore often inter-municipal models can achieve economies of scale, with facilities costing less per tonne at higher capacities until they reach a certain level. There is then a trade-off between the cost of transporting the waste to the site and the savings achievable in a lower cost-per-tonne facility.

The strength of inter-municipal models is cost efficiency in investment and operation that allows smaller municipalities to enjoy benefits from services they would not have been able to achieve on their own. Another gain is the way they can learn from each other and build on each other's experiences. Considerable economies of scale can be realized by such approaches, but smaller municipalities may lose decision-making power to their larger colleagues and creating such systems is complex and time-consuming, requiring high levels of goodwill from all sides.

---

77 Soos et al. (2013a).

78 In high-income countries, significant effort has gone into extending solid waste collection services also to rural areas; many countries now achieve 100% collection coverage across both urban and rural areas.

79 See Case Study 8 on waste management policy in Flanders and Case Study 6 on SYSAV, an inter-municipal company serving municipalities in the Malmö region of Sweden.

New regional institutional structures are needed and require having a very detailed understanding of precisely how the regional structures will operate. This is because the regionalization shifts the balance of costs between landfill and collection, between capital costs and operating costs and between large municipalities and small municipalities. Common important decision points are whether a common regional tariff will apply (covering both collection and disposal) and how existing arrangements (particularly collection services) will be accommodated in the new system.

BOX 5.7 EXAMPLES OF INTER-MUNICIPAL COOPERATION FOR WASTE MANAGEMENT

CIGRES Inter-Municipal Consortium for waste treatment in Brazil[80]

In 2001 in the state of Rio Grande De Sol, southern Brazil, 13 local authorities got together to form the Inter-Municipal Consortium of Solid Waste Management (CIGRES). The Consortium has grown ever since, encompassing 30 municipalities in 2012. The motivation of the municipalities was to improve treatment while sharing costs, but also to reduce unfair market practices of the private sector that kept waste management tariffs artificially high. Brazil has a long-standing tradition of agricultural cooperatives, so there was experience to build on.

The investment is public and the operation is performed by the consortium. The technical treatment is not perfect, in that instead of composting, the treatment is only a mechanical step to stabilize the waste. The facility was designed as a clean material recovery facility, but source separation and selective collection of waste has not yet been implemented by the municipal partners. The costs of the facility are shared by the member authorities, with each authority paying a fee according to its population; in order to ensure affordability for the smaller municipalities, the fees are cross-subsidized. With 88,050 inhabitants living in the partner municipalities and 1,300 tonnes received per month, the cost of treatment and final disposal at the regional facility is 50 EUR per tonne and 0.74 EUR per capita per month, and the cost to be covered by the municipalities after sale of recyclables is 29 EUR per tonne and 0.43 EUR per capita per month. Besides collecting local taxes or fees, the municipalities receive a transfer of funds from the state government to cover their costs related to waste management.

Regional strategy through a state-level nodal entity (Gujarat, India)[81]

Gujarat recognized the need for an integrated approach to MSW management in 2005, and adopted a state-wide, regional (clusterization) approach to meet requirements for safe treatment and disposal.

- **Institutional structure:** Gujarat Urban Development Company Limited (GUDCL) was designated as a nodal agency by the state government's Department of Urban Development in September 2005 and was mandated to develop a state-wide MSW management programme within a five-year period, using available 12th Finance Commission and state grants. GUDCL's in-house team then contracted private planning and technical design companies to provide technical inputs required for the implementation of treatment and disposal facilities.
- **Strategic Plans:** 159 urban local bodies (ULBs or municipalities) in Gujarat have a combined population of around 7.7 million and generate about 900,000 tonnes of MSW annually. The project study (2008) established that for processing and disposing of waste, the ULBs would have to spend 1,200 INR (25 USD[82])/tonne if each developed their own facilities, which would decrease to 450 INR (9.4 USD)/tonne if they worked together in clusters of ULBs, with decentralized treatment at the ULB level and regional landfills (with a minimum 100 tonne per day capacity) for disposal of treatment rejects to achieve cost effectiveness. Initially 51 clusters were identified, but after an in-depth assessment and study this was reduced to 36 regional landfill sites (clusters). The site search for landfills was constrained by the availability of suitable sites and by a maximum haul distance for any ULB of 50km.
- **Progress:** As of March 2015, vermicomposting plants had been constructed for 93 out of 159 ULBs for decentralized treatment and seven regional landfills catering for 37 ULBs had been constructed. These facilities were either already in operation or their operation and management contracts were in progress. In addition, six regional landfills are in the process of being refurbished in order to meet the national standards.
- **Governance arrangements:** GUDC in the role of facilitator has been mandated by the state government to identify and acquire land, hire technical consultants, conduct waste characterization studies and other technical studies, manage the bid process and facilitate PPP for construction of the landfills and vermicomposting plants, conduct awareness programmes and engage in asset management, among other responsibilities. Project activities have been jointly funded by the state and central governments. The development of regional landfills and ULB-level vermicompost plants has been done on a PPP basis. The land for treatment and disposal operations has been provided by the state government under a lease period of 20 to 30 years, and the cost of developing landfills and vermicomposting plants has been borne by central government and state grants.
- **Operational arrangements:** The compost facilities at 58 out of 93 ULBs are operated and maintained by NGOs paid directly by the state government. In return, NGOs sell the compost at Rs.2/kg and share half of their total revenues with the state government. Other compost plants are operated and maintained by the ULBs themselves. The regional landfills will be operated and maintained by private operators in return for a gate fee which will be paid by the ULBs. The tender documents are under preparation. In view of the limited paying capacity of the ULBs, the state government is in the advanced stages of considering Rs.100/tonne[83] as a standard gate fee, with any deficit (viability gap) being covered by the state government through a one-off contribution towards the capital costs.

80 Schmidt (2013), listed in Annex A, Chapter 5, Service delivery and private sector participation in SWM

81 Prepared primarily from information provided by Mr. Saurin Dave, DGM(P), GUDCL. Some information is also available on the web at http://www.udd.gujarat.gov.in/projects_SWM.php and at http://www.gudcltd.com/municipal-solid-waste-project.asp. A general guidance document on regional cooperation for municipal solid waste management has also been published by India's Ministry of Urban Development. See India MoUD (2011), listed in Annex A, Chapter 5, Service delivery and private sector participation in SWM

82 Conversion at the 2008 exchange rate of 48 INR = 1 USD.

83 This figure is to be reviewed every three years.

5.7 MSWM FINANCING MODEL: REVENUES

Referring to *Figure 5.2*, this section discusses alternative models for raising the revenues to pay for MSWM services.

5.7.1 Introduction

The revenues as discussed here refer to the annual revenue requirements needed to cover the net costs of providing an MSWM service.[84]

System costs include operation and maintenance (O&M) costs, costs for street sweeping, landfill closure and aftercare costs, depreciation, debt service costs and cost of capital. When services are delivered by the private sector, revenues should be sufficient to cover financial contractual obligations the client needs to pay to the private operator. When any resource recovery activity (such as composting or operating an MBT or EfW plant) is operated at a net cost, rather than as a net revenue generator, then any costs associated with resource recovery, become part of the cost items that need to be covered by the overall revenue.

As with other municipal activities, revenues in the waste management sector generally come from public finance sources (Section 5.7.2). A special case of this is where a direct charge (or tax) is made specifically for the waste management service (Section 5.7.3). In the absence of municipal services, private-to-private service provision may arise with direct charging by the provider to the user. Where direct charges are made, a decision needs to be made as to what proportion of the total costs of providing the service is to be recovered in this way (Section 5.7.4). One important criterion in making such decisions is the affordability of both the charges, and indeed of the overall SWM service, to the citizens as the service users (Section 5.7.5).

The responsibility for revenue collection for MSWM varies widely among countries. In some world regions, responsibility is generally with the municipality (e.g. in much of Western Europe and Asia), while in others it is generally with the waste service provider (e.g. in the US, and much of Eastern Europe and sub-Saharan Africa). Where direct charges are being levied on the service user, the authorities usually have better mechanisms at hand for enforcement of payment. Sometimes fees are collected together with other public utilities, such as water or electricity, especially if these are also a municipal responsibility. Advantages include lower administrative costs for collecting the fees and higher payment rates, as there is a more visible consequence of non-payment.

Because many municipalities, particularly in developing countries, struggle to raise revenues sufficient to cover the cost of their current MSWM, they find the prospect of covering the increased costs of an upgraded and environmentally sound waste management system rather daunting. Any alternative sources of revenues are therefore extremely attractive. Section 5.7.6 covers in particular profit sharing, extended producer responsibility, revenue-generating economic instruments, climate finance and output-based financing.

5.7.2 Public financing

Public financing may take the form of a user charge, a local tax, a national transfer of public funds or some combination of these. In each case, the 'general public' or 'other public income' is the ultimate source of the revenues that are used to pay for the MSWM service. Additional income sources directly related to waste management include licence fees for operators and littering fines; however, these represent minor revenue sources, which are usually sufficient to cover only part of the cost of enforcement.

Revenue collection policies for SWM vary widely among cities and countries. As an example, *Table 5.4* divides the 20 cities studied in a recent UN-Habitat study into four broad categories according to revenue collection mechanism. Around half of the cities have some form of direct charging, but equally half use alternative indirect revenue sources, either in whole or in part. The level of cost recovery targeted varies widely, as does the revenue collector and the collection mechanism chosen, which in turn influences the payment rate.

---

84 The net cost is the gross cost of the MSWM service, less income received. Income may be derived from resource recovery activities or from gate fees charged to private waste generators using the MSWM system.

Table 5.4 Revenue collection mechanisms for formal waste services to households

DIRECT CHARGING VIA A WASTE BILL OR A UTILITY BILL (U)	DIRECT WASTE FEE + PROPERTY TAX	NO DIRECT FEE (FINANCED VIA PROPERTY TAX)	NO DIRECT FEE (FINANCE FROM GENERAL SOURCES)
Adelaide, Australia	Bamako, Mali	Belo Horizonte, Brazil	Ghorahi, Nepal
Canete, Peru	Bengaluru, India	Curepipe, Mauritius	Quezon City, Philippines
Kunming, PRC		Delhi, India	
Lusaka, Zambia		Dhaka, Bangladesh	
Moshi, Tanzania		Managua, Nicaragua	
Nairobi, Kenya (U)		Sousse, Tunisia	
Rotterdam, Netherlands (U)			
San Francisco, US			
Tompkins County, US			

Note: Such comparative data are not routinely collected; this data set for 20 cities comes from a recent UN-Habitat study.[85] The symbol 'U' indicates collection of MSWM fees via another utility bill (water or electricity).

5.7.3 Direct charging

The payment rate of fees or taxes is never 100%; a good sustainable payment rate for MSWM is considered to be around 90%. Collecting revenues from citizens may work better in the form of a direct user charge or tax depending on how it has been traditionally done and on the local culture, for example in places where there is a tradition of service delivery through community-based systems, or in places with strong decentralization where cost-based transparent methods are used for establishing the charges. Some places breed a culture of user charging that emerges as private-to-private arrangements in cities or districts that are not covered by communal services. Examples include part of Bishkek, Kyrgyzstan,[86] several islands in the Maldives, and cases in Africa. Private sector participation may make it politically easier to introduce fees, because that means of introduction could be seen as less likely to threaten election results.

Collecting direct user charges may work well in cases in which there is an efficient enforcement mechanism in place, as for example when SWM fees are collected with other utility bills or phone bills and an effective sanction can be made in the case of non-payment. For example, linking waste fee collection to electricity bills was successfully implemented in Greece, Grenada and Jordan but contested in court and overruled in Egypt. Other enforcement measures include administrative penalties, such as refusal to issue a certificate of good fiscal standing that is often needed to purchase assets or apply for a job. In some of the Newly Independent States, the former Soviet Union administration's ability to sequestrate goods from non-payers is still applied as a form of enforcement. Entrusting private operators with user charge collection may be detrimental to payment rates, exactly because private operators lack an effective means of enforcement, as cases in southern Italy have shown.

Collecting waste management fees through local taxes that are earmarked for the waste management budget is widely practised and may be a cheaper and more effective way of collecting the revenue than a separate waste management user charge. For example, the waste management tax charged in Belo Horizonte, Brazil covers 40% of waste management costs and is collected with 95% efficiency. The reason for the high collection rate is believed to be the efficient and transparent revenue collection system, through a local tax called the Urban Cleansing Tax that has a long-standing tradition in the city and is trusted by the citizens.[87] Similarly in the city of Surat, India, the sanitation tax collection rate is 92%; this is collected together with the property tax and amounts to a 5% surcharge on that tax.[88] One disadvantage of a direct user charge from the point of view of the city is that, if there is a change for the worse in the municipal service, then users may choose to withhold payment for the service. This was the case in Tunis, Tunisia after the revolution in 2011, when in some areas

85 Scheinberg, Wilson & Rodic (2010), on behalf of UN-HABITAT, listed in Annex A, Chapter 1, Waste management; also Wilson, Rodic, Scheinberg et al. (2012).
86 Sim, N., D. Wilson, C. Velis, & S. Smith (2013). Waste management and recycling in the former Soviet Union – Case study of Bishkek, Kyrgyz Republic (Kyrgyzstan). *Waste Management and Research*, 31 (10 Supplement), 106-125.
87 Scheinberg, Wilson and Rodic (2010).
88 Soos, et al. (2013a).

the payment rate dropped from nearly 100% to 73%.[89] In such cases a vicious circle may arise, as poor payment performance leads to a further deterioration of service quality that may ultimately lead to the failure of the system.

Direct charges to households are usually a flat rate, or are based on the number of persons in the household, property size or value, road frontage (if collected with the property tax) or to water, sanitation or electricity usage, and in these cases collected simultaneously with the relevant utility bill. Charges that are related to the volume or weight of waste collected, and which thus reflect the actual cost of providing the service, are sometimes referred to as pay-as-you-throw (PAYT) charging schemes.[90] PAYT is attracting increasing attention especially in the developed countries of Western Europe and the US but also in developing countries.

Person in a household paying for collection services, Nicaragua

Interestingly, the larger waste generators are usually the first ones to opt for a charging system by the volume or quantity of waste generated. They generally rent a container, pay a 'pull charge' each time the container is emptied or removed/replaced, and often also have to pay the gate fee for landfilling their waste. Where they choose to use municipal rather than business-to-business services, the larger waste generators such as small businesses, shops, restaurants, hotels and institutions are regularly the ones who pay a higher fee for MSWM, cross-subsidizing those users who face affordability issues.

5.7.4 Cost recovery and its challenges

Where direct charges are made to households for a MSWM service, a decision needs to be made on the proportion of the total cost to be recovered in this way. On the one hand, it can be argued that the householders are the waste producers, so in accord with the 'polluter pays principle' they are responsible for the costs. On the other hand, MSWM is often seen by the citizens as an essential public service, which they should receive 'for free', as they have 'already paid for it' in their general taxes. One example of this situation is found in Chile, where 80% of the population is exempt from paying taxes on waste services.

Often international financial institutions (IFIs) and other donors make their funding for improvements in waste management in developing countries contingent on improving cost recovery, usually by increasing direct charging. Where a private operator is tendering for a service delivery contract, they need to balance the requirements of their municipal client who wants to have affordable tariffs and a clean city, and their shareholders who want to increase tariffs and ensure healthy profit margins. As a general principle, any increase in fees needs to be gradual and connected to real improvements in the waste management system that are well communicated to the citizens and actually perceived as such by them. High payment rates alone are not a guarantee for full cost recovery. Cost recovery from user fees alone is more likely in high-income countries, simply because fees set at a level allowing full cost recovery are more affordable in those countries.

Traditionally in many cities, usually where local governance is weak and decision making is centralized, both direct fees and other revenues intended for the solid waste service flow instead into a general municipal account where they tend to be absorbed by overall expenditures, rather than applied to the intended purpose of solid waste management. Tracing public funds becomes even more difficult when locally collected fees and revenues are transferred to the central government before being redistributed to the local level.

Transparency and traceability in the source and use of funds is important for accountability. The absence of a direct linkage between revenues and the actual level of service provision tends to undermine the accountability of local waste management institutions and remove their incentive to improve and/or extend service.[91] In light of this problem, there has been a major focus on improving cost recovery and extending it to cover all costs (full cost recovery) and this has often become a requirement of donor financing. While the pressure for increased

89 Personal communications with several municipalities of Greater Tunis, June 2014.
90 See Section 4.5.1.
91 See Sections 3.5 and 3.6 in the work by Schübeler et al. (1996), listed in Annex A, Chapter 2, Integrated sustainable waste management.

transparency in revenue collection is positive, the evidence that full cost recovery is either possible or necessary for the sustainability of waste management service provision is not conclusive.

Maputo, Mozambique, has used a World Bank loan to extend its collection service to slum areas. The city did its best to fulfil the World Bank's requirements for cost recovery and eventually achieved full cost recovery through long-term planning that included a step-wise reduction of public financing and increases in user charges as well as other cost recovery mechanisms. The municipality re-negotiated its contract with the World Bank and convinced donors to finance a small share of the running costs of part of the extended collection service by showing them a long-term plan that ensured gradual raising of the user charges to reach full cost recovery.[92]

Public meeting to raise awareness of the new waste collection service, and to foster a willingness to pay, Maputo, Mozambique

A recent World Bank study[93] investigated the cost recovery of waste management services in four cities in the PRC and found that in 2012, rates of cost recovery through user fees were between 28 to 45%, dropping further from an already low 60% cost recovery average in 2007. At the same time, financing solid waste management is not a particular strain on the local budget, as total waste management costs make up 4 to 9% of the total urban budget in the studied cities, and much of this cost is met by transfers from municipal and ultimately central government funds. So the issue does not seem to be one of affordability and is not linked to low payment rates or to the way user charges are collected.

This further strengthens previous findings that local policies and preferences based in the local tradition and context will determine the most workable approach for financing and cost recovery. In high-income countries the user's fee is more likely to be calculated based on the real costs incurred than in countries with lower levels of income. Full cost recovery is unlikely to be achievable in the short term in most low- and lower-middle income countries.

5.7.5 Affordability

Local authorities need to balance the twin objectives of cost recovery and affordability. From the point of view of an external financial institution, cost recovery is the single most important factor for financial sustainability, as it is a prerequisite for recovery of investments with a profit margin or for repayment of loans with interest. But from the point of view of the local authorities and the service users, waste management services have to be affordable to the citizens of the city and ultimately to the country.

It has been suggested that a practical upper limit (or simple 'rule of thumb') for affordable waste management costs is 1% of the per capita income level. In low-income countries (with an income level below about 1,000 USD per capita per year) this works out to 10 USD or less per capita per year. *Table 5.1* used this upper limit of affordability to calculate an affordable total net cost per tonne for waste management, for comparison with estimated costs of alternative unit operations that could form part of an environmentally sound management (ESM) system. The table shows clear affordability constraints in the lower income countries, and also that services become more affordable with rising income levels. Thus, it should come as no surprise that some low- and middle-income countries set their user charges below full cost recovery rates. For example Belo Horizonte, Brazil, and Chile as a whole, deliberately keep fees low and affordable to everyone, while at the same time striving to provide 100% waste collection coverage and getting all users to pay at least some contribution towards the costs.

In principle all people should pay something for waste management at the level that is equitable and affordable to them and proportional to the amount of waste they generate. Sometimes keeping fees low stems from political rather than affordability considerations. For example, Bengaluru, India, and Managua, Nicaragua, also keep fees low but do not apply punitive measures for non-payers even though the payment rate is low at 40-

92 See Box 5.6 above and Stretz, J. (2013) Economic Instruments in Waste Management, Case study of Maputo, Mozambique. http://www.giz.de/fachexpertise/downloads/giz2012-en-economic-instruments-mozambique.pdf

93 Ren & Hu (2014), listed in Annex A, Chapter 5, General reading.

50%. However, these cities' MSW management systems are also financially less sustainable and may fall into the 'vicious circle' of declining fee collection rates bringing down revenues and causing service rates to decline.

Citizens are able and willing to spend different amounts of money for waste management depending on their income level. Often tariff policies incorporate cross-subsidizing, whereby higher-income citizens pay higher fees and the poor pay lower fees. Examples include Moshi, Tanzania; Curepipe, Mauritius; and perhaps unexpectedly Rotterdam, Netherlands. In these cities, households that are classified as poor do not pay at all, or they pay a relatively small amount. In some places such as Ghorahi, Nepal, and Quezon City, Philippines, no waste management fee has been charged to households until recently. There are certain payment mechanisms that are suitable for cross-subsidizing. For example linking user charges to other utility bills make it easy to single out larger users and require them to pay a higher rate per volume of collected waste than small users. Similarly, collecting a percentage of property tax automatically ensures a higher tax from those who have more assets. When waste management service fees are paid via local taxes or governmental transfers, the public funds are usually not traceable, but this could also be seen as a form of cross-subsidy, as funds from other sources such as income, profit and trade taxes are used to cover the costs of waste management.

There is evidence that most users are willing to pay for reliable collection service to serve their neighbourhoods, even in the poorest countries.[94] However, the benefits of disposal or treatment that is environmentally sound are not obvious to most service users (except perhaps to those living close to a dumpsite), so it is difficult to convince them to pay. It should be noted that securing a loan to build a facility from, for example, a foreign development aid agency, is not necessarily a guarantee of financial sustainability. To that end, financing operation and transportation costs to the facility is essential, as these may be prohibitively high for potential distant disposal sites, as is the case in Bamako, Mali.

5.7.6 Other revenue sources

Alternative and innovative ways to raise revenues from sources other than government sources or direct user charges are likely to be extremely attractive to municipalities. The options considered in this section are profit sharing, extended producer responsibility, revenue-generating economic instruments, climate finance and output-based (also called results-based) financing.

Profit sharing is often practised in concession-type service contracts or public-private partnerships, where the municipality is the owner or has invested in facilities, equipment or land that is being used for operations. In such contracts the obligations of the operator may include, among other things, profit sharing with the municipality. This creates a revenue stream that can be used by the municipality for contributing to waste management activities that are not part of the contract, such as closing old landfills or carrying out awareness raising campaigns.

Extended producer responsibility (EPR) is one of a number of market-based instruments which is of particular interest to municipalities because it has the potential to shift the physical and/or (at least part of) the financial responsibility for managing the end-of-life products contained in municipal solid waste back to the producer who placed the product on the market. EPR has been applied in particular to packaging and to WEEE and is attracting much interest in developing countries.[95] Ultimately, those who consume the products pay for the management of these special waste streams, as the cost of recycling borne by the producer under an EPR scheme is internalized in the price of the product.

Revenue-generating economic instruments[96] include a range of stimulants that aim to improve the environmental performance of the private sector. The most common forms are those that provide fiscal relief or grants for investments in green technologies. Beside the direct user charges discussed in the sections above, green taxes[97] are also worth a mention. These create a disincentive by charging extra for the polluting products while also, if well-designed, making it possible to earmark the tax revenues and reinvest these into smart design, cleaner production or recycling, which work on the prevention side. Products that are subject to green taxation may include plastic bags, batteries, packaging, tires, waste oils, lubricants and refrigerators.

94 An example is the case presented earlier of Maputo, Mozambique, which is also discussed in Box 5.6.
95 See Section 4.5.2 on EPR and Topic Sheet 13, found after Section 4.5, on EPR initiatives outside of the 'usual' OECD countries.
96 See Section 4.5.1 on economic (market-based) instruments.
97 Green taxes are also known as eco taxes, with product charges as one example.

Climate finance, carbon finance or emissions trading is another market-based instrument that is used for generating additional revenues in this sector. Under the Clean Development Mechanism (CDM) introduced under the Kyoto Protocol, carbon credits (which could be counted towards buyers' greenhouse gas emission reduction obligations) were traded on the market and monitored, verified and registered on the UN registry. Most of the improvements in waste management in developing countries financed through these mechanisms were related to the capture and use of landfill gas. Because the carbon credits were directly linked to greenhouse gas emission reductions actually achieved and verified, there was a direct financial incentive for cities to operate their donor-funded sanitary landfills as designed. The annual carbon credits provided in effect a revenue stream to pay for both operations and the interest charges on the capital the following year. For this reason, CDM was actively promoted by a number of international financial institutions as an early form of 'output-based financing' (see discussion below).

As the Kyoto Protocol, in its present form, expires at the end of 2015, the funding mechanisms created under the Protocol are also coming to an end. These mechanisms were also criticized for the heavy bureaucratic burden on the project developer and implementer in terms of setting baselines, monitoring and verifying impacts. In an effort to reduce transaction costs, a replacement climate financing mechanism, the sector-based Nationally Appropriate Mitigation Actions (NAMA) approach, has been designed to allocate financing to much larger programmes, rather than individual projects, and with a less stringent methodology for monitoring results.[98]

NAMAs are at different levels of preparedness across the world, while the respective governments, private sector actors and project developers are waiting for developments related to the measuring, reporting and verification guidelines for this still-evolving mechanism and seeking financing from the limited funds that are put through various NAMA facilities. As of the end of 2014, total pledges to the Green Climate Fund total 10 billion USD across all sectors, and the UN is urging developing countries to prepare their projects. An example of a NAMA project under preparation in 2014 is provided in *Box 5.8*.

BOX 5.8 PROPOSED NAMA FINANCING FOR THE WASTE SECTOR IN COLOMBIA[99]

The goal of the solid waste NAMA is to transform Colombia's waste sector by creating greater certainty for investments in waste management projects that divert waste from landfill disposal towards more productive uses. Diversion avoids methane emissions from landfills while promoting sustainable economic growth and improving the living conditions of informal recyclers.

The Colombian government proposes, among other initiatives, to reform waste tariff regulations (gate fees) so that diverting waste for alternative uses such as recycling, composting or waste-to-energy plants can compete economically with landfill disposal. The proposed financial mechanism is geared towards building in-country experience with integrated waste management technologies. The government proposes the creation of a national revolving NAMA equity fund to overcome investors' hesitation to invest due to a lack of familiarity with the waste technology and processes. The equity fund would contribute equity capital on a concessional basis to help build waste treatment facilities on a municipal level, contributing to projects in multiple cities and attracting other equity and debt investors to finance the facilities.

Repayment of equity from project developers will remain in the fund and be available for equity investments in future projects. Over time, the contribution from the NAMA equity fund will be reduced as investors become more knowledgeable about the waste treatment technology and the associated operational and financial risks, with the eventual goal of not needing concessional support for integrated waste management projects.

98 See the NAMA database for examples in this sector. http://www.nama-database.org/index.php/Waste

99 Information taken from Mitigation Momentum (ECN, ECOFYS, GIZ, CCAP, TERI, UNEP, OECC), Annual Status Report on NAMAs (2013). Available at http://www.mitigationmomentum.org/downloads/Mitigation_Momentum_Anual_Status_Report_2013.pdf The proposal can be found in the NAMA database at http://www.nama-database.org/index.php/Recycling_Program_NAMA

Output-based financing is a form of donor financing which can be considered revenue financing, because no financing changes hands until a concrete output is first achieved. International development partners generally only finance investment costs, which means that the city must be able to pay for the operating cost as well as the cost of capital in the case of a loan. Operation costs usually increase with investments into upgrading, extension and modernization of waste management, and generally make up around 70% of total costs.[100] Full cost recovery of all waste management costs is usually already a challenge to cities and becomes increasingly so as systems are upgraded and become more sophisticated. It is thus no surprise that, in the past, numbers of low-income cities have borrowed money to build a new waste management facility, and then not been able to afford the operational costs to run it properly. Even the best-designed sanitary landfill will soon revert under such circumstances to an uncontrolled disposal site.[101]

For such reasons, there has been considerable interest among donors in the SWM field in financing mechanisms that are linked to performance and outputs, to help ensure that cities can afford to operate their new facilities as planned. The old form of climate finance under the Kyoto Protocol provided such incentives specifically for landfill sites (see above). The Global Partnership for Output Based Aid (GPOBA) started its activities in 2003 and is being financed by multiple donors. Its mission is to develop and test viable output based-financing methods. In waste management GPOBA has so far successfully supported projects for various improvements in the sector. One example for Nepal is shown in *Box 5.9*. Other successes can be found in Indonesia, Jamaica, Malaysia, Mali, the PRC, Tanzania and the West Bank.[102]

BOX 5.9 OUTPUT-BASED AID (OBA) FOR MSWM IN NEPAL[103]

The project aims to improve the quality and financial sustainability of SWM services in participating municipalities through an output-based financing scheme. Given the generally poor financial situation of the SWM sector, the project is designed to provide incentives to enable the gradual development of a beneficiary charging mechanism for SWM services in order to enhance financial sustainability, improve service quality, and enable expansion of SWM service coverage over a four-year period starting in 2011. The project has three components as described below:

Component 1 – Service delivery financing (3 million USD): This component finances output-based service delivery subsidies for each participating municipality over a four-year period, to cover the gap between the operating and maintenance costs of delivering SWM service improvements and the revenues collected through fees. The cost recovery gap is financed by aid, provided that the services meet agreed minimum performance criteria. When applying for this financing, a municipality commits upfront to improving the financial sustainability of its service according to a time-bound plan (but does not necessarily commit to full cost recovery).

Component 2 – Technical assistance for implementation support to participating municipalities (0.6 million USD): Participating municipalities benefit from technical assistance for successfully planning service improvements, developing manuals and training in landfill operations and management, as well as from mechanisms for improved billing and revenue collection, monitoring, evaluation and performance management. This component also includes increasing awareness of the principle of the 3Rs and assistance in selecting operator models and operators for different service components.

Component 3 – Project management, monitoring, evaluation and communications (0.7 million USD): This component covers the project management costs involved with the monitoring and verification of outputs and performance and baseline and beneficiary assessments. This is a key component in ensuring the performance-based nature of the project.

100 See Section 5.2.1 on financial and economic costs and benefits.

101 One such case study is described in Rouse, J.R. (2006). Seeking common ground for people: Livelihoods, governance and waste. *Habitat International* 30(4): 741-753.

102 World Bank (2014), listed in Annex A, Chapter 5, General reading.

103 Information taken from: http://www.tdf.org.np/new/custom/uploads/cms/56_9782.pdf

5.8 MSWM FINANCING MODELS: INVESTMENT FINANCE

Referring to *Figure 5.2*, this section focuses on various models for raising the investment finance required for MSWM services.

5.8.1 Investment needs

The investment needs in the waste management sector are considerable in all parts of the world. Taking into account current investment trends, needs and waste generation rates, various attempts have been made to estimate the need for investment. The World Bank *What a Waste* publication estimated the total global cost for investment and operation of waste management for the period 2010-2025 at 375 billion USD.[104] A similar estimation was carried out to arrive at a global 10-year investment demand for MSWM infrastructure of 330 billion USD, i.e. 33 billion USD per annum.[105] Both of these estimates look rather low compared to the data presented in Section 3.5.4, which show the total value of proposed worldwide waste management investment projects in the period January 2013 to December 2014 to be around 300 billion USD, of which around 85 billion USD was for MSW projects.[106] These latter values are an overestimate as they indicate the total for projects 'active' over a two-year period, and include many projects that will never be approved, financed or built. The majority of these investments are planned for high-income countries (particularly the UK for MSW projects and the US for total investment), with the PRC, India and the Russian Federation each accounting for 3 to 10% of the total proposed MSW investments.

While none of these estimations is an exact science, the important point made is that there is a very significant investment need in the waste management sector worldwide in the years to come.

5.8.2 Financing sources and instruments

Investments in the sector may be financed by national or local governments, international financing institutions and other donors, the private sector or a combination of these. The financing may come as a grant, free of cost and not requiring reimbursement; as commercial financing; or a financing instrument that falls somewhere between a grant and commercial financing. Financial sustainability depends on having access to a predictable flow of funds to finance initial asset purchases as well as future maintenance and replacement needs. Grants have a role to play here, as by supporting initial asset purchases they enable service charges to be kept at affordable levels while the services are improved and financial sustainability increases. The logic is that national income will grow in real terms and will allow municipalities to maintain affordability while achieving financial sustainability.

Commercial financing may be sourced from the private investor's own capital (equity), from commercial banks or other commercial financial institutions. Financing instruments are usually provided by international financing organizations as an alternative to grants and these include soft or preferential loans with grace periods, long-term reimbursement requirements, low interest rates and government guarantees. Soft loans from international financial institutions may be granted, for example, with a few years' grace period, for 20 to 30 years with interest rates as low as 3%. Commercial banks, depending on their appetite and their trust in the sector, may provide loans at 8-12% interest rates.[107] However, in many SIDS and other developing countries where the banking sector is not well-established, local authorities may have reached their debt limit, or their credit worthiness may be weak, and in such cases there is a high likelihood that commercial loans will not be available. Larger municipalities with a good financial record may have the option to raise capital by issuing municipal bonds. The World Bank estimates that of the 500 largest cities in the developing world only 4% are deemed credit worthy in international capital markets and 20% in local markets.[108] In this sense financial planning for developing country cities becomes an integral part of waste management planning and vice versa; without financial resources, waste management planning may become a futile exercise.

104 Hoornweg & Bhada-Tata (2012), listed in Annex A, Chapter 3, Collated data sources.

105 Whiteman A. & R. Soos (2011). Investing in Resources & Waste Management: Policy Context & Challenges, A Background Paper for the UN CSD Intersessional Conference on Building Partnerships for Moving towards Zero Waste, 16-18 February 2011, Tokyo, Japan.

106 AcuComm's Waste Business Finder database. http://acucomm.net/

107 Correct at the time of writing, 2014, with international interest rates at an all-time low (central bank rates around 0.0-0.5%).

108 See http://www.worldbank.org/en/news/feature/2013/10/24/financing-sustainable-cities-africa-creditworthy

5.8.3 Local and national government financing

As cities typically struggle with making ends meet in the waste management sector, especially in developing countries, public and grant financing are always welcome. Local government financing means the city has to raise the funds from local sources of income. The sources of income for the local government are for the most part local taxes, fees and fines such as property taxes and parking fees. Some municipalities are richer and manage their assets more effectively and are able to earn income through renting out assets or participating in different public-private partnerships.

When significant local funds are allocated to a project, the sense of ownership of the project by the local government tends to be high and significant effort typically goes into planning and participative decision-making. One example of this is the locally financed landfill in the city of Ghorahi, Nepal, a small town in a low-income country. The town authorities understood the functions of various aspects of landfill design, so rather than design a landfill using technical specifications copied from abroad, they instead engaged Nepalese experts who identified an appropriate site with thick deposits of natural, undisturbed clay to function as landfill bottom liner and thus provide an adequate level of environmental protection.[109] The keys to success in this case were likely that the project was a local initiative; the objective was to improve landfilling and reduce local environmental impacts; and investment funds were secured from both local and national public financing.

National government funding is a very important source of investment finance. For example, for the PRC, MSWM infrastructure has been a priority since 2000. In India, the Jawaharlal Nehru National Urban Renewal Mission launched by the central government envisages an investment of 20 billion USD.

5.8.4 International financial institutions

A recent study for ISWA has documented the contribution of international financial institutions (IFIs) to the financing of SWM in developing countries.[110] The results are summarized graphically in *Figure 5.5*.

An estimated 4 billion USD was committed to development cooperation in SWM between 2003 and 2012, rising over the period from around 180 million USD to 700 million USD per annum. While the proportion (3-year average) of overall development finance directed to SWM has risen from 0.12% to 0.32% over the same period, this is still only a tiny proportion of the total. Considered in terms of the population of the countries receiving SWM development finance in 2012, this equates to just 0.09 USD per capita. This compares with per capita levels of 2.43 USD in the water and sanitation sector and 31 USD of development finance overall.

The majority (70%) of this support for SWM has been in the form of lending from development banks, amounting to 2.8 billion USD over the 10 years from 2003 to 2012. This has provided access to capital for selected middle-income countries to develop MSWM infrastructure, particularly collection systems and engineered landfills. The distribution of SWM loan funding has been very uneven: one country (the PRC) received 12 loans with a total value of 510 million USD (18% of total development finance lending for SWM over the 10-year period). The top ten countries for SWM-focused development finance are all middle-income countries, and they account for over two-thirds of the total value of both grants and loans over 4 million USD.[111]

Overall, low-income countries appear to have received significantly less financial assistance – only ten sub-Saharan countries received grants or loans of more than 4 million USD, together accounting for less than 5% of the total. This uneven distribution may be because middle-income countries are better able to access and absorb development finance but it is certainly an issue that needs to be considered carefully to ensure that development finance for SWM is targeted appropriately.

109 Scheinberg, Wilson and Rodic (2010), pp. 62 and 204

110 Lerpiniere et al. (2014), listed in Annex A, Chapter 5, Service delivery and private sector participation

111 In descending order, these are the PRC, India, Morocco, Turkey, Azerbaijan, Vietnam, Venezuela, Ukraine, Tunisia and Argentina.

Figure 5.5 International development finance for solid waste management

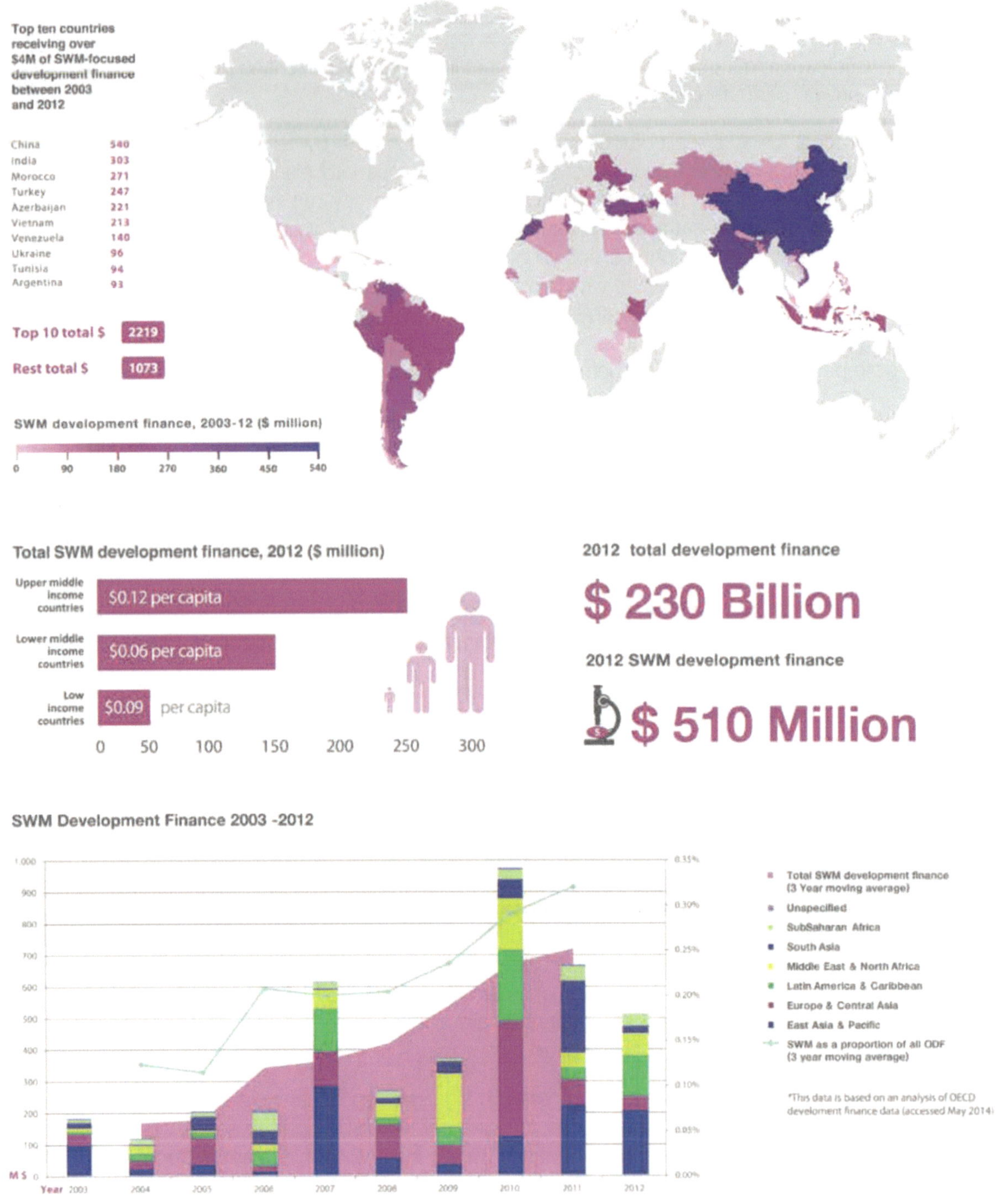

Source: Copyright ISWA. Background research: David Lerpiniere, University of Leeds. Graphics: D-Waste. Reproduced as is from Lerpiniere, D. et al. (2014). *Review of International Development Co-operation in Solid Waste Management.* Report for ISWA.

Grant-funded support is the other key element of development cooperation, amounting to an estimated 1.2 billion USD between 2003 and 2012 across more than 3,000 grants. Around three-quarters of total grant funding has been used to increase local skills and capacity and to provide other technical assistance in areas such as the informal recycling sector, private sector participation, cost recovery, awareness raising and climate change. The remaining grant funding has been used to fund the purchase of refuse collection vehicles and containers and to provide SWM in the aftermath of natural disasters or as part of conflict-related relief efforts.

From the point of view of recipient local authorities, grant financing is similar to national public financing in that it is free of cost. The equipment and facilities purchased or built through grant financing remain in the ownership of the public authority, allowing more flexibility in choosing the remaining elements of the financing model and enabling more control over tariffs, choice of treatment and performance standards. However, donor financing often comes with requirements. These might include for example requiring full cost recovery or the use of public-private partnerships or other forms of private financing to co-finance the integrated waste management system in question. Experience also shows that along with 'free money' may come certain pitfalls that cities need to be aware of. 'Free money' may create the incentive to over-size facilities, making operation costs higher than necessary. Investment projects may involve bringing in technologies that are inappropriate for the climate or the waste composition in the recipient country or that require spare parts or know-how in operation and maintenance that is not available locally. Often this type of financing is intended by the donor as a one-time intervention to provide much-needed equipment. However, for this to be a sustainable long-term solution, the recipient needs to put in place a careful plan and management system to build up the funds that will be needed to maintain the equipment and replace parts as they wear out, and eventually to replace the equipment at the end of its useful life.

BOX 5.10 PRIVATE VERSUS GRANT FINANCING FOR THE DEVELOPMENT OF SANITARY LANDFILLS IN EASTERN EUROPE[112]

In the Eastern European accession countries and pre-accession countries to the EU, a major amount of investment financing has been made available for modernizing waste management infrastructure and services through the EU financing programs. In countries such as Romania, Bulgaria and Serbia there was a big push over the years to close old open dumpsites and open sanitary landfills. In spite of the available public funds some cities, such as Oradea, Romania and Duboko, Serbia, opted for private financing raised by a private investor from commercial banks.

An analysis of these cases show that while the private financing is more expensive, in general the implementation of these projects is faster, risk allocation is better, and there is greater focus on cost recovery and resource efficiency. The impact on the gate fee is not significant, as in the analysed cases of grant and private financing the gate fee at the sanitary landfill worked out to be 11-13 EUR per tonne, with this being slightly higher for the privately financed facility. Private financing locked the municipalities into long-term contracts with the same operator, which can be a disadvantage if the contract is not flexible enough to cope with significant changes in circumstances.

In terms of saving on the cost of capital, soft loans or preferential loans from IFIs are the next most attractive option after grants. That said, these loans are associated with quite significant transaction costs in terms of preparing feasibility studies. They also require meeting preconditions or conforming to performance standards in terms of credit worthiness, debt capacity, environmental health and safety practices, social impacts, resettlement policy and the like. Local authorities are rarely powerful and often their incomes are unreliable or are highly dependent on central budgets, and their debt capacity is limited because they also need to finance infrastructure for other local priorities, including health, education, transport and water.

IFIs most often provide grants or loans to central governments and sometimes to municipalities. Many encourage their clients to enter into public-private partnerships, whereby a private company operates the service. Private companies seeking finance for a waste facility to provide business-to-business (B2B) services[113] may in some cases be able to seek finance from the International Finance Corporation (IFC)[114] or from a number of specialist bilateral donor funds, but it should be noted that this is the exception, not the rule.

112 Soos R. & C. Rus (2010). Modernizing Landfills through PPPs: A comparison of Romania, Bulgaria and Serbia. ISWA Beacon Conference, Novi Sad, Serbia.
113 See Section 5.5 on B2B financing models for waste management.
114 A case study showing IFC involvement with the recycling industry in the Balkans appears in Box 5.3.

5.8.5 Private sector participation in investment

Private sector operation of MSWM services was discussed in Section 5.6.3. Here the focus is on attracting additional private investment financing for improvements in the waste management sector. This implies both creating the conditions under which an investor can recover its investment and earn a profit and providing some certainty to the investment community. Whether the investment is being made via a foreign or a local company, in principle the commercial requirements are similar. Private sector entities may source financing from their own capital (equity) or from financial institutions such as commercial banks or specialized funds. Specialized funds are created for specific (development) sectors. They combine private, commercial and/or grant financing and are run by professionals who are highly familiar with the sector and therefore may facilitate the development of project pipelines better than the average commercial bank.

The private sector may participate in B2B models[115] or may cooperate with the public sector in service provision for MSWM. Some contractual options are illustrated in *Box 5.5*. Those which include the provision of investment finance may take the form of concessions, design-build-finance-operate-transfer (DBFOT) type contracts, or involve setting up a special purpose vehicle[116] for providing the service. These contracts, being service contracts, can be concluded for individual services such as collection, treatment or landfilling, or for the entirety of the services within the entire waste management system.

The private partner always aims at earning back its investment and that takes time. Therefore the duration of this type of contract tends to be long. The length increases both with the amount invested and with the working life of the physical investment. If new vehicles are specified in a collection contract, then the contract duration will likely need to be at least 5 to 7 years. For a 50 to 100 million USD treatment facility, the likely contract duration will be 20 to 30 years. Municipalities in developing countries are likely to need technical assistance that provides the professional advice required to negotiate such large and long-term contracts, even if these are limited to waste collection and sanitary landfills.

Private finance for large waste infrastructure investments is generally linked to long-term contracts, fixed waste amounts and relatively high gate fees. Thus it is inherently more suited to higher-income cities. If the contractual and financial arrangements are not flexible, these facilities may cause problems for the city over the long term if the city is successful in reducing waste generation or in diverting waste to recycling.[117]

Although private financing requires the payment of interest, it often offers great help to cities facing urgent waste management problems. In the Philippines, for example, a privately-built high-temperature incinerator for high-risk hazardous health-care waste is being used by more than 200 medical centres and hospitals.[118]

Private operators and their investors are usually looking to reduce risks, and since for treatment/disposal facilities they are paid by the tonne, they are looking for guaranteed waste input amounts to secure revenues. As full cost recovery for treatment and disposal facilities is rarely possible from sale of outputs such as recyclables, compost, refuse-derived fuel (RDF) or energy, the municipality and those bringing their waste to the facility pay a fee to the private investor at the gate.

The gate fees at these privately-financed facilities tend to be high as the technologies are generally relatively expensive. They are negotiated and agreed between the contractor, the private finance provider and the municipality, but in such cases the national government in fact often bears the responsibility of last resort should the city default on payment. Mostly such projects are implemented in high- or upper-middle income countries, regularly serving urban areas where both the density of population and the calorific value of the waste is relatively high. Finding examples of energy-from-waste (EfW) facilities that have successfully raised private finance for investment is not so easy, particularly in developing countries. The feedback from banks is typically that they consider EfW to be an infrastructure project with a time horizon that is too long. At the same time, it is not the most obvious choice for public-private partnership financing.[119] *Box 5.9* shows an example of one of the earlier privately-financed EfW facilities in the PRC, which relied on national government guarantees. *Box 5.10* shows a contrasting case in Surat, India, which is the exception to the rule of high gate fees; here the private contractor provided a service free of charge to the municipality. However, this was likely at the expense of unaccounted social and/or environmental costs, and the facility closed after five years due to financial losses.

115 See Section 5.5 on B2B financing models for waste management.
116 A special purpose vehicle is a joint venture company set up for the purposes of implementing a specific project.
117 An example of the potential problems is provided by the case study of Stuttgart in Box 4.4 in Section 4.2.2.
118 Modak (2011). Chapter 3 in UNEP's *Towards a Green Economy: Pathway to sustainable Development and Poverty Eradication*, listed in Annex A, Chapter 1, Precursors of the GWMO.
119 ISWA Energy Recovery Working Group, personal communication.

BOX 5.11 PUBLIC-PRIVATE PARTNERSHIP FOR FINANCING AND OPERATING AN EARLY ENERGY FROM WASTE PLANT IN THE PEOPLE'S REPUBLIC OF CHINA

In the PRC there has been a policy push to increase energy from waste (EfW), and by 2014, the PRC had 150 EfW plants.[120] The electricity produced in these plants is bought at a subsidized price by the government, making the investment more attractive. Until recently the environmental norms on various emissions and pollutants relevant for the sector were quite loose, adding to the economic viability of the earlier plants at the time of their development. The EfW plants are also exempt from corporate income tax in their first five years of operation.[121]

This favourable policy context made it possible for private companies to invest in EfW facilities. For example in the city of Wenzhou, a local private contractor entered into a contract in 2003 with the municipality to finance, build, operate and transfer a waste to energy incineration plant. Recovery of the investment was expected to take 12 years, while the contract duration is 25 years, after which time the company would transfer the facility to the municipality. The total investment was 90 million CNY[122] and revenues come from the sale of electricity and a gate fee of 73.8 CNY per tonne[123] paid by the municipality and ultimately by the central government.[124] These tariffs were comparable to other EfW plants in the region that were developed through private investment about the same time.

There is an ongoing debate about the environmental quality standards of the EfW plants. NGOs and civil society participating in public consultations blame the early EfW plants for air and water pollution impacts.[125] It is understood that air emission control requirements for new EfW plants are comparable to international standards.

BOX 5.12 – PUBLIC-PRIVATE PARTNERSHIP FOR FINANCING AND OPERATING A MATERIAL RECYCLING FACILITY IN SURAT, INDIA[126]

In Surat, India, the municipality, which was very keen on improving treatment, issued a tender using the public-private partnership model for solutions that would divert waste away from landfilling at minimal cost. The incentive used by the municipality was to provide infrastructure and land to the successful bidder for a token price. An Indian company from Mumbai won the bid by proposing a solution at no cost to the municipality. The required input amount of waste for the facility was 400 tonnes per day, which represented about 25% of the city's waste stream.

The MBT facility, fully operational since 2009, was extracting and selling metals and producing a compost-like output and refuse-derived fuel (RDF). At the time of the research carried out in this city in 2011, the investor was planning an expansion of the business. However, more recent information suggests that the plant closed in December 2013 due to financial losses incurred over five years of operation.[127] The same source states that a similar plant operated by the same company in another city closed at the same time after eight years of operation, while another three plants in other Indian cities have recently suffered catastrophic fires.[128]

The advantages of this investment from the viewpoint of the municipality were clear: it was private money engaging in resource recovery at no cost to them. At the same time, there were some issues to consider: the contract term was 30 years and locked the waste streams to an MBT facility processing mixed wastes, while better solutions may become feasible before the contract term expires. And, although the service was provided free of charge to the municipality, there may be some externalities that were not being adequately paid for upstream or downstream. For example the facility buying the RDF may not have had adequate environmental standards for using RDF and therefore was able to offer a better price for this fuel, or the compost was being prepared from mixed waste and therefore would contain some contaminants, or the informal sector that had formerly recycled this waste stream may have been displaced, since the facility was capturing a significant part of the waste stream and indeed relied for its income at least partly on separating materials for recycling.

120 See Box 3.7 on alternative technologies for resource recovery from waste.

121 Yang Z., H. Zhang, D. Van Den Bulcke, China Consulting Projects (CCP) (2009). Green means clean: Investing in China's Municipal Solid Waste Industry. http://www.antwerpmanagementschool.be/media/355505/ecc_green_means_clean.pdf

122 About 11 million USD, calculated at the average exchange rate for 2003. Note that this investment cost is very low compared to the 'norms' shown in Table 5.1.

123 About 12 USD (as of August 2015).

124 Nie, Y. (2008). Development and prospects of municipal solid waste (MSW) incineration in China. *Frontiers of Environmental Science and Engineering in China,* Volume 2, Issue 1, pp. 1-7, March 2008.

125 Yoshida, A. (2012). Japan Consulting Institute (JCI), CDM Center, Validation Report for Linjiang Erqi MSW Incineration for Power Project. http://cdm.unfccc.int/filestorage/3/t/H7MNXZF0IPOQB4DALJU19KTVWGC568.pdf/Validation%20Report.pdf?t=ODB8bjh0bmNnfDCEyuto4Q54PORlxndDYsfi

126 Gupta (2013), listed in Annex A, Chapter 5, Service delivery and private sector participation

127 https://grasshopperfiles.files.wordpress.com/2014/08/report-of-swm-in-gujarat.pdf

128 http://timesofindia.indiatimes.com/city/nagpur/Are-fires-at-Hanjer-plants-smokescreens/articleshow/45075666.cms

Potential private investors also include companies using secondary products and/or energy from waste, who thus have a business interest in investing in waste management. One example, given in *Box 5.1*, is an Indian conglomerate that invested in waste collection in order to secure reliable supplies of recovered paper as a raw material for its business. Another example is in Cape Town, South Africa, where a composting business was launched more than 10 years ago, driven by a requirement from farmers wanting to sell table grapes fertilized by compost not using any chemical additives. This requirement in the market created a demand for compost from source-separated green wastes, and the biggest composter in Cape Town in 2015 is a farmer who uses the compost to fertilize his own vineyards.[129] Quite similar is the situation of certain cement plants in the Middle East that have become cut off from their regular fuel supply by the spread of the Islamic State and are now looking to invest in RDF generation in the region. Co-management can be beneficial for other waste streams as well, such as co-digestion of the organic fraction of the municipal solid waste stream with sewage sludge or agricultural waste.

Although private sector finance is suitable for more than just large-scale infrastructure, small and micro-scale enterprises often have difficulty accessing the finance they need for working capital or equipment purchases. Experience shows that supporting smaller scale enterprises, cooperatives or individuals working as door-to-door waste collectors, waste pickers or itinerant waste buyers, or those operating recycling or composting facilities, can be beneficial to both the local economy and the service users. Micro-financing or smaller-scale financing made available to the home-grown businesses in the service and recycling chains have strengthened these chains and allowed them to step up and offer better services and increase and diversify their businesses.[130] The public authorities can play an important role here by designing policy that protects and facilitates small business development.[131]

© Oscar Espinoza

Selective collection in Peru with ex-informal recycler

129 This farmer relies on the composting of 2 million m³ of green waste per annum. Personal communication, May 2015.
130 See Box 5.3 in Section 5.3 for an example of a successful recycling linkages project.
131 See Section 4.8.1 on the possible roles of governmental institutions.

BOX 5.13 LIVELIHOOD FINANCE: SEWA'S APPROACH TO LIFT INFORMAL SECTOR WORKERS OUT OF POVERTY[132]

SEWA (Self Employed Women's Association), registered in 1972,[133] is a central trade union organizing poor, self-employed women workers in the informal economy. SEWA started in Ahmedabad, India, with a vision of full employment and self-reliance, mainly with women workers from textile mills and associated activities, and soon increased its membership to include more diverse occupations within its fold. As of 2015, SEWA has organized 1.8 million women workers[134] in 14 states of India including around 74,000 waste recyclers and cleaners. Members are organized into cooperatives, rural producers' groups, social security organizations, savings and credit groups or federations.[135]

SEWA Gujarat has about 1.0 million members of which about 60,000 are women waste recyclers and cleaners organized in Ahmedabad. These waste recyclers and cleaners have been organized by SEWA to undertake activities like door-to-door collection, street sweeping and recycling paper to make stationery, in order to link traditional waste recyclers inclusively with the solid waste management system by closely working with local authorities and other government agencies as required. Since 1995, three waste and recycling cooperatives have been operating, currently with 1,400 members.[136] As an example, the Gitanjali Cooperative with 400 members originally engaged in door-to-door waste collection and segregation. This resulted in sustainable income for its members from 3,000 to 3,500 INR[137] per month for five hours of work daily.[138] However, due to privatization and mechanization, these workers were again displaced. The immediate requirement for these waste recyclers was livelihood. In 2013, the Gitanjali Cooperative turned towards stationery making with support from SEWA. Among the products of the cooperative are a wide range of paper products like designer covers, notepads, pens made of paper, spring files etc. made from waste or recycled materials according to market demand and catering to international brands like Staples, CISCO, Giftlink, Mindtree and many more. Notepads and spring files made by members are also supplied to companies in the US and the UK.

The initial capital was raised by contribution of 100 INR[139] per member and small donor grants while SEWA provided essential technical and administrative support. The monthly profits are shared among the members in proportion to their time and labour contributions. Besides this, each member of the Gitanjali Cooperative is a member of SEWA bank. SEWA bank lays great emphasis on savings and provides a wide range of loans[140] according to its members' needs. However, the basic prerequisite to apply for and receive a loan is to have at least a one-year record of consistent savings with the bank. This would act as security in the absence of any other traditional collateral available from the applicant. The members get a consistent livelihood and income, a family-like environment, safe and sanitary work conditions, financial and social security through the facility of loans, accidental insurance benefits and dignity in social circles and at work. This not only provides the women with income and livelihood but also gives them skills, self-respect and status in society. This improves their standard of living and they can send their children to school.

The two major approaches adopted by SEWA are livelihood finance and partnerships. In the area of livelihood finance, SEWA strongly believes that in most situations, microcredit as a one-time intervention is not enough to pull a poor household out of poverty. Hence in all its projects, SEWA incorporates a complete livelihood finance package comprising the following constituents: a) organizing and institution building, intended to build collective strength and bargaining power; b) credit and other financial services (such as insurance and pensions) that lead to asset building; and c) productivity improvement services that include capacity building[141] and fostering market linkages, marketing services and human development services, specifically in health, nutrition and education. In order to enhance the economic self-reliance of its members' cooperatives, producers' groups, and other organizations, SEWA fosters partnerships with the private sector.

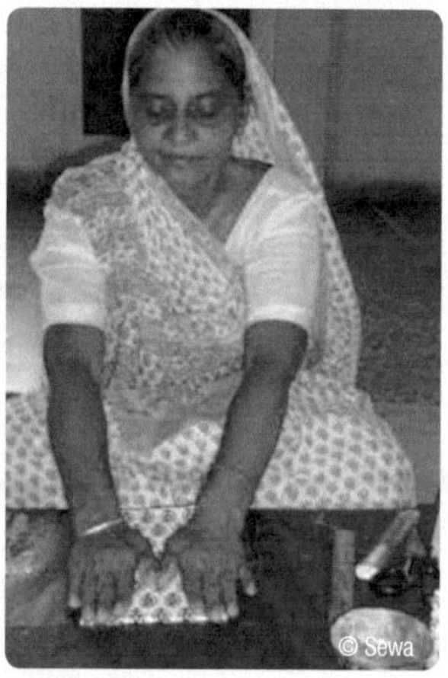

One of the members, Parvatiben Solanki, says, "For more than 15 years, I could not afford to have a door in my hut. But I started working with the Gitanjali Cooperative making stationery, so I have a regular income now, which enabled me to take a loan and get a door for my home. That was the first night I could sleep. I felt so happy."

---

132 Prepared based on inputs from Manali Shah, SEWA Ahmedabad (2012). Growth from the Bottom Up. Presented by Reema Nanavaty, SEWA in Workshop – Growth and Inclusion: Theoretical and Applied Perspectives, January 2012, and http://www.sewa.org/

133 The Labour Department refused to register SEWA due to the absence of a recognized employer, saying that the workers were self-employed and thus had no one to organize themselves against. SEWA argued that a trade union was not necessarily an organization working against an employer, but rather an organization working for the unity of the workers. SEWA finally became registered in April 1972.

134 This includes women workers with occupations such as hawkers; vendors; home-based workers like weavers; potters; food production workers; ready-made garment workers; and manual labourers and service providers such as agricultural labourers, construction workers, contract labourers, handcart pullers, head loaders, domestic workers and laundry workers, waste pickers and cleaners.

135 Each woman can be a member of one or more groups.

136 SEWA Gujarat supports 84 cooperatives with 11,610 members. These include dairy cooperatives, artisan cooperatives, service and labour cooperatives, land based cooperatives and trading and vending cooperatives.

137 Approximately 47 to 54 USD (as of August 2015).

138 Otherwise the average income of waste recyclers is between 25-50 INR (0.40 to 0.80 USD) per day, on which they barely survive.

139 Approximately 1.60 USD (as of August 2015).

140 For more information see http://www.sewa.org/images/pdf/new/product.pdf

141 SEWA undertakes capacity building exercises both on its own and through linkages with other expert institutes to make its members self-reliant and aware about legal issues. Training programmes cover a wide gamut of activities ranging from the sorting of waste, that is, the collection of waste, which will fetch a higher price upon disposal, to the making of paper plates and bowls from waste, to measures for recycling waste, to ensuring alternative employment opportunities for the workers, such as stitching ready-made garments and computer operations.

5.9.1 The case for taking action now on waste management

Accounting for direct waste management costs is itself a challenge, but estimating or accounting for the economic costs and the externalities that result from unsound waste management is even more complex. However, the evidence and data collected from qualitative and quantitative evaluations of both the positive and negative impacts clearly indicate that:

- Lack of waste management or inadequate waste management harms human health and damages the environment and ecosystems both locally and globally;
- The health and environmental impacts are most severe in densely populated urban areas and slums populated by the urban poor, where urban infrastructure and services are often non-existent or inadequate;
- The cost estimates that have been made suggest strongly that the economic costs to society of inadequate waste management are much greater than the financial costs of environmentally sound waste management;
- There are in addition significant benefits to society and the economy from sound waste and resource management including among others resource security, job creation and sustainable livelihoods; and
- The economic case to prioritize waste management now is already strong and does not need to wait for more evidence on the monetary costs and benefits.

In summary, bringing waste under control today will cost much less than cleaning it up tomorrow.

As the availability of financing and refinancing for waste and resources management infrastructure is limited, it is important to choose those financing models that are most likely to secure revenue and investment capital while delivering reliable services.

5.9.2 When is B2B appropriate?

There is no real alternative to the business-to-business (B2B) financing model for larger businesses and industry. If enforcement of environmental legislation is strong and it can be reasonably expected that the private players will pay for and implement high-standard solutions, this option is able to generate the necessary private financing. Enforcement takes a lot of effort and resources, but it also encourages innovation in the industry that will be seeking cleaner technologies, greater resource efficiency and further innovation in waste treatment and management technologies. For smaller businesses, opting into the MSW system may be an appropriate solution. This may also result in cross-subsidizing the citizens, benefiting both categories of users and taking better advantage of the benefits of economies of scale.

5.9.3 Know where you stand in terms of MSWM finances

The MSWM sector faces a chronic financing gap, ever-greater amounts of waste and the need to meet increasingly stringent environmental and public health standards. In terms of the financing model given in *Figure 5.2*, a financing gap is found in both revenues and investment finance. This section focuses on the gap between revenues and the total costs of operating the waste management service.

Managing well with limited resources requires a thorough understanding of the costs that need to be paid and the revenues available. The GWMO attempts to collect and collate up-to-date cost and revenue information for waste management around the world, and it has confirmed a series of important gaps in data and knowledge on waste management costs and revenues. One reason for this may be that cities often function as cost centres within regional or national budgets, and do not depend on locally collected revenues; in such cases, there may be little incentive for management to understand the true costs of the various activities that are part of the municipal budget. This may also contribute to the common situation that the costs of providing a public service are considered 'confidential'. The result is that relatively little information on costs is available. What is available is not comparable across cities due to different accounting practices and different ways of aggregating the data, and specific activity-based costs in waste management are still poorly understood. Greater transparency, better benchmarking and more standardized information would facilitate tighter financial management with better control of technical inefficiencies and fraud.

Understanding the drivers behind individual waste and resource management activities, including waste collection, recycling, resource recovery, energy recovery (EfW) and disposal, would facilitate improved financial management at the municipal level. While there is a clear demand for some activities, such as primary collection or extracting the resource value of secondary materials, other activities such as safe disposal are driven by policy requirements. Private companies will finance activities only when investments are recoverable and ideally deliver a healthy return, while authorities' main concerns will remain ensuring reliable service, protecting public health and the environment, and maintaining affordable fees. Understanding the financial and economic costs of future activities is equally important.

Recommended actions to be performed in a financial and economic analysis prior to deciding on improvements through investments:

- Know the full costs of services and service options before undertaking the investment, factoring in downtime of plants and periodic revamping costs
- Establish the annual revenue requirement
- Define an indicative affordable revenue stream from users
- Establish a cost recovery policy – the recurrent cost components to be recovered from users
- Analyse the twin opposing constraints of affordability and financial sustainability
- Assess possible requirements for receiving grant funding
- Recognize the need for a balance between grant funding, soft loans and commercial (private sector) finance

Also, given the impure public good nature of waste management and the blurred property rights of waste, it has been noted that in the case of dry materials the public service of waste collection delivered in exchange for a tariff or fee cannot compete with the services offered by itinerant informal buyers who are willing to purchase those materials from the waste generators. However, experience shows that source separation and systems that integrate the practice of buying or accepting recyclables free of charge while collecting mixed waste for a fee may be the solution. It is increasingly important to understand and manage the interface between the public service and the value chain, including the informal sector, to increase the cost efficiency of both activities and arrive at win-win solutions.

Given that these complexities cannot be avoided, they need to be managed through a delineation of rights and responsibilities, clarifying the drivers and responsibilities for each of the stakeholders. Because of all the different economic interests, robust integrated systems can only be built through participative decision-making. There are examples of good practices for stakeholders to learn from but they need to be adapted to local situations and adopted by the people and organizations involved in waste management.

5.9.4 Private sector participation in MSWM

It is commonly thought that the private sector can help with capital financing, cutting the costs of services, managing and recovering value from waste streams and even moving towards prevention through integrated services. Yet public-private participation or 'privatization' is not necessarily a universal answer in all circumstances, and even when it is appropriate, it needs to be carefully managed. Local authorities have no easy task in selecting the appropriate financing and operator models to ensure the effective delivery of all the various services that make up an MSWM system.

A study of examples of successful models across a variety of model types in different geographical regions shows that local needs, priorities and beliefs are very important in selecting a model. Selecting an appropriate model is an iterative process that starts with understanding characteristics of the local situation, problems and needs, setting goals and priorities, and defining strategies for addressing those. Some of the issues identified in the local situation may not be directly linked to waste management, but if they are a priority for the city and can be addressed through the sector, they are equally important when selecting the model. Addressing social problems of poverty and unemployment is one notable example.

These goals may be translated into more concrete objectives, which then become selection criteria. After that, a shortlist of several options that may work well for the city can be created through a detailed analysis of the four components of a financing model – the client, the operator (the entity delivering the service), revenue collection and investment finance (cf. *Figure 5.2*) – and of the local technical, economic, policy, social and cultural conditions. So for example before deciding to attract private financing, it is important to assess the market, see if there are operators with relevant investment capacity, understand the conditions under which they are interested in participation and determine if those are attractive to the city. For revenue collection, it is important to assess if the municipality has the capacity to collect user charges and enforce payment. It may be that outsourcing this function to an operator, a utility or another third party would be a better fit.

An important observation is that most cities in low- and middle-income countries operate more than one 'operator model' in parallel. Thus a variety of approaches may be appropriate for different components of the overall MSWM system, such as street sweeping; extending primary collection services to unserved, low-income neighbourhoods; recycling; providing secondary collection and transport to a perhaps distant treatment or disposal facility; resource recovery; and final disposal at an environmentally sound landfill site. Even within the same city, it may be the case that there is no 'one size fits all' approach.

Policies may strongly favour one model or another. For example a public model may be exempt from value added tax, or public financing may be available for regionalization. These can constitute decisive factors in choosing a particular model, all other things being equal. It is easier to navigate among the different model types for each component of the overall system by deciding first if it is going to be a public or private model, and if the objectives include achieving economies of scale through regionalization. This limits the options and therefore simplifies the analysis.

5.9.5 Selecting a financing model

The recent GIZ study on 'Operator Models' for the delivery of MSWM services has resulted in a guidance paper that provides the framework for this analysis and maps out the most common operator models for each physical component of the SWM system.[142] The guidance paper takes the reader through the steps of assessment and decision-making. The assessment of problems and framework conditions gives an analytical framework with a comprehensive long list of possible problems and issues around the three physical elements and three governance strategies in integrated sustainable waste management (ISWM; see *Figure 1.2*). These framework conditions define the decision-making space of the public authorities for selecting a model.

After establishing the local objectives, the decision-makers are provided with another tool in the form of a checklist to identify conditions and capacities that are suitable for different operators, contract types and financing types, together with the benefits and drawbacks of the options. Last but not least, 42 common operator models, each linked to one or more components of the overall MSWM system, are identified and described to help the user navigate among the options. While the guidance paper is a useful document for making informed choices, the human factor in decision making should not be underestimated. The embedded beliefs about the efficiency and benefits of one model over another will play an important role in the choice.

The steps for selecting an Operator Model have been adapted in *Figure 5.6* to reflect the slightly different steps involved in selecting a Financing Model, where investment financing is part of the selection process as a key consideration, alongside the client, operator and revenue collection functions and their interaction.

142 Soos et al. (2013b).

Figure 5.6 Steps in selecting an appropriate financing model for MSWM[143]

Step 1 Establish the framework conditions

- Examine the current political, legal, institutional, economic and cultural structures and constraints
- Investigate potential for economies of scale, the regional/municipal nexus and institutional constraints
- Clarify the sources and availability of investment funding options
- Define the role and scope of the informal sector

Step 2 Define the objectives for improvements in the waste management system

- Define the improvements needed in order of priority within the service chain
- Define the scope and desirability for introducing materials recovery systems

Step 3 Assess Conditions and Capacities

- Examine the capacities and experience of the public authority, operator and revenue collector with different operator models
- Examine institutional, economic & policy conditions that influence choice of models

Step 4 Select the model
Assess advantages and drawbacks

- Examine the potential role of, and constraints on, contracting with the private sector
- Examine the inherent advantages and disadvantages of possible financing models
- Establish revenue requirements, tariffing policy and revenue collection methods

IF the model is favourable: Implement

IF the model is NOT favourable: Go back to step 3

While it is clear that certain models are completely unsuited to certain conditions, it is also clear that there is no one perfect choice for a given situation. More than one model can work well if implemented with care and equally any model may fail if not managed properly. The analysis is important to create awareness of the strengths, weaknesses, costs and benefits of each model type so that these can be given proper attention during implementation.

143 The summary of the assessment and decision-making that needs to be taken in each of the steps has been adapted from those in the GIZ guidance paper on selecting an operator model. See Soos et al. (2013b).

5.9.6 What is an appropriate level of cost recovery in MSWM?

Where direct charges are made to householders for the MSWM service, a decision needs to be made as to what proportion of the total cost is to be recovered in this way. This question has already been explored in Section 5.7.4, and that discussion has not been included again here. However, it is worth repeating that full cost recovery is unlikely to be achievable in the short term in most low- and lower-middle income countries.

5.9.7 Selecting the appropriate sources of investment finance

The main sources of investment finance for MSWM improvements are the municipalities' own budgets, national budgets, IFIs and other donors, private investors or a combination of these. The public financing and grants from donors generally do not require reimbursement and are directed towards certain types of projects and conditionalities. Private financing comes at a cost and investors are looking to recover their investment. IFIs may offer soft loans that come at lower costs and have better repayment terms compared to commercial loans, but, IFIs also impose stringent conditions on their lending, which appears to have limited the availability of funds for SWM projects and also restricted funding to those emerging economies that have already developed the governance structures that allow them to meet the conditions.

There are benefits and drawbacks to each of these options. Non-reimbursable or low-cost investment finance is much needed in low-income countries and would allow a broader range of options and better control over service quality, but if the provision of such finance is predicated on full cost recovery, it may in effect be unaffordable to society. Private financing is costly but it may be accompanied by cost efficiency in operation, as private investors are very keen on recovering costs. Also, in some countries accessing private funds may be faster than going through the application procedures for obtaining public financing.

Experience shows that strong local authorities who have the capacity to anticipate the strengths and weaknesses of different models and incorporate risk reduction elements in their management will be better at attracting and using funds. This would mean for example that a strong municipality would try to negotiate contract terms with donors to obtain conditions that are suitable for the local situation. Similarly if private financing is sought, the municipality would anticipate the profit-seeking behaviour of private businesses and build in conditions in the cooperation agreement that would allow flexibility in fulfilling future environmental obligations and in phasing in increases in tariffs.

Choosing the source of financing is part of a more complex puzzle and there is no one best solution. The right choice will depend on local objectives, economic and policy conditions and capacities. The pressing problem seems to be not so much the selection of the source of financing but the limited options available, as municipal financing is increasingly limited in the context of increasing amounts of waste to manage; the availability of donor financing for SWM is very low compared to other sectors and has been restricted largely to the more advanced emerging economies; and the economic downturn has curbed the availability of private financing.

On the positive side, in recent years there has been more and more work done to compile data in this area reflecting the great variety of local conditions, including examples of both successful and unsuccessful financing choices. This can be expected to help to improve decision-making by local authorities, who often lack capacity in this area.

SYSAV – ENERGY FROM WASTE (EFW) FACILITY INTEGRATING RECYCLING AND REUSE WITH GREAT SUCCESS IN MALMÖ, SWEDEN[1]

CASE STUDY 6

Context and background

Historically, Sweden has shown strong commitment to environmental protection initiatives and policies, particularly in the area of waste. Over the last one-and-a-half decades, Sweden has seen some major legislative interventions which have led to changing patterns in waste management options. A landfill tax was introduced in 2000[2] to discourage waste from being disposed in landfills. This was followed by a ban on landfilling combustible waste in 2002 which was amended in 2005 to include organic waste. With these interventions, waste going to landfills decreased dramatically and the preferred waste management options in Sweden shifted to incineration and material recycling. Municipal solid waste quantities have been almost equally diverted between these two waste treatment paths with slightly more waste going to incineration.

As can be seen in *Table 1*, the amount of waste going to landfill has drastically decreased over the past decade or so. It also shows how there has been an increasing proportion of recycling as well as energy from waste (EfW). A study on EfW production shows that Sweden continues to have the highest rate of energy recovery from waste incineration in Europe. There are 32 plants for incineration of household waste in Sweden. Imported waste used for energy recovery has increased in recent years and is now close to 831,400 tonnes annually. This case study focuses on one such plant which integrates the (older) approach of EfW with a more recent focus on recycling and reuse and ensures that EfW goes hand-in-hand with prevention, reuse and recycling.

Table 1 Change in breakdown of how municipal solid waste is managed in Sweden[3]

WASTE MANAGEMENT OPTION		2001	2013
		%	%
EfW		41	50.3
Recycling		37	49
	Biological treatment	9	16
	Material recycling	28	33
Landfill[4]		22	0.7
Energy generation (GWH)			
	Heat	7.8	13.8
	Electricity	0.3	1.8

1 Inputs from Anders Persson, CEO, Sysav Development LtD and Gunilla Carlsson, Communications Manager, Sysav.

2 51 EUR per tonne in 2015.

3 Avfall Sverige (Swedish Waste Management).

4 Municipal waste management in Sweden Prepared by Leonidas Milios, ETC/SCP February 2013. EEA project manager: AlmutReichel. The Topic Centre has prepared this working paper for the European Environment Agency (EEA) under its 2012 work programme as a contribution to the EEA's work on waste implementation.

About the project

Sysav is a private-public company[5] owned by 14 municipalities in the south of Sweden serving a population of about 700,000. In 2013 Sysav dealt with some 905,900 tonnes of waste, from both households (about 43%) and industries (about 57%).[6] Municipalities are responsible for the collection and the transportation of household waste. Municipalities charge about 200 EUR per household annually for waste management (collection, transportation and treatment). Sysav has a long-term contract with municipalities for treating household waste and is responsible for investing in and operating recycling centres for bulky waste and treatment, by charging gate fee to its clients which includes municipalities and other businesses, including industries. The investments in treatment and disposal made by Sysav have been as follows: energy from waste – more than 2.2 billion EUR (total, 2003-2008); pre-treatment plant for food waste 10 million EUR (2010) and 15 recycling centres wherein investment for setting up each recycling centre is around 2 million EUR (1992-2015). Besides these, periodic investments are made in the creation of new landfills, the closing and care of old landfills and establishing treatment plants for hazardous waste. Revenues come from treatment charges,[7] energy revenues, material sales and others.

Sysav's overall goal since its foundation in 1974 has been to maximize the recovery and recycling of material and energy while landfilling as little as possible. This mission is carried out through a combination of treatment methods. At the time of its inception, the era of oil crises, Sysav incinerated half its waste (the incineration plant was connected to the district heating network in the Malmö area to enable the generation of energy from waste) and the other half was landfilled. In the beginning of the 1990s with the introduction of new regulations, the types of waste allowed to be landfilled were substantially reduced. Sysav kept pace with this and landfills less than 2%[8] of incoming waste as of 2015.

The aim is always to close the eco-cycle in the waste stream as far as possible (keeping in mind the needs of detoxification and avoiding down-cycling), from collecting the waste to when the recycled materials and energy are put on the market again, in order to bring waste management into the circular economy.

The eco-cycle model *(Figure 1)* summarizes how different waste streams (which include both industrial and household waste) that enter the company are treated. EFW accounts for approximately 63% of recovery at the moment but is decreasing, and other ways of recovery discussed hereafter such as source separation, recycling or composting account for approximately 35%, leaving less than 2% for landfill. The eco-cycle approach includes the following components:

1. **Source separation** is conducted in order to have clean waste streams for various processes. Households have up to 11 different fractions[9] they collect at their home. Hazardous waste from households and commercial areas is collected in separate collection schemes and constitutes about 4% of the total waste collected.[10] Even though source separation has been practised in Sweden since the 1970s, Sysav has undertaken regular communication activities for more than 20 years to ensure that source separation stays as the main agenda.

2. **Reuse**[11] is promoted by providing containers for reusable furniture at all recycling centres which are taken away and reused/treated mostly by charitable organizations. Sysav transports reusable building and construction materials to a shop selling such materials that is jointly run by the Malmö municipality and Sysav. Since this material is not handled at the Sysav plant, it is not considered in the mass balance for Sysav.

3. **Material recycling** includes recycling of all suitable materials, for instance incineration bottom ash, plastic, wood, textiles, soil, paper, cardboard, bulky waste and materials from the disassembly of waste electrical and electronic equipment (WEEE). All this together constitutes about 25% of the total waste received at the plants. The recycled material is sold to various companies.

4. **Energy from waste** for creating heat and electricity uses only residual household and industrial waste after source separation and also some source-separated materials which do not meet buyers' quality standards. Besides this, imported fuel waste is used to match the additional waste requirements; Sysav uses imported burnable waste from the UK and Norway. Thus about 63% of total waste received at Sysav goes to EfW. The produced heat is delivered to the nearby district heating company 'E.ON', which provides Malmö and Burlöv with district heating. Sysav's heat is the baseload on the net and provides 60% of the district heating demand. The electricity is sold on the open spot-market.[12]

5 Sysav was formed as a private company in 1974 by the owner municipalities for them to be able to fulfil the new laws on waste management. The municipalities are responsible for the waste from households and they collect the waste either on their own or through contractors. Sysav is responsible for the treatment and landfilling of the waste. The municipalities pay a gate fee when they deliver the waste to the company. Otherwise there is no financial exchange. The benefit for the municipalities is that the company can keep the gate fees as low as possible.

6 Industrial waste is received from an open market and a part is imported.

7 Sysav charges local authority a gate fee of about 60 euros per tonne of waste. Private businesses and industries are charged according to waste types.

8 Residuals such as asbestos and ashes from both household and industrial wastes.

9 The containers shown in Figure 2 are used for 8 fractions. Three more fractions include garden waste in a separate bin (brown) and batteries and light bulbs. Batteries and lightbulbs use a special arrangement that can be put on bin with four fractions.

10 The EfW plant and landfill have a permit and the capacity for some types of hazardous wastes and this waste goes for either of the two options.

11 This component is directly dealt with by other agencies and does not undergo treatment. Hence it is not included in the mass balance of the eco-cycle.

12 This also includes heat and energy generated at the landfill (see point 6 below).

Figure 1 The Eco-cycle approach

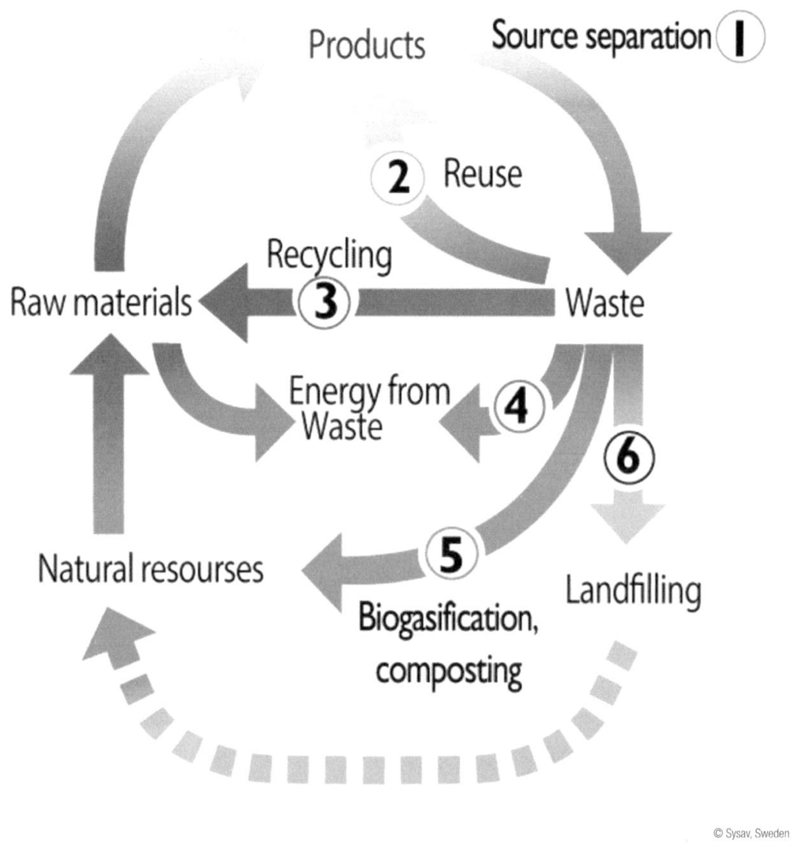

5. **A. Garden waste** (about 6%) is composted and prepared as garden soil.

B. Food waste (about 4%) is digested anaerobically for producing biogas (methane), which is liquefied and used as an automotive fuel to replace gasoline, which is sold to transport companies and has a much higher added-value market than is usual for methane from anaerobic digestion. A bio fertilizer is also produced from this food waste, closing the loop between the city (consuming food) and the countryside (producing food) and is delivered to farmers.

6. Only **residues** such as asbestos and fly ash are sent to landfill (approximately 2%). Landfill gas (methane gas from current and old landfill) is collected and used for heating purposes and to produce electricity. Leachate is treated in multiple stages to meet stringent discharge standards.

Enabling environment

The company structure, a private company with public owners, has made it possible to build capital-intensive infrastructure, as it can be used for long-term contracts with municipalities that ensure return on financial investments and effective waste handling.

Benefits

The most important benefits from the model implemented are related to the climate. Source-separated waste streams make it possible to get high-quality products that can be used again by the market and these separated waste streams also ensure good raw materials. Sysav provides for 60% of the heating demand in the Malmö and Burlöv area, enabling the owner of the net (E.ON) to avoid using fossil fuels for heating purposes. Sysav also produces 25,000 tonnes of bio fertilizer annually, as well as biogas corresponding to 2 million litres of gasoline, 10,000 tonnes of compost, various metals (including precious metals) and so on.

By using different methods that are environmentally sound (food waste to biogas fuel and efficient EfW, producing heat and electricity), Sysav reduces greenhouse gas emissions by at least 200,000 tonnes of CO_2 equivalent every year.[13]

The profit Sysav earns is used for necessary investments and to keep the treatment fees as low as possible for the municipalities, a win-win solution for municipalities, the inhabitants and the environment.

13 Calculations can be done in different ways. An overview for Swedish waste handling is available at http://www.wasterefinery.se/sv/project/projects/perspektivpaframtidaavfallsbehandling/Documents/PFA_eng_130703.pdf

Key challenges

The challenge for many years has been to make inhabitants and industry aware of the importance of sorting waste at source, ahead of collection. The municipalities are the drivers in this ongoing long-term communication awareness project. Another aspect of the challenge is to get people to change their behaviour and make them aware that when they buy things, they cause damage to the environment, including the climate. It is best to buy only what you need and to buy high-quality, environmentally-sound products. Sysav have started an awareness campaign on waste reduction that takes up this theme. The campaign 'Don't Buy Junk' won second prize in 2014 in Rome at the European Waste Reduction Week's annual contest.

Source separation by households into eight fractions

A recent challenge is adjusting to the new EU Directives, which increase the targets for material recycling and decrease the amount of waste that will go toward energy recovery. Sysav's own initiatives in campaigning for people to reduce waste may also compound this challenge by further decreasing the waste quantities available for EfW. However, the company estimates that for the coming 10 to 15 years, the quantity of waste is not expected to be a major problem for this plant, because the large quantities of waste that are still disposed of in landfills in Europe can be imported and treated by Sysav as a medium-term solution. However, such imports also pose competition for those in exporting countries trying to obtain financing for EfW plants of their own.

Lessons learned and the way forward

Sysav is one of many publicly-owned inter-municipal companies, providing municipal solid waste management services to a regional grouping of municipalities. This has resulted in economies of scale that help provide modern solid waste management facilities that meet high environmental standards and operate in a cost-effective manner. However, the term 'cost-effective' is relative: waste management charges to households are around 200 EUR per annum.

The initial solution decided upon in the 1970s was based around EfW, achieving high energy efficiency through supplying baseload heat to the city's district heating network. Achieving high recycling rates through separation at source of both dry recyclables and of organic wastes for composting, and the generation of biogas for use as a transport fuel, came later.

The more recent developments of waste reduction, reuse and recycling means that the EfW plant is now oversized, and Sysav is actively seeking to import refuse-derived fuel and/or residual municipal solid waste from other European countries so that the plant can continue to work at capacity for its remaining useful life of 10-15 years and can continue to meet its contractual obligations to provide heat for district heating. Such import could be regarded as controversial, in that it arguably makes it more difficult for other countries to obtain financing for their own EfW facilities. On the other hand, there are still huge amounts of treatable waste going to landfill in the EU.

INTERNATIONAL AID FLOWS FACILITATING AND SUPPORTING TAILORED LOCAL SOLUTIONS – BO CITY WASTE MANAGEMENT PROGRAMME

CASE STUDY 7

Context and background

Bo, the second largest city in Sierra Leone (after Freetown) with a population of approximately 167,000,[1] is the headquarters of Bo District and the provincial capital of the Southern Region. It is one of the leading financial, educational and commercial centres of Sierra Leone. With a rapidly increasing urban population, mainly as a result of economic growth and the establishment of agri-businesses and mining companies, existing waste management systems set in place by the Bo City Council (BCC) were unable to cope with the over 120 tonnes of waste generated daily within the city limits in spite of approximately 30% of its budget going toward waste management. Through the support of different donor-funded waste management programmes, the BCC has built up additional capacity in recent years.[2] While donor-funded projects often focus on either asset creation or capacity building, the solid waste management pilot project in the city of Bo[3] highlights a very basic but very important aspect for the effective utilization of international fund flows. The city's programme has strengthened and sustained the involvement of local stakeholder groups, and it has used local solutions designed around local strengths and local priorities together with long-term comprehensive planning to ensure that measures are not disjointed and that facilities do not turn non-functional after the support is withdrawn.

1 As per the 2004 population and housing census.

2 A decade of international development support in Bo City: In 2007, the UK Department for International Development (DFID) supported officers from Warwickshire Council in the UK to provide initial technical support on solid waste management issues to the city of Bo in Sierra Leone. Between 2008 and 2011, the UNDP grant funded approximately 365,000 USD of technical support to improve waste management in the cities of Bo and Makeni. UNDP's involvement has since been followed up by further grant funding and assistance from DFID. DFID established an 8 million USD Water, Sanitation and Health (WASH) challenge fund in Sierra Leone. This has funded 36 projects including two focused on solid waste management, in Bo and Makeni (see Lerpiniere et al. [2014], listed in Annex A, Chapter 5, General reading). Following the pilot project, in 2013 DFID committed a further 5 million USD to continue the project. The new phase of the project spans from February 2014 to January 2018.

3 Besides this pilot project, DFID simultaneously funded a Waste to Wealth programme, which was assisting small recycling businesses in Makeni. See http://wastetowealth.livingearth.org.uk/

The pilot phase

In the pilot phase, the project 'Support to local public and private sectors for solid waste management of Bo City and the environments' was implemented from March 2013 till February 2014.[4] The goal was to develop a strategy and design implementation activities that would guide the improvement of the solid waste management system up to 2020. The pilot project pivoted on three key components, namely:

- Creation of the 'Waste Management Plan 2020' for Bo City.
- Increased levels of waste separation and pilot waste recycling initiatives for improving the value chain for waste-derived products while creating livelihood opportunities.
- Effective and efficient collection and transportation of solid waste through private sector and community-based organization involvement.

In line with the goal of the pilot project, the new phase (February 2014 to January 2018) aims to scale up solid waste management activities by building upon the experiences gained in the pilot project and strengthen the BCC waste management sector, including its institutional setting and financing mechanisms.

4 The project, with an overall budget of 322,580 USD, is one of the projects funded through the DFID-established WASH challenge fund and has been implemented through a consortium led by a German NGO, Welthungerhilfe (WHH), with the support of the BCC, Klin Salone, One World Link and WASTE.

Creation of the 'Waste Management Plan 2020' for Bo City

The waste management plan for the city was prepared keeping strategic planning as a fundamental requirement for conducting waste management as an integral issue. The planning process utilized modern tools and incorporated visions for the short, medium and long term. After a long and rigorous process that involved research and baseline data collection in the city and extensive consultations with various stakeholders and experts based in the Netherlands (INGO WASTE), a waste management plan for Bo City 2014-2020 was prepared and launched in February 2014.

Increased levels of waste separation and development of pilot recycling activities that improve the value chain for waste-derived products and create livelihood opportunities

The project supported local waste recycling ideas that were mobilized through a Waste-To-Wealth Business Ideas Competition in July 2013. Each participant was given 30 minutes to present its ideas to a panel composed of BCC staff members and the local Waste Management Steering Committee, which included members of civil society and representatives of important social and governmental institutions in Bo City. The Competition used a voting system to select the six best-ranked business ideas, which would then be supported and up-scaled through the project. This transparent selection process drew a lot of attention and set a notable example of good governance in Sierra Leone.

The selected waste-to-wealth business ideas included: producing handbags (plastic sachets), making pots (aluminum cans/scrap metal), producing organic fertilizer (compost), making slippers (used tyres), producing energy-saving stoves (scrap metal) and making sieves and graters (scrap steel). To up-scale these ideas, the project helped in the preparation of investment and procurement plans in a participatory manner, provided business training and facilitated the opening of bank accounts for the selected waste-to-wealth entrepreneurs. Assistance was also extended in terms of finance and advisory support in procurement processes, including production machines, tools, improvements to workshops, raw material, protective equipment, signboards etc. All six SMEs are operational with a good market for their products and the entrepreneurs are earning their living through these reuse and recycling activities.

Only the compost-making initiative is struggling slightly due to low demand in the local market, but there are plans to make contracts with bigger agricultural companies.

Figure 1 Business ideas presented at the competition

Basket made out of plastics | Pots et al. made out of aluminium/scrap metal | Slippers made out of used tyres | Energy-saving stoves made out of scrap metal | Plastic weaving | Pave-stones made out of plastics | Necklaces made out of paper | Table Mats made out of Stoppers | Compost | Handbags made out of plastics/sachets

Figure 2 Training and capacity building of youth groups

Effective and efficient collection and transportation for solid waste through private sector and community-based organization involvement

In the area of effective and efficient door-to-door waste collection and transportation, the project through one of its partners Klin Salone[5] did a thorough scrutiny of over 30 youth groups that showed interest in waste collection at the household level and chose the best 10. The selected youth groups were trained in business management, financial management systems, banking procedures, waste management, tricycle operation and management, and waste collection from homes for an agreed fee. To ensure the sustainability of the intervention, a consistent revenue system was created for the youth groups. They have several revenue sources, specifically, collection fees from households willing to join the scheme (see *Figure 3* for the price list of different services); waste materials with market value, such as glass and plastic bottles, slippers, rubber, metals and cans, which are separated by the youth groups and then sold to small- and medium-sized enterprises involved in reuse and recycling activities and to local waste traders; and special assignments, such as private contracts for compound cleaning or cleaning services after large social events.[6]

Each youth group started with six members.[7] They were provided with a start-up kit at the beginning of the process, which included protective gear, a tricycle, wheelbarrows, etc. These youth groups function as small business entities under a community-based organization[8] called 'Klin Bo Services'. Each group services one of 10 operational areas in the city (excluding the central business district, where BCC waste collectors operate) and they manage their daily activities autonomously, including the management of all running and maintenance costs of their operations, such as salaries, fuel and tricycle maintenance. The City Council and WHH monitor their operations on a daily basis to ensure that the profits are managed rationally to ensure a self-sustainable scale-up of their businesses.[9]

Enabling environment

Having a local authority with able and visionary leadership that kept waste management in focus was the key aspect that enabled consultation with a broad spectrum of people, bringing together funds, ideas and local and international expertise. In all of its activities, the project has endeavoured to 'tap into the brains' of other cities' stakeholders and share experiences with them. In addition two successful intercity exchange workshops with BCC's counterparts in other cities and with councils in Sierra Leone were organized in Bo to help participants share and learn from each other's experiences. A Waste Management Steering Committee comprised of major local stakeholders in the waste business from civil society and the City Council was formed to act as an advisory body to ensure locally relevant waste management initiatives, arbitration in the case of conflicts, execution of the action plan for waste management and monitoring and evaluation of the system.

A consortium of partners that included, besides the funding agencies, international NGOs with a strong local presence, community-based organizations and international experts provided a well-grounded, participatory and all-embracing project concept. The

5 'Klin Salone' is a network of youth groups that have provided door-to-door waste collection services for households in Freetown since 2007. This pilot project used their expertise and know-how to replicate their successful system in Bo (Klin Salone was an implementing partner).

6 The project's overall implementation logic takes a market-based approach. Therefore it aims to create markets and value chains for those waste streams that currently do not have market value to increase the rate of separation (separation will then naturally follow, as the project has experienced thus far in its day-to-day work).

7 One focal person, one marketer/accountant, four persons in the Operational Unit including tricycle riders and waste workers. As of 2015, they have scaled up to seven to eight members per group. The size of the groups will be increased as required.

8 There are plans to scale them up to a network of fully-fledged private entrepreneurs that have effective and efficient finance and administration with an economically sustainable business.

9 The youth group called 'Cross Border Association' (CROBDAN) is now operating with a mini-van (instead of tricycle) that was financed with their own profits.

Figure 3 Price list for different services

SIZES AND SCHEDULES

SMALL (65 LITRES)

Collection Frequency	Monthly service fee	Weekly service fee
1x per week	LE 6,500	LE 1,600
3x per week	LE 19,500	LE 4,600
Daily (Mon-Sat)	LE 29,000	LE 6,800

MEDIUM (100 LITRES)

Collection Frequency	Monthly service fee	Weekly service fee
1x per week	LE 12,000	LE 2,800
3x per week	LE 29,000	LE 6,800
Daily (Mon-Sat)	LE 55,000	LE 12,800

LARGE (200 LITRES)

Collection Frequency	Monthly service fee	Weekly service fee
1x per week	LE 22,000	LE 5,200
3x per week	LE 55,000	LE 12,800
Daily (Mon-Sat)	LE 105,000	LE 24,500

OTHER SIZES AND SERVICES ARE NEGOTIABLE!

BIN OPTIONS

Basket

Rubber

KLIN BO CAN PROVIDE YOU WITH HALF-DRUM BINS!

HALF DRUM: 50.000 Le (One-Time Payment) OR 4.500 Le (Monthly)*

Half-drum

Rubber

KLIN BO CAN PROVIDE YOU WITH FULL-DRUM BINS!

FULL DRUM: 100.000 Le (One-Time Payment) OR 9.500 Le (Monthly)*

Full-Drum

* After you rent for 12 months, you owe the container. Or use your own container approved by Klin Bo.

Note: *1 USD= 4350 Le (Sierra Leonean Leone) in April 2015

existence of Waste Management Plan 2020 with its detailed actions helped to direct the activities of the Council.

Separate and well-drafted MoUs for waste-to-wealth entrepreneurs and the youth groups, including general terms and conditions of cooperation, the roles and responsibilities of each partner and project duration, gave a high level of clarity to the participating stakeholders.

Key challenges

Each of the three key components faced challenges during implementation. Key limitations encountered during the development of the 'Waste Management Plan 2020' were the lack of existing data, baselines and other information at both the ministries and the City Council and also the limited human resources available at the City Council. To combat this, as well as to increase the general acceptance of its implications and build the capacities of all stakeholders involved, the project chose a highly participatory approach in the drafting process of the document. Several workshops were held with the most important local stakeholders in solid waste management to ensure that the plan was discussed and debated in detail.

The project formalized all small- and medium-sized enterprises involved in reuse and recycling activities. These processes to formalize informal business entities proved to be challenging. For example, the winners of the business idea contest in Bo needed national ID cards in order to open a bank account but in fact the entrepreneurs behind those medium-sized business entities did not readily have them. In this way, creating the necessary preconditions for scaling up their capacities was time-consuming but very important to ensure the sustainability of the intervention. The entrepreneurs' level of knowledge regarding business development, administrative tasks and financial management was very low and further training is called for, as it continues to be a bottleneck for the overall success of the project's efforts. Long-term access to land for production purposes is another important factor that needs thorough consideration. But the most important challenge is the promotion and marketing of new products. Creating the demand for such products in local markets that have a low level of purchasing power is the key factor in success or failure.

The project also faced several challenges in setting up the door-to-door waste collection services provided by the youth groups. Citizens in Bo are not used to paying for such services. Awareness-raising campaigns, PR and the creation and enforcement of by-laws[10] are

10 Bo City prepared three city by-laws in 2008 which have not been approved at the national level. The project aims to organize a local referendum on solid waste management by-laws to ensure increased legitimacy and political pressure to adopt by-laws at the national level.

important factors to ensure the long-term success of such an intervention. There also needs to be strong political will locally to ensure an enabling environment for setting up waste collection and transportation services that are provided by the private sector. Low cohesion and conflicts among youth groups, coupled with low levels of business development, administrative and financial management skills, resulted in the need for additional training in leadership and team building. Youth groups also had to receive basic mechanical training in order to maintain the waste collection tricycles in a sustainable manner. Through ongoing mentoring, mediating and coaching, these challenges have been largely overcome.

Benefits

- Increased household door-to-door waste collection through the involvement of youth groups in Bo City. The coverage rate increased from 0%[11] to about 13% (1,688 households) of all households in Bo City.
- Increased willingness of the community to pay for improved waste collection services.
- Lowered youth unemployment through the door-to-door waste collection scheme; enhanced income levels for those involved. By the end of the pilot phase (February 2014), 60 jobs had been created through Klin Bo Services. That number then increased to 75 employed youths as of March 2015. The majority of them were previously either unemployed or had no steady income and now earn about 250,000 and 300,000 Sierra Leonean leones per month.[12]
- Enhanced capacities of already-existing entrepreneurs in reuse and recycling value chain activities (in total, 25 sustainable jobs were created in this sector).[13]
- Enhanced capacities of the City Council staff and improved awareness within the local council on waste management and governance within the pilot phase through training opportunities and workshops.
- Increased awareness on environmental pollution related to poor waste management practices and demonstrated activities that help to prevent pollution.
- Progress with infrastructure development according to the Waste Management Plan 2020. (This will be further scaled up and mainstreamed through the next phase of the project.)

Lessons learned and the way forward

The success of the project is related to the consistent and ingenious use of international aid, coupled with comprehensive, locally sustainable solutions guided and spearheaded by committed local leaders. The identification of local needs as well as the provision of support across a wide range of different areas including technical and project support, finance for infrastructure, capacity building and awareness raising have all contributed to the gradual establishment of a system that can be improved and sustained over time.

Two further lessons to be learned from the experience in Bo are worth highlighting:

- Sustainable solutions need to built locally; here the parallel involvement of youth groups in providing collection services and of local entrepreneurs in developing income-generating resource recovery businesses was a key element.
- Bo benefited significantly from sharing experiences with cities with similar waste management issues, through both national and international intercity exchange programmes.

© Klin Bo Services

11 Before the project intervention, there was only a skip collection system operated by the Council. Door-to-door waste collection services for households did not exist.

12 Approximately 54 to 64 USD (as of August 2015)

13 The project is in process of investing in 9 new SMEs which will further add to employment in the reuse and recycling sector.

CASE STUDY 8

WASTE MANAGEMENT POLICY IN FLANDERS – FROM 18% TO OVER 70% SEPARATE COLLECTION IN 20 YEARS[1]

Context and background

Waste management in Belgium is implemented by the regions (Brussels Capital Region, Flanders and Wallonia). It is the responsibility of each region to develop and implement waste policy and report to the central government independently. In the case of Flanders the entity responsible for waste management policy is the Public Waste Agency (OVAM).

Flanders, the northern federated state of Belgium, has 308 municipalities across a land area of 13,522 km^2, and as of 2015 was home to 6.43 million people. The municipal solid waste (MSW) generated in Flanders represents 60% of the MSW total for Belgium. This region has achieved the highest waste diversion rate in Europe, moving from a separate collection rate close to zero in the early 1980s to 18% in 1991 to over 70% in 2013, of which 48% is material recycling and 22% is composting.

This case study aims to showcase briefly the history of waste management and how Flanders has moved forward to achieve the desired results.

Waste management in Flanders

Flanders developed its first Waste Decree in 1981. This decree established the basis for developing regional waste plans every four or five years and set clear goals and targets to drive the desired approach. The focus of these plans transitioned over the years from the improvement of solid waste disposal to increasing source separation and recycling (in accordance with the European waste management hierarchy) and emphasizing waste prevention. The most recent regional waste management plan continues to emphasize prevention as the highest priority, sets up actions for sustainable consumption and production, and tries to consolidate the good results achieved through separation at source, composting and recycling.

Several factors have contributed to the successful local implementation of regional plans by Flemish municipalities, as explained below:

Bottom-up approach from local to regional level

In Flanders, while the regional authorities are responsible for the development and implementation of the waste policy and management and prevention plans, local governments have responsibility for organizing the collection and treatment of household waste. This responsibility includes guaranteeing the establishment of an adequate network of treatment facilities.[2]

When regional plans are drawn up, probably the most important factor is the strong bottom-up approach, based on a shared experience between the regional administration and the local authorities who will implement them locally. Representatives from local authorities are involved throughout the entire planning process. Those representatives can rely on a strong network of inter-municipal organizations and municipal civil servants. This interactive process in the preparation and the implementation of the plans creates mutual ownership over the waste management plans between the regional and the local governments.

Inter-municipal cooperation

Since the 1960s, when Flanders gradually became a very densely populated area, the municipalities, responsible for handling municipal waste, started working together to establish inter-municipal cooperation that would allow them to manage shared landfills. The number of municipalities

1 Text provided by Christof Delatter, Member of Staff of VVSG vzw, Association of Flemish Cities and Municipalities.

2 In some cases, municipalities also collect the waste from small- and medium-sized enterprises, because this waste is often comparable to household waste. For industrial waste, the responsibilities are defined in a more general manner, laying the responsibility for collection and safe treatment upon the waste generator.

involved in such cooperation has increased over the years and currently almost all[3] of them have grouped in some way to provide collective waste management services. This high participation rate demonstrates that the municipalities recognize advantages to inter-municipal cooperation, including the following:

- Local authorities remain the decision-makers, ensuring that democratic control and management are more local.
- The different municipalities are represented by people affiliated with a variety of political parties and local opposition group representatives also participate. They all have to cooperate, which excludes party politics.
- Cooperation creates more financial possibilities.
- Municipalities obtain more know-how in the long run.
- Vertical integration becomes possible at the local level.
- A larger scale leads to greater efficiency.
- A scale not too large leads to greater effectiveness.
- These 'intercommunals', or cooperative arrangements among municipalities, can play an important role in policy making at the regional level because of their know-how and experience.

In this way, inter-municipal cooperation has created on one hand, a good balance between a regional framework of standards, targets and minimal obligations, and on the other, the ability for local authorities to make various choices within this regional framework. Two political levels play both an important role in creating a long term waste management policy through this creation of political capacity both on the regional and on the local level.

Either the municipalities themselves, or multiple municipalities in the case of inter-municipal cooperation, can decide if they want to involve the private sector and in what manner, such as in deciding whether to contract for short-term or long-term arrangements. In order to do so a range of aspects are analysed before a decision is taken:

- What scale is necessary to achieve efficiency?
- Are residents satisfied with the level of quality of the service now being provided?
- What is the capacity of the municipality in terms of both financial and human resources?
- What are the strategic targets to be met?
- What role does the municipality want in the decision-making process?
- What level of control does the municipality want to maintain over the organization of the service?
- What level of flexibility is desired?
- How can a certain level of service quality be ensured?
- What costs are acceptable?
- How can continuity be ensured?
- How can transparency be ensured?
- To what degree is long-term predictability of the costs important?

Figure 1 Map of inter-municipal organizations in Flanders: 308 municipalities have formed 26 inter-municipal groups

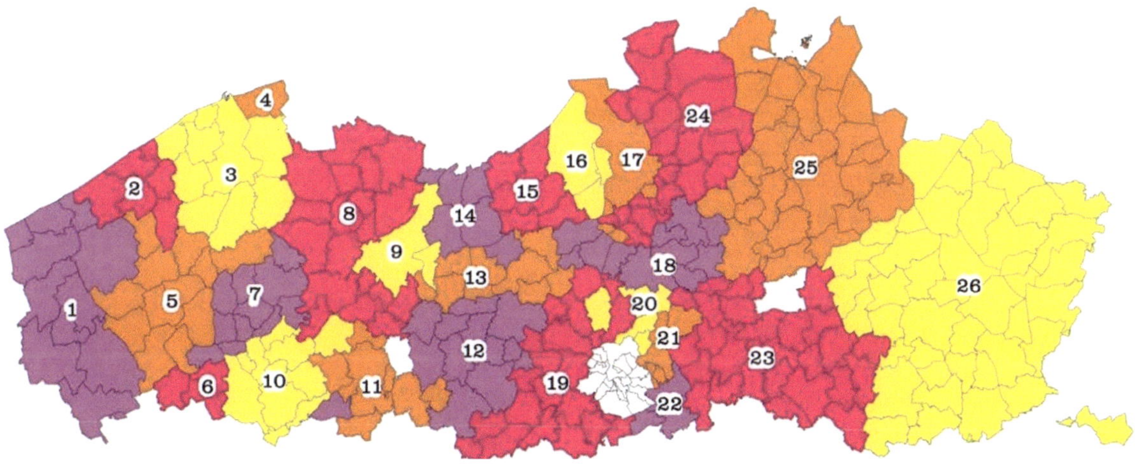

3 Recognizing that that there is no single concept that is ideal for all circumstances, every municipality decides on its most appropriate solution in light of its own local circumstances.

These aspects interact. If for example cost efficiency is the priority for a particular local authority, then it will have to accept the fact that a strong market-driven organization implies less flexibility and less direct control by the municipality. If the existing scale is not large enough to be efficient but municipalities want to retain a large degree of control, inter municipal cooperation will be the best solution. A high degree of privatization will be an attractive option if a municipality wants to decide only on the overall course of policy and does not want to determine the execution of this policy. Over the years, Flanders has learned that there is no single ideal option that guarantees the best results at the lowest cost. On the contrary, having different approaches within one region turns out to be informative, as this situation allows a continuous benchmarking of the results and efficiencies of different management options, both public and private.

Regional government support to municipalities

Looking for ways to accelerate the development of municipal environmental policy in the early 1990s, the Regional Government of Flanders provided the option of creating voluntary environmental agreements or 'covenants' between itself and municipalities. In exchange for a commitment to achieving a series of environmental goals set in accordance with the regional policy and sometimes even going beyond the legal requirements, municipalities received technical support and funding for specific investments from the Regional Government. The technical support was provided in response to questions from local authorities. Information was placed on the government website and brochures were prepared that could be ordered and used by the local authorities. When signing an agreement, each municipality could choose one of two levels, with Level 2 defining higher goals to be achieved, oftentimes going further than those set in legislation, and making the municipality eligible for greater amounts of funding than what a municipality opting for Level 1 could receive. From the beginning, the environmental covenant with its accompanying funds proved to be a very appealing instrument, as over 95% of the 308 municipalities participated. This covenant system is no longer in use as most municipalities have already achieved a very advanced level of waste management, but the Flemish Government has signed similar covenants with municipalities in other policy areas, including employment and mobility.

In addition, Flanders has a system of funds and subsidies for specific investments that support waste prevention and separate collection. The funds provided by the Regional Government cover a percentage of purchase or investment costs, depending on the nature of the investment. For example, subsidies cover 50% of the purchase price of home composting vessels or reusable diapers purchased by households. Investments in infrastructure for separate collection and volume reduction equipment are funded at a level of 30% of the total cost, making a crucial difference as municipalities acquire this infrastructure. Consequently separate collection services have been greatly enhanced for household hazardous waste, organic waste, paper, glass, plastics and metal waste streams. Funds formerly available to municipalities to finance the construction of incineration plants have meanwhile been abolished in order to promote the adoption of policies focused on waste prevention and separate collection.

Both the covenant system and the system to subsidize specific investments were used in parallel with other policy instruments, including communication, good policy development processes, landfill tax and the legal obligation of separate collection for certain waste streams. Furthermore, both measures provided more than just financial support as they were both linked to quality criteria and technical cooperation. The combination has proved to be a successful mix of instruments to achieve the envisaged policy goals.

Flemish approach to waste prevention

Flanders is one of the few European regions to have set a target for the amount of residual waste[4] that should be generated per capita. It was first introduced as a policy goal in its environmental covenant in 1995 and from there it was adopted into the Flemish Waste Management Plans. Today, this target is set at a maximum of 150 kg[5] of residual waste per capita per year as an average on the Flemish level and 180 kg as a maximum for individual municipalities. This target contributes to the waste prevention goal that drives Flemish waste policy overall. In order to achieve this target the following initiatives have been implemented.

Approach to waste prevention

The municipalities encourage their residents to prevent waste through a broad variety of actions that include home and community composting and 'chicken projects'[6] to reduce the amount of organic waste; supplying durable products such as breadboxes; providing feedback from the local government to the citizens and assigning waste mediators to civic amenity centres.

---

4 Residual waste is defined as waste that goes to landfill or incineration (with or without energy recovery) and it includes streams like bulky waste and litter waste.

5 The target was developed on the basis of an academic analysis of waste generated by households, including the types and composition of waste, the potential for prevention, and the potential for separate collection and recycling of different waste streams.

6 Municipalities distribute free chickens (three per family) or work out a special trade discount system with local pet shops for the distribution of chickens, which are then kept in gardens and eat organic kitchen and garden waste. People are often requested to register the amount of organic waste they have fed to the chickens. The success of these projects has – in some cases – been dazzling.

A typical view of the home composting units that are distributed among the households

Composting boxes used in larger gardens

Reuse centres

Flanders has a network of reuse centres, run as not for profit organizations, which collect all types of reusable goods. These centres are established to (i) achieve waste reduction through the repair and reuse of discarded products; (ii) sell reusable goods at discount prices; and (iii) create jobs for the poorly skilled. Therefore the reuse centres aim to break the short 'from product to waste' cycle through the extension of the service life of products by repairing them and selling them again at a low price.[7] They all have contracts with municipalities, from which they receive a fee that is just below the disposal cost of the collected goods. In this way, municipalities make a small saving while also supporting the reuse of goods. Reuse centres can rely on a repair shop for those goods that need small repairs or cleaning-up.

In 2013, Flanders had 118 reuse centres, with over 2,500 workers. They collected around 10 kg of reusable goods per resident, of which 45% was sold again as reusable products. The shops served over 2.5 million customers that year.

A reuse centre shop in Flanders

Separate collection system and civic amenity sites

Flemish municipalities generally organize doorstep collection of many waste streams. The collection, provided by the municipality, by an inter-municipal organization or in cooperation with private companies, can be at fixed interval periods for residual waste (weekly) and recyclables (e.g. paper and cardboard, glass and packaging waste at least monthly, garden waste at least four times a year or metals at least two times a year);[8] or upon demand (e.g. reusable materials and bulky waste). Specific collections of hazardous waste are also organized in many municipalities on a neighbourhood basis.

A typical street view on a packaging waste collection day

7 Waste electrical and electronic equipment (WEEE) is an important stream of waste for the reuse centres as it is among the most wanted items. WEEE is very often discarded when it is still quite recent and only needs minor repairs, and the associated training for repairing WEEE often enables the staff of the reuse centres to rejoin the regular labour market later. Flemish reuse centres have their own quality label for repaired WEEE and provide a one-year warranty on the products they sell.

8 Glass is also collected in bottle banks (1 for every 700 residents) and textiles may also be collected in street containers.

Collection of hazardous household waste

Every municipality also has a *civic amenity site* where the inhabitants can dispose of their waste during certain opening hours, under the supervision of municipal personnel. These sites also accept hazardous household waste. The 308 municipalities provide more than 340 civic amenity sites.

Civic amenity site of the city of Lier in Flanders

Taxes on the citizens for waste management

In years past, residents only paid a municipal waste tax, but as of 2012, 303 of the 308 municipalities were using a combination of a fixed tax and a variable tax or fee. Revenues from the fixed tax, topped up with financing from the general budget, cover the costs of the basic services that each municipality is obliged to provide. The fixed tax is not related to the amount of waste generated. It is used for financing the costs of awareness raising campaigns, cleaning services or the collection of separately collected waste streams. Complementing this is a variable tax or fee, which is an application of the 'polluter pays principle' in that it covers the costs related to the collection and treatment of residual waste and therefore varies depending on the amount of residual waste a household presents for collection.

The pay-as-you-throw system for the collection and treatment of residual waste[9] is charged through different systems:

- The compulsory use of a household waste bag of a given volume. These bags are sold at a certain price.
- To weigh and register the household waste container when it is emptied and bill based on the exact weight of the waste presented for collection.
- To count the number of times that a container is emptied and charge accordingly.
- To establish a system whereby households can 'subscribe' in order to have services provided for a container of a given volume. The fee is related to the volume of the container.

A similar system is also used for doorstep collection of organic waste, but for this waste stream, the charges cover only a portion of the actual cost of collection in order to keep it cheaper than the residual waste collection and give an incentive not to dump organic waste in the residual waste bin. Charging provides an incentive for home composting.

The result is a fee structure which always makes the failure to sort waste the most expensive option for households. Residual waste collection is always more expensive to a household than the separately collected waste streams.

In parallel to this system of differentiated taxes, municipalities have tried to make it easier for residents to reduce the amount of waste whose collection is taxed. Examples include a campaign in which home composting units were distributed for free or at low cost and the organization of additional separate collections.

It is also important that direct charging be accompanied by additional enforcement action. Household waste that was illegally dumped was examined for clues to ascertain who had owned the waste, using, for example, postal envelopes and other items with an address on them.

All these measures have successfully promoted waste prevention and reuse while also halting the growth in waste per capita, as shown in *Figure 2*. This Figure shows that steady year-on-year growth of total waste generation stopped in 2002 and that the separate collection of municipal solid waste for recycling[10] and composting increased by over 50% from 1991 to 2013. In this regard Flemish policy can be regarded as having been very effective for the separate collection of organic waste, paper and cardboard and WEEE, but waste streams like plastics still occur in significant quantities within residual waste.

9 It is required under the law to apply some kind of variable fee for the collection of bulky waste, rather than to finance the collection from the fixed tax.

10 Not including incineration with energy recovery.

Figure 2 Changes in household waste generated per capita per year

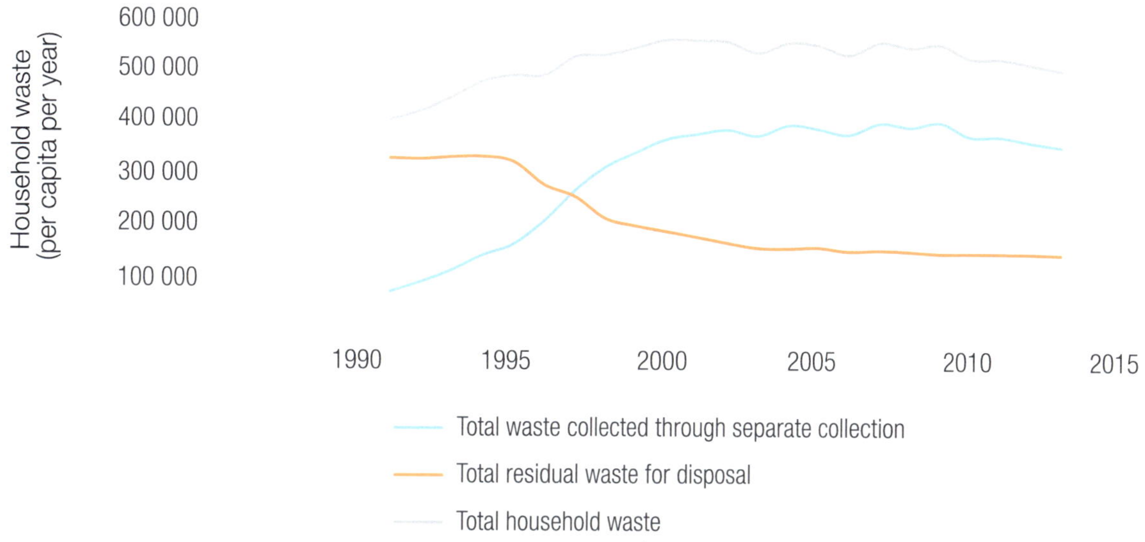

Note: Total waste collected through separate collection includes waste streams such as packaging, glass, paper and cardboard, metals, plastics, organic waste, drink cartons, textiles, construction and demolition waste, wood waste, tyres, WEEE, animal waste, diapers, hazardous waste and reusable waste. Total residual waste for disposal includes the residual waste generated by the residents, litter waste and bulky waste that is not sent for recycling.

Source: OVAM (Public Waste Agency of Flanders).

Lessons learned

- The **continuous use of a combination of different policy instruments** (legal, social and financial) is crucial to steer the behaviour of those who generate waste. These instruments tend to work best when they are used together with waste management/prevention targets.
- The organization of household waste management should be the **responsibility of the local authorities**, as they are closest to the field and can make the necessary policy decisions based on the local circumstances.
- Where there is a large number of relatively small municipalities, **inter-municipal co-operation is very effective** in achieving economies of scale, particularly in the treatment and disposal of residual wastes.
- There is **no ideal organizational approach (whether private or public)**, but having different approaches in one region allows for continuous benchmarking of results and costs.
- The **involvement of the citizens and other relevant stakeholders** in waste management (e.g. consultation when shaping the programmes) makes them feel ownership for waste prevention/management programmes.
- The introduction of a **clever pay-as-you-throw system**, in combination with prevention campaigns and good service for separate collection is a strong driver in helping households act as desired.
- Effective **coordination within and between various government agencies** at various levels is necessary. and
- The government needs to have the means for dedicated, **ongoing monitoring of the implementation of the policy instruments**.

CHAPTER

6

GLOBAL WASTE MANAGEMENT – THE WAY FORWARD

This final chapter focuses on synthesis, bringing together key points that have been made at different points of the GWMO, to form a series of 'key messages'. So the purpose is NOT to repeat or to present new analysis and discussion, but rather to pull together the evidence, experiences and insights that have been presented earlier. When read with Chapter 1, this chapter is intended to provide a higher-level, extended summary of the GWMO.

The initial focus is on what needs to be done to address waste management in a specific national or local situation (6.2). Attention then turns to the tools available to make waste management happen (6.3), regarding the various aspects of governance: responsibilities and partnerships, proactive policy and sound institutions, money matters and the data revolution. The overall recommendations are collated both under these headings and in terms of how to select appropriate actions at a national or local level and priorities for follow-up work (6.4). The chapter concludes with a Global Call for Action on waste management (6.5).

© Waste Concern

© Waste Concern

6.1 THE HOLISTIC APPROACH

Before focusing on the key messages from the GWMO and the tools available for activating them, it is necessary to recap what is meant by 'waste'. [C2.2.1][1] Natural processes generate 'residuals', but these are generally broken down to regenerate raw materials that can go back into the cycle of nature. Human population growth, urbanization, intensive agriculture and industrialization have interfered with these natural cycles, both increasing the sheer volume of residuals and increasing their complexity and their content of hazardous chemicals. Drawing an analogy between the lifecycle of materials and products and the human body, traditional 'end-of-pipe' waste management plays inter alia the role of the 'kidney', removing the contaminants from the materials flow through the 'body'. Taking the analogy one step further, it is important to avoid overloading your kidneys by controlling what you eat and drink; so an important component of waste management is to control the flow or generation of wastes by appropriate up-stream actions on waste prevention.

The UN defines residuals as one of the six components of a comprehensive set of environmental statistics. [C2.2.1] This is then sub-divided into three parts: emissions to air, generation of wastewater and generation of waste.[2] But the residuals from air pollution control and from wastewater treatment concentrate the contaminants into yet more 'waste' which is mainly disposed of to land. In other words, actions to reduce the impacts of the discharge of residuals to the atmosphere and water bodies concentrate the contaminants as waste to be dealt with elsewhere later. So waste management cannot be viewed in isolation – **a holistic approach is required for residuals management, pollution control and environmental management.** For such a holistic approach to work, policy and regulatory control, and the agencies in charge of each, need to be integrated. While there are often 'separate' environmental control regimes for air, water, land and (solid) waste, the interfaces between them need to the strong. Otherwise, it could result in moving pollution from one receiving medium to another.

6.2 WHAT NEEDS TO BE DONE TO ADDRESS WASTE MANAGEMENT

The focus here is on the physical aspects – what needs to be done to manage waste in a specific national or local situation. Four clear priorities are identified, beginning with the most critical, the public health imperative to ensure access for all citizens to proper waste management services. However ALL four priorities are necessary for achieving sound waste management. The way to move forward, tackling the priorities in parallel, will vary from country to country according to where that country stands and the local circumstances.

6.2.1 The health imperative of extending proper waste management services to all citizens

Extending sound waste management services in rapidly growing cities is a global public health priority which requires a coordinated national and international response and action. A regular and reliable waste collection service needs to be extended to all citizens, and uncontrolled dumping and open burning need to be eliminated.

Synthesis

Waste management sits alongside potable water, sanitation, energy, transport and communications as an essential utility service that underpins society in the 21st century, particularly in urban areas. Like food and shelter, it is a basic human need to live in a clean and healthy environment. Waste management can thus be considered a 'basic human right'. However, the public and political profile of (solid) waste management is often lower than other utility services.

1 This synthesis is underpinned by evidence presented earlier in the GWMO. To avoid repetition or lengthy cross-references, an abbreviated style has been adopted, with a cross-reference to evidence presented in, say, Figure 4.7 in Chapter 4.5 written as [C4.5/F4.7. B] denotes a box, [C] a chapter, [CS] a case study, [F] a figure, [T] a table, and [TS] a topic sheet.

2 Such 'waste' is generally but not always in the form of solid or sludge, although it may also include gas and liquid not suitable for standard air pollution control and discharge to air (e.g. pressurized gas in cylinders) or wastewater treatment and discharge to water (e.g. contaminated non-aqueous solvents); it is often referred to relatively loosely as 'solid waste'.

The costs of inaction – of doing little, or even doing nothing, about waste management – can be very high both for society and the economy as a whole. A waste generator will generally choose the cheapest available disposal option. In the absence of waste regulations, rigorously implemented and enforced, that often means, for household waste, dumping it in the street, on vacant ground, into drains, streams or watercourses; or burning it to 'reduce' the perceived nuisance of accumulated piles of waste. [C1.1] On a larger scale, when significant quantities of waste from cities and/or industries are dumped or burned in the open, the adverse impacts on public health and on air, land and water, including the marine environment, can be severe. [TS2]

Uncontrolled waste is not 'managed' and thus not measured, so estimating the size of the problem, and then estimating its costs for society, are both more than difficult. But the evidence gathered for the GWMO, [C5.2/T5.2] suggests that the costs to society and the economy are perhaps 5 to 10 times the cost per capita for proper waste management in a middle- or low-income city. The impacts of uncontrolled disposal include:

- *Public health:* Not having a waste collection service has a direct health impact on residents, particularly children; uncontrolled burning of waste creates emissions containing fine particulates (including black carbon) and complex organic compounds (including both carcinogenic and persistent organic pollutants [POPs]), which are highly damaging globally and locally. Accumulated waste and blocked drains both encourage vectors to breed, leading to increases in infectious diseases including cholera and dengue fever. They are also a major cause of flooding. Uncontrolled dumpsites, and in particular hazardous waste mixed with other waste, can cause disease in neighbouring settlements, as well as among waste workers. Such contaminated sites can continue to be a problem long after they have stopped receiving waste, and clean-up costs can be astronomical. [C1.1/B1.1]
- *Environmental pollution:* Dumpsites on land can cause pollution of both surface and groundwater. Sites are often alongside rivers or the sea, so can directly contribute to river and marine pollution. Erosion from coastal dumpsites is one source of marine litter. Among other potential costs are tourism losses from polluted beaches and damage to fisheries. [C1.1/B1.2/TS2/C5.2/T5.2/TS9]

Figure 6.1 collates the evidence on two key indicators for municipal solid waste (MSW) management: collection coverage and controlled disposal rates. *Figure 6.2* collates data on worldwide waste arisings, on the link between municipal solid waste generation per capita and income levels, and on expected growth in the quantities of municipal solid waste in major cities in the poorest countries in Africa and Asia, where the quantities requiring proper management are likely to double in the next 20 years.

The issue of uncollected waste and of uncontrolled dumping and burning is most obvious in low- and middle-income countries. But the residual issue of 'litter' is ubiquitous; street sweeping and litter management is an integral part of waste management services in all countries, and still absorbs a significant proportion of the budget in high-income countries.

Waste burning, Nepal.

Figure 6.1 The challenge in extending waste management to all citizens [C3.4]

Solid waste management is an essential utility service

Waste collection coverage

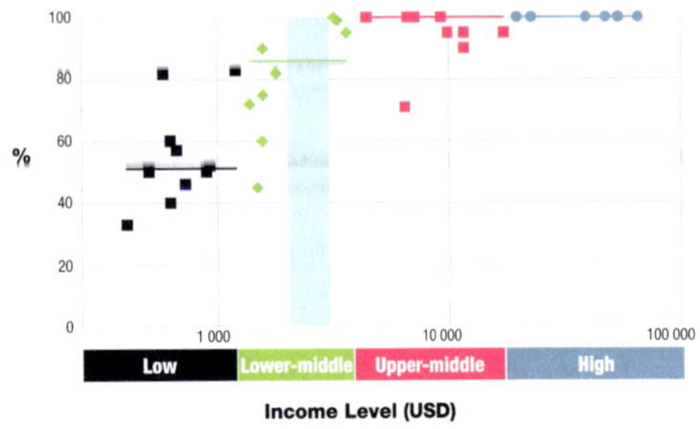

Note: 2012 data for selected cities

Developing countries have made significant progress since the 1990s, when **average collection coverage was around 50%**

2 billion people

without access to solid waste collection

Public health priority

Extending municipal solid waste collection to **100%** of the urban population

Controlled disposal rate

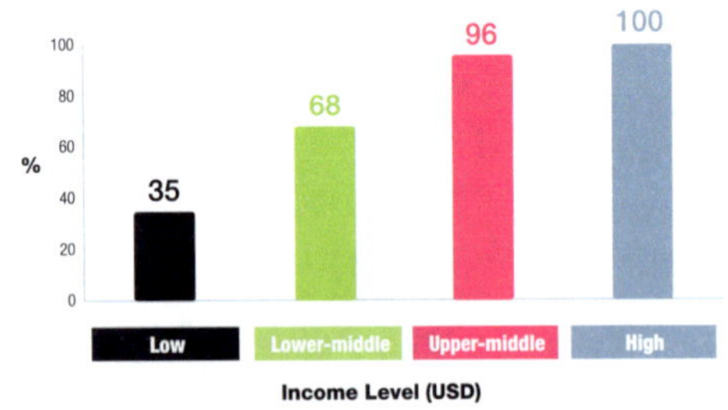

Note: 2012 data for selected cities

Developing countries have made significant progress since the 1990s, when **controlled disposal rates were often 0%**

3 billion people

lack access to controlled waste disposal facilities

Environmental priority

Achieve **100%** controlled disposal Eliminate open dumping and burning

Figure 6.2 The challenge of growing waste quantities in the poorest countries

Municipal solid waste per capita increases with income level

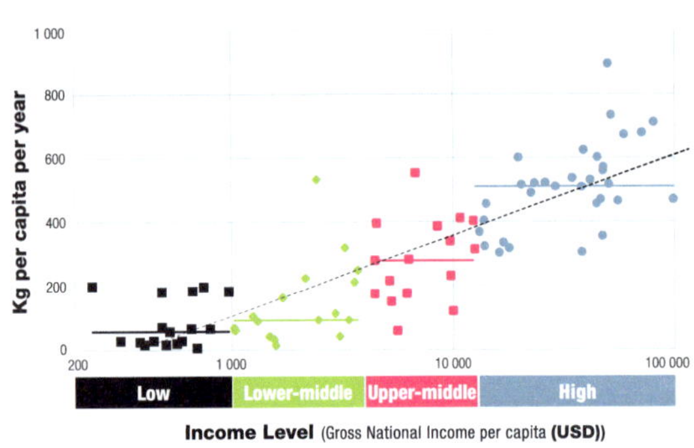

Note: Data for selected countries

2 billion tonnes per year of municipal solid waste

7-10 billion tonnes of 'urban' solid waste from households, commerce, industry and construction

Worldwide quantities increasing

Lower income cities in Africa and Asia will double their municipal solid waste generation within 15-20 years

Developing countries

- Population continues to grow
- Waste per capita is rising as economies develop
- Migration from rural to urban areas continues
- Number and size of cities increases

Developed countries

- Per capita rates doubled 1970-2000
- Rates stabilized since 2005
- 50% of total worldwide waste in 2012

Globalization results in industrial and hazardous waste generation shifting to developing countries

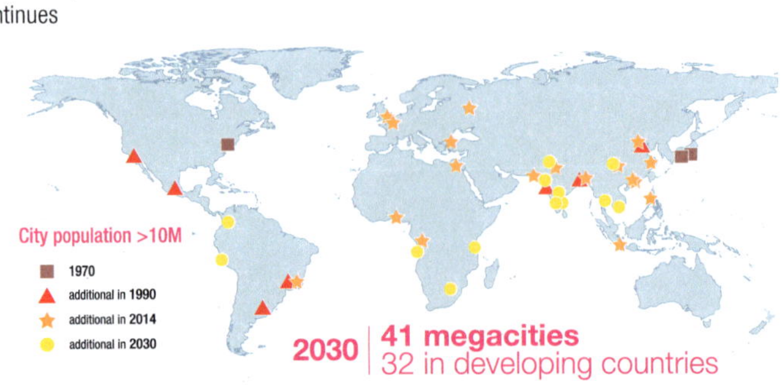

2030 | **41 megacities** 32 in developing countries

Conclusions

Public health is always the first objective of any waste management system. Accumulating waste poses potential risks to the health of all citizens of a city, rich as well as poor. The costs of inaction are much higher than the costs of providing a proper service. So there is a clear case for extending waste collection to cities in their entirety, including informal settlements and peri-urban areas; and also for phasing out all uncontrolled disposal. [C1.1/B1.1/ C3.4]

An example here reinforces the need for a holistic approach. Extending basic services to informal settlements and peri-urban areas requires different agencies to join hands, as the provision of sanitation and solid waste management services needs to be co-ordinated. To take a simple example, a first step in providing sanitation is to eliminate open defecation, which often means providing pit latrines. But if there is not a regular waste collection service, then the pit latrines becomes an 'easy' dumping ground for the waste, which makes emptying them and managing the sludge even more problematic.

The GWMO's data suggests that today, around half of those living in cities in lower income countries still do not have solid waste collection. More than 2 billion people worldwide lack access to this essential utility service (see *Figure 6.1*). Much of the waste in low-income countries is also still being dumped or burned without any control, with more than 3 billion people worldwide not served by controlled disposal facilities (*Figure 6.1*). As many of the poorest cities in Africa and Asia are set to double their waste generation within 20 years (*Figure 6.2*), the problem is set to get worse rather than better. **The global challenge of extending waste management to all citizens in every city can and must be addressed, and needs to be recognized as a priority by both national governments and the international community.**

Relying on developing a greater number of environmentally sound disposal facilities to cope with this rapidly increasing waste mountain, even if combined with a substantial increase in recycling, will put a severe strain on institutions which are already struggling financially. Section 6.2.3 below also therefore highlights the need for a **parallel focus on waste prevention**.

These issues can be tackled using some of the practical tools for responsibilities and partnerships, policy and institutions, financing and data that are discussed in Section 6.3. Primary responsibility must rest with individual cities and their national governments, but the international community also has a role and an obligation to help the lowest-income countries to address this challenge, not least through capacity building support as well as investment funding for the necessary infrastructure.

Action imperatives

(With a particular focus on cities and national governments in low- and middle-income countries, and on the international community that needs to support them)

- Health imperative – keep the city clean – extend waste collection to ALL who live in urban areas, irrespective of income level
- Eliminate the open dumping and burning of waste as an essential first step towards the environmentally sound management of waste
- **Proposed Goal: Ensuring access for all to adequate waste collection services; elimination of uncontrolled dumping and burning; and environmentally sound management of all waste by 2030.**

6.2.2 Bring hazardous wastes under control

Waste management concentrates pollutants and contaminants from the material life-cycle which would otherwise be discharged to air, water or land. It is therefore essential to take an integrated approach to the management of waste and of hazardous chemicals.

Synthesis

Much more general attention needs to be given to waste that is classified as 'hazardous'. Its improper disposal has resulted in massive costs to clean up 'the sins of the past', including numerous prominent cases in the US and Europe in the 1970s. The initial priority was to ensure the environmentally sound management of such waste, with a focus on treatment that destroys the hazardous components, for example through the high temperature combustion of organic pollutants including polychlorinated biphenyls (PCBs) from electrical transformers and the chemical treatment of cyanide wastes. More recently the focus has moved further up the waste hierarchy, to their reuse, recycling and energy recovery; however, for hazardous wastes the first concern in establishing what is the environmentally sound management route for a particular waste stream is the need to protect human health and the environment.[3]

One consequence of the introduction of expensive new controls over hazardous wastes in the high-income countries was the proliferation of unlawful groups, who saw the opportunity to make money by organizing the illegal dumping of waste. When controls over such illegal disposal were strengthened in the countries of origin, some criminal operators sought to export hazardous wastes to developing countries that did not have such controls. This scandal outraged the world and led to the agreement of the Basel Convention in 1989, which now has 183 Parties, a fact that clearly indicates the priority given to this particular issue.[4] While the scale of the abuse has hopefully decreased, elimination is difficult: globalization has resulted in a tremendous amount of closed-container trade, and it is not possible to inspect the many tens of thousands that arrive at ports around the world daily to check that they only contain what is stated on the shipping manifest. [C4.3.7/ C5.3.4]

Controls over hazardous waste management have been in place in the 'old' OECD countries since the 1970s and 1980s, and facilities for environmentally sound management of the hazardous wastes generated by industry have also been available since that time. Hazardous waste quantities in these countries may now be beginning to decline, partly through waste prevention and cleaner technologies but also due to globalization of the world economy. Globalization has resulted over the last few decades in the effective 'out-sourcing' of much resource extraction and processing, heavy industry and product manufacturing to emerging economies. In the BRICS countries (Brazil, Russia, India, the People's Republic of China and South Africa) and other rapidly industrializing emerging economies, hazardous waste management systems have been developed and are gradually being implemented, including the provision of facilities, as the quantities of industrial hazardous wastes in some places already provide the necessary economy of scale. Notwithstanding the progress made, there is still much to be done to ensure that fully compliant sound management facilities are available in all regions within these countries.

However, ALL countries generate hazardous waste. An obvious example is hazardous healthcare waste, whose separation and separate management is essential to the safe and environmentally sound management of ALL waste (both non-hazardous and hazardous). Yet this is only feasible when the facilities exist for their environmentally sound management. The quantities of hazardous wastes are often relatively low and the investment costs are high, so relatively few middle- and low-income countries yet have such facilities in place.

Not all waste which is hazardous is covered within the specific 'hazardous waste' control regime – some is covered under separate regulatory regimes specific to the industrial or infrastructure activity concerned. A common and particularly relevant case here is waste from the mining and extraction of metals and other minerals – mine tailings in particular are generated in very large amounts and their physical form requires careful management. [C4.7/B4.23] Other examples may include waste from shipping, from dredging and from laboratories and medical facilities.

---

3 For example, under the Stockholm Convention (Article 6(1)d(iii)), the recycling of persistent organic pollutants (POPs) is not permitted. Under the Minamata Convention on mercury (Hg), Article 11(3)(b) prohibits the recycling of mercury wastes unless allowed to a Party under that Convention.

4 The notification of transboundary movement of hazardous waste under the Basel Convention is an excellent tool to help ensure the environmentally sound management of exports and imports of hazardous waste. The competent authorities in the export, transit and import countries are all involved, which facilitates controls during transport, treatment and final disposal of the hazardous waste.

© David C. Wilson

Similarly, it is not just waste that meets the defined criteria for hazardous waste that needs to be considered here. The number of chemicals and the range of products in which they are being used are both increasing. As the ‘kidney’ function of waste management accumulates such chemicals in waste, [C2.2] low concentrations of chemicals increasingly occur in all waste. It is therefore important to integrate the management of hazardous chemicals with the management of waste in general and hazardous waste in particular. Many products in daily use contain hazardous chemicals, for example batteries, cleaners, oils, paints, pesticides and small electrical and electronic products. [C3.3.2] A particular recent focus has been on hazardous waste prevention, for example through the substitution of non-hazardous alternatives for hazardous materials in many products (e.g. cadmium, lead and mercury). [C4.3.6]

Such small-volume, commonly used materials are often termed ‘household hazardous waste’. In principle these should be segregated at source, collected separately and managed within the hazardous waste system. However many high-income countries are still in the process of implementing such controls, and the lack of environmentally sound facilities for hazardous waste management is a major barrier in low- and middle-income countries. So achieving this ideal worldwide is likely to remain a long-term aspiration.

Conclusions

The recent UNEP Global Chemicals Outlook shows that chemicals production and emissions are growing, particularly in emerging and developing countries, and both the number of chemicals and the range of products in which they are being used are increasing. So hazardous waste management is becoming a growing issue in all countries, and the presence of hazardous waste disposed of alongside municipal and industrial waste adds additional urgency to the global challenge of eliminating the uncontrolled dumping and burning of waste in the lowest-income countries. [C6.2.1] This issue is thrown into stark focus by the 2014 Ebola crisis in West Africa, where proper management of hazardous healthcare waste was an essential part of the solution.

Three key elements in addressing the problem are to raise awareness in government, industry and the general public, to build institutional and human capacity and to develop the required environmentally sound facilities. Given the high costs of sound treatment facilities, attention also needs to be paid to interim solutions for managing hazardous wastes, including temporary or final waste storage (landfill) facilities.

Globalization means that consumers in high-income countries can now enjoy the benefits of many products, while the burden of the pollution and of managing the hazardous wastes resulting from their production falls to generally lower-income developing countries, an issue which is important within the context of corporate social responsibility (CSR). Similarly, the hazardous waste generated in many low-income countries arises largely from imported products such as batteries, electrical appliances, electronic equipment and vehicles. Both international development assistance, extended producer responsibility and corporate social responsibility would thus appear to be appropriate mechanisms to assist emerging, middle- and low-income countries tackle their hazardous waste management problems, to help develop the necessary capacity and to ensure that environmentally sound management facilities are available and affordable in those countries.

It is also recommended that international organizations should consider a future study on wastes from the mining and extraction of metals and other minerals, with a particular focus on mine tailings.

Action imperatives

- Separate hazardous waste, and in particular hazardous healthcare waste, from non-hazardous waste at source, and manage the three types separately in environmentally sound facilities, in order to protect human health and the environment. Raise awareness both in business and the general public. *This is a particularly important issue in those lower-income countries where uncontrolled dumping and burning of mixed waste is still commonplace. Extending sound management to small-quantity hazardous waste, e.g. household hazardous waste, is still an ongoing challenge in many high-income countries.*
- *(Developing countries, the international community and multi-national companies).* Enforce the Basel Convention, adequately finance its monitoring and compliance aspects, build capacity and ensure that environmentally sound management facilities for either national or 'regional' hazardous wastes are available to developing countries.
- *(Governments and industry).* Take a holistic approach to managing all residuals: pollution control concentrates contaminants from air emissions and wastewater into (often hazardous) waste.

6.2.3 Tackle the problem at source – prevent waste: reduce, reuse, re-manufacture

Waste needs to be properly managed in order to protect public health and the environment. But that does not imply an exclusive focus on treatment and disposal: the best way to manage waste is as a resource, and by avoiding materials becoming 'waste' in the first place. The role of waste managers needs to be expanded to incorporate more visibly the upstream prevention actions.

Synthesis

Traditional waste management has dealt with waste after it has been discarded. A more effective life-cycle approach suggests that the focus should shift upstream, aiming to tackle the problem at the source through the options at the 'top' of the waste hierarchy and through sustainable consumption and production practices: design out waste to prevent its generation; reduce both quantities and the use of hazardous substances; minimize; repair and reuse; prepare end-of-life products for reuse, including through 're-manufacture' where required; and where residuals do occur, keep them concentrated and separate to preserve their intrinsic value for recycling and recovery and/or to prevent them from contaminating other waste that still has economic value for recovery. The idea is to move from 'waste disposal' to 'waste management' AND from 'waste' to 'resource' – **hence the updated terminology 'waste and resource management' and 'resource management', as part of the 'circular economy'**; [C4.2/B4.1/TS4] Here the GWMO interfaces with the earlier Sustainable Consumption and Production (SCP) Global Outlook. [TS3 SCP]

A fundamental challenge is to change a mindset sometimes encountered in the worlds of politics or business that more waste is somehow an 'indicator' of greater growth and prosperity and is thus in some sense a 'good thing', so long as it is properly managed. On the contrary, recent work on the circular economy by reputable financial analysts has accumulated a substantial evidence base which shows that the efficient use of resources and waste prevention can save businesses hundreds of billions of dollars each year through no-cost and low-cost improvements. So the efficient use of resources and waste prevention are not only good for the economy, but also essential for continued growth and prosperity. [C5.2/T5.3]

The scope of waste prevention and resource efficiency covers the whole economy – all material flows and all products used by a nation, from their respective 'cradles' to their ultimate 'graves'. [Figure in TS4] Thus, a comprehensive waste prevention programme needs to encompass the mining sector and productive industries, designers and service providers, and public and private consumers in addition to the waste management sector. Consequently, all economic sectors need to be stakeholders in waste prevention, and a systematic approach to waste prevention can also yield substantial greenhouse gas reductions, not just in the waste sector but right across the economy. [TS1]

An important component of resource management is reuse, including the repair and refurbishment of products, which has been traditional for millennia [C2.3] and is still an active economic sector in many low- and middle-income countries. Building on such traditional repair and reuse practices makes good sense. [CS1] The revival of 'reuse' in higher-income countries has also been accompanied by a new focus on re-manufacture.

Conclusions

Waste prevention has sat proudly at the top of the waste hierarchy since the concept was first outlined in the 1970s, but it has only recently emerged as a major focus of concrete actions in some high-income developed countries. The reasons for this delay were largely pragmatic: early priority was given first to phasing out uncontrolled and poorly controlled disposal practices; and then to increasing environmental standards and rebuilding recycling rates. [C2.3] The recent focus on prevention is driven in part by sharp increases in some commodity prices after 2000 and concerns about supply security for some critical materials [T5.3] and also by efforts to reduce the hazardous content of products as another dimension of waste prevention. [C4.3.6]

The major priority in low- and middle-income countries is to ensure universal access to waste collection services, to eliminate uncontrolled disposal and burning and to move towards environmentally sound management facilities for all waste. [C6.2, F6.1] Achieving this challenge is made even more difficult by forecasts that amounts of waste in major cities in the lowest-income countries are likely to double in the next 20 years. [F6.2] Relying just on increasing recycling, and on developing more and more environmentally sound disposal facilities, to cope with this rapidly increasing waste mountain will put a severe strain on institutions and budgets which are already struggling to cope. It follows that **developing effective policies and practices for waste prevention in order to stem the inexorable increase in waste per capita in low- and middle-income countries needs to become an increasing priority**. Building on traditional repair and reuse practices is one place to begin.

BOX 6.1 FOOD WASTE PREVENTION – AN INTEGRAL COMPONENT OF FOOD SECURITY

- 1.3 billion tonnes of edible food – one-third of the total produced for human consumption in the world – is lost or wasted per year
- In developing countries, the major challenge is upstream, improving storage to prevent spoilage between farm and plate
- In developed countries, the issue is mainly wastage of food that could have been eaten, at the household or retail stage
- The total amount of global food wastage in 2007 generated about 3.3 billion tonnes of CO_2 equivalent in greenhouse gas emissions. Were that a country, it would be the third-largest greenhouse gas emitter in the world, behind only the US and the People's Republic of China.

BOX 6.2 END-OF-LIFE PRODUCTS

The composition of waste, and of municipal solid waste in particular, has changed significantly over the past 30 years. A notable change has been the increasing contribution of end-of-life products, due both to the growth in consumerism and to the IT revolution. It is not just a question of the sheer volume of, inter alia, packaging and e-waste. Their increasing complexity (hazardous content, broad range of substances, composite materials, very small sizes of components, etc.) also makes them difficult to manage. [C3.4.2/TS3]

This development has raised a new set of options as to who is responsible for managing waste – previously it was the waste generator or the municipality. 'Extended producer responsibility' (EPR) now extends that potentially to the brand owner/'producer' of the product (defined variably between countries, but essentially placing responsibility for the management of products at the end of their [first] life firmly on the entire supply chain; an alternative term, 'product stewardship', is commonly used in North America). A major aim of extended producer responsibility is to incentivize reduction, reuse, recycling and recovery. [C4.5.2/ TS13]

In the globalized world, the design, resource extraction, manufacture and consumption of products are all in different countries. A major challenge is to develop forms of extended producer responsibility that work across national boundaries and also provide appropriate incentives for waste reduction and reuse. [TS3/CS1]

Action imperatives

- *(High-income countries)* Move the focus of 'waste management' upstream, to become 'resource management' as an integral part of a circular economy. Reducing waste saves everyone money. Design out waste and hazardous waste. Maximize repair, reuse and remanufacture.
- *(The international community working with middle- and low-income countries)* A short-term priority is to assist these countries in developing practical and innovative actions in order to stem the inexorable growth in waste per capita as their economies develop.
- *(All stakeholders, all countries)* Prevent food waste – as an integral part of wider efforts to promote food security, feed the world and reduce greenhouse gas emissions.
- *(Governments and industry)* Companies need to assume responsibility for the management of their products at the end of their useful life, when they would otherwise become waste, either voluntarily or based on legally binding instruments, through extended producer responsibility schemes and reverse logistics. Companies also need to ensure the environmentally sound management of waste generated upstream within their supply chains, particularly when material extraction and processing and product manufacture take place in developing countries.

6.2.4 Close the material cycle: Ensure recycling and energy recovery are compatible

A key part of the circular economy is to build 'feedback loops'. Recycling is part of natural processes, and re-establishing this circularity is a key component of overall sustainable development. There are limits to recycling, at which point energy recovery needs to be considered.

The general priority order of the waste hierarchy is: prevention, minimization, reuse, recycling, other recovery including energy recovery, and final disposal. [C2.4/F2.4] The previous section covered the first three of these, including reuse; here the focus is first on recycling and then on energy recovery.

Recycling

It is useful to distinguish between recycling 'dry recyclables' and 'wet' organics. For most of the dry materials commonly found in waste such as metals, glass, paper, plastics and textiles there are well-established industrial value chains. Recycled organics return to the agricultural value chain as either animal feed or soil conditioners (including compost and anaerobic digestate). [C3.5.3/T3.1/B3.6] The drivers for recycling include the market value of the material as well as policies and legal obligations; the latter have been established due to various concerns about 'market failure', including unstable markets, resource scarcity, environmental impacts and the like.

RECYCLE

RECYCLE

Key factors in determining the feasibility of recycling include the concentration and the quality (purity levels) of collected materials. In **source separation**, keeping materials clean and separate is critical, as a concentrated, clean material is a resource, while the same material mixed with general waste will at best be a contaminant and may even turn the mixture into a hazardous waste. [C3.5.2] For example, many countries restrict the use of compost or anaerobic digestate for food production to materials prepared from clean, source segregated feedstock for this reason. [B3.6] As another example, the supply of a number of 'rare earth' and other metals is regarded as being 'critical', so the case for their recycling should be self-evident; yet their recycling rates from consumer electronics are often quite low. Many devices contain 50 or even 70 different materials, so achieving sufficient quantities and concentrations is a challenge. [F3.14]

- Designing products for future durability, repairability, dismantling and recycling as explicit criteria is important. See Figure 6.3. [C4.2/B4.1/TS3 SCP/TS4]
- Waste generators (particularly in business and industry, but also householders) should try to keep different waste streams separate from each other in order to facilitate reuse and recycling. [C3.5.2].

Separation and collection for recycling only makes economic sense if the material is actually recycled, which depends on there being a market for the material. Some markets are relatively local, for example for compost as a soil conditioner; others may be national or regional, e.g. for glass; many secondary metals are globally traded commodities. Prices can fluctuate widely, and are often determined by competition with virgin raw materials, and influenced by fluctuations in the price of oil (for example, both oil prices and some secondary materials markets collapsed more or less simultaneously at the beginning of 2015). [C5.3.3/ F5.1] The global market for recycled plastics and paper is rather more recent than that for metals. The PRC is now the major importer. [C3.6] The long-term acceptability and sustainability of these markets depends on there being a '(more) level playing field' in terms of environmental standards.

The economics of recycling varies. In developed countries, recycling has been 'rediscovered' since the 1980s, as an alternative 'sink' to alternative treatment and disposal methods. [C2.3/F2.2] This has been driven by policies such as landfill tax aimed at diverting waste from landfill, at least in part to mitigate greenhouse gas emissions. **Recycling, along with waste prevention and the environmentally sound management of waste, has very substantial potential for reducing a country's overall GHG emissions by perhaps 15-20%.** [TS1]

In developing countries, recycling is generally performed by the 'informal sector', and therefore a purely private sector activity, undertaken only if the prices received for the separated materials cover all of the direct costs of the recycler and yield some profit. [C4.7/TS14] For some materials, and at times in the cycle when prices of other materials are relatively high, certain components in the waste stream become more sought after, leading to competition. [B5.1]

All of this is much easier with clean material cycles. So separating dry recyclables from wet organics at source is a central component of many initiatives to improve the working conditions and the livelihoods of informal sector recyclers and to integrate them within the mainstream waste and resources sector. [C3.5.2 / TS14]

The applicability of recycling demands that it should make economic, social and environmental sense. It must be carried out in a sound manner, protecting both human health and the environment. Such considerations are key when judging the relative position of recycling in the waste hierarchy – for example, exporting ships for ship-breaking on the beaches of an ill-equipped developing country is not equivalent to sound recycling in a modern, permitted and inspected shipyard; and the long-term sustainability of emerging country markets as a destination for plastics and other recycled materials from high-income countries depends on the success of national government efforts to eliminate the low-tech, uncontrolled and unregulated recycling facilities which still operate alongside modern environmentally sound facilities.

Figure 6.3 Making the life-cycle of materials and products circular

The role of design is crucial to the move from a linear, TAKE - MAKE - USE - WASTE, economy to a more sustainable circular economy. [F2.1]

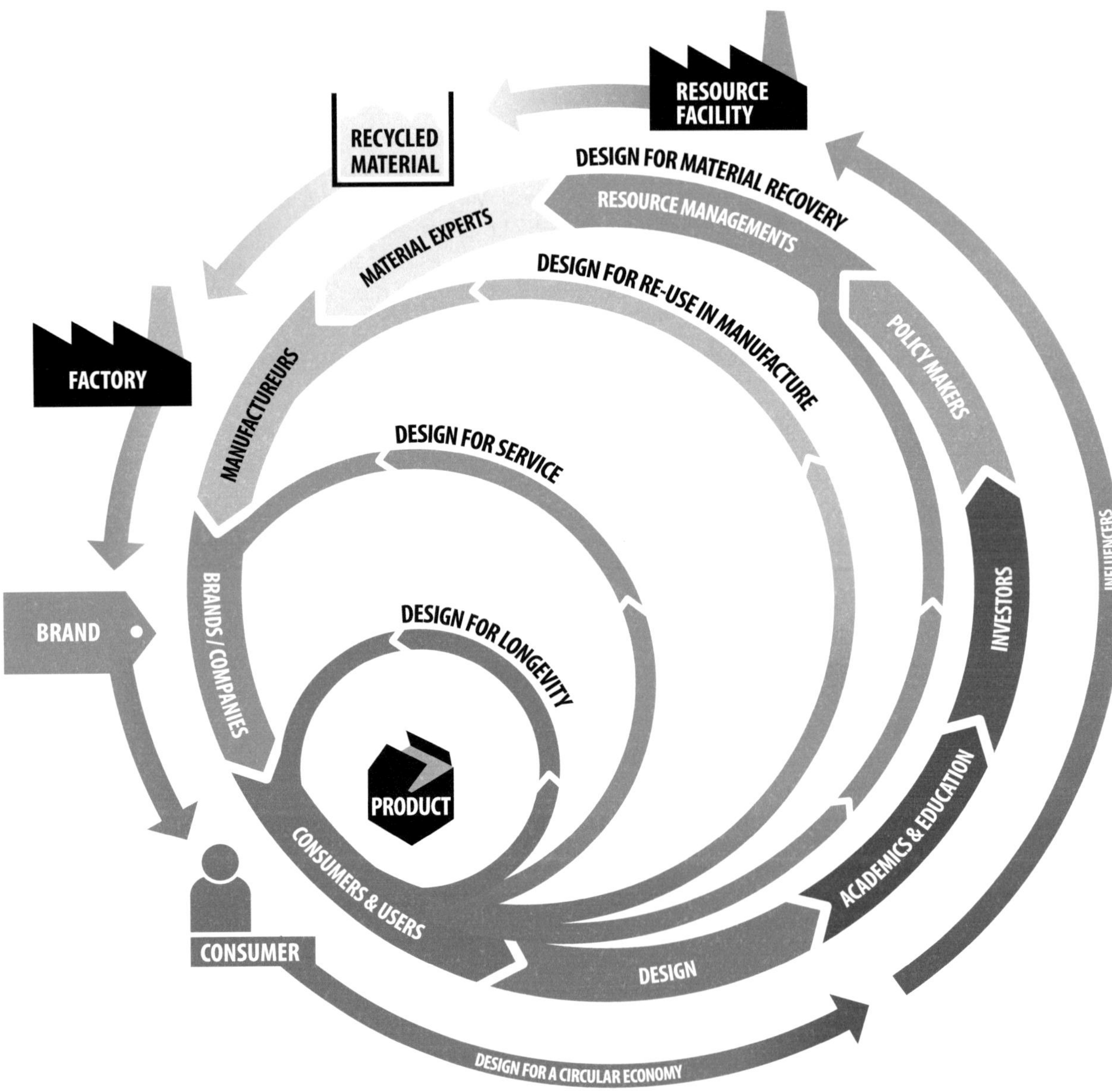

Source: RSA – The Four Designs Models http://www.greatrecovery.org.uk/

Energy recovery from waste

What is the optimum level of recycling in a solid waste management system? This will depend on how the balance is drawn between the often-conflicting drivers of policy, economics, environmental impact and sometimes resource scarcity. One 'rule of thumb' is to maximize sustainable recycling, and to recover energy from any residual waste. Again, there is no clear 'dividing line'. From an environmental point of view, the use of a specific life-cycle assessment to consider the particular local circumstances is recommended. [C2.4]

Biomass is one of the major renewable energy sources available as an alternative to fossil fuels. So for many types of waste of natural origin (so-called 'biogenic waste'), including agricultural crop residues, forestry waste and wood waste from industry, energy recovery from either thermal processing or from anaerobic digestion has clear benefits, although seasonal variations in supply may be an issue – again, life-cycle assessment should be used to compare recycling the nutrients and organic matter with energy recovery in any specific case. [C2.4]

The relative merits of recycling and energy recovery for mixed waste, including commercial and industrial waste and municipal solid waste, are perhaps less clear-cut. The calorific content of residual waste remaining after sustainable recycling has been maximized is partly of biogenic and partly of fossil fuel origin (largely from plastics). Only the former help to mitigate greenhouse gas emissions. But at some point, a life-cycle assessment could show that the environmental benefits 'switch' to favour energy recovery by thermal processing for the remaining fraction of the waste which cannot be sustainably recycled, composted or anaerobically digested. [C3.5.3/B3.7]

A wide variety of technologies for energy-from-waste (EfW) have been developed in high-income countries, and can meet high environmental standards when operated as designed – at a cost. Priority should be given to high energy efficiency. This favours combined heat and power, or heat recovery, over electricity only. Waste can be used directly, or after basic pre-processing to prepare refuse-derived fuel (RDF); or after more pre-processing to prepare a quality-controlled manufactured fuel (secondary-recovered fuel – SRF). Both SRF and RDF can be stored and transported for use by industry in thermal processing energy generation facilities, but for this to be environmentally sound management, it is important that high emission control standards are met by the industrial user. [C3.5.3/T3.1/B3.7] Many countries have developed policies to support the development of renewable energy as an alternative to fossil fuels, including inter alia guaranteed feed-in tariffs. Such policies should extend to energy-from-waste, including anaerobic digestion, thermal processing and landfill gas.

The applicability of energy-from-waste will vary between countries. Key issues to explore in middle- and low-income countries, which may constrain the applicability of energy from the thermal processing of waste, include: (1) the local waste characteristics (will the waste burn unsupported by additional fuel?); (2) the potential 'competition' with informal sector recyclers for materials in the waste; (3) energy needs and the availability of local markets for selling the heat; (4) affordability issues and their interaction with (5) the need to install and operate expensive, high-technology gas cleaning equipment; and (6) the local institutional capacity to permit and regulate facility operations and monitor emissions. [C4.2.2/4.9.2/B4.36]

Environmentally sound management of residual waste

Ideally, every effort should be made to reduce, reuse, re-manufacture, recycle and recover energy – but once that has been done, then it is necessary to use environmentally sound management (ESM) technologies to treat and dispose the final residual waste. Indeed, for some hazardous wastes, this requirement may take priority over the 'standard' waste hierarchy. [C6.2.2]

Progress in high-income countries has come in steps. [C2.3] For example, environmental controls were first introduced in developed countries in the 1970s. By 1990, most uncontrolled disposal had been phased out, but as shown in Figure 6.4, 85% of municipal solid waste in the EU was still going to landfill. Subsequent steps have focused on diversion from landfill, primarily through increased recycling but also through an increase in combustion with energy-from-waste (incineration). These changes were largely driven by policy and by explicit targets in the EU, while the US, for example, has relied mainly on market forces (with higher recycling rates driven by the high cost of state-of-the-art landfill and energy-from-waste facilities as alternative 'sinks'). [C2.3] As noted above, the feasibility of energy-from-waste technologies may be limited by various factors in many low- and middle-income countries, so it is likely that a well-engineered landfill will often represent the 'best practicable environmental option' for waste that cannot be recycled economically, at least until industry succeeds in designing out such waste.

Figure 6.4 Trends in MSW fate in the EU-27 from 1990-2020

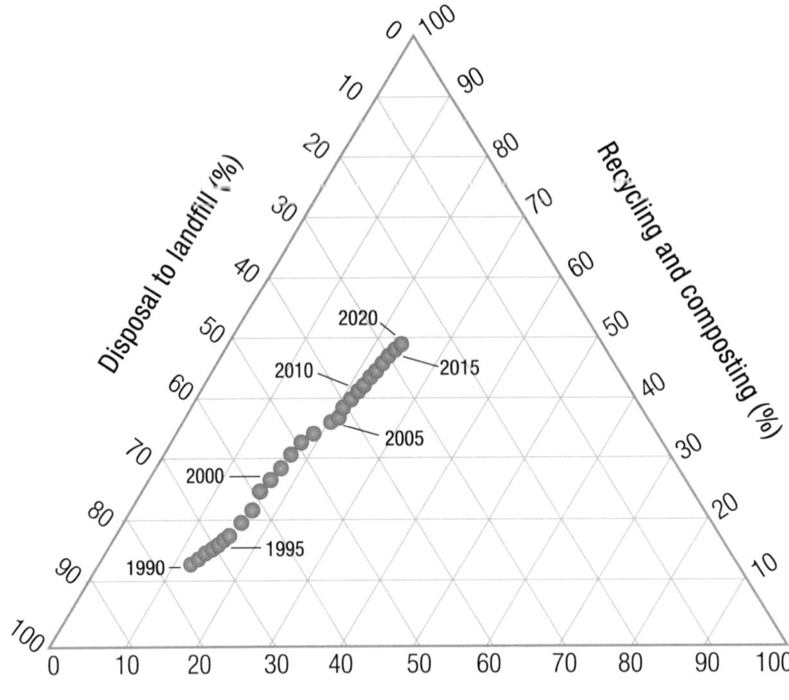

Notes: EU policy is to divert waste from landfill. Priority is given to recycling; waste that cannot be recycled should go to energy recovery. Recycling rates will have increased from 13% to 50% by 2020; waste-to energy from 12% to 24%; while landfill will have decreased from 85% to 28%.

Source: Bartl, A. (2014) Moving from recycling to waste prevention: A review of barriers and enables. Waste Management & Research. 32(9) Supplement 3–18.

Action imperatives

- *(All waste generators)* Where practicable, keep materials separate and/or ensure segregation at source so as to reduce contamination and facilitate recycling.
- *(Developing countries)* Build on existing 'bottom-up', small-scale entrepreneurial recycling by integrating the informal recycling within the mainstream waste management sector.
- *(High-income countries)* Maximize sustainable recycling, including composting and anaerobic digestion (AD) from clean, source-separated feedstock – consider high efficiency energy-from-waste (EfW) for any residual waste. Include energy from waste (anaerobic digestion, thermal processing and landfill gas) in national policies to support the development of renewable energy.
- *(Lower-income countries)* Maximize recycling, including organics recycling; in parallel, develop facilities for the proper management of any residual waste. In the short term, the priority is to replace uncontrolled disposal with controlled facilities, and in the longer term to strengthen standards to meet full 'environmentally sound management' criteria. In either case, a well-controlled or well-engineered landfill may represent the 'best practicable environmental option'.

6.3 TOOLS TO MAKE WASTE MANAGEMENT HAPPEN

So far the discussion has focused on the physical aspects – what needs to be done to manage waste. Attention turns now to how to do it – the tools to make it happen – addressing in turn key aspects of waste governance. Waste governance starts from the strategic goals and guiding principles. Here, we address in turn responsibilities and partnerships (involving all the stakeholder groups); proactive policies and sound institutions; money matters (financial sustainability); and the data revolution (ensuring the availability of reliable and timely data to underpin decision-making).

6.3.1 Responsibilities and partnerships

It is critical for the success of a waste management system that all the relevant stakeholders are fully 'on board'. This requires awareness, then knowledge and then commitment. It is essential to assign responsibilities clearly and to spend time creating effective partnerships, by building citizen and stakeholder engagement into policy-making processes.

Waste management involves a multitude of stakeholders (or actors). Every citizen and every business has an interest because we all generate waste, and we all potentially live near to current or future waste management facilities. [C4.7/F4.5]

Establishing strategic goals and guiding principles

Waste governance starts from strategic goals and guiding principles. It is necessary to formulate explicit and clear goals and to develop plans for how to achieve them. The goals express the very purpose of the system, and thus determine its elements and the way they interact. [C4.2] Universal access to waste management services, waste prevention and environmentally sound management are all important goals to be considered.

A basic principle is that waste management is the responsibility of the generator. Concerns over public health in the 19th century led to this responsibility for municipal solid waste (MSW), from a multitude of relatively small generators, being taken up by municipal governments to ensure the health of their citizens. Political commitment and keen leadership are indispensable for tangible results. Strategic planning for waste management requires a long time horizon – it is essential to ensure continuity beyond the current mayor's term of office. Future waste management transcends both party political lines and health and environment issues. The ideal situation is to develop a collegiate approach across political parties and with other relevant sectors and ministries. [CS8]

But municipalities cannot tackle waste management entirely on their own. For example, they must work within a national regulatory framework; large waste generators need to take responsibility for their own waste; and the 'producers', who place on the market the products and packaging that at the end of their useful life make up an increasing proportion of municipal solid waste, also need to take up their share of responsibility, along with the other stakeholders in the supply chain.

It follows that active engagement and building partnerships with ALL the actors in the system is necessary to make sure that those who are key to the success of the system are on board. Among other things, this generally means spending time on building citizen and stakeholder engagement into policy-making processes. [C4.7/F4.5]

Involving waste generators – achieving behaviour change

Behaviour change is key to most initiatives aiming to improve waste management. This is true when fostering a sense of responsibility for delivering waste into the proper management system, for example:

- for waste generators to store waste and present it for collection rather than dumping it in the street, in a watercourse or on nearby vacant land; or
- for people to place waste in a bin rather than dropping litter on the street at times when waste is generated 'on the go'.

It is equally true when undertaking efforts to move waste management 'up the hierarchy', for example:

- for product designers and the supply chain to design out and reduce waste and to facilitate recycling; or
- for consumers/waste generators to make choices to prevent waste generation including during the selection of products and when deciding whether to repurchase or to reuse or repair; or
- for waste generators, to segregate waste, store it separately at source and present it separately for collection in order to facilitate recycling.

Initiatives to change behaviour can be led by governments, by municipal authorities, by community-based organizations (CBOs) or by non-governmental organizations; any of these can be assigned the important role of a 'change agent'. [C4.7.1/ F4.5]

A useful tool for designing effective interventions to achieve behaviour change is the '4Es' model. [C4.5/F4.4] A successful intervention is likely to select elements appropriate to the local circumstances from each of four categories:

1. *Enable* - make it easy for people to practice the behaviour required to achieve the goals, e.g. by making available both information and instructions together with the required facilities/services;
2. *Engage* - get people involved, e.g. through public awareness and community participation;
3. *Encourage* - give the right signals, e.g. through both 'carrots' (incentives to make the behaviour attractive financially or in some other way) and 'sticks' (penalties for non-compliance); and
4. *Exemplify* - lead by example, e.g. through pilot projects to show how the behaviour can be changed and the benefits it brings; and also through government institutions being seen achieving the goals in the management of their own waste.

There is a range of tools available for engaging with service users. [C4.7] Participatory planning involves users in the design of the waste management system. [C4.2] One approach is to set up a 'waste management platform' to bring together a range of local stakeholders. [C4.7/B4.22] Such mechanisms will certainly not avoid all local opposition to the siting of new waste management facilities, but they do allow for more effective communication between different stakeholder groups and a better understanding of the importance to society of environmentally sound waste and resource management. Waste prevention is important in this context, so that people can link the need for facilities with the (increasing) amounts of waste that they themselves generate.

Social inclusion

Waste management can provide a tool for promoting social inclusion.

The absence of municipal solid waste management services in shanty towns and peri-urban areas poses a public health and environmental pollution risk to everyone living in a city. Access for all to municipal solid waste management services is a basic human need.

Providing proper waste management services is an important source of employment – in both the formal and in the 'community' and 'informal' sectors. (See *Box 6.3* below.) Street sweeping and waste collection have been traditionally labour intensive in low-income countries; informal sector repair, reuse and recycling has for centuries provided an important source of livelihood for the poor; the community sector in many high-income countries has focused on social benefit, using reuse and recycling as a means of creating jobs for disadvantaged groups. The transformation of waste management into resource management holds the promise for creating many new 'green' jobs. [T5.3] Similarly, better integration of the community and informal recycling sectors into mainstream waste management has the potential to enhance recycling rates; and (where it is necessary) to upgrade the quality of existing jobs by improving working conditions, phasing out child labour and providing access to healthcare, education and social benefits; and to create new jobs and better livelihoods. Ensuring better livelihoods within the formal economy is an important part of any programme to eliminate poverty in the lowest-income developing countries. [C4.7/ TS14]

BOX 6.3 EXAMPLE ESTIMATES OF EMPLOYMENT IN WASTE AND RESOURCE MANAGEMENT

DEVELOPING COUNTRIES	EUROPE	CIRCULAR ECONOMY
15-20 million informal sector workers worldwide	Doubled between 2000 and 2010, to more than 2 million Over the same time, average recycling rates in Europe increased from 20% to 35%	Worldwide potential for new jobs: 9 to 25 million

Due to the gender inequalities existing in waste management, there is a need for gender mainstreaming in national and local planning. The inclusion of a gender perspective will help in designing more efficient and inclusive waste management systems, ensure equitable access to livelihoods and enhance the role of women as economic and political actors. [TS15]

Creating partnerships to deliver effective services

Municipal governments are responsible for municipal solid waste management to protect public health and the environment for all the citizens. Service delivery can be by the public and/or the private sector, and the latter by either the local or international formal private sector, or by local community-based organizations and/or the informal or small-scale entrepreneurial private sector. Many larger and/or private sector waste generators will either manage their own waste or engage the private sector to do so on their behalf. So it is clear that the municipal, commercial and 'community' waste sectors need to work together, to ensure that *all* waste generated in a city is properly managed in a cost effective manner. [C5.6]

There can be advantages in involving the private sector in service delivery for municipal solid waste management. These may include better access to capital for financing, lower costs for services and greater experience in operating modern technologies. Similarly, involving the local community and small-scale entrepreneurial (or 'informal') sectors can be advantageous in terms of social inclusion, providing livelihoods and making the best use of their understanding and familiarity with local conditions. The use of public service provision can also be appropriate. So there is not a universal answer in all circumstances, and even within the same city, it is common to find several different models in use for providing different aspects of the overall service. [C5.6/F5.3]

Where public-private partnerships are used, it is important that the city develops the capacity – both technical and managerial – to develop, tender, negotiate, manage and supervise the contracts. [C4.8.3/B4.31 Important aspects include using performance-based contracting, including proportionate penalties for non-performance, capping fees, building up funds for replacement and maintenance and incorporating a degree of flexibility into the contractual conditions to enable the municipality to react to future changes (e.g. in waste composition or national policy). There needs to be a balanced partnership between the city as the client and the private contractor(s). [C5.6/F5.4]

The international community is another important partner for improving waste management in low- and middle-income countries. Although responsibility rests with national and city governments, a case can be made for providing international development assistance when requested, particularly to help build local capacity to enable the local actors to understand and solve their own particular problems. Assistance with investment funding in low-income countries can also be important, provided that it is aimed firmly at locally sustainable solutions. [C5.8.4/F5.5]

Action imperatives

- *(Government, municipalities, stakeholders)* Ensure that all sections of society have affordable access to waste management services
- *(Government, municipalities, stakeholders)* Work together to establish clear strategic goals and guiding principles for waste management
- *(Government, municipalities, stakeholders)* Work together to secure political commitment to those goals and their implementation, and to ensure continuity beyond political terms of office
- *(All stakeholders)* Ensure public participation in decision making so that stakeholders can express their concerns, provide ideas and knowledge, and get involved and develop a sense of 'ownership'
- *(All stakeholders)* Once a new system has been agreed, ensure that waste generators know what is required of them for its implementation, and are facilitated to change their behaviour in a sustainable way
- *(Government, municipalities, other stakeholders)* Establish mutually beneficial partnerships to deliver effective and sustainable waste management services. The type of partnership should be selected and tailor-made to suit local conditions.
- *(Government, municipalities, other stakeholders)* Ensure the inclusion of the community and informal sectors within an integrated waste and resource management system in the city.

6.3.2 Proactive policy and sound institutions

Sound waste management as we know it in some parts of the world only exists as a result of proactive policy, actively implemented and enforced by sound institutions.

An important component of this is direct regulation, which serves to protect society's common interests, such as public health and the environment. A combination of legislation (laws and derived regulations) and its credible and consistent enforcement has resulted in the waste industry as we currently know it – otherwise waste would be dumped or burned at the lowest cost. [C4.3]

A basket of policy instruments

Experience has shown that effective waste management systems use multiple types of policy instruments in a coherent and balanced mix. Thus direct regulation is complemented by both economic instruments, providing incentives and disincentives for specific waste practices; and 'social' instruments, based on communication and interaction with stakeholders. [C4.2.1/F4.1/F4.2]

- The basket of policy instruments that is selected needs to match the goals they seek to achieve and the local circumstances under which they will be applied. Financial sustainability is critical; beyond that, the reasons for relative success in a particular country are primarily social, cultural and political, rather than technical.
- Early policy instruments tended to focus on waste generators and the formal waste sector; now they are increasingly focusing on 'producers' (including manufacturers, brand owners, importers and others in the supply chain), as well as end-consumers in consideration of the entire life cycle of materials and products. This is part of a broader societal trend toward sustainable consumption and production.

Direct regulations for waste management first addressed public health issues, and later environmental protection. Relatively recent additions address the recovery of resources from waste, in terms of materials, nutrients and energy, as well as waste reduction; this has been as much a response to increasing costs of waste management as to the value of resources recovered. Some good practices with respect to the use of direct regulations are summarised in *Table 6.1* below.

Table 6.1 Some good practices in the use of direct regulations [C2.3/C4.3]

A	Laws and regulations define basic concepts such as 'waste' and 'hazardous waste', set standards for the environmental performance of facilities and operations, may establish targets such as recycling targets, and set forth sanctions in cases of non-compliance and violation.
B	Direct regulations need to allocate responsibilities clearly among stakeholders. It is particularly important that each government authority has clear responsibilities and duties and that overlaps among them are avoided. [C4.8]
C	Direct regulation has often been introduced gradually, in stages, with progressively more stringent standards, to allow the actors in the system to develop expertise and raise necessary financial resources. This remains a valid recommendation for countries that currently seek to introduce such laws and regulations.
D	Direct regulation needs to set out the requirements for collecting and regular reporting of timely and reliable data. Good data reporting facilitates...
E	.... consistent and fair enforcement, which also requires regular monitoring and inspection, which are costly in terms of the institutional capacities required. Resources should be focused on situations involving high risks and serious consequences of non-compliance, such as handling hazardous waste and illegal, uncontrolled disposal (waste crime), which undermines the compliant waste industry.

Economic instruments serve to steer stakeholders' behaviours and practices towards strategic goals through market-based incentives and disincentives. [C4.4] For example, taxes on landfilling or incineration have been imposed to discourage opting for these methods; and fiscal benefits have been used to encourage investment by private companies in solid waste management. When in doubt whether such instruments are appropriate, look at the system as a whole.

- If one of the main goals is to get disposal under some level of control, then taxes on controlled disposal are certainly out of place. In contrast, pay-as-you-throw (PAYT) schemes, in which charges are directly linked to the quantities and types of waste generated, may turn out to be beneficial to raise awareness and encourage segregation at source – so long as enforcement systems are in place to prevent 'wild disposal'.
- Extended producer responsibility (EPR) will hold producers, importers and others in the supply chain accountable for products placed on the local market at the end of the product's life.

'Social' instruments focus on communications, raising the awareness of the public and businesses, and encouraging the active participation of communities and other stakeholder groups. The aim goes beyond just providing information to promoting active engagement and leading by example on an ongoing basis, as behaviour change takes time. Voluntary agreements with groups of producers and waste generators, sealed by signing a formal Memorandum of Understanding, can also be used to prompt action. [C4.3.8]

Institutional framework

Effective waste governance will also depend on the institutional framework in place. Institutional capacities to prepare legislation and – particularly – to enforce it, are of paramount importance. For effectiveness and credibility in enforcement, it is much better to keep separate the two roles of legislator and regulator (enforcer). As government, keep learning new skills. [C4.8]

Waste management is an integral component of public health and environmental management, and it works best if the institutional arrangements for air and water pollution control and for waste management are integrated, with a strong ministry setting policy and with an effective and well-resourced environmental regulator. The increasing emphasis on waste prevention necessarily draws in an even wider range of affected industries and associated ministries. By its nature, waste management is relevant to the areas of responsibility of a number of ministries. A natural 'home' might be within the remit of the Environment Ministry, but other relevant ministries include those for local government, industry, health, finance, planning, infrastructure, agriculture, energy and natural resources.

It is important that one ministry or agency has *responsibility* for waste management policy at the national level, and the authority and autonomy to ensure proper implementation and coordination, and to avoid possible contradictions, overlaps or gaps between different agencies. Such institutional coherence is also important at the local level in order to achieve clean cities.

In some countries, it may be appropriate for implementation of the waste management system to be decentralized to regional-level governments. In such cases, experience shows that it is still beneficial to have *some form* of a central agency so as to (1) establish the same interpretation of national laws across regional entities, (2) have some overarching level of control to ensure consistency of approach throughout the country, (3) facilitate better cooperation and use of facilities around the country, and (4) facilitate communication and exchange of experiences and enable collective learning. [C4.8.2]

Implementation

Putting in place national legislation is merely the beginning of a process, not an end in itself. The first step is often preparing a framework law, which may then need to be brought into force through a series of subsidiary regulations, which in turn may require guidance documents on how to interpret and implement them in practice.

New legislation often focuses on raising the standards of the facilities (infrastructure) for the handling of waste – perhaps replacing uncontrolled disposal with controlled facilities, or strengthening the environmental standards that need to be met. This raises an 'implementation conundrum' – waste generators cannot comply with the new legislation until the new, compliant facilities are available, and the waste industry cannot raise the finance to invest in (expensive) new facilities until they have the 'regulatory certainty' that they will not have to compete against the existing non-compliant and thus cheaper facilities. Unfortunately history is littered with examples of new facilities going out of business because waste generators were still able to use cheaper, existing facilities, which makes financial institutions wary to invest in new ones. Paying due attention during this transition is therefore critical; it is important to allow time for new instruments to be implemented, and for change to take hold; a phased approach to implementation often works best; but equally, all stakeholders need to know when existing, non-compliant facilities will finally be closed down by the regulator. [C4.4]

An important tool here is *strategic planning*, a process which needs to begin prior to developing the legislation, and the timeframe for which needs to encompass future changes to legislation as environmental standards are gradually strengthened. [C4.2.2] Technology selection is an important part of strategic planning – it is important that the key principles here are matters of governance, not only matters of technical management. It is important to understand the function and purpose of technological options, rather than just their features, and study their performance and real costs before making a selection. [C3.5.3/C4.2.2/B4.37/T5.1] The availability of reliable data on waste generation (both quantities and characteristics) is also critical to planning the necessary infrastructure for waste and resource management. The system needs to be stable enough to provide 'regulatory certainty', but it also needs to be flexible enough to be able to adapt to future changes.

Monitoring and evaluation is key to successful strategic planning and implementation: how is the implementation going? Is the legislation working and is it being enforced? Is the situation manageable for all the actors? Have there been any unforeseen consequences? An important part of this is to measure the performance of the waste management system across a wide range of criteria. [C2.5]

The high cost of environmentally sound management for all waste continues to give rise to illegal activity, even in high-income countries, so strong and proactive enforcement is essential. Outright waste crime is a particular problem in countries with a widespread 'mafia' (e.g. Italy but also many other countries) or paramilitary presence (e.g. the island of Ireland). International waste trafficking remains an issue. Two examples where it can 'piggy-back' on legitimate trade are the export of recyclable paper or plastics from the US or EU for recycling in Asia, and the export of used computers or mobile phones for reuse in Africa, often by charities. In both cases, criminals hide either untreated municipal solid waste, or e-waste, in containers behind a surface 'layer' of recyclable materials, in order to export waste for low-cost uncontrolled disposal rather than the declared recycling or reuse. [C4.3.7/5.3.4]

Action imperatives (This is necessarily a summary. Many aspects and steps are outlined in C4)

(Particular focus on national governments and municipalities, and on the international community to support them in developing capacity)

There is a need to ensure that a number of minimum requirements are in place:

- a basket of policy instruments, including direct legislation, economic and social instruments, appropriate to the local context;
- a national framework law for waste management (in place and enforced);
- each government authority knows its responsibilities and duties and takes care to avoid gaps and overlapping among them;
- a well-resourced Waste Management Department, with the appropriate level of authority and autonomy, both in the national (environment) ministry and in city administrations;
- a well-resourced environmental regulator, with sufficient authority to enforce regulations in a consistent and effective manner; and
- strategic waste management planning to provide an agreed, long-term and stable framework within which the necessary facilities and infrastructure for environmentally sound management of waste and resources can be provided.

Sufficient resources and high levels of expertise need to be in place in the national, regional and city waste management departments as well as at the environmental regulator. Capacity development may be required.

Embed within working practices the importance of monitoring and evaluation, measuring progress, knowing what works, learning from experience, drawing lessons and accumulating local knowledge.

6.3.3 Money matters

The available evidence on both direct and indirect costs is often weak, but actions need to be taken NOW in parallel to efforts to improve the evidence base. In low- and middle-income countries, affordability is likely to be a key constraint, and securing sustainable sources of revenue to improve the level of service provided is likely to be challenging. But even in high-income countries, raising the necessary investment finance for new, environmentally sound waste and resource management facilities is still a challenge.

The economic case for taking action now on waste management

The direct costs of providing a well-suited solid waste management service will be less than the long-term cost to society of doing nothing (when the current situation includes uncontrolled disposal).

Lack of waste management or inadequate waste management harms the health of people and damages the environment and ecosystems. The indirect economic costs of not addressing waste management problems in developing countries (the 'costs of inaction') are difficult to quantify, but the available evidence suggests that they (greatly) exceed the direct financial costs of environmentally sound waste management for both municipal solid waste and hazardous waste. [C5.2/T5.2] This means that action on waste management is an urgent political priority – there is no excuse to wait for better evidence.

The evidence base on the costs of inaction is strongest in the case of small island developing states (SIDS). The high economic value of their coastal aesthetics and tourism, combined with scarce land and expensive options for transport, recycling, treatment and disposal, has arguably made them more extensively studied, while their small size has made it easier to quantify the indirect costs. [TS5] However, similar arguments apply to any city – a clean city is attractive to inward investment and to visiting business people and tourists. In order to get the waste management system working well, particularly in lower-income and peri-urban areas, all of the many aspects of governance must also be addressed – so the cleanliness of a city can also be used as proxy indicator for good governance. [C1.3]

Good waste and resource management bring many wider benefits to society and the economy, including green jobs and improved livelihoods, improved profitability for businesses through resource efficiency and waste prevention, reduction in greenhouse emissions and improved food security. [C1.3/ C5.2/T5.3]

So the case for taking action on solid waste management is strong, but the challenges in overcoming inertia are also great:

- Collection coverage is often already high in the central business district and in rich residential areas of a city, so it is necessary to make the case for the rich to pay for extending services to the poorer areas. These issues are not new – in the modern era, waste management first came onto the political agenda in Europe in the 19th century in response to cholera epidemics. Since infectious diseases affected even the rich, it was accepted that providing sanitation and solid waste management services to all citizens is a 'public good'. [C2.3]
- The costs of proper waste management need to be paid now, while the (much higher) economic costs of inaction are longer-term costs, generally falling beyond politicians' current term of office.

Financing models for waste management

Affordability is a significant constraint on municipal solid waste management services in lower-income countries. Short-term solutions must be financially sustainable, and ambitions must be tailored to what is affordable.

Low- and lower-middle income countries can often barely afford current municipal solid waste collection costs, so even the first steps of extending collection coverage and eliminating uncontrolled disposal and burning will raise affordability issues. [C5.2/T5.1/C5.7.5]

Financing models for waste management can be analysed in terms of four components: the relationship between the 'client' and the 'operator' and the sources of 'revenue' and 'investment finance'. Different combinations give almost an infinite number of permutations. There is no 'right' or 'wrong' answer for every case. Each local situation requires a tailor-made solution. [C5.4/F5.2]

One approach to extending services to unserved, often very poor, areas of cities in low-income countries is through community-based initiatives, providing livelihoods for local people and generating income from reuse and recycling. [C4.7.3/B4.27/B4.28/B5.6/TS15 /TS14]

Whatever model is used, a basic requirement is to understand the costs that need to be paid and the revenues available. While this may sound self-evident at first, in fact many cities in developing countries appear not to keep detailed accounts that bring together all of the costs of providing the municipal solid waste management service, especially since the costs may be spread among a number of cost centres. This results in a lack of accountability and control. So basic management data are often not available, an example being comparative unit costs for providing basic elements of the municipal solid waste management service. Greater transparency, more consistent accounting procedures and better benchmarking would ensure 'tighter' financial management with better control over technical inefficiencies and fraud. [C5.2]

Sources of revenue

Municipal solid waste management is a core utility service provided by or on behalf of a city to its citizens, and can be considered as a 'basic human right'. The costs are often met from the municipality budget; the municipality may raise the income required from a user charge, a local tax or a transfer from national funds, or a mixture of these. **All of these revenue sources can work well, so long as the system is transparent and fits with local custom and tradition and the service fits customer needs.** [C5.7]

Raising service standards increases costs. As a result, as standards improve, pressure intensifies on cities to increase direct cost recovery from service users. This is possible where there is a demand for a service and a tangible benefit to the service user (e.g. primary waste collection which cleans up the neighbourhood and enables the local children to avoid coming into contact with uncollected or burning waste). However this is substantially more difficult if activities are policy driven, such as a transition to more environmentally sound treatment and disposal options. It is possible to increase payment rates and cost recovery through smart enforcement mechanisms and by providing support for those who cannot afford to pay. Full cost recovery is more affordable to the users when income levels are higher, even though the absolute costs also increase. [C5.7]

Waste management charging is partly a political issue. The 'polluter pays principle' suggests that larger waste generators should assume financial responsibility for their own waste and pay the full economic cost of environmentally sound management. [C5.5] It is important that environmentally sound disposal be priced. In the early stages, care will be required to ensure that this does not serve as a driver for illegal disposal, but the price signal is necessary to incentivize the prevention and sorting of waste by the generator. The implementation of pay-as-you-throw (PAYT) schemes, in which non-recyclable municipal solid waste is charged more than recyclable waste fractions, has led to impressive results in those high-income regions where it has been applied. [C4.4.1] Relative facility pricing in neighbouring countries is affected by differences in policy, particularly where landfill and incineration taxes are in use – export of waste for disposal is generally banned under the Basel Convention, but export for recycling and energy recovery is permitted – an interesting case has been a huge rise in export of refuse-derived fuels from the UK to Europe since 2010, driven by high UK landfill tax. [C3.6/CS6]

Innovative financing instruments can enhance the efficiency of funds directed to waste management. For example, output based financing has shown positive results. [C5.7.6/B5.9] Economic instruments such as extended producer responsibility (EPR) and product charges can be a very strong instrument to provide municipalities with an additional income stream to cope with the rising quantities of end-of-life products entering municipal solid waste. Experience in the EU shows that extended producer responsibility schemes work and are affordable and help to finance the system in a just manner. [C4.4.2]

Investment financing

Raising the capital required to provide infrastructure for the environmentally sound management (ESM) of all waste streams is still a major challenge in both developing and developed countries.

Cities in developing countries struggle to access the investment funding required, and often look to national governments or international donors for assistance with capital costs. Existing revenue streams often barely cover operating costs for current inadequate services, so establishing appropriate cost-recovery mechanisms for the operation of any new facility will be an important part of project preparation. The lack of capacity to absorb investments has been a major barrier to international development financing in solid waste management.

An estimated 4 billion USD was committed to development cooperation in solid waste management between 2003 and 2012, rising over the period from around 180 million USD to 700 million USD per annum. The proportion of overall development finance directed to solid waste management has risen over the same period from 0.12% to 0.32%, which is still only a tiny proportion of the total. Considered in terms of the population of the countries receiving solid waste management development finance in 2012, this equates to just 0.09 USD per capita. Most of this funding has gone to just 10 middle-income countries, which arguably have less 'need', but which were better able to absorb the investment. [C5.8.4/F5.5]

So the challenge to the international community is both how best to assist the lowest income countries to address their waste management challenges, and how to increase the flow of investment finance available. A new approach is needed, to open up faster, better funded, more flexible credit lines which recognize the need to deliver rapid improvements to waste management systems on the ground. Such sustainable financing can be linked to wider sector reform, and would concentrate on delivering investments in infrastructure AND capacity development alongside short-term operational financing, within the framework of a sector policy and/or strategic plan. The focus should be on delivery of basic service needs of citizens; generating local business and employment opportunities; maximizing waste reduction (expenditure-reducing) and reuse, recycling and recovery (income-generating) opportunities; and fostering a healthy environment for the private sector to invest in. These stand in opposition to the quick-fix, technology-fix offers that bombard local authorities in developing countries and which threaten to stall the development process of the waste sector for a generation[5]. As a funding mechanism under the framework of the UNFCCC, the Green Climate Fund has a clear role to play here, but additional actions specific to waste management are also required.

Commercial financing to cover investment needs in municipal solid waste is often not possible because of the poor banking service coverage in many parts of the developing world and/or the high level of indebtedness of the cities. The World Bank estimates that of the 500 largest developing country cities, only 4% are deemed

5 Work is required to design at the macro-level, a coordinated programmatic approach that meets the needs of both loan and grant financing organizations AND national governments and local authorities, and is linked both to carbon, but also to public health and improving the living environment: and at the micro-level, procurement processes that are robust and transparent, but also work in the context of a fast moving developing economy situation, and appropriate key performance indicators (KPIs) and financing triggers.

credit worthy in international capital markets and 20% on the local markets. One possible solution here would be a revolving capital fund for MSW investments, administered by one or more of the international financial institutions.

Obtaining commercial finance for environmentally sound higher technology infrastructure for waste and resource management in high-income countries is still a major challenge. This is particularly a problem when a new facility will operate to a higher environmental standard, and thus requires a higher gate fee, than existing facilities – the lender will ask if is there 'regulatory certainty' that the existing, non-compliant facilities will be closed down by the regulatory authorities. Statistics on current investment projects in the pipeline show a high volume (around 300 billion USD, of which 85 million USD is for MSW projects), of which more than 40% is in just two countries which one might expect to have already put in place their infrastructure for the environmentally sound management of waste – the US and the UK. However, it needs to be noted that NOT all of these facilities will be built, due in part to the problems in raising private finance. [C3.5.4]

Potential access to private capital for investment is one of the attractions for involving the private sector in delivering services for municipal solid waste management. However, this is proving a challenge even in high-income countries; for developing countries, lenders are likely to require a very high level of guarantees from national governments, which may pose difficulties. Privately financed treatment and disposal facilities usually have a higher capital cost and operate at higher gate fees, so are more affordable in upper-middle or high-income cities or countries; they may also involve long term, inflexible contracts for guaranteed waste amounts. [C5.8.5]

Access to finance is an issue not only for large-scale investments in new infrastructure. When small and medium enterprises (SMEs), community-based organizations (CBOs) and the informal sector become involved in waste service delivery, they require access to relatively small amounts of funding to cover working capital (e.g. to pay up-front for materials separated out for recycling) and initial investments (e.g. tricycles or handcarts for collection). Facilitating access to micro-financing is a relatively simple intervention that can help communities to help themselves and improve the livelihoods of urban poor and marginalized people. [C5.8.5]

Action imperatives

- *(Municipalities and all waste generators)* Know your costs: understand the costs that need to be paid and the revenues available
- *(Governments, municipalities, citizens)* Someone has to pay for municipal solid waste management. Find the appropriate financing model.
- *(Governments, municipalities and larger waste generators)* Larger waste generators should assume financial responsibility for their own waste and pay the economic cost of environmentally sound management
- *(Government, municipalities and the waste industry)* Ensure disposal is priced, to provide an incentive to the waste generator to reduce waste quantities
- (*Municipalities*) There is no 'right' or 'wrong' financing model – each local situation requires a tailor-made solution. Select a model that is transparent and fits with the local custom and tradition and select a service that fits customer needs.
- (*Municipalities*) Aim to increase cost recovery gradually. Begin by making a direct charge where there is a clear demand for the service and a tangible benefit to the service user (e.g. primary waste collection). Ensure support is available to those who cannot afford to pay.
- (*Governments, municipalities and producers*) Consider extended producer responsibility (EPR) as a means to transfer the costs of managing end-of-life products in municipal solid waste from the municipality to the producer and other appropriate stakeholders along the supply chain.
- (*Governments, international financial institutions, commercial banks*) Create financial instruments to provide access to capital for cities for investment in municipal solid waste management and infrastructure
- (*Governments, international community*) Increase the flow of Official Development Finance (ODF) to help improve solid waste management in the lowest-income countries. Increase substantially the share of funding spent on improving solid waste management from the current meagre level of 0.32% of total funding.

6.3.4 Data revolution

'Data are the lifeblood of decision-making and the raw material for accountability. Without high-quality data providing the right information on the right things at the right time, designing, monitoring and evaluating effective policies becomes almost impossible.'[6]

This is the opening statement in the first report of the UN Secretary-General's *Independent Expert Advisory Group on a Data Revolution for Sustainable Development's*, which addresses the challenge of integrating the recent explosion of data from new technologies with traditional data to produce high-quality information to foster and monitor sustainable development. Key recommendations include developing a global consensus on principles and standards for data; new resources for capacity development to improve data in the lowest-income countries; and a UN-led 'Global Partnership for Sustainable Development Data'.

Improving waste and resource management data

A recurrent theme throughout this GWMO has been both a lack of data and the poor reliability of available data on waste (and resource) management, and not just in low- and middle-income countries. These issues are not unique to waste management, but they are recognized as being more severe than in many other environmental areas. An international comparison of the environmental performance of countries is now published every two years, using the Environmental Performance Index (EPI) that has been developed over 15 years and six iterations. The EPI includes component indicators measuring performance across 20 environmental issues; a further nine issues are explicitly listed as important, but currently excluded from the EPI[7] as 'global data are still lacking' – these include municipal solid waste management, recycling rates and toxic chemical exposures. Hazardous waste and any other waste indicators are also excluded. [C2.5]

The *Data Revolution for Sustainable Development* focuses on the need for high-quality and reliable data, but also on 'issues (which) are still barely covered by existing data'. According to the EPI, waste management is one of those 'barely covered' issues; the opportunity of co-ordinated international action to improve data for sustainable development MUST be taken to facilitate the STEP-CHANGE that is urgently needed in the availability and reliability of waste and resource management data. Some of the elements to be considered in future work to improve waste and resource management data are included in *Table 6.2*.

© Jeroen IJgosse

© Jeroen IJgosse

© Jeroen IJgosse

Waste composition analysis in Bengaluru, India

6 Independent Expert Advisory Group on a Data Revolution for Sustainable Development (2014). *A World that Counts: Mobilising The Data Revolution for Sustainable Development.* p.2 http://www.undatarevolution.org/report/

7 http://epi.yale.edu/our-methods

Table 6.2 Elements to be considered in future work to improve waste and resource management data

- ***Definitions.*** The Basel Convention controls transboundary transport of hazardous wastes, so requires a standard definition. Where wastes are managed at the national (or sub-national) level, definitions will vary – but some degree of standardization needs to be agreed to facilitate the international sharing and comparison of data.
- At a minimum, data are required on the ***quantities and composition*** of municipal solid waste, commercial and industrial waste, construction and demolition waste and hazardous waste; and on the ***relative quantities going to the main management methods***. Data are required on quantities generated, not just on quantities collected or disposed of.
- ***Regular measurement and reporting is important***. Collecting basic data should be part of initiatives to extend collection coverage, eliminate uncontrolled disposal and develop environmentally sound management facilities in middle- and low-income countries. Waste quantities are not all-important vis-à-vis other metrics – but in the absence of actual measurement using a weighbridge, any data are likely to be unreliable. Regular measurement of waste composition, using a standardized methodology, is also important.
- ***Coordination and communication*** among the local, regional and national levels is essential.
- ***Need a fundamental review of waste metrics***. Waste quantities are important, but so also are metrics that focus on impacts (e.g. waste toxicity, health impact and environmental sensitivity) and on management methods (e.g. material composition, biogenic content, moisture content, calorific value).
- It is necessary to ***measure waste prevention*** as well as waste generation – this is even more challenging! Further research is required to develop suitable protocols for waste prevention metrics, which should then be incorporated into national data systems.
- Alongside waste management, ***material and resource flows through the economy need to be measured***. [C2.5.4]
- ***Innovation is necessary*** – how can data from new technologies (mobile phones, social media etc.) be harnessed to plug some of the existing gaps in waste and resource management data?
- ***Availability and transparency***. The ideal is that waste and resource data should be publicly available, on-line, on at least a quarterly basis. Even in high-income countries, this is currently the exception rather than the rule.
- ***Mining & quarrying and agricultural & forestry wastes*** are often outside of national waste legislation. Nevertheless, some estimation of waste quantities and properties is desirable. This is essential for some higher risk waste streams – a particular example is mine tailing from the extraction of heavy and precious metals.

Measuring waste that is generated and actively managed is important, but not in itself enough. The first challenge is to capture the waste in the first place, by eliminating uncontrolled disposal and burning, and by bringing waste into the formal control system. Understanding the current situation is an important input to planning such improvements, but such 'losses from the system' are, almost by definition, never measured. A useful tool is to carry out an approximate mass balance, estimating the flows through and losses from, for example, the combined formal and informal sector municipal solid waste management system in a city. [C2.4.2]

Benchmarking the performance of waste management systems

Coordinated international efforts to improve the 'hard data' available on waste management are essential, but are just part of what is needed. A major objective of the GWMO has been to: *'Assemble sets of standardized performance indicators on waste management that would allow benchmarking exercises and so facilitate better analysis of the state of waste management around the world and also provide a standardized means to monitor progress over time'*. Assessing performance of the waste management system of a city or a country as a whole, including both the 'hard' physical and 'soft' governance aspects has been addressed by a number of international teams over the last few years, and some of the practical indicator sets available have been used in the GWMO. Some of these combine the available data with expert judgement on a number of clearly defined aspects of system performance to come up with an overall assessment of how a city's solid waste management system is performing. The aim is NOT to come up with one definitive index number, but rather to allow assessment across a range of indicators, to allow both strengths and weaknesses to be identified, and to assist in setting priorities for the deployment of the limited resources available to address the most pressing weaknesses. These systems now need to be used routinely in a wide variety of situations, and the results used to revise the indicator sets, with a view to coming up with internationally agreed sets of performance indicators, suitable for use in different situations. [C2.5.3]

Waste and the Sustainable Development Goals (SDGs)

The *Data Revolution for Sustainable Development* focuses on data to monitor the targets within the action plan for the Post-2015 Development Agenda. Waste management is well embedded within the Sustainable Development Goals, with all of the components of the Global Waste Management Goals set out in the GWMO's Global Call for Action being included within the Goals. (See Section 6.5 below.) Indeed, waste management goals are explicit or implicit in more than half of the 17 higher-level Sustainable Development Goals. [C1.3/T1.1] **This emphasizes the strategic importance of waste management: actions here will contribute to progress against targets for a wide range of SDGs.** Setting and monitoring targets for waste management is thus an important part of meeting the Sustainable Development Goals.

Action imperatives

- *(International community)* Ensure that waste and resource management data are actively included within wider international action as part of the data revolution to improve data for sustainable development.
- *(International community)* Set and monitor goals and targets for waste management, as an important element in meeting the Sustainable Development Goals.
- *(All stakeholders)* Local, national and international action is required to improve several-fold the availability and reliability of waste and resource management data. Build reliable databases.
- *(International community, all stakeholders)* Roll out the newly available performance indicators for solid waste management across a wider range of cities, and assess the results with a view to their further development and potentially their more widespread use
- *(Municipalities)* Benchmark performance of your city's municipal solid waste management system using available indicators and highlight areas for improvement
- *(International community, all stakeholders)* Take advantage of investment in new services and facilities for solid waste management to institute routine collection of the data needed to monitor and further develop services
- *(All stakeholders)* Be transparent – make waste generation and monitoring data available online

6.4 RECOMMENDATIONS

Many recommendations to address both the priority action points – what needs to be done to manage waste – and also the tools to make it happen have already been included as ***Action Imperatives****. To avoid repetition, this section collates the overall recommendations from four different perspectives. (1) What needs to be done to develop a sustainable waste management system at the local or national level. (2) How to select the appropriate actions at the governance level to make those changes happen. (3) Waste management as an entry point to achieve sustainable development. (4) Priorities for further work to follow-up on the GWMO.*

6.4.1 What needs to be done at the local and national level

Figure 6.5 below summarizes the recommendations on what needs to be done, grouped according to the four main physical aspects of waste management as discussed in Section 6.2.

In selecting what needs to be done in a particular country or city, it is clear that the answers will vary according to the starting point – the current baseline. This is reflected in the grouping of the priority actions presented in Section 6.2 and summarized in Figure 6.5; these are NOT sequential steps, as one cannot afford to wait until one 'problem is solved' before beginning to address the next. But equally, it is not possible to do everything at once when your resources are limited – it is clear that the high-income countries have reached their current, relatively sophisticated waste management systems via a series of intermediate steps over a period of 30 to

50 years. [C2.3] But due to the recent step change in communications as a result of the Internet and mobile phones, people can now learn from each other at an unprecedented scale in time and space, something that was not possible 50 years ago.

Figure 6.5 What needs to be done at the local and national level

This Figure summarizes schematically the Action Imperatives set out in Section 6.2.

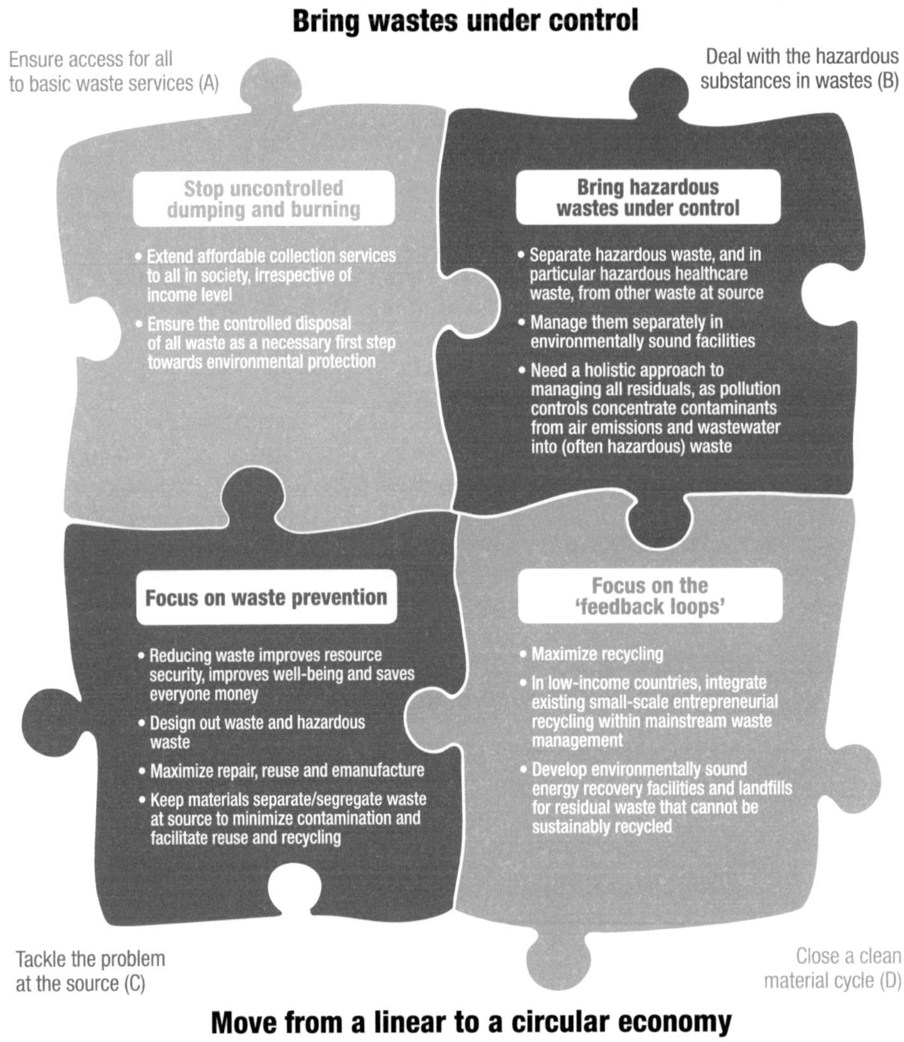

A paradigm shift is in process in some high-income countries, moving from 'waste management' to 'resource management' – controls over illegal disposal (point (A) in *Figure 6.5*) and hazardous waste (B) clearly need to be maintained, but the focus is moving to implementing the 'circular economy' through (C) and (D). The task facing middle- and low-income countries is enormous: they often have major challenges implementing (A) and (B) and they need to take actions on (C) to 'decouple' waste generation from economic development, to avoid 'forever running to stand still' as waste quantities continue inexorably to rise. Building on existing, informal recycling systems by integrating them more into the mainstream waste management sector (D) often forms a relatively low-cost, win-win component of an integrated solution.

Achieving what is a substantial step-change in waste management in low- and in some middle-income countries is going to require national and local political commitment to specific national and local solutions. **However, making it happen by 2030 is also going to require financial support from the international community. International and bilateral donors need to prioritize waste management and to increase significantly the percentage of overall international development funding directed to meeting this challenge.**

6.4.2 How to make it happen – selecting the appropriate actions

The focus here is on how to select the appropriate actions at the governance level to make the required physical changes happen in practice, in order to take the next appropriate steps in developing (your own) specific waste management system. The recommendations are addressed primarily to a national minister or city mayor, but some are also applicable to waste generators, producers of products that become waste and the waste industry.

A series of action imperatives was set out in Section 6.3, including the main stakeholder groupings on which each action falls. These are summarized in *Figure 6.6*.

Figure 6.6 Summary of the recommended action imperatives on waste governance

This Figure summarizes schematically the Action Imperatives set out in Section 6.3.

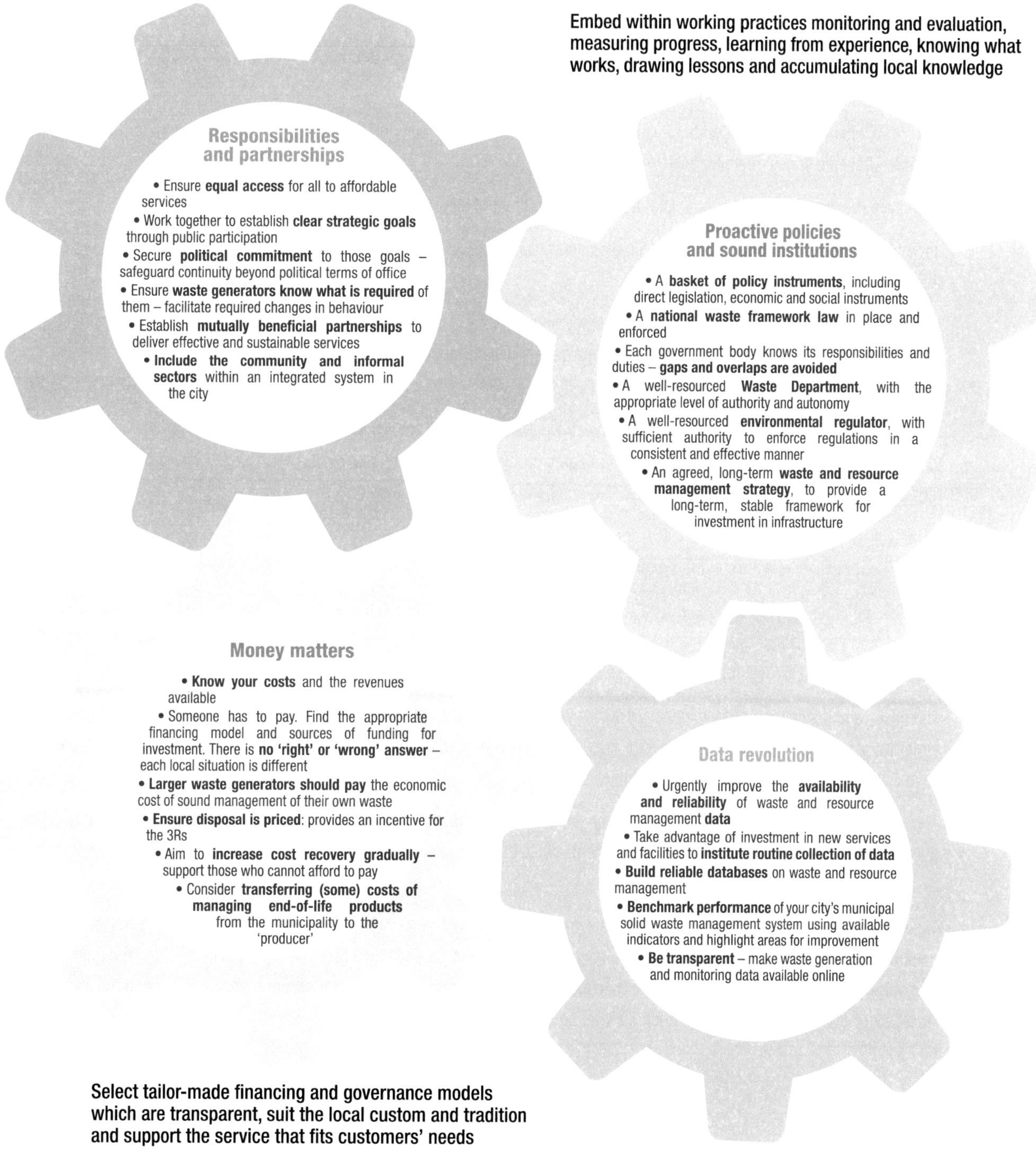

Table 6.3 brings together some of the points summarized in *Figure 6.6* with other recommendations made throughout the GWMO, to provide a set of general principles for decision makers to bear in mind. The focus here is on the actions that focus on governance rather than the physical aspects and on selecting the tools to use to make the physical changes happen.

Table 6.3: Some general principles for selecting the next appropriate steps in developing a waste management system

GENERAL ADVICE
• Need strong leadership and courage. Political commitment makes a huge difference. • Solutions need to be adapted to local conditions – no 'one size fits all'. • Formulate explicit and clear goals; plan how to achieve them. Formulate goals carefully – the goals express the very purpose of the system, and thus determine its elements and the way they interact. • It usually works more easily if social consensus exists among stakeholders about waste issues and waste management principles and goals. • Start from where you are – understand your baseline – build on what already exists – build gradually and work towards each ultimate goal as a series of steps. • Reconfirm the waste prevention principle, and confirm the role of all waste generators, including the public. • Need long-term thinking and good strategic planning. Investment requires 'regulatory certainty' – decisions should not be subject to change after each election. • The reasons for relative success are often primarily social, cultural and political, rather than technical or even economic. Invest in information, dialogue, education, communication, collaboration.
PROACTIVE POLICIES AND SOUND INSTITUTIONS
• Experience has shown that effective waste management systems combine multiple types of policy instruments – direct regulation, economic and social instruments – in a coherent and balanced mix. • Introduce framework legislation, to allow for regulations to be implemented in stages, with progressively more stringent standards, to allow the actors in the system to develop expertise and raise necessary financial resources. • Effective enforcement, by a strong, independent and well-resourced regulator, with sufficient authority to enforce the regulations in a consistent and effective manner, is critical to create a 'level playing field' for proper waste management. There need to be penalties in place for non-compliance. • The policy and institutional system needs to evolve within, and for, the local situation and get rooted, supported and 'owned' under the local circumstances. This is a long-haul learning process, which takes time, effort and commitment, and it requires developing resources, expertise and capacities. • The passing of stable legislation that allows businesses and local governments to plan their operations and investment ahead will increase the effectiveness of the system. While being stable, such legislation also needs to be flexible enough to accommodate future changes. • Take charge of technology selection as a matter of governance, not as a matter of technical management. Technological solutions need to support the goals and match the local situation, needs and capacities. Take time to learn about the function and purpose of various technologies, rather than just their features, and study their track record, performance and real costs. As a rule of thumb: If a solution looks too good to be true, it probably is.

RESPONSIBILITIES AND PARTNERSHIPS

- Open up as government, go broader than traditional waste management, work with other ministries, with producers and importers and with civic society, to PREVENT waste at the design stage and through a product's lifecycle, and to move toward a circular economy.
- Build citizen and stakeholder engagement into policy-making processes. Communicate, facilitate involvement, engage with the actors in the system. Spend time on consulting all affected stakeholders, private or public. Make sure that those who are key to the success of the system are on board. Ensure that all parties know what is required of them – facilitate required changes in behaviour.
- Establish mutually beneficial partnerships to deliver effective and sustainable waste management services. The type of partnership should be selected and tailor-made to suit local conditions, including both public and private sectors as appropriate. Ensure the inclusion of the community and informal sectors within an integrated waste and resource management system in the city.
- Future waste management transcends both party political lines and health and environment issues – develop a collegiate approach across political parties and with other relevant sectors and ministries.

AFFORDABLE AND COST-EFFECTIVE WASTE MANAGEMENT

- Work on and communicate about the economic dimension of waste problems, strategies and actions (policy).
- Affordability is a significant constraint on municipal solid waste management services in lower income countries. Short-term solutions must be financially sustainable. Tailor ambitions to what is affordable.
- Understand your costs and revenues. More transparency, more consistent accounting procedures and better benchmarking allow tighter financial management with better control over technical inefficiencies and fraud.
- Aim to gradually increase cost recovery. Begin by making a direct charge where there is a clear demand for the service and a tangible benefit to the service user (e.g. primary waste collection). Ensure support is available to those who cannot afford to pay.
- Ensure disposal is priced – implement the 'polluter pays principle' and provide an incentive to reduce, reuse and recycle (the 3Rs)
- There is no 'right' or 'wrong' financing model – each local situation requires a tailor-made solution. Select a model which is transparent and fits with the local custom and tradition, and select a service that fits customer needs.
- Consider extended producer responsibility (EPR) as a means to transfer at least some of the costs of managing end-of-life products in municipal solid waste from the municipality to the producers and other stakeholders in the supply chain who put the products on the market.
- Consider potential investment opportunities from private players in markets for secondary products and energy and the potential financial benefits of co-management of specific municipal waste streams with other material or waste streams.

EFFECTIVE MANAGEMENT INFORMATION SYSTEMS (THE 'DATA REVOLUTION')

- Aim to develop fit-for-purpose data collection and management information systems, to allow performance monitoring and sound strategic planning.
- Take advantage of investment in new waste management services and facilities to institute routine collection of the data needed to monitor and further develop waste management services.
- Utilize the newly available performance indicators for municipal solid waste management to benchmark the city's performance, highlight areas for improvement and monitor improvements over time.

A key message from the GWMO is that there is no **'one size fits all' answer** – each local situation is different, and each solution needs to be tailored to the specific social, cultural and political situation at both the local and the national levels. However, the GWMO has highlighted some tools that can be used in a particular local situation to assist decision makers.

Table 6.4 shows one example that uses the '4Es' framework for designing an intervention for behaviour change. [C3.5/F3.4] It addresses the challenge set out in point (A) of *Figure 6.5*, developing a programme to ensure access for all to basic waste services by extending waste collection and eliminating uncontrolled dumping and burning. This is only a hypothetical example and would need to be worked up for each specific local situation. This tool can also be applied to the other priority action needs identified in the GWMO and summarized in *Figure 6.5*.

Table 6.4: Ensure access of all to waste services *(Point (A) in Figure 6.5)*

As an illustration, this Table shows one example that applies the '4Es' framework to designing an intervention for behaviour change. This example addresses the challenge set out in point (A) in *Figure 6.5*, namely the development of a programme to extend waste collection services to all the citizens of a low-income city and to stop uncontrolled dumping and burning. It assumes a baseline of 50% of households having a regular waste collection service; an informal recycling rate of 25%; and 50% of the collected and non-recycled waste being disposed of in controlled facilities.

ENABLE – MAKE IT EASY TO 'DO THE RIGHT THING'	ENGAGE – GET PEOPLE INVOLVED
• Ensure that a regular and reliable municipal solid waste management collection service is available to all citizens irrespective of income level • Phase out existing uncontrolled dumps, upgrade to or replace with controlled disposal facilities • Work to build on existing repair workshops and reuse systems, as part of a concerted approach to waste prevention • Work to increase the coverage of 'itinerant waste buyers' who collect source-separated materials for recycling • Ensure whenever possible that national policy and strategy is fully supportive of the local initiatives	• Raise awareness of waste generators (households, businesses, institutions, industry) regarding the health and environmental impacts and high costs to society and to business resulting from uncontrolled dumping and burning • Educate and inform people so that they know what is the 'correct' thing to do • Work with CBOs, NGOs, MSEs and the informal sector to provide primary collection services in low-income areas • Work with informal repair and reuse shops and recyclers to Integrate them into the solid waste management system – to improve working practices, increase reuse and recycling rates and divert waste from (expensive) collection and disposal by the city
ENCOURAGE – GIVE THE RIGHT SIGNALS	**EXEMPLIFY – LEAD BY EXAMPLE**
• 'Carrot' (Provide a meaningful incentive or reward): For example, facilitate clean up campaigns, so that neighbourhoods are clean, watercourses are clear and flood risks are reduced. This then provides an incentive to communities to maintain this • Institute affordable charges for primary collection of municipal solid waste; assist those who cannot pay. • 'Stick' (Ensure that there is a clear penalty for undesirable behaviour): Implement and enforce legislation against uncontrolled dumping and burning • Ensure that the regulatory agency is adequately resourced to enforce the legislation to stamp out both casual evasion and organized waste crime • Communicate achievements and success stories to all stakeholders, to encourage and keep up engagement	• Implement pilot and demonstration projects to show that the intended goal can be achieved • Work with larger (public sector) waste generators to demonstrate that they themselves are already using, and paying for, controlled disposal • Work with larger (private sector) waste generators to ensure that they too are using, and paying for, controlled disposal

6.4.3 Waste management: An 'entry point' to sustainable development

Waste management is a cross-cutting issue impacting many aspects of society and the economy. As shown in *Figure 6.7*, the political case for action can be significantly strengthened **if waste management is viewed as an 'entry point' to address a range of other (often 'difficult to reach') sustainable development issues.**

A major driver for waste management is the protection of public health and the local environment (although some 'local' impacts such as litter may extend globally if they reach the oceans). However, the particular focus at this point is rather on one particular aspect of global environmental protection, specifically the very significant contribution that good waste and resource management can make to mitigating anthropogenic climate change. [C1.3/TS1]

It has been proposed that the cleanliness of a city can be used as a proxy indicator for good governance – which is both a major objective of development programmes and extremely difficult to measure. Improving waste and resource management practices can be used as a catalyst for creating sustainable livelihoods and promoting economic development, in both developing and developed economies. [C1.3]

Figure 6.7 Waste management: An 'entry point' to sustainable development

Waste management has strong linkages to a range of other global challenges: health, climate change, poverty reduction, food and resource security, sustainable production and consumption. The political case for action can be significantly strengthened when waste management is viewed as an entry point to address a range of sustainable development issues, many of which are difficult to tackle.

Climate change

Potential impact of improved waste management on reducing GHG emissions across the economy: 15-20%

Diversion from disposal of biodegradable wastes prevents emissions of methane, a powerful greenhouse gas (GHG)

Reduction, reuse and recycling all displace virgin materials and products, and the GHG emissions in their manufacture

A clean city

- Where the solid waste management service works well
- A holistic approach is taken to managing all residuals

A successful city

- A healthy, pleasant and safe place to live
- A good place to do business and visit as a tourist
- Fosters a sense of community and belonging

Good governance

- The cleanliness of the city can be used as a proxy indicator of good governance

Enterprise and creating sustainable livelihoods

'Waste to wealth' projects in Africa have demonstrated that new waste services can be used as **a catalyst for sustainable livelihoods and economic development** in poor neighbourhoods of some of the world's poorest cities

15-20 million people working in the small-scale entrepreneurial 'informal' waste sector worldwide

2000-2010 in Europe **employment in waste and resource management doubled: > 2 million**

Estimate of worldwide potential for new jobs in the circular economy: **9 to 25 million**

Section 6.5 below sets out priority Global Waste Management Goals, all of which are already included within the Post-2015 Development Agenda. To meet these goals would contribute to more than half of the 17 sustainable development goals (SDGs). [C1.3/T1.1] This again underlines the strategic importance of waste management.

6.4.4 Beyond the GWMO – priorities for future work

Recommendations for follow-up work to the GWMO are collated here. These are directed to UNEP, ISWA and other international institutions, to national and city governments, and to academia, international NGOs, the waste industry and other stakeholder groups. They are organized into three tables:

- *Table 6.5* collates recommendations for the coverage of the Regional Waste Management Outlooks, which UNEP and ISWA are planning as a direct follow-up to the GMWO.
- *Table 6.6* brings together recommendations on improving data and on testing and further developing performance indicators.
- *Table 6.7* lists a number of priority studies required, either to prepare guidance to assist with the implementation of some of the earlier action imperatives, or for research to fill some of the evidence gaps identified during the course of the GWMO.

Table 6.5 Recommendations for the coverage of the Regional Waste Management Outlooks (RWMOs)

UNEP and ISWA are planning a series of RWMOs as a direct follow-up to the GMWO

- Collate regional data. Include within the scope municipal solid waste, commercial & industrial and construction & demolition wastes; hazardous wastes; wastes particularly relevant to the region (e.g. ship-breaking in Asia); and make an attempt to collate data or make estimates for mining & quarrying and agricultural & forestry wastes.
- Collate regional data on the other major residual streams, namely wastewaters and emissions to air. Take a holistic approach, examining the interfaces between the environmental control regimes for air, water, land and wastes.
- Decide on what priority is appropriate to give within each Regional Outlook to mining & quarrying and agricultural & forestry wastes
- Use the newly available performance indicators to benchmark and compare the performance of the municipal solid waste management systems in a sample of regional cities. Also examine the use of suitable indicators at a national level.
- Develop regional comparisons of MSW collection coverage, controlled disposal rates and recycling rates across a wide range of cities and countries.
- Focus on 'domestic' (i.e. 'in-country') generation and management of hazardous wastes within the region – how far are hazardous wastes routinely separated from non-hazardous wastes for proper management? How widespread is the availability of facilities for the environmentally sound management of hazardous wastes?
- What is the status of waste prevention in the region? What practical steps can be taken to decouple waste generation per capita from economic growth?
- Each RWMO should include a particular focus on food waste generation and prevention, and construction & demolition waste.
- Focus on documenting existing repair, reuse and recycling across the region. Include both formal and informal sector recycling. Document case studies, including examples of successful integration of informal sector reuse and recycling into mainstream solid waste management, and examples where the resource value of waste has been used as one income stream to help extend services to unserved communities.
- Focus on markets for recovered products – dry recyclables, organics and energy – at local, national, regional and global levels. Examine the balance between developing local/national markets and relying on export.
- Examine regional experiences of working with producers of end-of-life products which contribute to the waste stream.
- Investigate and establish data on willingness to pay for services and level of affordability.
- Develop a wide range of short examples and longer case studies on selected individual countries, to investigate the policy instruments and financing arrangements discussed in GWMO and how they have been adapted to the local situation when replicated elsewhere. The longer case studies should follow a common methodology and be elaborated and analysed in-depth, so as to allow also comparative analysis among them.

Table 6.6 Recommendations on improving data and on testing and further developing performance indicators

- Ensure that waste and resource management data are actively included within wider international action as part of the 'data revolution' to improve data for sustainable development.
- Set and monitor targets for waste management, as an important element in meeting the Sustainable Development Goals.
- Need local, national and international action to facilitate the urgently needed STEP-CHANGE in the availability and reliability of waste and resource management data.
- As part of this, ISWA and UNEP to develop, by 2018, globally recognized and internationally-agreed methodology and protocols for collecting waste data at a national and city level. This should consider the elements identified in ***Table 6.2*** and design a recognized data base hosting the updated information.
- Develop suitable protocols for waste prevention metrics, which should then be incorporated into national data systems.
- Commission research on forecasting future rates of municipal solid waste generation per capita. Institutionalize the regular publication and updating of the resulting forecasts, in the same way that the UN currently publishes regular forecasts of world population and urbanization prospects.
- Roll out the newly available performance indicators for solid waste management across a wider range of cities.
- Based on experience in using indicators in the regional guidelines and elsewhere, further work should be carried out to develop standardized indicators, taking into account the existing ones, so as to facilitate benchmarking and better analysis of the state of waste management around the world and also provide a standardized means to monitor progress over time. The long-term aim should be to allow comparison of indicators from a local to a national to a regional to a global level.

Table 6.7 Recommendations for priority studies required as a follow-up to the GWMO

a. Guidance to assist with the implementation of some of the earlier action imperatives

- Guidance on a national framework law for waste management
- Guidance on waste prevention for middle- and low-income countries. Practical and innovative actions are required to stem the inexorable growth in waste per capita as their economies develop. Include some practical examples and case studies, e.g. based on the existing repair and reuse sector, and build on existing guidance focusing on sustainable consumption and production.
- Guidance on low-cost reuse and recycling ('waste to wealth') technologies – including those which involve minimal or low capital investment and which produce products for a local market. Compile case studies. Include barriers and constraints and how to address them. Include 'how to' kits to encourage replication. Coverage should include municipal solid waste, e-waste and potentially other waste streams as well as the necessary health & safety and environmental protection measures to protect both the workers and society.
- Guidance on minimum environmental and health & safety protection standards for waste and resource recovery operations. A particular priority would be those facilities importing materials for recycling.
- Practical guidance on separate handling of household and other relatively small-quantity hazardous waste, particularly in middle- and low-income countries.
- A municipal solid waste management accounting and budgeting tool, and/or practical guidance on financing models, for local government in low- and middle-income countries:
  - Include guidance on budgeting for the 'invisible' costs of modernization – training, customer satisfaction, awareness raising campaigns, managing contractors etc. – and on the need for transparency in both costs and revenues.
  - Provide benchmarks for operating costs and capital costs of commonly used equipment, technologies and facilities for municipal solid waste collection, transfer, treatment, resource recovery and disposal.
  - Provide benchmarks for productivity of labour and commonly used equipment, land-take, energy and fuel consumption etc.

Table 6.7 Recommendations for priority studies required as a follow-up to the GWMO (continued)

b. Research to fill some of the evidence gaps identified during the course of the GWMO
• Research to understand better the dissipation of value when materials, and products at their end-of-life, become waste. Identify points of intervention to mitigate or prevent loss of value. • Economics of waste – with a focus on decoupling waste generation from economic development and quality of life • A better evidence base linking waste and health • A better evidence base on the costs of inaction • Research on the better integration of sanitation and solid waste management in the lowest-income countries • Research on how to achieve behaviour change, to move from a position of uncontrolled disposal and widespread littering in low- and middle-income countries, to universal waste collection, controlled disposal and environmentally sound management. Providing access to services is one necessary component, but the continued prevalence of littering even in high-income countries shows that this on its own is not sufficient. • More effective approaches to international development financing for solid waste management that recognize the need to deliver rapid improvements to waste management systems on the ground. Open up faster, better funded, more flexible credit lines.[8] • Compilation of an evidence base for successful financing of waste management and resource recovery infrastructure, identification of good practices and development of good practice recommendations, aimed at the waste industry, at potential users of secondary raw materials and waste-derived fuel products, and at banks and other financial institutions.[9] The work should include financing models suitable for application to countries of all income levels, and highlight the circumstances more likely to favour each model. – Financing models available, and case studies showing both positive and negative experiences – Detailed analysis of barriers, constraints and risk factors – Identification and analysis of innovative financing instruments that match the needs for guarantees and working capital – Differentiation between financing models that are more suitable for 'waste management' (which rely primarily on a 'charge' for collection or a 'gate fee' for environmentally sound disposal) and those for 'resource recovery' (which are based on the revenue generation potential from recovered materials and/or energy in cases where that is significant) • A better evidence base linking improvements in waste and resource management to mitigation of greenhouse gas (GHG) emissions and climate change. More research is needed, inter alia, on black carbon emissions from the open burning of waste and also on the climate benefits of improved solid waste management in developing countries – both would assist in mobilizing financial assistance for improved waste management from the Global Climate Fund and other potential sources. • Development of more effective approaches for the producers of products and other stakeholders in the supply chain to take more responsibility for waste management associated with their products and waste in developing countries. Suitable approaches should be explored for implementing extended producer responsibility (EPR) for end-of-life products in developing countries, including both voluntary and more legislative approaches, direct cooperation between the producers and the informal sector recyclers who are or could be involved in reuse and recycling of the products, as well as the role of consumers in an EPR system. Another area looks upstream, to ensure the environmentally sound management of waste generated within producers' supply chains, particularly where material extraction and processing and product manufacture take place in developing countries. • Study of experiences with implementation of UNEP/IETC guiding documents in individual countries • Exploration of the evidence base on whether waste management can be used as a reliable proxy-indicator for good governance • An international study on mine tailings, which are of particular interest due to a special combination of factors including their high volume, content of hazardous substances and physical form

8 More detailed recommendations on the work required here are provided in Section 6.3.3.

9 This project has elements of both research and guidance. The Terms of Reference need to ensure that the results are useful both to developed and developing countries – clearly the challenges, barriers and constraints, and risk factors will differ depending on income level and some models will be more applicable to one or the other situation.

6.5 GLOBAL CALL FOR ACTION ON WASTE MANAGEMENT

A key message from the GWMO is that waste is a global as well as a national and a local issue. We therefore conclude with recommendations for a set of Global Waste Management Goals, and a Global Call to Action to achieve those goals.

6.5.1 Global Waste Management Goals

Table 6.8 presents a set of priority Global Waste Management Goals. As noted in the table, all of these goals are already explicit or implicit within the Post-2015 Development Agenda. **Addressing waste management as a priority will facilitate early progress against more than half of the sustainable development goals (SDGs).**

Table 6.8 The Global Waste Management Goals and their relation to the SDGs [C1.3/T1.1]

GLOBAL WASTE MANAGEMENT GOALS			RELATED SDGs	
Ensure by 2020	W.1	Access for all to adequate, safe and affordable solid waste collection services	3 – Health for all	11 – Safe cities
	W.2	Stop uncontrolled dumping, open burning	3 – Health for all 11 – Safe cities 12 – Sustainable consumption and production (SCP)	6 – Clean water and sanitation 14 – Marine resources 15 – Terrestrial ecosystems
Ensure by 2030	W.3	Achieve sustainable and environmentally sound management of all waste, particularly hazardous waste	12.4 – Managing all waste 13 – Climate change	7 – Access to energy
	W.4	Substantially reduce waste generation through prevention and the 3Rs (reduce, reuse, recycle) and thereby create green jobs	12.5 – The 3Rs 8 – Growth & employment	1 – End poverty 9 – Sustainable industry
	W.5	Halve per capita global food waste at the retail and consumer levels and reduce food losses in the supply chain	12.3 – Food waste	2 – End hunger; food security

* Cross-reference is made to Sustainable Development Goals. More detail of the linkages between the Global Waste Management Goals and the specific targets within each of the 17 SDGs are provided in Table 1.1.

A food waste redistribution centre ('food bank') in Belo Horizonte, Brazil

6.5.2 Global call for action

The priority recommendations from the GWMO form a GLOBAL CALL FOR ACTION to national governments and the international community to address these pressing goals.

FOCUS ON DEVELOPING COUNTRIES

- Mobilize international aid and environmental and climate funds to assist the poorest countries. Increase the level of funding on waste management by a factor of 10, from the 0.3% achieved since 2000 to an average of 3% of total international aid funding in the period from 2015 to 2030.
- As an initial step, aim to:
  - achieve 100% collection coverage in all cities with a population more than 1 million,
  - eliminate open burning of municipal solid wastes and similar wastes, and
  - close large open dumps, replacing them with controlled disposal facilities.
- Develop a holistic approach to managing all residuals. In particular develop the integration of sanitation and solid waste management services.
- Build on existing recycling systems while eliminating hazardous resource recovery practices as a means to achieve sustainable livelihoods and reduce costs for the cities.
- Manage hazardous wastes safely. Enforce, and adequately finance domestic enforcement of, the Basel Convention and ensure that sound facilities are available within developing countries for their own wastes.
- Promote producer responsibility programmes to ensure that international companies take more responsibility for waste management associated with their products and wastes in developing countries.
- Establish/strengthen wide-reaching capacity development programmes for administrative, technical and business sectors. Involve developed country cities through twinning and promote collaboration.

FOCUS ON ALL COUNTRIES

All countries still have some way to go to meet the 2030 goals

- Improve access to financing for sound waste management facilities and operations.
- Reduce waste at source, engage citizens, industries and other stakeholders – move from linear waste management to the circular economy.
- Improve substantially the availability and reliability of waste and resource management data – if you don't measure it, you can't manage it.

The 21st century will witness an enormous increase in both quantity and composition of waste. A proactive leadership is needed to turn this growing problem into a resource for sustainable development. Some of the key actions needed by this leadership include: a framework legislation; a priority in the national development plan; a strategy and a plan with full engagement of stakeholders; and a strong networked research and development programme taking full advantage of the data revolution. Sound management of waste is a key component of the circular economy and will make a significant contribution to the Post-2015 Development Agenda.

Annex A: Further Resources

The convention for referencing in the GWMO is that direct footnotes are used in the text, either to cross reference to this list of 'Further Resources' which has been arranged by chapter; or to reference a specific source which is not of more general interest; or to add some explanatory comments. This Annex both provides the reader with a selective list of 'Further reading' and collates the more important references used earlier.

CHAPTER 1 AND OVERARCHING REFERENCES

Precursors of the GWMO in other sectors

UNEP (2011c). *Towards a Green Economy: Pathways to Sustainable Development and Poverty Eradication*. www.unep.org/greeneconomy Green Economy Report. Waste Chapter available at http://www.unep.org/greeneconomy/Portals/88/documents/ger/8.0_Waste.pdf

UNEP (2012a). *Global Environment Outlook (GEO-5)*. http://www.unep.org/geo/pdfs/geo5/GEO5_report_full_en.pdf

UNEP (2012b). *Global Outlook on SCP Policies: taking action together*. http://www.unep.org/pdf/Global_Outlook_on_SCP_Policies_full_final.pdf

UNEP (2013b). *Global Chemicals Outlook - Towards Sound Management of Chemicals*. http://www.unep.org/chemicalsandwaste/Portals/9/Mainstreaming/GCO/Rapport_GCO_calibri_greendot_20131211_web.pdf

Waste management

Abarca-Guerrero, L., G. Maas & W. Hogland (2013). Solid waste management challenges for cities in developing countries. *Waste Management*, 33 (1), 220-232. doi:10.1016/j.wasman.2012.09.008

Huda, K. M. N. (2008). *Municipal Solid Waste Management - Bangladesh Perspective*. Dhaka: Academic Press and Publishers Library, www.applbooks.com

ISWA (2014b). Globalisation and Waste Management. Final report from the ISWA Task Force. Mavropoulos, A., D.C. Wilson, B.A. Appelqvist, C.A. Velis et al. Vienna: ISWA. http://www.iswa.org/fileadmin/galleries/Task_Forces/TFGWM_Report_GWM_LR.pdf

Modak, P. (2010). "Municipal Solid Waste Management: Turning Waste into Resources," in *Shanghai Manual - A Guide for Sustainable Urban Development in the 21st Century*. United Nations Bureau International de Expositions Shangai 2010 World Exposition Executive Committee. http://www.un.org/esa/dsd/susdevtopics/sdt_pdfs/shanghaimanual/Chapter%205%20-%20Waste_management.pdf

Onibokun, A. (Ed.). (1999). Managing the Monster – Urban Waste and Governance in Africa. International Development Research Centre (IDRC), Ottawa, Canada. http://www.idrc.ca/EN/Resources/Publications/Pages/IDRCBookDetails.aspx?PublicationID=346

Scheinberg, A., D.C. Wilson & L. Rodic (2010). *Solid Waste Management in the World's Cities: Water and Sanitation in the World's Cities 2010*. Prepared on behalf of UN-HABITAT. London: Earthscan. http://www.waste.nl/en/product/solid-waste-management-in-the-worlds-cities

Vergara, S.E. & G. Tchobanoglous (2012). Municipal solid waste and the environment: A global perspective. *Annual Reviews of Environment and Resources*, 37: 277-309.

Wilson, D.C., L. Rodic, A. Scheinberg et al. (2012). Comparative analysis of solid waste managment in 20 cities. *Waste Management & Research*, 30(3), 237-254.

Worrell, W.A. & P.A. Vesilind (2012). *Solid Waste Engineering*, 2nd edition. Stanford, California: Cengage Learning.

Zhu, D., P.U. Asnani, C. Zurbrügg et al. (2008). *Improving Municipal Solid Waste Management in India – A Sourcebook for Policy Makers and Practitioners*. Washington DC: The World Bank. https://openknowledge.worldbank.org/handle/10986/6916

Strategic planning for solid waste management

ABRELPE/ISWA (2013). *Solid waste: guidelines for successful planning*. Available in Portuguese and English. http://www.abrelpe.org.br/_eng/manual_apresentacao.cfm

IJgosse, J., J. Olley, V. de Vreede et al. (2003). Waste Keysheets for Municipal Waste Management Planning. www.wastekeysheets.net

UNDG (2011). *Results-Based Management Handbook. Harmonising RBM concepts and approaches for improved development results at country level.* http://www.undg.org/docs/12316/UNDG-RBM%20Handbook-2012.pdf

UNEP (2009c). *Developing Integrated Solid Waste Management Plan – Training Manual Vol. 1: Waste Characterization and Quantification with Projections for Future; Vol. 2: Assessment of Current Waste Management System and Gaps therein; Vol. 3: Targets and Issues of Concern for ISWM. Vol. 4: ISWM Plan*. http://www.unep.org/ietc/OurWork/WasteManagement/Capacitybuilding/tabid/56256/Default.aspx

UNEP, UNITAR (2013). *Guidelines for National Solid Waste Management Strategies: Moving from challenges to opportunities.* http://www.unep.org/ietc/Portals/136/Publications/Waste%20Management/UNEP%20NWMS%20English.pdf

Wilson, D.C., A. Whiteman & A. Tormin (2001). Strategic planning guide for municipal solid waste management. World Bank. Washington, DC, U.S., www.worldbank.org/urban/solid_wm/erm/start_up.pdf

Topic Sheets

Waste and climate

Bond, T.C., S.J. Doherty, D.W. Fahey et al. (2013). Bounding the role of black carbon in the climate system: A scientific assessment. *Journal of Geophysical Research: Atmospheres, 118*(11), 5380-5552. doi: 10.1002/jgrd.50171

Christensen, T. (Ed.). (2009). Fundamentals in Greenhouse Gas Accounting: Concepts and mechanisms. *Waste Management & Research*, 27(8): 693-836.

Climate and Clean Air Coalition to Reduce Short-Lived Climate Pollutants (CCAC). (2014). CCAC MSW Knowledge Platform. http://waste.ccac-knowledge.net

Dehoust, G., D. Schüler, R. Vogt et al. (2013). *Climate Protection Potential in the Waste Management Sector - Examples: Municipal Waste and Waste Wood*. Germany, UBA and BDE. http://www.umweltbundesamt.de/publikationen/klimaschutzpotenziale-abfallwirtschaft

Gentil, E., T.H. Christensen & E. Aoustin (2009). Greenhouse gas accounting and waste management. *Waste Management & Research*, 27(8): 696-706.

Giegrich, J. & R. Vogt (2009). KfW & GIZ SWM-GHG calculator. Tool for calculating greenhouse gases (GHG) in solid waste management (SWM). Frankfurt, Germany. Prepared on behalf of KfW Development Bank (German Financial Development Cooperation) and GIZ (German Technical Development Co-operation). http://www.ifeu.org/abfallwirtschaft/pdf/Klimarechner/SWM-GHG%20Calculator_blanko_2010_01.xls

Hansen, J.A. (Ed.). (2009). Applied Greenhouse Gas Accounting: Methodologies and cases. *Waste Management & Research*, 27(9): 837-936.

IPCC (2013). Climate Change 2013: The Physical Science Basis. Contribution of Working Group I to the Fifth Assessment Report of the Intergovernmental Panel on Climate Change. Stocker, T.F., D. Qin, G.-K. Plattner et al. (Eds.). Cambridge, United Kingdom and New York, NY, U.S. http://www.ipcc.ch/report/ar5/wg1/

ISWA (2009). *White Paper on Waste and Climate Change*. Vienna, Austria: International Solid Waste Association. http://www.iswa.org/fileadmin/user_upload/_temp_/Small_GHG_white_paper_01.pdf

UNEP (2010). *Waste and Climate Change: Global Trends and Strategy Framework*. Osaka/Shiga: Division of Technology, Industry and Economics. International Environmental Technology Centre. http://www.unep.or.jp/ietc/Publications/spc/Waste&ClimateChange/Waste&ClimateChange.pdf

UNEP/ISWA (2012). Joint thematic paper on waste and climate change. http://www.iswa.org/index.php?eID=tx_iswaknowledgebase_download&documentUid=3145

USEPA (U.S. Environmental Protection Agency). (2013). Clean Energy & Climate Change — Waste Management. *How waste management impacts greenhouse gas emissions.* http://www.epa.gov/region9/climatechange/waste.html

The world's 50 biggest dumpsites

Waste Atlas Partnership (2014). Waste Atlas: The World's 50 Biggest Dumpsites, 2014 Report. http://www.d-waste.com/d-waste-news/item/263-the-world-s-50-biggest-dumpsites-official-launching-of-the-2nd-waste-atlas-report.html

ISWA (2015b). Wasted Health – The tragic case of dumpsites. A. Mavropoulos ISWA STC Chair, with contributions from D. Newman, ISWA President. Vienna.

CHAPTER 2: BACKGROUND, DEFINITIONS, CONCEPTS AND INDICATORS

Waste history and drivers

Marshall, R.E. & K. Farahbakhsh (2013). Systems approaches to integrated solid waste management in developing countries. *Waste Management,* 33, 988-1003.

Melosi, M.V. (2000). *The Sanitary City: Urban Infrastructure in America from Colonial Times to the Present (Creating the North American Landscape)*. Baltimore: Johns Hopkins University Press.

Strasser, S. (1999). *Waste and Want: A Social History Of Trash*. Owl Books, New York.

Velis, C., D.C. Wilson & C. Cheeseman (2009). 19th century London dust-yards: A case study in closed-loop recycling and zero waste. *Waste Management* 29: 1282-1290.

Wilson, D.C. (2007). Development drivers for waste management. *Waste Management & Research,* 25(3): 198-207.

Woodward, D. (1985). "Swords into Ploughshares": Recycling in pre-industrial England. *Economic History Review 2nd Series*, 38, 175–191.

Integrated sustainable waste management

Schübeler, P., K. Wehrle & J. Christen (1996). Conceptual framework for municipal solid waste management in low-income countries. SKAT, St Gallen, Switzerland, SKAT. UMP/SDC Collaborative Programme on Municipal Solid Waste Management in Developing Countries. Urban Management Programme (UMP) Working Paper Series, No. 9., pp. 59. www.skat.ch

Van de Klundert, A. & J. Anschütz (2001). Integrated Sustainable Waste Management - the Concept. WASTE, Gouda, The Netherlands. http://www.waste.nl/en/concept-tool-iswm

Wilson, D.C., C.A. Velis & L. Rodic (2013). Integrated sustainable waste management in developing countries. Proceedings of the Institution of Civil Engineers, Waste and Resource Management, 166, WR2, 52-68. http://www.icevirtuallibrary.com/content/article/10.1680/warm.12.00005

LCA and other assessment tools for waste and resource management

Allesch, A. & P.H. Brunner (2014). Assessment methods for solid waste management: A literature review. *Waste Management & Research*, 30 (6), 461-473.

Brunner, P.H. & H. Rechberger (2004). Practical Handbook of Material Flow Analysis. Boca Raton: CRC Press LLC.

Gentil, E., A. Damgaard, M. Hauschild et al. (2010). Models for waste life cycle assessment: Review of technical assumptions. *Waste Management*, Volume 30, Issue 12, December 2010, Pages 2636–2648.

European Topic Centre on Sustainable Consumption and Production (n.d.). Life-cycle assessment and life-cycle thinking in resource and waste management. http://scp.eionet.europa.eu/themes/lca

Laurent, A., J. Clavreul, A. Bernstad et al. (2014a). Review of LCA studies of solid waste management systems – Part I: Lessons learned and perspectives. Waste Management, Volume 34, Issue 3, March 2014, Pages 573–588.

Laurent, A., J. Clavreul, A. Bernstad et al. (2014b). Review of LCA studies of solid waste management systems – Part II: Methodological guidance for a better practice. Waste Management, Volume 34, Issue 3, March 2014, Pages 589–606.

Zurbrügg, C., M. Caniato & M. Vaccari (2014). How Assessment Methods Can Support Solid Waste Management in Developing Countries—A Critical Review. *Sustainability 2014*, 6(2), 545-570; doi:10.3390/su6020545

Waste data and indicators

Armijo, C., A. Puma & S. Ojeda (2011). A set of indicators for waste management programs. 2nd International Conference on Environmental Engineering and Applications. *IPCBEE* vol.17 (2011). Singapore: IACSIT Press.

Bringhentia, J.R., E. Zandonadeb, & W.M.R. Günther (2011). Selection and validation of indicators for programs selective collection evaluation with social inclusion. *Resources, Conservation and Recycling*, 55, 876–884.

CEPT University - PAS (2010). *Performance Measurement Framework for Urban Water and Sanitation. Volume I: Approach and Framework. Volume II: List of Indicators and Reliability Assessment.* http://pas.org.in/Portal/document/ResourcesFiles/pdfs/Performance%20Measurement%20Framework%20Report_Vol%20I%20&%20II.pdf

Forfás (2010). Waste Management In Ireland. Benchmarking Analysis and Policy Priorities: Update 2010. http://www.forfas.ie/media/forfas101005-Waste_Management_Benchmarking_Analysis_2010.pdf

GCIF (2012). Global City Indicators Facility. www.cityindicators.org

Greene, K. & D. Tonjes (2014). Quantitative assessments of municipal waste management systems: Using different indicators to compare and rank programs in New York State. *Waste Management*, 34(4), 825-836.

Hotta, Y. (Ed.). (2014). 3R Policy Indicator Factsheets - Discussion Paper. Asia Resource Circulation Policy Research Group. Kanagawa, Japan: Institute for Global Environmental Strategies (IGES). [Online] http://pub.iges.or.jp/modules/envirolib/upload/4977/attach/3RIndicator_B5report_web.pdf (Accessed 23 April, 2014)

Huang, Y.T., T.C. Pan and J.J. Kao (2011). Performance assessment for municipal solid waste collection in Taiwan. *Journal of Environment Management*, 92, 1277-1283.

IEAG (2014). *A World that Counts - Mobilising the Data Revolution for Sustainable Development*. First report of the UN Secretary General's Independent Expert Advisory group on a Data Revolution for Sustainable Development's, November 2014. http://www.undatarevolution.org/report/

India, Ministry of Urban Development (MOUD). (2010). *Handbook of Service Level Benchmarking*. http://jnnurm.nic.in/wp-content/uploads/2010/12/SLB-Handbook.pdf.

Jasch, C. (2000). Environmental performance evaluation and indicators. *Journal of Cleaner Production*, 8(1), 79-88.

Mendes, P., A.C. Santos, F. Perna et al. (2012). The balanced scorecard as an integrated model applied to the Portuguese public service: A case study in the waste sector. *Journal of Cleaner Production*, 24, 20-29.

Menikpura, S.N.M., S.H. Gheewala, S. Bonnet et al. (2013). Evaluation of the effect of recycling on sustainability of municipal solid waste management in Thailand. *Waste and Biomass Valorization*, 4(2), 237-257.

Munizaga, J.A. & A.L. Garcia (2012). Propuesta para la Evaluación Integral de Sistemas de Gestión de Residuos Domésticos [Proposal for Comprehensive Evaluation of Municipal Solid Waste Management System]. In Spanish. Conama2012 - Congreso Nacional del Medio Ambiente.

Munizaga, J.A. & A.L. Garcia (2013). Garbometer: A methodology for comprehensive evaluation of municipal solid waste management systems. Proceedings of ISWA World Congress 2013, 7-11 October, Vienna, Austria. Vienna: International Solid Waste Association (ISWA).

Polaz, C.N.M. & B.A.N. Teixeira (2009). Indicadores de sustentabilidade para a gestão municipal de resíduos sólidos urbanos: Um estudo para São Carlos [Indicators of sustainability for municipal solid waste management: Case study of the city of São Carlos]. In Portuguese. *Engenharia Sanitaria e Ambiental*, 14(3), 411-420.

Regions4Recycling (2014). *Regions for Recycling: R4R Toolkit and Methodology*. http://www.regions4recycling.eu/R4R_toolkit/R4R_methodology

Romualdo J.C. (2014). Development and testing of an indicator set to benchmark the performance of a national hazardous waste management system (MSc thesis). Imperial College London.

Ronconi, M. (2001). The Development of Waste Indicators at European Union Level: Some Recent Eurostat Experiences. Working paper 29 to a joint ECE/ Eurostat Work Session on Methodological Issues in Waste Statistics. Statistical Commission and Econonmic Commission for Europe (ECE), Commission of the European Communities, Conference of European Statisticians and Eurostat. http://www.unece.org/fileadmin/DAM/stats/documents/2001/10/env/wp.29.e.pdf

SWEEPNet (2014). Challenges and Opportunities for Solid Waste Management in the Mashreq and Maghreb Region. Sweepnet - The Regional Solid Waste Exchange of Information and Expertise network in Mashreq and Maghreb countries. http://www.sweep-net.org/sites/default/files/REGIONAL%20RA%20ANG%20WEB.pdf

UNSD (United Nations Statistics Division). (2013). *Framework for the Development of Environment Statistics 2013*. http://unstats.un.org/unsd/environment/fdes.htm

Wilson, D.C., L. Rodic, M.J. Cowing et al. (2015). 'Wasteaware' benchmark indicators for integrated sustainable waste management in cities. *Waste Management*, 35 (1), 329-343. doi:10.1016/j.wasman.2014.10.006

Wilts, H. (2012). National waste prevention programs: Indicators on progress and barriers. *Waste Management & Research*, 30((9) Supplement), 29-35.

Zaman, A.U. & S. Lehmann (2013). The zero waste index: A performance measurement tool for waste management systems in a 'zero waste city'. *Journal of Cleaner Production*, 50, 123-132.

Materials flow analysis

Baccini, P. & P.H. Brunner (2012). *Metabolism of the Anthroposhere - Analysis, Evaluation, Design*. 2nd Edition. Cambridge, Massachusetts, U.S.: The MIT Press.

Bringezu, S. (1997). Material flow indicators. In Moldan, B. and S. Billharz (Eds.) *Sustainability indicators*. New York: John Wiley & Sons Ltd., 168-176.

Cleveland, C.J. & M. Ruth (1998). Indicators of Dematerialization and the Materials Intensity of Use." *Journal of Industrial Ecology* 2(3), 15-50.

Hinterberger, F., E. Luks & F. Schmidt-Bleek (1997). Material flows vs. "natural capital": What makes an economy sustainable? *Ecological Economics* (23), 1-14.

Loppolo, G., R. Heijungs, S. Cucurachi et al. (2014). Urban Metabolism: Many Open Questions for Future Answers. In Salomone, R. & G. Saija (Eds.) *Pathways to Environmental Sustainability*, Springer, 23-32.

Pincetl, S., P. Bunje & T. Holmes (2012). An expanded urban metabolism method: Toward a systems approach for assessing urban energy processes and causes. Landscape and Urban Planning, 193-202.

Schmidt-Bleek, F. (1994). Revolution in resource productivity for a sustainable economy - a new research agenda. Fresenius Environmental Bulletin (2): 245-490.

UNEP and CSIRO (2011). *Resource Efficiency: Economics and Outlook for Asia and the Pacific*. http://apps.unep.org/publications/pmtdocuments//pdf/Resource_Efficiency_EOAP_web.pdf.

UNEP and CSIRO (2013). *Recent Trends in Material Flows and Resource Productivity in Asia and the Pacific.* http://www.unep.org/pdf/RecentTrendsAP(FinalFeb2013).pdf

West, J., H. Schandl & S. Heyenga (2013). Resource Efficiency: Economics and Outlook for China. UNEP, CSIRO and IPM. ISBN 13:9789280733181. http://www.unep.org/pdf/China_Resource_Efficiency_in_Chinese_2013.pdf

Topic Sheets

Waste prevention

Brook Lyndhurst for DEFRA (2009). *WR1204 Household Waste Prevention Evidence Review: L1m1 – Executive Report. A report for Defra's Waste and Resources Evidence Programme*.

Cox, J., S. Giorgi, V. Sharp et al. (2010). Household waste prevention - A review of evidence, *Waste Management and Research,* 28(3): 193-219.

EEA (European Environment Agency). (2013b). Waste prevention in Europe - the status in 2013, *EEA Report No 9/2014*. http://www.eea.europa.eu/publications/waste-prevention-in-europe-2014

Sabogal, N. (2013). Cartagena Declaration on the Prevention, Minimization and Recovery of Hazardous Wastes and Other Wastes, Proceedings of the Eighth International Conference on Waste Management and Technology, Towards Ecological Civilization, Shanghai, China, October 23-25, 2013.

UK, DEFRA (UK Department for Environment, Food & Rural Affairs). (2012). Business Waste Prevention Evidence Review – WR1403. http://randd.defra.gov.uk/Default.aspx?Menu=Menu&Module=More&Location=None&Completed=0&ProjectID=17499

Wilson, D.C., D. Parker, J. Cox et al. (2012). Business waste prevention: A review of the evidence, *Waste Management and Research*, 30(9 SUPPL.1): 17-28.

Wilts, H., G. Dehoust, D. Jepsen et al. (2013). Eco-innovations for waste prevention – Best practices, drivers and barriers. *Science of the Total Environment,* 461-462: 823-829.

Sustainable consumption and production

Clement, S. (Ed.). (2007). The Procura+ Manual. A Guide to Cost-Effective Sustainable Public Procurement. 2nd edition. ICLEI – Local Governments for Sustainability. http://archive.iclei.org/index.php?id=796

European Commission (EC). (2011b). Proposal for a Directive of the European Parliament and of the Council on Public Procurement, COM(2011) 896 Final, Brussels, 2011. http://www.europarl.europa.eu/meetdocs/2009_2014/documents/com/com_com(2011)0896_/com_com(2011)0896_en.pdf

Georghiou, L., J. Edler, E. Uyarra et al. (2013). Policy instruments for public procurement of innovation: Choice, design and assessment, *Technological Forecasting and Social Change*. http://dx.doi.org/10.1016/j.techfore.2013.09.018

Lorek, S. & D. Fuchs (2013). Strong sustainable consumption governance – Precondition for a degrowth path? *Journal of Cleaner Production* 38: 36-43

McDonough, W. & M. Braungart (2002). *Cradle to Cradle: Remaking the Way We Make Things*. North Point Press, New York, U.S.

McDonough, W. & M. Braungart (2013). *The Upcycle: Beyond Sustainability – Designing for Abundance*. North Point Press, New York, U.S.

Tukker, A., S. Emmert, M. Charter et al. (2008). Fostering change to sustainable consumption and production: An evidence based view, *Journal of Cleaner Production* 16(11): 1218-1225.

UNEP (2009b). *Design for sustainability: Step-by-step approach.* http://www.d4s-sbs.org Various other interesting publications on design for sustainability can be found at : http://www.unep.org/resourceefficiency/Business/SustainableProducts/DesignforSustainability/tabid/78845/Default.aspx.

UNEP (2012d). Global Outlook on Sustainable Consumption and Production Policies. Taking action together. http://www.unep.org/pdf/Global_Outlook_on_SCP_Policies_full_final.pdf. Various other useful UNEP publications on this topic can be found at http://www.unep.org/resourceefficiency/.

UNEP/Wuppertal Institute Collaborating Centre on SCP, with the Stockholm Environment Institute (2010). What public policy framework is required to encourage sustainable consumption business strategies? http://www.encourage-sustainable-lifestyles.net/fileadmin/content/downloads/CSCP_Sustainable_Consumption_Policy_Instruments.pdf

USEPA (U.S. Environmental Protection Agency). (n.d.). The Revised (Draft) Guidelines for Product Environmental Performance Standards and Ecolabels for Voluntary Use in Federal Procurement. http://www.epa.gov/epp/draftGuidelines/index.html.

Small Island Developing States (SIDS)

ADB (2014) Solid Waste Management in the Pacific – Financial Arrangements http://www.adb.org/sites/default/files/publication/42656/solid-waste-management-financial-arrangements.pdf and http://www.environment.gov.ki/?page_id=37

Secretariat of the Pacific Regional Environment Programme (SPREP). Regional Strategies on Hazardous waste and solid waste, Case studies. http://www.sprep.org/case-studies/

SIDS Accelerated Modalities of Action (S.A.M.O.A.) Pathway http://www.sids2014.org/index.php?menu=1537

Stock, P. (2014). *Island Innovations – UNDP and GEF: Leveraging the Environment for the Sustainable Development of Small Island Developing States.* UNDP. http://www.undp.org/content/dam/samoa/docs/UNDP_WS_IslandInnovations_UNDP_GEF_Leveraging_the_Env.pdf

UNEP (2012f). *SIDS-Focused Green Economy – An analysis of challenges and opportunities.* http://www.unep.org/pdf/Green_Economy_in_SIDS.pdf

UNEP (2014b). *GEO Small Island Developing States Outlook.* http://apps.unep.org/publications/pmtdocuments/-Global%20Environment%20Outlook:%20small%20island%20developing%20states-2014GEO_SIDS_final.pdf

CHAPTER 3: GLOBAL STATUS OF WASTE MANAGEMENT

Collated data sources

Chalmin, P. & C. Gaillochet (2009). *From waste to resource: World waste survey*. Paris: Economica.

D-Waste with University of Leeds, ISWA, SweepNet, and WTERI. Waste Atlas. http://www.atlas.d-waste.com/

EEA (European Environment Agency). (2013a). *Managing municipal solid waste - a review of achievements in 32 European countries*. http://www.eea.europa.eu/publications/managing-municipal-solid-waste

EEA (European Environment Agency). (2015). *2015 State and Outlook of the Environment Report*. This includes both a European assessment on waste: http://www.eea.europa.eu/soer-2015/europe/waste and a European cross-country comparison on waste: http://www.eea.europa.eu/soer-2015/countries-comparison/waste

Hoornweg, D. & P. Bhada-Tata (2012). *What a Waste: A Global Review of Solid Waste Management*. Urban Development Series Knowledge Papers. World Bank. http://siteresources.worldbank.org/INTURBANDEVELOPMENT/Resources/336387-1334852610766/What_a_Waste2012_Final.pdf

IFC (2014). Waste in Russia: Garbage or Valuable Resource? http://www.ifc.org/wps/wcm/connect/6895a9004392026fb0dbb8869243d457/IFC+Waste+in+Russia+Report_rus.pdf?MOD=AJPERES

India, Ministry of Urban Development (MOUD). (2012). *Service levels in urban water and sanitation sector, Status report 2010-2011*. http://moud.gov.in/sites/upload_files/moud/files/SLB%20National%20Data%20Book_0.pdf Karak, T., R.M. Bhagat & P. Bhattacharyya (2012). Municipal solid waste generation, composition, and management: The world scenario. *Critical Reviews in Environmental Science and Technology* 42(15): 1509-1630.

Russia, Ministry of Natural Resources and Environment (2014). National Report "On State and Protection of the Environment in the Russian Federation in 2013". Prepared in accordance with the Resolution No. 966 of the Government of the Russian Federation. http://www.mnr.gov.ru/regulatory/list.php?part=1101

UNDESA (2010). *Trends in Sustainable Development - Chemicals, Mining, Transport, Waste Management 2010-2011*. http://www.un.org/esa/dsd/resources/res_pdfs/publications/trends/trends_Chemicals_mining_transport_waste/ch4_waste_management.pdf

UNDESA (2012). *World Population Prospects: The 2011 Revision*. http://esa.un.org/wpp/

UNDESA (2013). *World Urbanization Prospects, 2012 Edition*. http://esa.un.org/unpd/wup/

UNDESA (2014). *World Population Prospects: The 2013 Revision*. http://esa.un.org/wpp/

UNDESA (2015). *World Urbanization Prospects, 2014 Edition*. http://esa.un.org/unpd/wup/

Zoi Environmental Network and GRIS-Arendal (2012). *Vital Waste Graphics 3*. Earlier editions also available. http://www.grida.no/publications/vg/

Municipal solid waste management (MSWM) – trends, composition, collection, disposal

Beigl, P., G. Wassermann, F. Schneider et al. (2004). *Forecasting Municipal Solid Waste Generation in Major European Cities*. Institute of Waste Management, BOKU – University of Natural Resources and Applied Life Sciences, Vienna, Austria. http://www.iemss.org/iemss2004/pdf/regional/beigfore.pdf

Coffey, M. & A. Coad (2010). *Collection of municipal solid waste in developing countries, 2nd edition*. UN-Habitat, Nairobi.

ISWA (2013a). Guidelines for Design and Operation of Municipal Solid Waste Landfills in Tropical Climates. Munawar, E. and J. Fellner. http://www.iswa.org/nc/home/news/news-detail/browse/6/article/new-publication-guidelines-for-the-design-and-operation-of-municipal-solid-waste-landfills-in-tropi/109/

Kurian, J., R. Nagendran, K. Thanasekaran et al. (n.d.). *Dumpsite Rehabilitation Manual*. Centre for Environmental Studies. Anna University, Chennai, India. http://www.elaw.org/system/files/Dumpsite+Rehabilitation+Manual_0.pdf

Hoornweg, D., P. Bhada-Tata & C. Kennedy (2013). Waste production must peak this century. *Nature, 502 (7473): 615-617. 31 October 2013*.

Hoornweg, D., P. Bhada-Tata & C. Kennedy (2015). Peak Waste: When Is It Likely to Occur? *Journal of Industrial Ecology*, 19 (1), 117-128. http://onlinelibrary.wiley.com/enhanced/doi/10.1111/jiec.12165/

ISWA (2010). *Landfill Operational Guidelines, 2nd Edition*. ISWA Working Group on Landfill. Vienna. http://www.iswa.org/index.php?eID=tx_iswaknowledgebase_download&documentUid=1449

ISWA (2011). *International Guidelines for Landfill Evaluation*. ISWA Working Group on Landfill. Vienna. http://www.iswa.org/index.php?eID=tx_bee4mememberships_download&fileUid=156

ISWA (2012a). *State of the Nation Report: Landfilling practices and regulation in different countries*. Working Group on Landfill. Vienna. http://www.iswa.org/index.php?eID=tx_iswaknowledgebase_download&documentUid=3141

Rushbrook, P. & M. Pugh (1999). Solid waste landfills in middle- and lower- income countries – a technical guide to planning, design and operation. Washington: World Bank Technical Paper No 426. http://documents.worldbank.org/curated/en/1999/01/2063254/solid-waste-landfills-middle-lower-income-countries-technical-guide-planning-design-operation

Slack, R.J,, M. Bonin, J.R. Gronow et al. (2007). Household Hazardous Waste Data for the UK by Direct Sampling. *Environmental Science & Technology*. 41: pp 2566-2571.

Recycling (including collection for recycling and segregation)

Bartl, A. (2014). Moving from recycling to waste prevention: A review of barriers and enables. *Waste Management & Research*. 32(9) Supplement 3–18

Dahmus, J.B. & T.G. Gutowski (2007). What gets recycled: an information theory based model for product recycling. *Environmental Science &Technology* 41: 7543–7550.

Regional 3R Forum in Asia (2013). The Ha Noi 3R Declaration - Sustainable 3R Goals for Asia and the Pacific for 2013-2023. https://www.env.go.jp/en/focus/docs/files/20130318-67.pdf

UNEP (2011). *Recycling Rates of Metals: A Status Report*. http://www.unep.org/resourcepanel/Portals/24102/PDFs/Metals_Recycling_Rates_110412-1.pdf

Velis, C.A. & P.H. Brunner (2013). Recycling and resource efficiency: it is time for a change from quantity to quality. *Waste Management & Research*. 31(6) 539–540. http://wmr.sagepub.com/content/31/6/539.short

Technologies for Resource Recovery

Administrative Staff College of India (2014). *Waste to Resources Projects in Municipal Solid Waste Management in India* (2 volumes).

Arena, U. (2012). Process and technological aspects of municipal solid waste gasification. A review. *Waste Management*, 32: 625–639.

Asian Development Bank (2013). Materials Recovery Facility Tool Kit. http://www.adb.org/sites/default/files/publication/30220/materials-recovery-facility-tool-kit.pdf

Basel Convention (2011). Technical guidelines on the environmentally sound co-processing of hazardous wastes in cement kilns. http://www.basel.int/Portals/4/Basel%20Convention/docs/pub/techguid/cement/tg-cement-e.pdf

Canada, Environment Canada (2013). Technical Document on Municipal Solid Waste Organics Processing, 208pp. http://www.ec.gc.ca/gdd-mw/3E8CF6C7-F214-4BA2-A1A3-163978EE9D6E/13-047-ID-458-PDF_accessible_ANG_R2-reduced%20size.pdf (Accessed 11 May 2015)

Chen, D.Z., L.J. Yin, H. Wang et al. (2015). Pyrolysis technologies for municipal solid waste: A review. *Waste Management*, 37, 116-136.

Christensen, T.H. (Ed.). (2011). *Solid Waste Technology and Management* (2 volumes). Chichester: Wiley.

Dube, R., V. Nandan, P. Bineesha et al. (2014). Status Paper on Utilisation of Refuse Derived Fuel (RDF) in India. Indo-German Environmental Partnership (IGEP). http://www.igep.in/live/hrdpmp/hrdpmaster/igep/content/e54413/e54441/e61172/e61173/20130425_RDFstatuspaper_final4.pdf

Eawag/Sandec & Waste Concern (2006). *Decentralised Composting for Cities of Low- and Middle-Income Countries - A User's Manual*. http://www.eawag.ch/fileadmin/Domain1/Abteilungen/sandec/publikationen/SWM/Decentralized_Composting/Rothenberger_2006_en.pdf

European Commission (EC). (2006). Integrated Pollution Prevention and Control Reference Document on Best Available Techniques for the Waste Treatments Industries. http://eippcb.jrc.ec.europa.eu/reference/BREF/wt_bref_0806.pdf

Hoornweg, D., L. Thomas & L. Otten (1999). Composting and Its Applicability in Developing Countries. Urban and Local Government Working Paper Series #8, World Bank, Washington, DC, U.S. http://www.worldbank.org/urban/solid_wm/erm/CWG%20folder/uwp8.pdf

Huang, Q., Y. Yang & Q. Wang (2012). Potential for Serious Environmental Threats from Uncontrolled Co-processing of Wastes in Cement Kilns. *Environmental Science & Technology*, 46: 13031–13032. http://pubs.acs.org/doi/ipdf/10.1021/es3042274

India, Planning Commission of India (2014). Report of the Task Force on Waste to Energy (Volume I). http://planningcommission.nic.in/reports/genrep/rep_wte1205.pdfISWA (2012b). Waste to Energy: State of the Art Report, 6th Edition. ISWA Working Group on Energy Recovery. Denmark, 2012. http://www.waste-management-world.com/content/dam/wmw/online-articles/documents/2013/ISWA_WtE_State_of_the_Art_Report_2012_08_FV.pdf

ISWA (2013b). ISWA Guidelines: Waste To Energy In Low And Middle Income Countries. Report prepared for ISWA by Working Group Energy Recovery. Kamuk, B. & J. Haukohl. Vienna: ISWA http://www.iswa.org/index.php?eID=tx_iswaknowledgebase_download&documentUid=3252

Leckner, B. (2015). Process aspects in combustion and gasification Waste-to-Energy (WtE) units. *Waste Management*, 37:13-25.

Mueller, C. (2007). Anaerobic Digestion of Biodegradable Solid Waste in Low and Middle-Income Countries. Eawag/Sandec. http://www.eawag.ch/forschung/sandec/publikationen/swm/dl/Mueller_2007.pdf

Rapport, J., R. Zhang, B.M. Jenkins et al. (2008). Current Anaerobic Digestion Technologies Used for Treatment of Municipal Organic Solid Waste. Report for the California Integrated Waste Management Board. http://www.calrecycle.ca.gov/Publications/Documents/1275/2008011.pdf

REN21 (2014). Renewables 2014 Global Status Report. http://www.ren21.net/Portals/0/documents/Resources/GSR/2014/GSR2014_full%20report_low%20res.pdf

Storey, D., L. Santucci, J. Aleluia et al. (2013). Decentralized and Integrated Resource Recovery Centers in Developing Countries: Lessons Learnt from Asia-Pacific United Nations Economic and Social Commission for Asia and the Pacific (ESCAP), Thailand. Paper presented at the 2013 ISWA Congress, 7-11 October 2013. http://www.unescap.org/sites/default/files/Paper_ESCAP%20paper%20on%20IRRC%20ISWA%20Congress.pdf

Terraza, H. & H. Willumsen (2009). Guidance Note on Landfill Gas Capture and Utilization, Inter-American Development Bank. http://idbdocs.iadb.org/wsdocs/getdocument.aspx?docnum=35447750

Umweltbundesamt (UBA)/Federal German Environment Agency (2013). Best practice municipal waste management. www.umweltbundesamt.de/en/topics/waste-resources/waste-management/waste-technology-transfer

UK, DEFRA (UK Department for Environment, Food & Rural Affairs). (2013). Mechanical Biological Treatment of Municipal Solid Waste. https://www.gov.uk/government/uploads/system/uploads/attachment_data/file/221039/pb13890-treatment-solid-waste.pdf

UNEP (2009a). Converting Waste Plastics into a Resource, Assessment Guidelines; and Converting waste plastics into a resource: Compendium of technologies. http://www.unep.or.jp/letc/Publications/spc/WastePlasticsEST_AssessmentGuidelines.pdf / http://www.unep.or.jp/letc/Publications/spc/WastePlasticsEST_Compendium.pdf

USEPA/ISWA (2012). International Best Practices Guide for Landfill Gas Energy Projects. Global Methane Initiative. https://www.globalmethane.org/documents/toolsres_lfg_IBPGcomplete.pdf

Velis, C.A., P.J. Longhurst, G.H. Drew et al. (2009). Biodrying for mechanical–biological treatment of wastes: A review of process science and engineering. *Bioresource Technology* 100 (11), 2747-2761.

Velis, C.A., P.J. Longhurst, G.H. Drew et al. (2010). Production and quality assurance of solid recovered fuels using mechanical—biological treatment (MBT) of waste: A comprehensive assessment. *Critical Reviews in Environmental Science and Technology*, 40 (12), 979-1105

Voulvoulis, N., R.J. Slack, & J.R. Gronow (2004). *Household Hazardous Waste: Establishing Sources and Disposal Routes*. The Waste Conference Ltd., Coventry.

Waste Consult (2014). Discussion paper on the use of Mechanical Biological Treatment (MBT) to treat mixed putrescible waste. West Australian Waste Authority. http://www.wasteauthority.wa.gov.au/media/files/documents/MBT_Paper_2014.pdf

World Bank (1999). *Technical Guidance Report: Municipal Solid Waste Incineration*. http://web.mit.edu/urbanupgrading/urbanenvironment/resources/references/pdfs/MunicipalSWIncin.pdf

World Energy Council (2013). *World Energy Resources: Waste to Energy*. http://www.worldenergy.org/wp-content/uploads/2013/10/WER_2013_7b_Waste_to_Energy.pdf

Global Secondary Materials Industry

Bureau of International Recycling (BIR). Website lists summary publications on world recycling markets for steel (updated annually); non-ferrous metals; paper (updated periodically); and a general market survey. Also a report on the environmental benefits of reccyling, and on recycling through the ages. http://www.bir.org/publications/brochures/ Specific publications used in Section 3.6 include:

- Global non-ferrous scrap flows 2000-2011
- Recovered paper market in 2012 (2014 report)
- World markets for recovered and recycled commodities - 2010, 2011
- World steel recycling in figures - six reports, from 2005-2009 through 2010-2014

Confederation of European Paper Inductries (CEPI). (2014). *Key Statistics - European Pulp and Paper Industry – 2013*. http://www.cepi.org/system/files/public/documents/publications/statistics/2014/Final%20Key%20statistics%202013.pdf

Crang, M.A., A. Hughes, N. Gregson et al. (2013). Rethinking governance and value in commodity chains through global recycling networks. *Transactions of the Institute of British Geographers*, 38 (1). 12-24.

De Angelis, A. (2013). *Global Recycling Industry*. http://www.companiesandmarkets.com/MarketInsight/Environmental/Global-Recycling-Industry/NI7749

FAO (2010). *Recovered paper data 2009*. http://www.fao.org/docrep/013/k9444e/k9444e00.pdf and http://www.fao.org/forestry/industries/en/

Kojima, M. & E. Michida (Eds.). (2011). Economic Integration and Recycling in Asia Institute of Developing Economies, 2011. http://www.ide.go.jp/English/Publish/Download/Report/2010/2010_431.html

Minter, A. (2013) *Junkyard Planet - Travels in the billion-dollar trash trade*. London: Bloomsbury.

Norris, L. (2013). Worn clothing - anthropologies of reuse and recycling. www.wornclothing.co.uk Provides an extensive list of publications (click on 'About' and 'Publications').

Rogich, D.G. & G.R. Matos (2008). *The Global Flows of Metals and Minerals from 1975-2005*. US Gelogical Service. http://pubs.usgs.gov/of/2008/1355/pdf/ofr2008-1355.pdf

UNEP (2013c). *Metal Recycling: Opportunities, Limits, Infrastructure: A Report of the Working Group on the Global Metal Flows to the International Resource Panel*. Reuter, M., C. Hudson, A. van Schaik et al; and *E-Book: International Resource Panel Work On Global Metal Flows*. http://www.unep.org/resourcepanel/Portals/24102/PDFs/Metal_Recycling_Full_Report.pdf http://www.unep.org/resourcepanel/Portals/50244/publications/UNEP_Synopsis_131008.pdf

Velis, C.A. (2014a). Global recycling markets - plastic waste: A story for one player – China. Report prepared by FUELogy and formatted by D-waste on behalf of International Solid Waste Association - Globalisation and Waste Management Task Force. ISWA, Vienna. http://www.iswa.org/fileadmin/galleries/Task_Forces/TFGWM_Report_GRM_Plastic_China_LR.pdf

Velis, C.A. (2015). Circular economy and global secondary material supply chains. *Waste Management & Research*. 33(6), 389-391. http://wmr.sagepub.com/content/33/5/389.full.pdf+html

World Steel Association (2012). *Sustainable Steel at the Core of a Green Economy*. http://www.worldsteel.org/dms/internetDocumentList/bookshop/Sustainable-steel-at-the-core-of-a-green-economy/document/Sustainable-steel-at-the-core-of-a-green-economy.pdf

Topic Sheets

Construction and demolition waste

BIO Intelligence Services (2011). Management of construction and demolition waste. European Commission (DG ENV). http://ec.europa.eu/environment/waste/pdf/2011_CDW_Report.pdf

Centre for Science and Environment (2014). Construction and demolition waste. http://www.cseindia.org/userfiles/Construction-and%20-demolition-waste.pdf

Edge Environment PtY Ltd. (2011). *Construction and demolition waste guide – recycling and re-use across the supply chain*. Australia, Department of Sustainability, Environment, Water, Population and Communities. http://www.environment.gov.au/system/files/resources/b0ao6oo1-1263-1d2b-b001-0be-f84b52b8/files/case-studies.pdf

Gratton, R. (2013). *Mission 2030 – Creating Sustainable Building Industry Change*. http://www.cricouncil.com/wp-content/uploads/2013/02/Mission-2030-General-Overview-6P.pdf

Guah, E.H. (2013). *Status and Technologies for Construction & Demolition Waste Management in Asia*. Prepared for 3R Knowledge Hub, Asian Institute of Technology. http://3rkh.net/index.php?option=com phocadownload&view=file&id=623:cad-waste-manual&Itemid=238

USEPA (U.S. Environmental Protection Agency). (2000). Building Savings: Strategies for Waste Reduction of Construction and Demolition Debris from Buildings. http://www.epa.gov/epawaste/nonhaz/municipal/pubs/combined.pdf

Hazardous waste

Basel Convention. Guidance Manuals, Technical Guidelines and Training Manuals. http://www.basel.int/Implementation/Publications/GuidanceManuals/tabid/2364/Default.aspx http://www.basel.int/Implementation/Publications/TechnicalGuidelines/tabid/2362/Default.aspx http://www.basel.int/Implementation/Publications/TrainingManuals/tabid/2363/Default.aspx

Basel Convention (n.d.). Reporting database. http://www.basel.int/Countries/NationalReporting/ReportingDatabase/tabid/1494/Default.aspx

Eurostat (2014). Statistics on Hazardous waste. http://epp.eurostat.ec.europa.eu/portal/page/portal/waste/key_waste_streams/hazardous_waste

ISWA, UNEP & Basel Convention (2002). *Training Resource Pack for hazardous waste management in emerging economies*. Wilson, D., F. Balkau & M. Thurgood. This is in the process of updating. The current version (Ed: Balkau, F.) is available at: http://www.trp-training.info

Slack, R.J., J.R. Gronow & N. Voulvoulis (2004). *Household Hazardous Waste: Establishing Sources and Disposal Routes*. Publisher: The Waste Conference Ltd., Coventry.

UNEP (2012a). *Compendium of technologies for treatment and destruction of health-care waste*. http://www.unep.org/ietc/Portals/136/News/Publication%20of%20Healthcare%20Waste%20compendium%20of%20technologies/Compendium_Technologies_for_Treatment_Destruction_of_Healthcare_Waste_2012.pdf

WHO (2014). *Safe management of wastes from health-care activities*. 2nd edition (New edition of the "Blue Book"). Edited by Chartier, Y., J. Emmanuel, U. Pieper et al. http://apps.who.int/iris/bitstream/10665/85349/1/9789241548564_eng.pdf?ua=1

WHO (n.d.). Training modules in health-care waste management. http://www.who.int/water_sanitation_health/healthcare_waste/training_modules_waste_management/en/

E-waste

Baldé, C.P., R. Kuehr, K. Blumenthal et al. (2015). *E-waste statistics: Guidelines on classifications, reporting and indicators*. United Nations University, IAS - SCYCLE, Bonn, Germany. http://i.unu.edu/media/ias.unu.edu-en/project/2238/E-waste-Guidelines_Partnership_2015.pdf

Baldé, C.P., F. Wang, R. Kuehr et al. (2015). *The global e-waste monitor – 2014*. United Nations University, IAS – SCYCLE, Bonn, Germany. http://i.unu.edu/media/unu.edu/news/52624/UNU-1stGlobal-E-Waste-Monitor-2014-small.pdf

Greenpeace (2008). Poisoning the poor: Electronic waste in Ghana. http://www.greenpeace.org/denmark/Global/denmark/p2/other/report/2008/poisoning-the-poor-electroni.pdf

International Labour Organization (2012). The global impact of e-waste: Addressing the challenge. http://www.ilo.org/wcmsp5/groups/public/---ed_dialogue/---sector/documents/publication/wcms_196105.pdf

Ongondo, F.O., I.D. Williams & T.J. Cherrett, (2011). How are WEEE doing? A global review of the management of electrical and electronic wastes. *Waste Management*, 31(4), 714-730.

Song, Q. & J. Li (2014). Environmental effect of heavy metals derived from the e-waste recycling activities in China: a systematic review. *Waste Management*, 34 (12), 2587–2594.

UNEP (2007). *E-waste Vol 1: Inventory Assessment manual. Vol. 2: E-waste Management Manual.* http://www.unep.org/ietc/Portals/136/Publications/Waste%20Management/EWasteManual_Vol1.pdf http://www.unep.org/ietc/Portals/136/Publications/Waste%20Management/EWasteManual_Vol2.pdf

UNEP (2013a). *E-Waste Vol. III – WEEE/e-waste "Take back System."* http://www.unep.org/ietc/Portals/136/Publications/Waste%20Management/Ewaste%20Manual%203.pdf

UNEP – Basel Convention Mobile Phone partnership Initiative (2012). Guidance document on the environmentally sound management of used and end-of-life mobile phones. Prepared by the Mobile Phone Working Group. http://www.basel.int/Implementation/TechnicalAssistance/Partnerships/MPPI/MPPIGuidanceDocument/tabid/3250/Default.aspx

UNODC (United Nations Office on Drugs and Crime). (2013). *Transnational Organized Crime Threat Assessment Report - East Asia and the Pacific*. https://www.unodc.org/documents/toc/Reports/TOCTA-EA-Pacific/TOCTA_EAP_c09.pdf

UNU-IAS SCYCLE. Solving the E-waste Problem (StEP) Initiative. Many publications related to E-waste at http://www.step-initiative.org/publications.html

Plastic Waste and Marine Litter

Cózar, A., F. Echevarría, J. González-Gordillo et al. (2015). Plastic debris in the open ocean. *PNAS* Vol 111 (28), 10239-10244. http://www.pnas.org/content/111/28/10239.full.pdf

European Commission. (2013). Green Paper on a European Strategy on Plastic Waste in the Environment. Brussels. http://ec.europa.eu/environment/waste/studies/pdf/green_paper_plastic.pdf

Global Underwater Awareness Association GUWAA (2013). http://whiteflagint.com/guwaa/index.php/guwaa-projects/finished/guwaa-2nd-world-cleanup-2013

Jambeck, J.R., R. Geyer, C. Wilcox et al. (2015). Plastic waste inputs from land into the ocean. *Science,* 347(6223), pp. 768-771.

UNEP (2014c). Valuing Plastics: The Business Case for Measuring, Managing and Disclosing Plastic Use in the Consumer Goods Industry.

U.S. Department of Commerce and National Oceanic and Atmospheric Administration (NOAA). (n.d.). Marine Pollution: Ocean Dumping. http://www.gc.noaa.gov/gcil_mp_ocean_dumping.html

USEPA (U.S. Environmental Protection Agency). (2001). Marine litter – trash that kills. http://www.epa.gov/owow/oceans/debris/toolkit/files/trash_that_kills508.pdfUSEPA (2009). Factsheet: Marine Debris. http://water.epa.gov/type/oceb/marinedebris/factsheet_marinedebris_debris.cfm

US NOAA (National Oceanic and Atmospheric Administration). (2014). Marine Debris Program Report: Entanglement of Marine Species in Marine Debris with an Emphasis on Species in the United States. http://marinedebris.noaa.gov/sites/default/files/mdp_entanglement.pdf

Velis, C.A. (2014b) Plastic waste in marine litter: Action now and at the source. *Waste Management & Research* 32(4), 251-3. http://wmr.sagepub.com/content/32/4/251.short

Disaster Waste

Suganuma, K. (2006). Recent Trends in Earthquake Disaster management in Japan. Science and Technology Trends. Quarterly Review No. 19. http://data.nistep.go.jp/dspace/bitstream/11035/2714/1/NISTEP-STT019E-91.pdf

UNEP (2012e). Managing post-disaster debris: The Japan Experience. http://postconflict.unep.ch/publications/UNEP_Japan_post-tsunami_debris.pdf

UNEP/OCHA (2011). Disaster Waste Management Guidelines. UNEP/OCHA Environment Unit. Switzerland. https://docs.unocha.org/sites/dms/Documents/DWMG.pdf

Food Waste

FAO (2011). Global Food Losses and Waste. Extent, Causes and Prevention. http://www.fao.org/docrep/014/mb060e/mb060e00.pdf

FAO (2013). Food wastage footprint. Impacts on natural resources. http://www.fao.org/docrep/018/i3347e/i3347e.pdf

Papargyropoulou, E., R. Lozano, J.K. Steinberger et al. (2014). The food waste hierarchy as a framework for the management of food surplus and food waste. *Journal of Cleaner Production*, 76. 106 – 115

Stuart, T. (2009). *Waste: Uncovering the Global Food Scandal*. Penguin Books.

CHAPTER 4: WASTE GOVERNANCE

Sustainability

Ceschin, F. (2013). Critical factors for implementing and diffusing sustainable product-service systems: Insights from innovation studies and companies' experiences. *Journal of Cleaner Production* 45: 74-88.

Ellen MacArthur Foundation (2012). Towards the Circular Economy: economic and business rationale for an accelerated transition. McKinsey/Ellen MacArthur Foundation. http://www.ellenmacarthurfoundation.org/business/reports

European Commission (EC). (2011a). A Resource-efficient Europe – Flagship Initiative under the Europe 2020 Strategy. Communication from the Commission to the European Parliament, the Council, the European Economic and Social Committee and the Committee of the Regions COM(2011) 21.

Geng, Y. & B. Doberstein (2008). Developing the circular economy in China: Challenges and opportunities for achieving 'leapfrog development'. *International Journal of Sustainable Development & World Ecology* 15: 231-239.

Grosse, F. (2010). Is recycling "part of the solution"? The role of recycling in an expanding society and a world of finite resources, *Sapiens,* 3(1).

Beuren, F.H., M.G. Gomes Ferreira & P.A. Cauchick Miguel (2013). Product-service systems: A literature review on integrated products and services, *Journal of Cleaner Production*, 47: 222-231.

ISWA (2015a). *Circular Economy: Resources and Opportunities.* Final report of the ISWA Task Force on Resource Management. Vienna, 2015. https://www.iswa.org/resourcemanagement

Jackson, T. (2009). Prosperity without Growth? The transition to a sustainable economy. Report of Economic Commissioner for the UK Sustainable Development Commission. http://www.sd-commission.org.uk/data/files/publications/prosperity_without_growth_report.pdf

McDonough, W. & M. Braungart (2002). *Cradle to Cradle: Remaking the Way We Make Things*. North Point Press, New York, U.S.

McDonough, W. & M. Braungart (2013). *The Upcycle: Beyond Sustainability – Designing for Abundance*. North Point Press, New York, U.S.

Robèrt, K-H., B. Schmidt-Bleek, J. Aloisi de Larderel et al. (2002). Strategic sustainable development - Selection, design and synergies of applied tools. *Journal of Cleaner Production*, 10(3): 197-214.

Rockström, J., W. Steffen, K. Noone et al. (2009). Planetary boundaries: Exploring the safe operating space for humanity. Ecology and *Society* 14(2). Prepared by the Resilience Alliance. http://www.ecologyandsociety.org/vol14/iss2/art32/

Sekulova, F., B. Rodríguez-Labajos, G. Kallis et al. (Eds.). (2013). Degrowth: From Theory to Practice. *Journal of Cleaner Production,* 38: 1-98.

Stahel, W.R. (2010). The Performance Economy (2nd ed.). Palgrave MacMillan.

Su, B.W., A. Heshmati, Y. Geng et al. (2013). A review of the circular economy in China: Moving from rhetoric to implementation. *Journal of Cleaner Production*, 42: 215-227.

UNEP (2011a). Decoupling natural resource use and environmental impacts from economic growth: A Report of the Working Group on Decoupling to the International Resource Panel. Fischer-Kowalski, M., M. Swilling, E.U. von Weizsäcker et al.

UNEP (2014a). Decoupling 2: technologies, opportunities and policy options. A Report of the Working Group on Decoupling to the International Resource Panel. Von Weizsäcker, E, J. de Larderel, K. Hargroves et al.

Economic instruments

Calcott, P. & M. Walls (2005). Waste, recycling, and “Design for Environment”: Roles for markets and policy instruments. *Resource and Energy Economics* 27(4): 287-305.

Cointreau, S. & C. Hornig (2003). Global Review of Economic Instruments for Solid Waste Management in Latin America. Inter-American Development Bank (IDB), Washington DC, U.S. http://publications.iadb.org/handle/11319/3483?locale-attribute=en.

Greyson, J. (2007). An economic instrument for zero waste, economic growth and sustainability. *Journal of Cleaner Production*, 15(13-14): 1382-1390.

GIZ (2015). *Concepts for Sustainable Waste Management. Economic instruments in solid waste management*. Results published as a series of five reports. The synthesis report is Applying economic instruments for sustainable solid waste management in low- and middle-income countries; four separate case study reports cover industrialized and Latin-American countries; Maputo, Mozambique; Bayawan, Philippines; and Bulgaria. http://www.giz.de/en/worldwide/15109.html

Harrington, W. & R.D. Morgenstern (2004). Economic Incentives versus Command and Control – What's the best approach for solving environmental problems? *Resources,* Fall/Winter: 13-17. http://www.rff.org/rff/Documents/RFF_Resources_152_ecoincentives.pdf

Helm, D.R. (2005). Economic Instruments and Environmental Policy. *The Economic and Social Review,* 36(3): 205-228. http://www.esr.ie/ESR_papers/vol36_3/02_Helm_article.pdf

UNEP (2004). The Use of Economic Instruments for Environmental Policy: Opportunities and Challenges. http://www.unep.ch/etb/publication/EconInst/econInstruOppChnaFin.pdf

Watkins, E., D. Hogg, M. Mitsios et al. (2012). Use of Economic Instruments and Waste Management Performances. Final Report to the European Commission, DG Environment, Unit G.4 Sustainable Production and Consumption. http://ec.europa.eu/environment/waste/pdf/final_report_10042012.pdf

Extended producer responsibility

Agrawal, S., R.K. Singh & Q. Murtaza (2015). A literature review and perspectives in reverse logistics. *Resources, Conservation and Recycling*, 97, 76–92.

Cahill, R., S.M. Grimes & D.C. Wilson (2011). Extended producer responsibility for packaging wastes and WEEE – a comparison of implementation and the role of local authorities across Europe. *Waste Management & Research*, 29(5) 455–479.

European Commission – DG Environment (2014). Development of Guidance on Extended Producer Responsibility (EPR); (Comparative analysis of European experiences). http://ec.europa.eu/environment/waste/pdf/target_review/Guidance%20on%20EPR%20-%20Final%20Report.pdf

Hickle, G.T. (2014). Moving beyond the “patchwork”: A review of strategies to promote consistency for extended producer responsibility policy in the *U.S. Journal of Cleaner Production,* 64, 266-276.

INSEAD Social Innovation Centre (2014). Extended Producer Responsibility: Stakeholder Concerns and Future Developments. White paper prepared with European Recycling Platform. http://centres.insead.edu/social-innovation/what-we-do/documents/ExtendedProducerResponsibility_StakeholderConcernsandFutureDevelopments_web_000.pdf

Institute for Global Environmental Strategies (IGES). (2012). Applying EPR in Developing Countries. Also other publications on EPR in developing countries and Asia in particular. http://pub.iges.or.jp/modules/envirolib/upload/3553/attach/rio_issue_brief_vol3_EPR_mar2012.pdf

ISWA (2014a). Extended Producer Responsibility. Key issue paper. http://www.iswa.org/index.php?eID=tx_iswaknowledgebase_download&documentUid=4202

Nash, J. & C. Bosso (2013). Extended producer responsibility in the United States: Full speed ahead? *Journal of Industrial Ecology,* 17 (2): 175–185.

OECD (various dates). Various documents on EPR available to read online at http://www.oecd.org/env/tools-evaluation/extendedproducerresponsibility.htm

OECD (2014). The State of Play on Extended Producer Responsibility (EPR): Opportunities and Challenges. Issues Paper prepared for a meeting in June 2014. http://www.oecd.org/environment/waste/Global%20Forum%20Tokyo%20Issues%20Paper%2030-5-2014.pdf

Product Stewardship Institute (n.d.). Current information on EPR in the U.S. http://www.productstewardship.us

Quoden, J. (2014). EPR from the Obligated Business Perspective in the European Context. http://upstreampolicy.org/epr-from-the-obligated-business-perspective-in-the-european-context/

Sachs, N. (2006). Planning the funeral at the birth: Extended producer responsibility in the European Union and the United States, *Harvard Environmental Law Review*, 30(1): 51-98.

Sheehan, B. (2014). Responsible production: How far is it possible to go? http://upstreampolicy.org/responsible-production-how-far-is-it-possible-to-go/

Walls, M. (2006). Extended producer responsibility and product design – economic theory and selected case studies. Discussion Paper 06-08. Resources for the Future, Washington, DC, U.S. http://www.rff.org/Documents/RFF-DP-06-08-REV.pdf

Social instruments

Darnton, A. (2008). Practical Guide: An Overview of Behaviour Change Models and their Uses. For the Government Social Research Unit. http://www.civilservice.gov.uk/wp-content/uploads/2011/09/Behaviour-change_practical_guide_tcm6-9696.pdf

Heath, C. & D. Heath (2012). *Switch. How to change things when change is hard*. Random House Publishing Group.

Jackson, T. (2005). *Motivating sustainable consumption. A review of evidence on consumer behaviour and behavioural change.* A report to Sustainable Development Research Network.

Shove, E. (2010). Beyond the ABC: climate change policy theories of social change. *Environment and Planning A*, 42, 1273-1285.

Sustainable Consumption Roundtable (2006). *I will if you will. Towards sustainable consumption*. UK National Consumer Council and Sustainable Development Commission. http://www.sd-commission.org.uk/data/files/publications/I_Will_If_You_Will.pdf

Inclusivity *(including informal sector recycling)*

Aparcana, S. & S. Salhofer (2013). Application of a methodology for the social life cycle assessment of recycling systems in low income countries: Three Peruvian case studies, *International Journal of Life Cycle Assessment* 18(5): 1116-1128.

Cohen, P., J. IJgosse & G. Sturzenegger (2013). Preparing informal recycler inclusion plans: an operational guide. Available in Spanish and English. Washington: IDB. http://publications.iadb.org/handle/11319/697?locale-attribute=en

Ezeah, C., J.A. Fazakerley & C.L Roberts (2013). Emerging trends in informal sector recycling in developing and transition countries. *Waste Management*, 33, 2509-2519.

Fergutz, O., S. Dias & D. Mitlin (2011). Developing urban waste management in Brazil with waste picker organizations. *Environment and Urbanization* 23(2): 597-608.

Gunsilius, E., S. Spies, S. García-Cortés et al. (2011). *Recovering resources, creating opportunities: Integrating the informal sector into solid waste management*. Eschborn, GIZ: Eschborn, Germany. http://www.giz.de/de/downloads/giz2011-en-recycling-partnerships-informal-sector-final-report.pdf

Medina, M. (2007). The world's scavengers: salvaging for sustainable consumption and production. AltaMira Press, U.S.

Owusu, V., E. Adjei-Addo & C. Sundberg (2013). Do economic incentives affect attitudes to solid waste source separation? Evidence from Ghana. *Resources, Conservation and Recycling,* 78: 115-123.

Petts, J. (2004). Barriers to participation and deliberation in risk decisions: Evidence from waste management. *Journal of Risk Research,* 7(2): 115-133.

Reed, M.S. (2008). Stakeholder participation for environmental management: A literature review. *Biological Conservation,* 141(10): 2417-2431.

Reed, M.S., A. Graves, N. Dandy et al. (2009). Who's in and why? A typology of stakeholder analysis methods for natural resource management. *Journal of Environmental Management,* 90(5): 1933-1949.

Velis, C.A., D.C. Wilson, O. Rocca et al. (2012). An analytical framework and tool ('InteRa') for integrating the informal recycling sector in waste and resource management systems in developing countries. *Waste Management and Research,* 30(9 SUPPL.1): 43-66. http://wmr.sagepub.com/content/30/9_suppl/43.full.pdf+html

Wilson, D.C., C.A. Velis & C. Cheeseman (2006). Role of informal sector recycling in waste management in developing countries. *Habitat International,* 30(4): 797-808.

Policy combinations

Bengtsson, M., Y. Hotta, S. Hayashi et al. (2010). Policy tools for sustainable materials management: appications in Asia. IGES Discussion paper SCP 2010-001. http://pub.iges.or.jp/modules/envirolib/upload/2885/attach/policy_tools_smm_dp_final.pdf

Braathen, N.A. (2007). Instrument mixes for environmental policy: How many stones should be used to kill a bird? *International Review of Environmental and Resource Economics*, 1(2): 185-235.

Bhuiyan, S.H. (2010). A crisis in governance: Urban solid waste management in Bangladesh, *Habitat International*, 34(1): 125-133.

Finnveden, G., M. Bisaillon, M. Noring et al. (2012). Developing and evaluating new policy instruments for sustainable waste management, *International Journal of Environment & Sustainable Development*, 11(1): 19-31.

Gunningham, N. (2009). Environment law, regulation and governance: Shifting architectures. *Journal of Environmental Law*, 21(2): 179-212.

Hou, D., A. Al-Tabbaa, P. Guthrie et al. (2012). Sustainable waste and materials management: National policy and global perspective, *Environmental Science and Technology*, 46(5): 2494-2495.

Howlett, M. & J. Rayner (2007). Design principles for policy mixes: cohesion and coherence in 'New Governance Arrangements', *Policy and Society*, 26(4): 1-18.

Nilsson, M., T. Zamparutti, J.E. Petersen et al. (2012). Understanding Policy Coherence: Analytical Framework and Examples of Sector-Environment Policy Interactions in the EU, *Environmental Policy and Governance*, 22(6): 395-423.

Probst, K.N. & T.C. Beierle (1999). *The Evolution of Hazardous Waste Programs: Lessons from Eight Countries*. Center for Risk Management, Resources for the Future. Washington, DC, U.S.

Taylor, C., S. Pollard, S. Rocks et al. (2012). Selecting Policy Instruments for Better Environmental Regulation: A Critique and Future Research Agenda, *Environmental Policy and Governance,* 22(4): 268-292.

Gender and waste

Espino, A. & N. Bidegain (2011). Hacia una gestión integrada de los residuos con inclusion social: Recomendaciones para la acción [In Spanish]. Centro Interdisciplinario de Estudios sobre el Desarrollo (Interdisciplinary Centre for Development Studies), Uruguay (CIEDUR).

Fredericks, R. (2008). Gender and the Politics of Trash in Dakar: Participation, Labor and the "Undisciplined" Woman. Thinking Gender Papers, UCLA Center for the Study of Women, UC Los Angeles, California, U.S.

Gonzenbach, B. & A. Coad (2007). Solid Waste Management and the Millennium Development Goals: Links that inspire action. Collaborative Working Group on Solid Waste Management in Low- and Middle-income Countries. CWG Publication Series No. 3.

GTZ (German Technical Cooperation). (2006). Study on gender issues in the management of municipal solid waste in the State of Mexico. State of Mexico: Tlalnepantla. German Agency for Technical Cooperation.

IDB (2013). Gender and Recycling: Tools for Project Design and Implementation: Regional Initiative for inclusive recycling. http://publications.iadb.org/handle/11319/6138?locale-attribute=en

Muller, M. & A. Schienberg (1999). Gender and Urban Waste Management. http://www.gdrc.org/uem/waste/swm-gender.html

Poswa, T.T. (2004). The importance of gender in waste management planning: a challenge for solid waste managers. Durban Institute of Technology (DIT), Department of Environmental Health. 8th World Congress on Environmental Health, South Africa.

CHAPTER 5: WASTE MANAGEMENT FINANCING

General reading on financing and economics of waste management

Cavé, J. (2014). Who owns urban waste? Appropriation conflicts in emerging countries, *Waste Management and Research,* 32 (9), 813-821.

Cavé, J. (2015). *La ruée vers l'ordure. Conflits dans les mines urbaines de déchets* [In French]. Presses universitaires de Rennes.

Chaturvedi, B. & V. Gidwani (2010). The Right to Waste: Informal Sector Recyclers and Struggles for Social Justice in Post-Reform Urban India. *India's new economic policy: A critical analysis*, Vol. 15, 125-153.

Lane, R. (2011). The Waste Commons in an Emerging Resource Recovery Waste Regime: Contesting Property and Value in Melbourne's Hard Rubbish Collections, Geographical Research, 49(4), 395-407.

Lerpiniere, D., D.C. Wilson, C.A. Velis et al. (2014). Review of International Development Co-operation in Solid Waste Management. Report prepared by University of Leeds and formatted by D-Waste on behalf of ISWA Globalisation and Waste Management Task Force. International Solid Waste Association, Vienna.

Lupton, S. (2011). *Economie des déchets : une approche institutionnaliste* [In French]. De Boeck, Bruxelles: Ouvertures Economiques. https://developpementdurable.revues.org/9636

Pfaff-Simoneit, W. (2013). *Entwicklung eines sektoralen Ansatzes zum Aufbau von nachhaltigen Abfallwirtschaftsystemen in Entwicklungsländern vor dem Hintergrund von Klimawandel und Ressourcenverknappung* [In German]. Universität Rostock, Darmstadt/ Rostock. http://rosdok.uni-rostock.de/file/rosdok_disshab_0000000936/rosdok_derivate_0000005003/Dissertation_Pfaff-Simoneit_2013.pdf)

Ren, X. & S. Hu (2014). Cost recovery of municipal solid waste management in small cities in inland China. *Waste Management and Research*, 32(4): 340-347.

Soos, R. (2010). Section 5.2 *Financial Sustainability*, in Scheinberg, Wilson and Rodic, 2010 (listed under Chapter 1, Waste Management), 164-183.

World Bank (2014). *Results-based financing for municipal solid waste management*. World Bank Urban Development Series Knowledge Papers. http://www.worldbank.org/en/topic/urbandevelopment/publication/results-based-financing-for-municipal-solid-waste

Estimating economic costs and benefits of waste and resource management and waste and health

Afon, A. (2012). A survey of operational characteristics, socioeconomic and health effects of scavenging activity in Lagos, Nigeria. *Waste Management and Research*, 30(7), 664-671.

Binion, E. & J. Gutberlet (2012). The effects of handling solid waste on the wellbeing of informal and organized recyclers: A review of the literature. *International Journal of Occupational and Environmental Health*, 18(1), 43-52. http://www.juttagutberlet.com/wp-content/uploads/2013/10/2012-IJOEH-Health-Binion-Gutberlet.pdf

Chatham-Stephens, K., J. Caravanos, B. Ericson et al. (2013). Burden of Disease from Toxic Waste Sites in India, Indonesia, and the Philippines in 2010. *Environmental Health Perspectives*, 121 (7), 791-796. [Online] http://ehp.niehs.nih.gov/1206127/

Ellen MacArthur Foundation *Towards the circular economy.* A series of reports. http://www.ellenmacarthurfoundation.org/business/reports

Eshet, T., O. Ayalon & M. Shechter (2005). Valuation of externalities of selected waste management alternatives: A comparative review and analysis. *Waste Management & Research*, 23, 487-504.

Hajkowicz, S., K. Tellames, J. Aitaro (2006). Economic cost scenarios for solid waste-related pollution in Palau. IWP-Pacific Technical Report. http://pacificwater.org/userfiles/file/IWRM/Toolboxes/financing%20IWRM/Palau.pdf

ISWA (2015b). Wasted Health – The tragic case of dumpsites. A. Mavropoulos ISWA STC Chair, with contributions from D. Newman, ISWA President. Vienna.

Jamaica, Planning Institute of Jamaica, Sustainable Development and Regional Planning Division (2007). *Management of hazardous & solid wastes in Jamaica.* http://pioj.gov.jm/Portals/0/Sustainable_Development/Management_of_Wastes.pdf

Kimani, N.G. in cooperation with UNEP (2007). Environmental Pollution and Impacts on Public Health, Implications of the Dandora Municipal Dumping Site in Nairobi, Kenya. http://www.kutokanet.com/Storage/UNEP_Dandora_Environmental_Pollution_and_Impact_To_Public_Health_2007.pdf

McKinsey & Company (2011). *Resource Revolution: Meeting the world's energy, materials, food, and water needs*. http://www.mckinsey.com/insights/energy_resources_materials/resource_revolution

Nkwocha, E., R. Ejejuru, E. Pat-Mbano et al. (2011). Proximity of municipal waste dumpsites to residential neighbourhoods and rate of hospitalization for malaria. *International Journal of Advanced Biotechnology and Research*. Vol 2, Issue 1, 159-167 http://bipublication.com/files/ijabarv2l120116.pdf

Phillips, W. & E. Thorne (2011). *Municipal solid waste management in the Caribbean – A benefit cost analysis*. United Nations UN ECLAC Subregional headquarters for the Caribbean. http://www.cepal.org/publicaciones/xml/3/45473/S22Municipalsolidwastemanagement.pdf

Sankoh, F.P., X. Yan, Q. Tran (2013). Environmental and Health Impact of Solid Waste Disposal in Developing Cities: A Case Study of Granville Brook Dumpsite, Freetown, Sierra Leone. *Journal of Environmental Protection*, 4, 665-670.

Scheinberg, A., M. Simpson, Y. Gupt, et al. (2010). *Economic aspects of the informal sector in solid waste management*. WASTE, SKAT, and city partners for GTZ - Deutsche Gesellschaft für Technische Zusammenarbeit and CWG - Collaborative Working Group on Solid Waste Management in Low- and Middle-Income Countries, Eschborn, Germany http://www.giz.de/fachexpertise/downloads/gtz2010-en-Economic-Aspects-WASTE.pdf. Summary published as CWG Publication Series No 5. http://www.giz.de/en/downloads/giz2011-cwg-booklet-economicaspects.pdf

SWEEPNet (2014). *Cost of environmental degradation due to solid waste management practices in Beirut and Mount Lebanon*. GIZ. http://www.sweep-net.org/sites/default/files/CASWD%20LEBANON.pdf

UNEP/DTIE Chemicals Branch (2010). First results on dioxins and Furans under the Stockholm Convention on POPs 2010, presented at Dioxin Symposium in Barcelona by Heidelore Fiedler. http://apps.thermoscientific.com/media/SID/IOMS/PDF/Dioxin_Symposium_Barcelona_2010/12-HFiedler_Barcelona_2010.pdf

UNEP (2005). Standardized Toolkit for Identification and Quantification of Dioxins and Furan Releases

USEPA (U.S. Environmental Protection Agency). (n.d.). Superfund. http://epa.gov/superfund/about.htm

Service delivery and private sector participation in SWM

Coad, A. (2005). *Private Sector Involvement in Solid Waste Management, Avoiding Problems and Building on Successes*. Eschborn: GTZ. http://www2.gtz.de/dokumente/bib/05-0412.pdf

Cointreau, S. (1994). *Private sector participation in Municipal Solid Waste Services in Developing Countries*. Urban Management Programme Paper UMP 13. The World Bank, Washington DC, U.S. http://ww2.unhabitat.org/programmes/ump/documents/UMP13.pdf

Cointreau, S., P. Gopalan & A. Coad (2000). *Private Sector Participation in Municipal Solid Waste Management*. World Bank Toolkit, 5 volumes. Washington: World Bank. http://ppp.worldbank.org/public-private-partnership/sector/solid-waste/toolkits

Gupta, S. (2013). Case Study Report for Surat, India. Prepared for GIZ Operator Models project. http://www.giz.de/en/mediacenter/publications.html, search the database for 'operator models', then select Annex 5.

Hall, D. (2010). Waste management in Europe: framework, trends and issues. Public Services International Research Unit (PSIRU). http://www.epsu.org/IMG/pdf/it_10_2010-02_Waste_trends-3.pdf

Hall, D. (2012). Re-municipalising municipal services in Europe. Public Services International Research Unit (PSIRU). http://www.epsu.org/IMG/pdf/Redraft_DH_remunicipalization.pdf

Hall, D., J. Lethbridge, E. Lobina et al. (2003). The UK Experience – privatised sectors and globalised companies. In CESifo (2003), *Privatisation Experiences in the EU*, Munich, 10-11 January, CESifo, Munich, Germany. http://www.cesifo-group.de/ifoHome/events/Archive/conferences/2003/01/2003-01-10-event-ConfCESifo/Programme.html

India, Ministry of Urban Development (MOUD). (2011). Guidance Note on Municipal Solid Waste Management on a Regional Basis. http://www.moud.gov.in/guidance_note

Mourdzhev, B. & R. El Shebriny (2013). Case Study Report for Qena, Egypt, prepared for GIZ Operator Models project. http://www.giz.de/en/mediacenter/publications.html Search the database for 'operator models', then select Annex 6.

Scheinberg, A., M. Simpson, Y. Gupt et al. (2010). *Economic Aspects of the Informal Sector in Solid Waste Management*; GTZ (German Technical Cooperation): Eschborn, Germany. Summary available at http://www2.gtz.de/dokumente/bib-2011/giz2011-0116en-informal-sector-solid-waste-management.pdf

Schmidt, T. (2013). Case Study Report for Rio Grande De Sol, Brazil. Prepared for GIZ Operator Models project. http://www.giz.de/en/mediacenter/publications.html Search the database for 'operator models', then select Annex 3.

Soos, R., A.D. Whiteman, D.C. Wilson et al. (2013a). *Operator Models. Respecting Diversity. Concepts for Sustainable Waste Management.* Eschborn: GIZ. http://www.giz.de/en/downloads/giz2013-swm-operator-models-sourcebook-en.pdf (Accessed 04 August, 2014).

Soos, R., A.D. Whiteman, D.C. Wilson et al. (2013b). *Operator Models – Respecting Diversity. Guidance Paper for Solid Waste Management Practitioners.* GIZ. http://www.giz.de/en/downloads/giz2013-swm-operator-models-guidance-paper-en.pdf

Stretz, J. (2013). Case Study Report for Maputo, Mozambique. Prepared for GIZ Operator Models project. http://www.giz.de/en/mediacenter/publications.html Search the database for 'operator models', then select Annex 7.

Wilson, D.C., J.B. Kangojera, R. Soos et al. (2013). Operator Models for Delivering Municipal Solid Waste Management Services in Emerging and Developing Countries. Proceedings of ISWA World Congress 2013, 7-11 October, Vienna, Austria. Vienna: International Solid Waste Association (ISWA).

Annex B: Data Sources Used for the GWMO

A number of databases have been put together for use in the GWMO, particularly in Chapter 3, which have been supplemented by other available sources. Waste and resource data have been accumulated at both the country and the city level. The GWMO master databases were compiled in May 2014 with limited updates thereafter.

1. Two *'master' databases* of the latest available official statistics were compiled for use in the GWMO by the Environmental Management Centre (EMC), LLP, India. *The country level database* primarily uses waste statistics submitted by the countries to three international statistical organizations: UN Statistics,[1] which covers all countries, and Eurostat[2] and the OECD,[3] which cover the EU and OECD member countries. For hazardous wastes, returns submitted under the Basel Convention[4] have also been used. The latest data available as of May 2014 varied between countries: 2012 data was available for some EU and OECD countries, but it was necessary to go back to 2010 to get reasonable coverage of the remainder. For non-EU, non-OECD countries, the latest available data is often older than 2010, and sometimes significantly older. Non-waste data on the countries, including population and gross national income (GNI)/gross domestic product (GDP), uses data published by the World Bank.[5]

2. *The city level database* of official estimates again uses UN Statistics, supplemented as appropriate by data compiled and published by the World Bank in 2012.[6] The latter was the latest available published analysis of municipal solid waste management statistics, but most of the data contained therein is from the period 2001-2005 (some older).

3. A second *country-level database* was prepared from the Waste Atlas.[7] As this uses both official data and data from various other published sources, ensuring consistency and reliability is more challenging than with the official sources. The database used for the GWMO was prepared in a project undertaken at the University of Leeds, which collated and cross-checked country level data from the Waste Atlas. This database is referenced herein as 'Waste Atlas - University of Leeds'.[8]

4. A second *city-level database* uses the outcomes from applying the Wasteaware ISWM indicators[9] to some 40 cities in all six inhabited continents. These indicators have been compiled at different times in projects undertaken for UN-Habitat, GIZ and Imperial College London. Again here, a project at the University of Leeds has collated and cross-checked all the available data as part of the preparatory work for the GWMO. This database is referenced herein as 'Wasteaware - University of Leeds'.[10]

5. *Population data* has been sourced from two United Nations publications, *World Population Prospects*, using both the 2013 and 2015 Revisions, and *World Urbanization Prospects*, using both the 2012 and

---

1 United Nations Statistics Division - Environment Statistics. http://unstats.un.org/unsd/ENVIRONMENT/qindicators.htm The United Nations Statistics Division of the Department of Economic and Social Affairs (DESA) has a data service for the global user community. It brings UN statistical databases within easy reach of users through a single entry point. This database was used to extract all the solid waste-related data and country classification data for all the countries across years. It was also used to compare data parameters for the same year and country presented by different sources in order to judge data quality and reliability.

2 Eurostat Statistics. http://epp.eurostat.ec.europa.eu/portal/page/portal/eurostat/home The main objective of this database is to make available the largest possible stock of past and present datasets, legislations, definitions, other national standards and methodologies in order to help users in the analysis of statistical data for the EU. This database was used to extract all the solid waste generation and treatment data for the EU countries across years.

3 OECD Statistics. http://stats.oecd.org/ The website includes data and metadata for OECD countries and selected non-member economies. It enables users to search for and extract data from across many databases under different themes such as agriculture and fisheries, development, economic projections, education and training, energy, environment, finance, globalization, health, national accounts, productivity, regional statistics, social and welfare. This database was used to extract all the solid waste-related data for the OECD member countries across years.

4 Basel Convention. http://www.basel.int/ The website publishes data on hazardous waste generation reported to the Basel Convention by contracting Parties, but there are issues with both consistency of the data and with incomplete reporting by the Parties.

5 World Bank Indicators. http://data.worldbank.org/indicator/ This website presents free and open access to data about development in countries around the globe. It was used for accessing the global and country-wise population data as well as for information on GNI and GDP per capita data for every country. Use is made in the GWMO of GNI per capita data derived using the Atlas method.

6 Hoornweg & Bhada-Tata (2012), *What a Waste*, World Bank, listed in Annex A, Chapter 3, Collated data sources

7 Waste Atlas. http://www.atlas.d-waste.com/. Waste Atlas is an online database, produced in collaboration between D-Waste, ISWA, Waste to Energy Research and Technology Council (WtERT), Solid Waste Exchange of Information and Expertise Network (SWEEP-Net), the Society of Solid Waste Management Experts in Asia and Pacific Islands (SWAPI), and the University of Leeds. The aim is to collate and standardize the wealth of MSW data available on the Internet. At the time of writing, the database had amassed waste data for 1773 cities and 162 countries.

8 Lawless, M. (2014). *Waste Atlas: Recycling and resource recovery around the world.* (MEng dissertation) Supervisors: Velis, C. and Z. Wadud. School of Civil Engineering, University of Leeds, Leeds, UK.

9 See Section 2.5.3 and Wilson et al. (2015), listed in Annex A, Chapter 2, Waste data and indicators.

10 Hickman, H. (2014). *MSW management in cities around the world: An overview of evidence based on revised UN-Habitat methodology.* (MEng dissertation) Supervisors: Velis, C. and Z. Wadud. School of Civil Engineering, University of Leeds, Leeds, UK.

2014 Revisions. These are biennial estimates and projections of the total population, and urban and rural populations, of all countries in the world and of their major urban agglomerations.[11]

6. India appears to be unique among emerging countries in the extent and availability of *waste data and indicators*. A database has been made available to the GWMO with solid waste indicators for 400 urban local bodies in the two Indian states of Gujarat and Maharashtra, compiled using the Performance Assessment System (PAS). This database is referenced here as 'CEPT University - PAS'.[12] In addition, the Ministry of Urban Development (MoUD) has initiated a service level benchmarking programme that provides city-level data with participation from over 1,400 municipalities and municipal corporations from 13 states. It includes indicators for performance related to access, effectiveness of operations and environmental sustainability and financial sustainability of operations.

7. Brazil's data on waste has been sourced from annual reports from ABRELPE (Brazilian Association of Urban Cleansing and Waste Management) (in Portuguese). ABRELPE is the representative of the International Solid Waste Association (ISWA) in Brazil.[13]

8. The National Bureau of Statistics of China has compiled annual city-level and region-level data on waste in its online Statistical Yearbook for the years 1990 through 2014. It contains data for stock, disposal and utilization of municipal solid waste, industrial solid waste and hazardous waste.[14]

9. Annual Reports prepared by the Ministry of Natural Resources and Environment of the Russian Federation in accordance with Resolution No. 966 of the Government of the Russian Federation on environmental data have been accessed to compile waste data from Russia.[15]

10. ISWA has recently made available to its members a regular Waste Business Monitor prepared by AcuComm[16] to track existing and planned investments in waste management infrastructure around the world. AcuComm has made available to the GWMO the underlying database on investment projects that have been active in the two-year period from January 2013 to December 2014.

11. A number of organizations publish regular data on international trade in secondary materials and on secondary raw material prices. These are referenced in the text as they are used. Particular mention here should be given to the international secondary materials trade association Bureau of International Recycling (BIR), which publishes a range of useful summary reports.

12. UN Comtrade,[17] a database of official international trade statistics, and data from the United Nations Conference on Trade and Development (UNCTAD)[18] have been used for trade statistics on secondary raw materials. UN Comtrade comprises annual trade statistics starting from 1962 and monthly trade statistics since 2010. UNCTAD has some data dating back to 1948.

13. These databases have been complemented by extensive research accessing other data sources and publications. The 2009 World Waste Survey was useful in providing more information than others on industrial and hazardous waste and trade in secondary materials, as well as on municipal solid waste.[19] A 2012 review by Karak et al. focused on municipal solid waste.[20] The work of Hoornweg et al. on forecasting future municipal solid waste generation around the world was used extensively;[21] a paper by Beigl et al. focused on forecasting MSW Generation in major European cities.[22] Other individual sources are cited as they are used.

11 United Nations, Department of Economic and Social Affairs. World Population Prospects: http://esa.un.org/wpp/ World Urbanization Prospects: http://esa.un.org/unpd/wup/

12 CEPT University- PAS (2010), listed in Annex A, Chapter 2, Waste data and indicators

13 Brazilian Association of Urban Cleansing and Waste Management (ABRELPE) www.abrelpe.org.br (in Portuguese)

14 China, National Bureau of Statistics. http://www.stats.gov.cn/tjsj/ndsj/. Data available in English.

15 State report *On the State and Environmental Protection of the Russian Federation* (in Russian). In the report on 2013 data, waste data 'ОТХОДЫ' is on page 63. http://www.mnr.gov.ru/regulatory/list.php?part=1101.

16 AcuComm's Waste Business Finder database http://acucomm.net/

17 http://comtrade.un.org/

18 http://unctad.org/

19 Chalmin & Gaillochet (2009), listed in Annex A, Chapter 3, Collated data sources

20 Karak et al. (2012), listed in Annex A, Chapter 3, Collated data sources

21 Hoornweg et al. (2013, 2015), listed in Annex A, Chapter 3, Municipal solid waste management

22 Beigl et al., (2004), listed in Annex A, Chapter 3, Municipal solid waste management

Annex C: Glossary of and Concepts

Glossary of guiding principles for waste governance[23]

Waste prevention – See Topic Sheet 4 on waste prevention, found after Chapter 2.

Duty of care[24] – A requirement that a person act toward others and the public with the watchfulness, attention, caution and prudence that a reasonable person would use in those circumstances. If a person's actions do not meet this standard of care, then the acts are considered negligent, and any damages resulting may be claimed in a lawsuit for negligence.

Polluter pays – The principle in environmental law of making the party responsible for producing pollution also responsible for paying the cost of the damage done to the natural environment. Principle 16 of the 1992 Rio Declaration[25] states: "National authorities should endeavour to promote the internalization of environmental costs and the use of economic instruments, taking into account the approach that the polluter should, in principle, bear the cost of pollution, with due regard to the public interest and without distorting international trade and investment."

Inclusivity and public participation – Principle 10 of the 1992 Rio Declaration states: "Environmental issues are best handled with participation of all concerned citizens, at the relevant level. At the national level, each individual shall have appropriate access to information concerning the environment that is held by public authorities, including information on hazardous materials and activities in their communities, and the opportunity to participate in decision-making processes. States shall facilitate and encourage public awareness and participation by making information widely available. Effective access to judicial and administrative proceedings, including redress and remedy, shall be provided."

Subsidiarity – The principle of devolving decisions to the lowest practical level in a political system so that decisions are taken at the closest possible level to the citizen. For example, in the EU it is the principle whereby the Union does not take action (except in the areas that fall within its exclusive competence), unless it is more effective than action taken at the national, regional or local level.[26]

Precautionary principle[27] – Caution practiced in policy-making in the context of (scientific) uncertainty. Many definitions of the precautionary principle exist, each including two key elements: (1) an expression of a need by decision-makers to anticipate harm before it occurs, (2) the establishment of an obligation, if the level of harm may be high, for action or policy to prevent or minimize such harm even when the lack of scientific certainty makes it difficult to predict the likelihood of harm occurring, or the level of harm should it occur. One of the primary foundations, and globally accepted definitions, of the precautionary principle comes from the 1992 Rio Declaration, Principle 15: "In order to protect the environment, the precautionary approach shall be widely applied by States according to their capabilities. Where there are threats of serious or irreversible damage, lack of full scientific certainty shall not be used as a reason for postponing cost-effective measures to prevent environmental degradation."

Cost recovery – See section 5.7.4.

Proximity[28] – The proximity principle advocates that wastes should be managed and disposed of as close as practicable to their point of origin. The principle is therefore aimed at ensuring efficient waste management practices, by minimizing the cost, resource use and emissions of transporting waste.

Self-sufficiency – Along with the proximity principle, aiming to achieve responsible self-sufficiency at a regional or subregional level. Where this is not possible, priority should be given to transportation by rail or water.

23 Referred to in Section 4.2.1, which highlights key messages on waste governance.
24 http://legal-dictionary.thefreedictionary.com/. Some footnotes (including this one) provide the full reference; for those which provide simply an author and date, full details are listed in Annex A, Chapter 4, Sustainability.
25 1992 Rio Declaration on Environment and Development:
http://www.unep.org/Documents.Multilingual/Default.asp?documentid=78&articleid=1163
26 http://eur-lex.europa.eu/legal-content/EN/TXT/?uri=uriserv:ai0017
27 http://encyclopedia.thefreedictionary.com/precautionary+principle
28 http://www.4980.timewarp.at/sat/ZeroWIN/wiki/index.php/Proximity_principle

Concepts for sustainable resource management

Various concepts have been proposed for sustainable development, addressing one or more of its three domains – environmental, social and economic. Some are formulated as *boundary conditions* or *conditions to be met*, such as planetary boundaries,[29] Cradle to Cradle and The Natural Step. Some propose strategies to achieve sustainable resource use, such as circular economy, performance economy, industrial ecology[30] and prosperity without economic growth.[31] Some concepts are formulated as goals to be reached in order to achieve sustainable development, such as resource efficiency, the decoupling of resource use from economic growth,[32] dematerialization and zero waste.[33] A number of indicators have been proposed to measure progress toward the goals, such as the intensity of material use, material input per unit of service, ecological rucksack and ecological footprint. A selection of these concepts is introduced here.

- "A **circular economy** is an industrial system that is restorative or regenerative by design. It replaces the 'end-of-life' concept with restoration, shifts towards the use of renewable energy, eliminates the use of toxic chemicals, which impair reuse, and aims for the elimination of waste through the superior design of materials, products, systems, and, within this, business models."[34] In a more limited meaning this concept is widely used in the People's Republic of China, in accordance with the Circular Economy Promotion Law of 2009.[35] The circular economy is also increasingly prominent in European strategies.[36]
- Starting from the premise that 'design is the first signal of human intention', **Cradle to Cradle** is a design concept that is based on three fundamental principles that are 'borrowed' from natural ecosystems: (1) Waste = Food, (2) Use current solar income (3) Celebrate diversity. In natural ecosystems, processes of individual organisms and species contribute to the health and vitality of the system as a whole. In those ecosystems, waste from one organism serves as nutritious food for another. Living systems thrive on solar energy. Natural systems function and thrive through diversity and complexity. These principles are translated and applied to the design of materials and products and to production processes.[37]
- Commonly referred to as **The Natural Step** principles, after the NGO that spearheaded the process of identifying them, these systems conditions for sustainability describe full sustainability on a societal level. In a sustainable society, nature is not subject to systematically increasing: (1) concentrations of substances extracted from the Earth's crust; (2) concentrations of substances produced by society; (3) degradation by physical means; and in that society (4) people are not subject to conditions that systematically undermine their capacity to meet their needs. These conditions are incorporated in a strategic planning approach for companies or communities, with back-casting as the central approach – a successful outcome is imagined in the future, then the actions are identified that are needed to reach that vision of success.[38]
- **Performance economy** is an economy in which consumers purchase certain performances (services) rather than physical products.[39] For example, consumers pay for the service of having light rather than for light bulbs, fixtures and electricity. Ownership of the goods (in this case, the light bulbs, electricity and so on) is retained by the producer rather than the consumer and therefore encourages the producer to reduce waste, minimize resource consumption and lengthen product life. 'Product-service systems' is the term used to denote the performance-based business models in which the manufacturer retains ownership of the items and equipment involved,[40] which is a fundamental change from the current practices.[41]

29 Rockström et al. (2009)
30 Introduction providing definitions: http://www.umich.edu/~nppcpub/resources/compendia/INDEpdfs/INDEintro.pdf
31 Jackson (2009); Journal of Cleaner Production, special issue (2013)
32 UNEP (2011a); UNEP (2014a)
33 http://www.zerowaste.org; http://www.zerowastescotland.org.uk/content/scotlands-zero-waste-plan-launched
34 Ellen MacArthur Foundation (2012)
35 http://www.fdi.gov.cn/1800000121_39_597_0_7.html; http://europesworld.org/2014/06/15/chinas-policies-and-instruments-for-developing-the-circular-economy/#.VBAMY0vGBaE; Su et al. (2013)
36 http://ec.europa.eu/environment/circular-economy/index_en.htm; http://www.satw.ch/publikationen/schriften/kreislaufwirtschaft/index_EN; http://www.green-alliance.org.uk/wasted_opportunities:smarter_systems_for_resource_recovery.php
37 McDonough & Braungart (2002); McDonough & Braungart (2013)
38 http://www.naturalstep.org; Robèrt et al. (2002)
39 Stahel (2010)
40 Ceschin (2013)
41 This topic is discussed from a business perspective in Section 5.5.3.

Acronyms

ACR+	Association of Cities and Regions for Recycling and Sustainable Resources
AD	Anaerobic digestion
AMSA	Azienda Milanese Servizi Ambientali (Milanese Environmental Services)
ANGed	National Waste Management Agency in Tunisia
ANR	Association of Recyclers in Colombia
APC	Air pollution control
ARB	Association of Recyclers of Bogotá
A&F	Agriculture & forestry
BAT	Best available technique
BC	Black carbon
BCC	Bo City Council
BIR	Bureau of International Recycling
BOO	Build, operate and own
BOOT	Build, operate, own and transfer
BOR	Build, operate and renew
BRICS	Brazil, Russia, India, the People's Republic of China, and South Africa
BSI PAS	British Standards Institute's Publicly Available Specification
B2B	Business-to-business
CBO	Community based organization
CDC	Centre for Development Communications
CDM	Clean Development Mechanism
CFCs	Chlorofluorocarbons
CIGRES	Inter-municipal Consortium of Solid Waste Management
CMC	Coimbatore Municipal Corporation
CNY	Chinese yuan (currency unit)
COP	Conference of the Parties
CO_2	Carbon dioxide
CSD	Commission on Sustainable Development
CSR	Corporate social responsibility
CWG	Collaborative Working Group on Municipal Solid Waste Management in Low- and Middle-Income Countries
C&D	Construction & demolition
C&I	Commercial & institutional
DALY	Disability adjusted life-year
DBO	Design, build and operate
DDT	Dichlorodiphenyltrichloroethane
DEFRA	Department for Environment, Food and Rural Affairs (UK)
DENR	Department of Environment and Natural Resources (Philippines)
DFID	Department for International Development (UK)
EACR	East African Computer Recycling
ECHA	European Chemicals Agency
EEE	Electrical and electronic equipment
EfW	Energy from waste
EHS	Environment, health and safety
EIA	Environmental impact assessment
EIP	Eco industrial park
EnBW	Energie Baden-Württemberg
EMC	Environmental Management Centre (India)
EPA	Environmental Protection Agency (U.S.)
EPI	Environmental performance index
EPP	Environmentally preferable purchasing

EPR	Extended producer responsibility
ESM	Environmentally sound management
ETS	European Emissions Trading System
EU	European Union
EUR	Euro (currency unit)
eq	equivalent
EXPRA	Extended Producer Responsibility Alliance
FAO	Food and Agriculture Organization of the United Nations
FGT	Flue gas treatment
FOEN	Federal Office for the Environment (Switzerland)
FSC	Food supply chain
FSP-K	Foundation of the Peoples of the South Pacific - Kiribati Branch
GAIA	Global Alliance for Incinerator Alternatives/ Global Anti-Incinerator Alliance
GBP	Great Britain pounds (Sterling; currency unit)
GDP	Gross domestic product
GEF	Global Environment Facility
GFN	Global FoodBanking Network
GHG	Greenhouse gas
GIS	Geographic information system
GIZ	Deutsche Gesellschaft fuer Internationale Zusammenarbeit (German Agency for Technical Cooperation. Former GTZ for International Cooperation)
GM	General Motors
GNI	Gross national income
GoG	Government of Gujarat (India)
GPOBA	Global Partnership for Output Based Aid
GRC	Glass Recycling Company
GUWAA	Global Underwater Awareness Association
GWP	Global warming potential
HBV	Hepatitis B virus
HCV	Hepatitis C virus
HDPE	High density polyethylene
HHW	Household hazardous waste
HIV	Human immunodeficiency virus
ICT	Information communication technology
IDB	Inter-American Development Bank
IETC	International Environmental Technology Centre
IFC	International Finance Corporation
IFIs	International finance institutions
IPCC	Intergovernmental Panel for Climate Change
IRRC	Integrated Resource Recovery Centers
ISRO	Indian Space Research Organization
ISWA	International Solid Waste Association
ISWM	Integrated sustainable waste management
ITC	Indian Tobacco Company
IWWG	International Waste Working Group
JICA	Japan International Cooperation Agency
JL	Jimbaran Lestari
JNNURM	Jawaharlal Nehru National Urban Renewal Mission
JPSPN	National Solid Waste Management Department (Malaysia)
KKPKP	Kagad Kach Patra Kashtakari Panchayat
kW	Kilowatt
LCA	Life cycle assessment
LCT	Life cycle thinking
LDC	Least developed countries
LFG	Landfill gas
LFHW	Love Food Hate Waste

LGA	Local Government Association
LGUs	Local Government Unit
MBS	Mechanical biological stabilisation
MBT	Mechanical biological treatment
MDGs	Millennium Development Goals
MEA	Multilateral environmental agreements
MENA	Middle East and North Africa
MEP	Ministry of Environmental Protection (PRC)
MFA	Material flow accounting
MHW	Ministry of Health and Welfare
MIPS	Material input per unit of service
MNCR	Movimento Nacional dos Catadores de Materiais Recicláveis (National Movement of Catadores of Recyclable Materials)
MNES	Ministry of Non-Conventional Energy Sources (Egypt)
MNRE	Ministry of New and Renewable Energy (India)
MoUs	Memorandum of understanding
MoUD	Ministry of Urban Development (India)
MPPI	Mobile Phone Partnership Initiative
MSEs	Micro- or small-enterprises
MSW	Municipal solid waste
MSWM	Municipal solid waste management
MW	Megawatt
M&Q	Mining & quarrying
N/A	Not applicable
NAMA	Nationally Appropriate Mitigation Action
NCM	Non compostable materials
NCPCs	Network of National Cleaner Production Centres
NEA	National Environment Agency (Singapore)
NEMA	National Environmental Management Authority (Kenya)
NEPA	National Environmental Protection Agency
NGO	Non-government organizations
NIMBY	Not in my backyard
NIS	New Independent States
NRC	National Recycling Coalition
OBA	Output-based aid
OECD	Organisation for Economic Co-operation and Development
OWG	Open working group
PACE	Partnership for Action on Computing Equipment
PAS	Performance assessment system
PAYT	Pay-as-you-throw
PCBs	Polychlorinated biphenyl
PCMC	Pimpri-Chinchwad Municipal Corporation
PET	Polyethylene terephthalate
POPs	Persistent organic pollutants
PPP	Public-private partnership
PRONAGDES	Programme National de Gestion des Déchets Solides (Tunisia) (National Solid Waste Management Programme)
PROs	Producer responsibility organizations
PSP	Private sector participation
PTEs	Potentially toxic elements
RDF	Refuse derived fuel
REACH	Registration, Evaluation, Authorisation and Restriction of Chemicals
RED LACRE	Red Latinoamericana de Recicladores (Latin American network for recyclers)
RMGC	Rosia Montana Gold Corporation
RoHS	Restriction of Hazardous Substances

ROT	Rehabilitate, operate and transfer
Rs	Indian rupees (currency unit)
SA	South Australia
SCP	Sustainable consumption and production
SDGs	Sustainable Development Goals
SEPA	State Environmental Protection Administration (PRC)
SERI	Sustainable Europe Research Institute
SIDS	Small Island Developing States
SLB	Service level benchmark
SLCP	Short-lived climate pollutants
SMEs	Small- and medium-sized enterprises
SPA	Singapore Packaging Agreement
SSO	Source segregation organic
SO2	Sulphur dioxide
SPREP	Secretariat of the Pacific Regional Environment Programme
SRF	Solid recovered fuel
SWAPI	Society of Solid Waste Management Experts in Asia and Pacific Islands
SWaCH	Solid waste collection and handling
SWEEP-Net	Solid Waste Exchange of Information and Expertise Network
SWM	Solid waste management
TPD	Tonnes per day
UAESP	Special Administrative Unit of Public Services (Colombia)
UK	United Kingdom
ULBs	Urban Local Bodies
UNCSD	United Nations Conference on Sustainable Development
UNDP	United Nations Development Programme
UNEP	United Nations Environment Programme
UNESCAP	United Nations Economic and Social Commission for Asia and the Pacific
UNFCCC	United Nations Framework Convention on Climate Change
UN-Habitat	United Nations Human Settlement Programme
UNIDO	United Nations Industrial Development Organization
US	United States of America
USD	United States dollars (currency unit)
USEPA	United States Environmental Protection Agency
VALIRE	Valorisation of incineration residues
WARM	Waste reduction model
WASH	Water, sanitation and health
WCO	World Customs Organization
WEEE	Waste electrical and electronic equipment
WHH	Welthungerhilfe
WHO	World Health Organization
WIEGO	Women in Informal Employment: Globalizing and Organizing
WIG	Women's Initiative the Gambia
WMC	Waste management component
WtE	Waste to energy
3Rs	Reduce, reuse and recycle
4Es	Enable, encourage, engage and exemplify
4Rs	Reduce, reuse, recycle and recover

About the UNEP Division of Technology, Industry and Economics

Set up in 1975, three years after UNEP was created, the Division of Technology, Industry and Economics (DTIE) provides solutions to policy-makers and helps change the business environment by offering platforms for dialogue and co-operation, innovative policy options, pilot projects and creative market mechanisms.

DTIE plays a leading role in three of the six UNEP strategic priorities: **climate change, harmful substances and hazardous waste, resource efficiency.**

DTIE is also actively contributing to the **Green Economy Initiative** launched by UNEP in 2008. This aims to shift national and world economies on to a new path, in which jobs and output growth are driven by increased investment in green sectors, and by a switch of consumers' preferences towards environmentally friendly goods and services.

Moreover, DTIE is responsible for **fulfilling UNEP's mandate as an implementing agency for the Montreal Protocol Multilateral Fund** and plays an executing role for a number of UNEP projects financed by the Global Environment Facility.

The Office of the Director, located in Paris, coordinates activities through:

> **The International Environmental Technology Centre** – IETC (Osaka), which promotes the collection and dissemination of knowledge on Environmentally Sound Technologies with a focus on waste management. The broad objective is to enhance the understanding of converting waste into a resource and thus reduce impacts on human health and the environment (land, water and air).

> **Sustainable Consumption and Production** (Paris), which promotes sustainable consumption and production patterns as a contribution to human development through global markets.

> **Chemicals** (Geneva), which catalyses global actions to bring about the sound management of chemicals and the improvement of chemical safety worldwide.

> **Energy** (Paris and Nairobi), which fosters energy and transport policies for sustainable development and encourages investment in renewable energy and energy efficiency.

> **OzonAction** (Paris), which supports the phase-out of ozone depleting substances in developing countries and countries with economies in transition to ensure implementation of the Montreal Protocol.

> **Economics and Trade** (Geneva), which helps countries to integrate environmental considerations into economic and trade policies, and works with the finance sector to incorporate sustainable development policies. This branch is also charged with producing green economy reports.

DTIE works with many partners (other UN agencies and programmes, international organizations, governments, non-governmental organizations, business, industry, the media and the public) to raise awareness, improve the transfer of knowledge and information, foster technological cooperation and implement international conventions and agreements.

For more information,
see **www.unep.org/dtie**